YAMAHA OUTBOARD

Volume III

V4 and V6
1984-1988

Tune-Up and Repair Manual

Includes
Jet Drive
Counterrotating Drive

MARINE MANUALS

Other books

by

Joan Lorton Coles

and/or

Clarence W. Coles

Glenn's Complete Bicycle Manual
The Book of Harry (Limited Edition)
Glenn's Flat Rate Manual -- American Cars
Seloc's OMC Stern Drive Tune-up and Repair Manual
Seloc's Mercury Volume I Tune-up and Repair Manual
Seloc's Mercury Volume II Tune-up and Repair Manual
Seloc's Mercury Volume III Tune-up and Repair Manual
Seloc's Yamaha Volume I Tune-up and Repair Manual
Seloc's Yamaha Volume II Tune-up and Repair Manual
Seloc's Yamaha Volume III Tune-up and Repair Manual
Seloc's OMC Cobra Stern Drive Tune-up and Repair Manual
Seloc's Mercruiser Stern Drive Tune-up and Repair Manual
Seloc's Volvo Penta Stern Drive Tune-up and Repair Manual
Seloc's Johnson/Evinrude Volume I Tune-up and Repair Manual
Seloc's Johnson/Evinrude Volume II Tune-up and Repair Manual
Seloc's Johnson/Evinrude Volume III Tune-up and Repair Manual
Seloc's Johnson/Evinrude Volume IV Tune-up and Repair Manual

SELOC'S YAMAHA OUTBOARD

Volume III

V4 and V6
1984-1988

Tune-Up
and
Repair Manual

Includes

Jet Drive
Counterrotating Drive

by
Joan and Clarence Coles

Cover
An original illustration
by
Tom Baus

Technical Director
Dani Fisk

Graphics
Barbara Flotow

Copyright © *1988*
by
Seloc Publications

All rights reserved. No part of this book may be reproduced or utilized in any form or by any means, electronic or mechanical including photocopying, recording, or by any information storage and retrieval system, without permission in writing from the publisher.

Inquiries should be addressed to:

Seloc Publications
10693 Civic Center Drive
Rancho Cucamonga, California 91730-3804

ISBN 0-89330-023-3

Printed in the United States of America

FOREWORD

This is a comprehensive tune-up and repair manual for Yamaha V4 and V6, 2-cycle outboard units manufactured from 1984 through 1988, including jet drive units and counterrotating units. The book has been designed and written for the professional mechanic, the do-it-yourselfer, and the student developing his mechanical skills.

Professional Mechanics will find it to be an additional "tool" for use in their daily work on V4 and V6 Yamaha units because of the many special helpful techniques described and not found in the factory "shop manuals".

Boating Enthusiasts interested in performing their own work and in keeping their unit operating in the most efficient manner will find the step-by-step illustrated procedures used throughout the manual extremely valuable. In fact, many have said this book almost equals an experienced mechanic looking over their shoulder giving advice.

Students and **Instructors** have found the chapters divided into practical areas of interest and work. Technical trade schools, from Florida to Michigan and west to California, as well as the U.S. Navy and Coast Guard, have adopted Seloc manuals as a standard classroom text.

Illustrations and procedural steps are so closely related and identified with **matching** numbers that, in most cases, captions are not used. The exploded drawings show internal parts and their relationship with each other.

Troubleshooting sections have been included in many chapters to assist the individual performing the work to quickly and accurately isolate problems to a specific area without unnecessary expense and time-consuming work. As an added aid and one of the unique features of this book, many worn parts are illustrated to identify and clarify when an item should be replaced.

Accurate comprehensive specifications and wiring diagrams are included.

ACKNOWLEDGMENTS

A sincere expression of indebtedness is extended to **YAMAHA MOTOR CORPORATION U.S.A,** particularly Ike Hirao, Engineer, for the assistance, permission to use certain illustrations, and for making outboard units available to us for photographic purposes. The exploded drawings and wiring diagrams are a positive contribution to this manual.

A special and personal thanks is expressed to John Alger, Technical Coordinator, and Ted Zahorski, Technical Advisor, both of Yamaha Motor Corporation, for their advice, expertise, and assistance in the preparation of the material in this book.

Joan & Clarence Coles
Seloc Publications

TABLE OF CONTENTS

1 SAFETY

INTRODUCTION	1-1
CLEANING, WAXING & POLISHING	1-1
CONTROLLING CORROSION	1-1
PROPELLERS	1-2
Diameter & Pitch	1-3
Selection	1-3
Cavitation	1-4
Vibration	1-4
Shock Absorbers	1-5
Propeller Rake	1-5
Progressive Pitch	1-6
Cupping	1-6
Rotation	1-6
High Performance Propellers	1-7
JET DRIVE	
Description & Operation	
Model Identification and Serial Numbers	1-8
Operation Hints	1-8
Powerhead Stall	1-8
Grille Blockage	1-9
FUEL SYSTEM	1-10
Taking on Fuel	1-10
Static Electricity	1-10
Fuel Tank Grounding	1-11
LOADING	1-12
HORSEPOWER	1-12
FLOTATION	1-12
EMERGENCY EQUIPMENT	1-14
COMPASS	1-16
MISCELLANEOUS EQUIPMENT	1-19
BOATING ACCIDENT REPORTS	1-20
NAVIGATION	1-21

2 TUNING

INTRODUCTION	2-1
TUNE-UP SEQUENCE	2-1
COMPRESSION	2-3
SPARK PLUG INSPECTION	2-3
IGNITION SYSTEM	2-4
TIMING & SYNCHRONIZING	2-5
BATTERY CHECK	2-5
FUEL & FUEL TANKS	2-7
CARBURETOR ADJUSTMENTS	2-7
Air/Fuel Mixture	2-8
Idle Speed	2-8
FUEL PUMPS	2-9
CRANKING MOTOR	2-9
INTERNAL WIRING HARNESS	2-9
WATER PUMP CHECK	2-10
PROPELLER	2-11
LOWER UNIT	2-11
JET DRIVE UNITS	2-12
BOAT TESTING	2-13

3 MAINTENANCE

INTRODUCTION	3-1
EMERGENCY TETHER	3-2
LUBRICATION -- COMPLETE UNIT	3-3
PRE-SEASON PREPARATION	3-4
SEALANTS, ADHESIVES, LUBS., HYD. FLUID & FUEL STABILIZER	3-7
FIBERGLASS HULLS	3-8
BELOW WATERLINE SERVICE	3-8
INSIDE THE BOAT	3-9
SUBMERGED ENGINE SERVICE	3-9
LOWER UNIT -- PROP DRIVE	3-11
LOWER UNIT -- JET DRIVE	3-12
PROPELLER SERVICE	3-15
WINTER STORAGE	3-17

4 FUEL

INTRODUCTION	4-1
GENERAL CARBURETION INFO.	4-1
CARBURETOR MODELS	4-3
FUEL SYSTEM	4-4
Leaded Gasoline & Gasohol	4-4
Removing Fuel from System	4-5
TROUBLESHOOTING	4-5
"Sour" Fuel	4-5
Fuel Pump Test	4-6
Fuel Line Test	4-7
Rough Engine Idle	4-8
Excessive Fuel Consumption	4-8
Engine Surge	4-10

4 FUEL (Continued)

CARBURETOR SERVICE 4-10
  Removal 4-10
  Disassembling 4-12
  Cleaning & Inspecting 4-18
  Assembling 4-19
  Installation 4-23
    Tachometers & Connections 4-26
FUEL PUMP
  Description & Operation 4-26
  Pressure Check 4-27
  Removal & Disassembling 4-29
  Cleaning & Inspecting 4-30
  Assembling & Installation 4-31

5 OIL INJECTION

DESCRIPTION & OPERATION 5-1
  Main Oil Tank 5-2
  Main Oil Tank Sensor 5-3
  Remote Oil Tank Sensor 5-3
  Oil Injection Pump 5-4
  Oil Feed Pump 5-4
  Control Unit 5-5
  Tilt Switch 5-5
  Emergency Switch 5-5
  Buzzer 5-5
  Warning Light Display 5-6
TROUBLESHOOTING 5-7
SERVICING SYSTEM 5-11
  Main Oil Tank Removal 5-11
  Oil Pump Removal 5-12
  Disassembling Oil Pump 5-13
  Cleaning & Inspecting Pump 5-14
  Assembling Oil Pump 5-15
  Installation -- Oil Pump 5-17
  Remote Oil Tank Removal
    and Oil Feed Pump Service 5-20
  Cleaning & Inspecting 5-22
  Assembling 5-22
  Installation 5-22
PURGING AIR FROM
  OIL INJECTION SYSTEM 5-23
ADJUSTING CONTROL LINK ROD 5-25
OPERATIONAL CHECK --
  WARNING SYSTEM 5-26

6 IGNITION

INTRO. & CHAPTER COVERAGE 6-1
SPARK PLUG EVALUATION 6-1
POLARITY CHECK 6-4
CAPACITOR DISCHARGE IGNITION 6-5
  Pulsar Coil Circuit 6-6
  Charge Coil Circuit 6-6
  Thyristor Circuits 6-6

  Secondary Circuits 6-6
  Timing Basics 6-6
  Timing Advance 6-7
  Charging System 6-7
TROUBLESHOOTING CDI SYSTEM
AND CHARGING SYSTEM 6-7
  Spark Plugs 6-7
  Compression 6-8
  Testing CDI Components 6-9
    Charge Coil Test 6-10
    Lighting Coil Test 6-11
    Rectifier/Regulator Test 6-12
    Primary Winding Test 6-12
    Secondary Winding Test 6-13
    CDI Resistance Tests 6-13
    Control Unit on V6 Units 6-13
FLYWHEEL AND
STATOR PLATE SERVICE 6-17
  "Pulling" Flywheel 6-17
  Cleaning & Inspecting 6-19
  Installation 6-20
CDI UNIT, REGULATOR/RECTIFIER
AND IGNITION COIL -- REMOVAL
FOR TESTING OR REPLACE. 6-22
CDI UNIT, REGULATOR/RECTIFIER
AND IGNITION COIL --
INSTALLATION 6-24
YMIS (YAMAHA MICROCOMPUTER
IGNITION SYSTEM)
  Description & Operation 6-26
    Control Unit 6-27
    Throttle Position Sensor 6-27
    Crank Position Sensor 6-28
    Knocking Sensor 6-28
    Thermo Sensor 6-28
  Engine Controls 6-29
    Engine Start Control 6-29
    Engine Warm-up Control 6-30
    Idle Stabilizing Control 6-30
    Ignition Timing Control 6-30
    Overheating Control 6-30
    Overrev. Control 6-30
    Knocking Control 6-30
    Reverse Direction Control 6-31
    Operating Voltage Control 6-31

TROUBLESHOOTING 6-31
  YMIS Bypass Procedure 6-31
  Resistance Tests 6-32
  YMIS Power Supply Test 6-32
  Check Crank Position Sensor 6-33
  Check Knocking Sensor 6-33
  Check Thermo Sensor 6-34
  Check Throttle Position
    Sensor 6-34
  Adjust Throttle Position
    Sensor 6-35

OPERATION TESTS
  Thermo Sensor   6-36
  Crank Position Sensor   6-37
  Knocking Sensor   6-37
  Throttle Position Sensor   6-38
  Thermo Switch   6-38
  Microcomputer   6-38

7 TIMING & SYNCHRONIZING

INTRODUCTION & PREPARATION   7-1
  Dial Indicator   7-1
  Timing Light   7-2
  Tachometer   7-2
  Flywheel Rotation   7-2
  Test Tank   7-2
ALL V4 AND V6 MODELS **NOT**
EQUIPPED WITH YMIS
  Timing Plate Alignment
    and Preliminary Adjustments   7-2
  Prelim. Link Rod Adjustments   7-3
  Static Timing --
    Fully Retarded   7-4
  Static Timing --
    Fully Advanced   7-5
  Static Timing --
    Pickup Adjustment   7-6
  Dynamic Timing --
    Fully Retarded at Idle   7-8
  Dynamic Timing --
    Fully Advanced   7-8
  Dynamic Timing --
    Pickup Adjusment   7-9
  Throttle Linkage Adjustment   7-10
  Idle Adjustment   7-11
  Pilot Screw Adjustment   7-12
ALL MODELS **WITH** YMIS   7-12
  Throttle Control
    Link Rod Adjustment   7-14
  Pickup Timing Adjustment   7-14
  Throttle Linkage Adjustment   7-15
  Idle Adjustment   7-15
  Pilot Screw Adjustment   7-16

8 POWERHEAD

INTRODUCTION AND
  CHAPTER COVERAGE   8-1
DESCRIPTION AND OPERATION   8-2
ACTUAL OPERATION   8-3
POWERHEAD SERVICE   8-4
  Removal   8-4
  Disassembling   8-7
    Lower Oil Seal   8-7
    Intake Manifold   8-8
    Exhaust Cover   8-9
    Thermostat   8-10
    Cylinder Head   8-11
    Upper Oil Seal Housing   8-11
    Crankcase Separation   8-12
    Crankshaft & Rods   8-13
    Pistons   8-15
    Crankshaft Disassembling   8-16
  Cleaning & Inspecting   8-17
    Reed Block   8-17
    Exhaust Cover   8-18
    Bleed System   8-19
    Crankshaft   8-19
    Connecting Rods   8-20
    Pistons   8-21
    Cylinder Block   8-24
  ASSEMBLING   8-29
    Crankshaft   8-29
    Pistons & Rods   8-30
    Upper Oil Seal Housing   8-33
    Crankshaft   8-34
    Cylinder Head   8-37
    Thermostat   8-39
    Exhaust Cover   8-40
    Intake Manifold   8-40
    Lower Oil Seal Housing   8-40
  INSTALLATION
    Powerhead   8-42
      Without YMIS   8-42
      With YMIS   8-43

9 ELECTRICAL

INTRODUCTION   9-1
BATTERIES   9-1
  Battery Ratings   9-2
  Battery Service   9-3
  Jumper Cables   9-6
  Storage   9-6
THERMOMELT STICKS   9-7
TACHOMETERS   9-7
ELECTRICAL SYSTEM
  GENERAL INFORMATION   9-7
CHARGING CIRCUIT SERVICE   9-8
CRANKING MOTOR CIRCUIT   9-9
  Description   9-9
  Troubleshooting   9-10
  Motor Removal   9-14
  Disassembling   9-15
  Cleaning & Inspecting   9-18
  Testing   9-19
  Assembling   9-20
  Installation   9-22
TESTING OTHER
  ELECTRICAL COMPONENTS   9-23
  Testing Thermo Switch   9-23
  Trim/Tilt Relay   9-24
  Trim/Tilt Switch   9-24
  Trim/Tilt Motor   9-25

9 ELECTRICAL (Continued)

Main Switch	9-25
"Kill" Switch	9-26
Choke Solenoid	9-27
Choke Switch	9-27
Neutral Safety Switch	9-27
Buzzer	9-28
Oil Injection Components	9-28

10 TRIM/TILT

INTRODUCTION AND CHAPTER COVERAGE	10-1
DESCRIPTION AND OPERATION	10-1
Trim Up	10-2
Trim Down	10-3
Tilt Up	10-3
Tilt Down	10-4
Shock Absorber Action	10-4
PURGING AIR FROM SYSTEM -- UNIT INSTALLED	10-4
TROUBLESHOOTING	10-5
SERVICING TRIM/TILT SYS.	10-7
Disassembling	10-7
Cleaning & Inspecting	10-11
Assembling	10-11
PURGING AIR FROM SYSTEM -- UNIT REMOVED	10-13

11 LOWER UNIT

DESCRIPTION	11-1
STANDARD AND COUNTER-ROTATING LOWER UNIT	
Description & Operation	11-2
STANDARD UNIT -- SHIFTING PRINCIPLES	11-3
COUNTERROTATING -- SHIFTING PRINCIPLES	11-4
EXHAUST GASES	11-5
TROUBLESHOOTING -- PROPELLER DRIVE LOWER UNIT	11-6
PROPELLER SERVICE	11-7
LOWER UNIT SERVICE -- PROPELLER DRIVE	
Removal	11-9
Disassembling	11-12
Water Pump	11-12
Shift Rod	11-12
Oil Seal Housing	11-13
Bearing Carrier	11-14
Propeller Shaft	11-17
Driveshaft	11-19
Cleaning & Inspecting	11-22
Assembling	11-26
Forward/Reverse Gear	11-26
Driveshaft Bearing	11-26
Pinion Gear	11-28
Driveshaft Oil Seals	11-30
Pinion Gear Depth	11-31
Clutch "Dog"	11-32
Bearings & Oil Seals	11-34
Short Prop. Shaft	11-35
Gear Pattern & Mesh	11-38
Water Pump Installation	11-42
Lower Unit Installation	11-44
Propeller Installation	11-45
JET DRIVE	
Description & Operation	11-47
Removal & Disassembling	11-48
Cleaning & Inspecting	11-52
Assembling	11-55
Installation	11-57
Adjustments	11-60

12 REMOTE CONTROLS

DESCRIPTION	12-1
DISASSEMBLING	12-2
CLEANING AND INSPECTING	12-6
ASSEMBLING	12-8

APPENDIX

METRIC CONVERSION CHART	A-1
ENGINE SPECIFICATIONS AND TUNE-UP ADJUSTMENTS	A-2
WIRING SCHEMATIC AND COLOR CODE IDENTIFICATION	
V4 Powerheads	A-7
V6 Powerheads **WITHOUT** YMIS	A-8
V6 Powerheads **WITH** YMIS	A-9
Other Seloc Manuals	A-10

1
SAFETY

1-1 INTRODUCTION

In order to protect the investment for the boat and outboard, they must be cared for properly while being used and when out of the water. Always store the boat with the bow higher than the stern and be sure to remove the transom drain plug and the inner hull drain plugs. If any type of cover is used to protect the boat, be sure to allow for some movement of air through the hull. Proper ventilation will assure evaporation of any condensation that may form due to changes in temperature and humidity.

1-2 CLEANING, WAXING, AND POLISHING

Any boat should be washed with clear water after each use to remove surface dirt and any salt deposits from use in salt water. Regular rinsing will extend the time between waxing and polishing. It will also give you "pride of ownership", by having a sharp looking piece of equipment. Elbow grease, a mild detergent, and a brush will be required to remove stubborn dirt, oil, and other unsightly deposits.

Stay away from harsh abrasives or strong chemical cleaners. A white buffing compound can be used to restore the original gloss to a scratched, dull, or faded area. The finish of your boat should be thoroughly cleaned, buffed, and polished at least once each season. Take care when buffing or polishing with a marine cleaner not to overheat the surface you are working, because you will burn it.

1-3 CONTROLLING CORROSION

Since man first started out on the water, corrosion on his craft has been his enemy.

Maximum enjoyment can only be realized when the boat is kept clean, the engine is performing at peak efficiency, and proper safety precautions are exercised.

1-2 SAFETY

The first form was merely rot in the wood and then it was rust, followed by other forms of destructive corrosion in the more modern materials. One defense against corrosion is to use similar metals throughout the boat. Even though this is difficult to do in designing a new boat, particularly the undersides, similar metals should be used whenever and wherever possible.

A second defense against corrosion is to insulate dissimilar metals. This can be done by using an exterior coating of Sea Skin or by insulating them with plastic or rubber gaskets.

Using Zinc

The proper amount of zinc attached to a boat is extremely important. The use of too much zinc can cause wood burning by placing the metals close together and they become "hot". On the other hand, using too small a zinc plate will cause more rapid deterioration of the metal you are trying to protect. If in doubt, consider the fact that it is far better to replace the zincs than to replace planking or other expensive metal parts from having an excess of zinc.

When installing zinc plates, there are two routes available. One is to install many different zincs on all metal parts and thus run the risk of wood burning. Another route, is to use one large zinc on the transom of the boat and then connect this zinc to every underwater metal part through internal bonding. Of the two choices, the one zinc on the transom is the better way to go.

All Yamaha outboard units have a zinc attached somewhere to the exterior of the lower unit. Each bank of cylinders also has two wegde shaped anodes in the water jacket around the cylinders. These anodes provide protection against corrosion where cooling water is routed inside the powerhead.

1-4 PROPELLERS

As you know, the propeller is actually what moves the boat through the water. This is how it is done. The propeller operates in water in much the same manner as a wood screw does in wood. The propeller "bites" into the water as it rotates. Water passes between the blades and out to the rear in the shape of a cone. The propeller "biting" through the water in much the same manner as a wood auger is what propels the boat.

All units covered in this manual are equipped, from the factory, with a through the propeller exhaust. With these units, exhaust gas is forced out through the propeller.

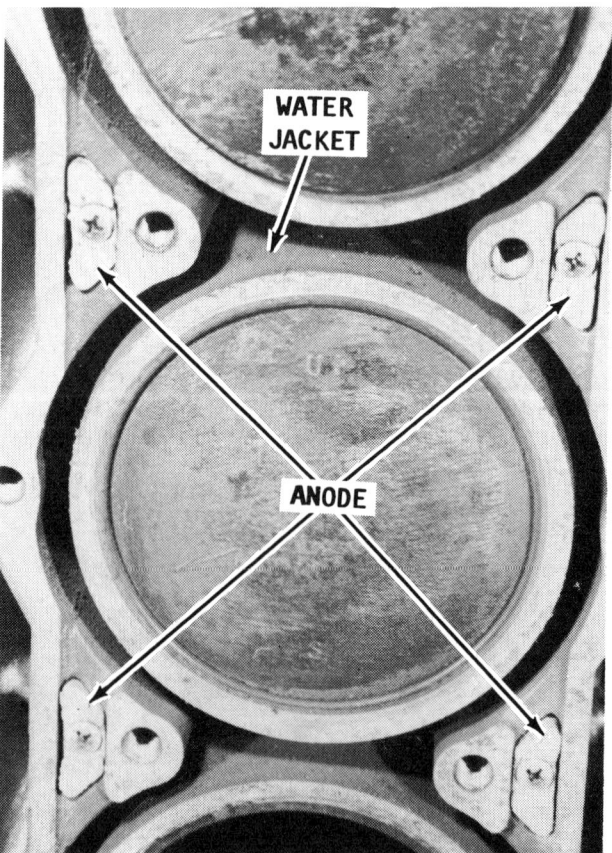

Anodes are installed in each cylinder bank as a corrosion prevention measure. The V4 powerheads have a total of four anodes and the V6 powerheads have eight.

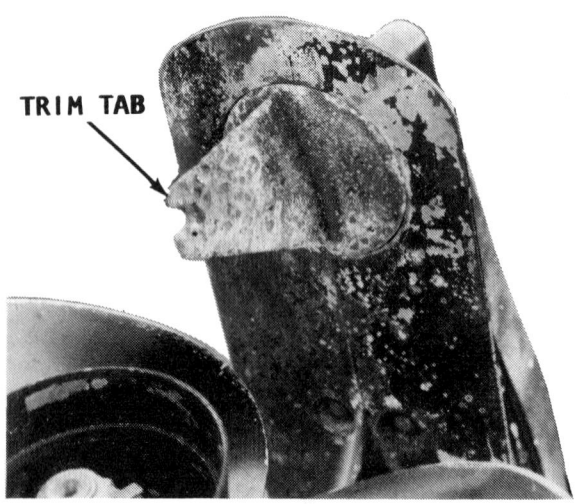

Zinc installation also used as the trim tab. The tab assists the helmsperson to maintain a true course without "fighting" the wheel.

Diameter and Pitch

Only two dimensions of the propeller are of real interest to the boat owner: The diameter and the pitch. These two dimensions are stamped on the propeller hub and always appear in the same order: the diameter first and then the pitch. Propellers furnished with the outboard by the manufacturer for the units covered in this manual have a letter designation following the pitch size. This letter indicates the propeller type. For instance, the numbers and letter 9-7/8 x 10-1/2 - F stamped on the back of one blade indicates the propeller diameter to be 9-7/8", with a pitch of 10-1/2" and it is a Type "F".

The diameter is the measured distance from the tip of one blade to the tip of the other as shown in the accompanying illustration.

The pitch of a propeller is the angle at which the blades are attached to the hub. This figure is expressed in inches of water travel for each revolution of the propeller. In our example of a 9-7/8 x 10-1/2 propeller, the propeller should travel 10-1/2 inches through the water each time it revolves. If the propeller action was perfect and there was no slippage, then the pitch multiplied by the propeller rpms would be the boat speed.

Most outboard manufacturers equip their units with a standard propeller having a diameter and pitch they consider to be best suited to the engine and the boat. Such a propeller allows the engine to run as near to the rated rpm and horsepower (at full throttle) as possible for the boat design.

The blade area of the propeller determines its load-carrying capacity. A two-blade propeller is used for high-speed running under very light loads.

A four-blade propeller is installed in boats intended to operate at low speeds under very heavy loads such as tugs, barges, or large houseboats. The three-blade propeller is the happy medium covering the wide range between the high performance units and the load carrying workhorses.

Propeller Selection

There is no standard propeller that will do the proper job in very many cases. The list of sizes and weights of boats is almost endless. This fact coupled with the many boat-engine combinations makes the propeller selection for a specific purpose a difficult task. Actually, in many cases the propeller may be changed after a few test runs. Proper selection is aided through the use of charts set up for various engines and boats. These charts should be studied and understood when buying a propeller. However, bear in mind, the charts are based on average boats with average loads, therefore, it may be necessary to make a change in size or pitch, in order to obtain the desired results for the hull design or load condition.

Propellers are available with a wide range of pitch. Remember, a low pitch

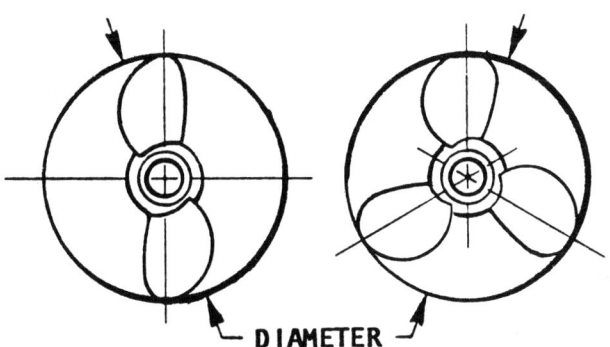

Diameter and pitch are the two basic dimensions of a propeller. The diameter is measured across the circumference of a circle scribed by the propeller blades, as shown.

Correct propeller selection can be made from charts developed considering, powerhead size, hull design, and intended boat performance.

takes a smaller bite of the water than the high pitch propeller. This means the low pitch propeller will travel less distance through the water per revolution. The low pitch will require less horsepower and will allow the engine to run faster.

All engine manufacturers design their units to operate with full throttle at, or slightly above, the rated rpm. If the powerhead is operated at the rated rpm, several positive advantages will be gained.

1- Spark plug life will be increased.
2- Better fuel economy will be realized.
3- Easier steering qualities.
4- Best performance received from the boat and power unit.

Therefore, take time to make the proper propeller selection for the rated rpm of the engine at full throttle with what might be considered an "average" load. The boat will then be correctly balanced between engine and propeller throughout the entire speed range.

A reliable tachometer must be used to measure powerhead speed at full throttle to ensure the engine will achieve full horsepower and operate efficiently and safely. To test for the correct propeller, make a test run in a body of smooth water with the lower unit in forward gear at full throttle. If the reading is above the manufacturer's recommended operating range, try propellers of greater pitch, until one is found allowing the powerhead to operate continually within the recommended full throttle range.

If the engine is unable to deliver top performance and the powerhead is properly tuned, then the propeller may not be to blame. Operating conditions have a marked effect on performance. For instance, an engine will lose rpm when run in very cold water. It will also lose rpm when run in salt water as compared with fresh water. A hot, low-barometer day will also cause the engine to lose power.

Cavitation

Cavitation is the forming of voids in the water just ahead of the propeller blades. Marine propulsion designers are constantly fighting the battle against the formation of these voids due to excessive blade tip speed and engine wear. The voids may be filled with air or water vapor, or they may actually be a partial vacuum. Cavitation may be caused by installing a piece of equipment too close to the lower unit, such as the knot indicator pickup, depth sounder, or bait tank pickup.

Vibration

The propeller should be checked regularly to ensure all blades are in good condition. If any of the blades become bent or nicked, this condition will set up vibrations

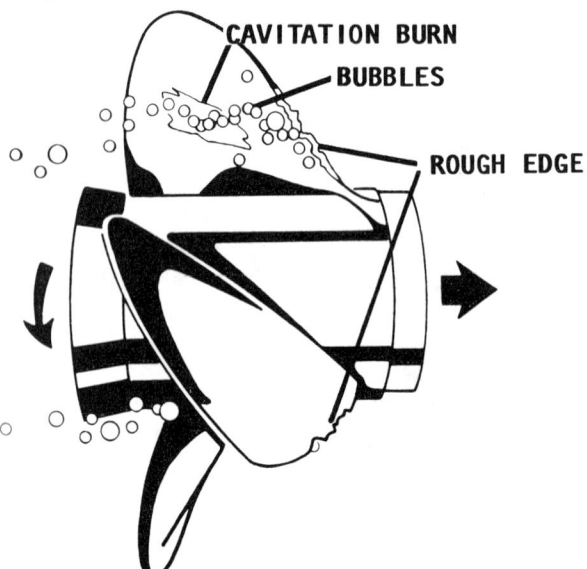

Cavitation (air bubbles) formed at the propeller. Manufacturers are constantly fighting this problem, as explained in the text.

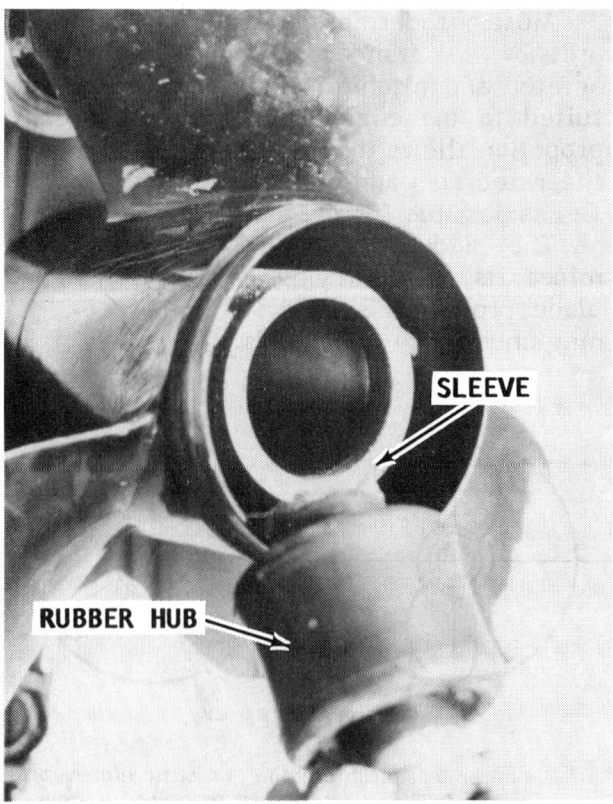

Rubber hub removed from the propeller because the hub was slipping in the propeller.

in the drive unit and the motor. If the vibration becomes very serious it will cause a loss of power, efficiency, and boat performance. If the vibration is allowed to continue over a period of time it can have a damaging effect on many of the operating parts.

Vibration in boats can never be completely eliminated, but it can be reduced by keeping all parts in good working condition and through proper maintenance and lubrication. Vibration can also be reduced in some cases by increasing the number of blades. For this reason, many racers use two-blade props and luxury cruisers have four- and five-blade props installed.

Shock Absorbers

The shock absorber in the propeller plays a very important role in protecting the shafting, gears, and engine against the shock of a blow, should the propeller strike an underwater object. The shock absorber allows the propeller to stop rotating at the instant of impact while the power train continues turning.

How much impact the propeller is able to withstand, before causing the shock absorber to slip, is calculated to be more than the force needed to propel the boat, but less than the amount that could damage any part of the power train. Under normal propulsion loads of moving the boat through the water, the hub will not slip. However, it will slip if the propeller strikes an object with a force that would be great enough to stop any part of the power train.

If the power train was to absorb an impact great enough to stop rotation, even for an instant, something would have to give, resulting in severe damage. If a propeller is subjected to repeated striking of underwater objects, it would eventually slip on its clutch hub under normal loads. If the propeller should start to slip, a new shock absorber/cushion hub would have to be installed.

Propeller Rake

If a propeller blade is examined on a cut extending directly through the center of the hub, and if the blade is set vertical to the propeller hub, as shown in the accompanying illustration, the propeller is said to have a zero degree ($0°$) rake. As the blade slants back, the rake increases. Standard propellers have a rake angle from $0°$ to $15°$.

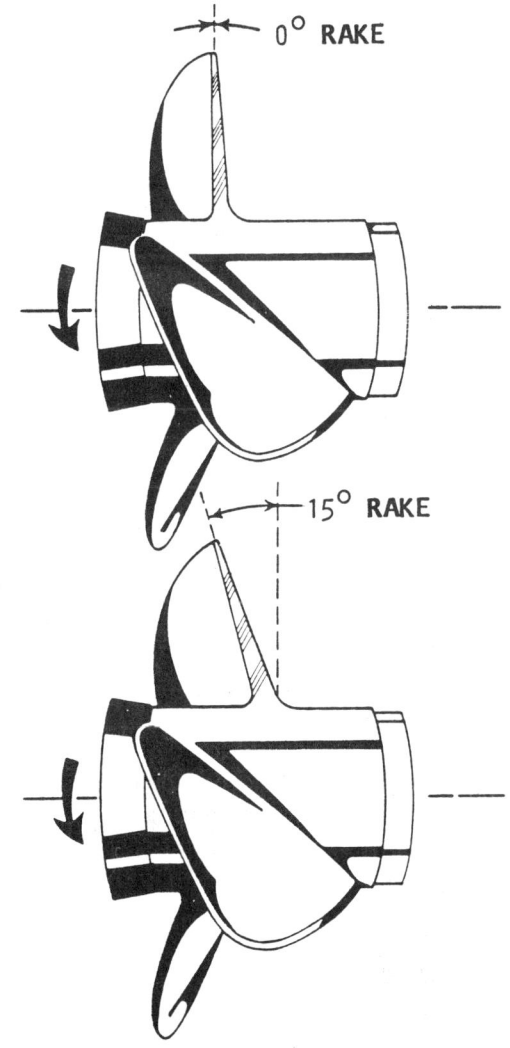

Example of a damaged propeller. This unit should have been replaced long before this amount of damage was sustained.

Illustration depicting the rake of a propeller, as explained in the text.

1-6 SAFETY

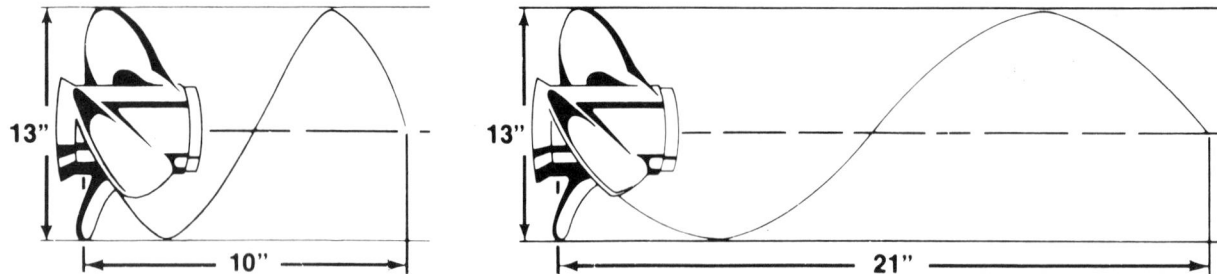

Diagram to explain the pitch dimension of a propeller. The pitch is the theoretical distance a propeller would travel through water if there were no friction.

A higher rake angle generally improves propeller performance in a cavitating or ventilating situation. On lighter, faster boats, higher rake often will increase performance by holding the bow of the boat higher.

Progressive Pitch

Progressive pitch is a blade design innovation that improves performance when forward and rotational speed is high and/or the propeller breaks the surface of the water.

Progressive pitch starts low at the leading edge and progressively increases to the trailing edge, as shown in the accompanying illustration. The average pitch over the entire blade is the number assigned to that propeller. In the illustration of the progressive pitch, the average pitch assigned to the propeller would be 21.

Cupping

If the propeller is cast with an edge curl inward on the trailing edge, the blade is said to have a cup. In most cases, cupped blades improve performance. The cup helps the blades to "HOLD" and not break loose, when operating in a cavitating or ventilating situation.

The cup has the effect of adding to the propeller pitch. Cupping usually will reduce full-throttle engine speed about 150 to 300 rpm below the same pitch propeller without a cup to the blade. A propeller repair shop is able to increase or decrease the cup on the blades. This change, as explained, will alter powerhead rpm to meet specific operating demands. Cups are rapidly becoming standard on propellers.

In order for a cup to be the most effective, the cup should be completely concave (hollowed) and finished with a sharp corner. If the cup has any convex rounding, the effectiveness of the cup will be reduced.

Rotation

Propellers are manufactured as right-hand rotation (RH), and as left-hand rotation (LH). The standard propeller for outboard units is RH rotation.

A right-hand propeller can easily be identified by observing it as shown in the accompanying illustration. Observe how the blade of the right-hand propeller slants from the lower left to upper right. The left-hand propeller slants in the opposite direction, from lower right to upper left.

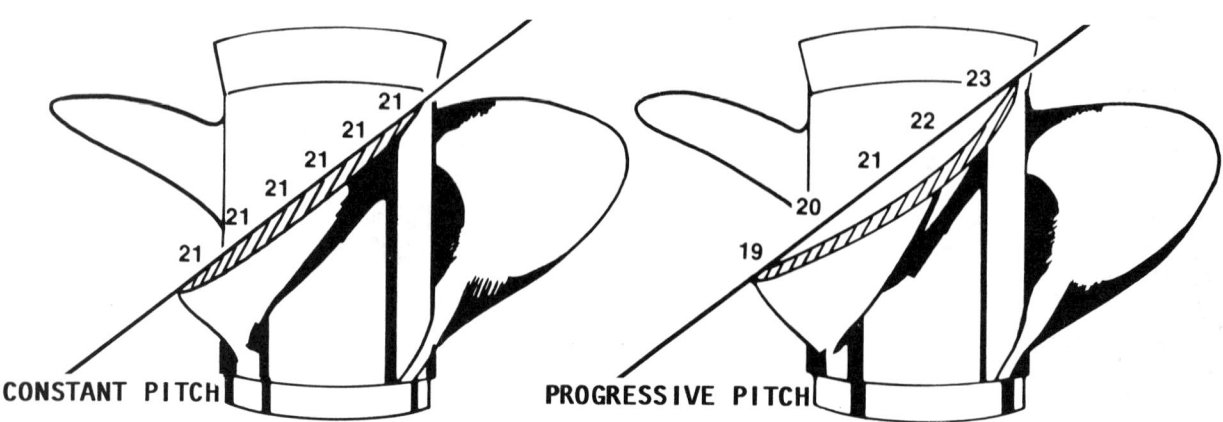

Comparison of a constant and progressive pitch propeller. Notice how the pitch of the progressive propeller, right, changes to give the blade more thrust and therefore, the boat more speed.

PROPELLERS 1-7

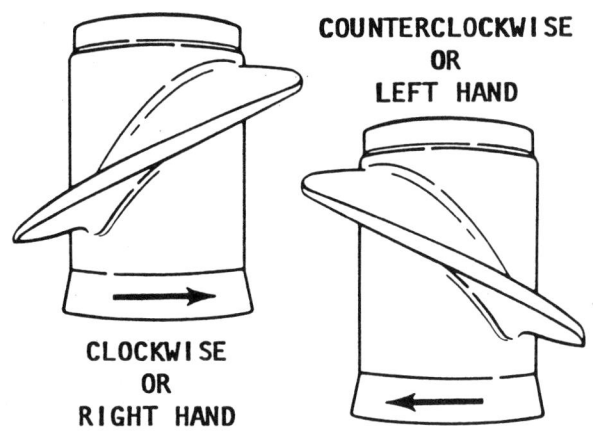

Right- and left-hand propellers showing how the angle of the blades is reversed. Right-hand propellers are by far the most popular for outboard units.

When the RH propeller is observed rotating from astern the boat, it will be rotating clockwise when the outboard unit is in forward gear. The left-hand propeller will rotate counterclockwise.

High Performance Propellers

The term "high performance" is usually associated with, or has the connotation of, something used only for racing. The Yamaha high performance propeller does not fit this category and is not considered an "aftermarket" item.

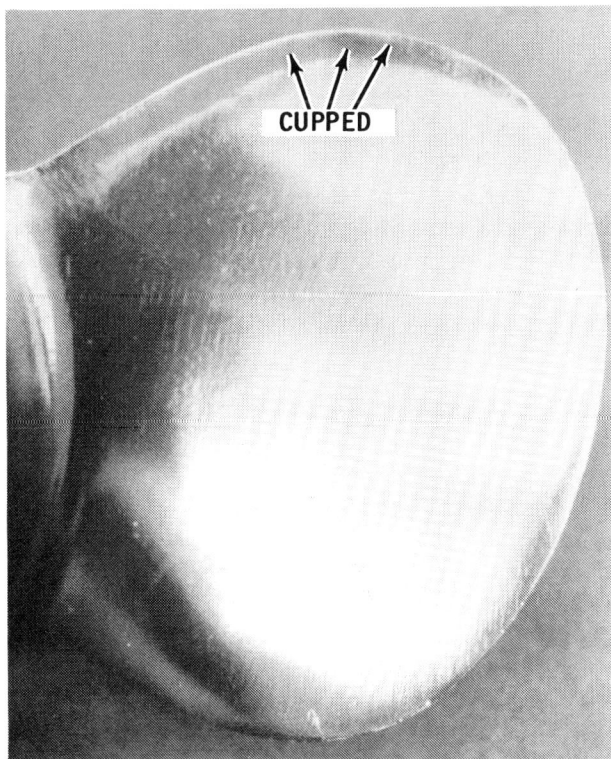

Propeller with a "cupped" leading edge. "Cupping" gives the propeller a better "hold" in the water.

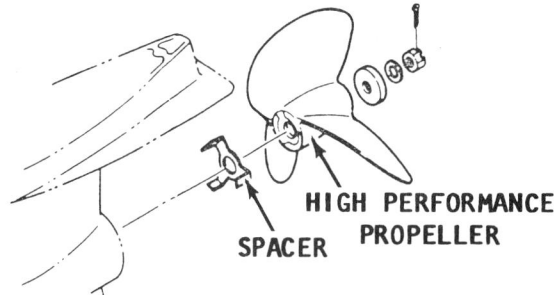

Arrangement of attaching hardware for a high performance propeller.

The Yamaha high performance propeller is made of stainless steel with sophisticated designed blades and carries an embossed **"P"** for positive identification.

The accompanying illustration of a high performance propeller clearly shows the unique design of the long blades and other features. This propeller is the weedless type, having extra sharp blades.

Installation of a high performance propeller requires raising the transom height, trim out after planing, and installation of a special design trim tab. Installation of a water pressure gauge is highly recommended, because raising the transom height will affect the amount of water entering the lower unit through the intake holes. An inadequate amount of water taken in will certainly cause powerhead cooling problems.

The high performance propeller has standard attaching hardware with the exception of the inner spacer which is a special three-pronged design, as shown in the accompanying illustration.

Example of a high performance propeller. Notice how the blades are more narrow with more pitch than a standard propeller of the same size.

1-5 JET DRIVE

Description and Operation

The jet drive unit is designed to permit boating in areas prohibited to a boat equipped with a conventional propeller drive system. The housing of the jet drive barely extends below the hull of the boat allowing passage in ankle-deep water, white water rapids, and over sand bars or in shoal water which would foul a propeller drive.

The jet drive provides reliable propulsion with a minimum of moving parts. Simply stated, water is drawn into the unit through an intake grille by an impeller driven by a driveshaft off the crankshaft of the powerhead, and then expelled under pressure through an outlet nozzle directed away from the stern of the boat.

As the speed of the boat increases and reaches planing speed, the jet drive discharges water freely into the air, and only the intake grille touches the water.

The jet drive is provided with a gate arrangement and linkage permitting the boat to be operated in reverse. When the gate is moved downward over the exhaust nozzle, the pressure stream is reversed by the gate and the boat moves sternward.

Conventional controls are used for powerhead speed, movement of the boat, and shifting.

Model Identification and Serial Numbers

A model letter identification is stamped on the rear, port side casing. A serial number for the unit is stamped on the starboard side of the casing as indicated in the accompanying illustration.

Operational Hints

Wear on the jet impeller and water intake grille will be greatly reduced if the intake of sand, gravel, and rocks, can be avoided.

When approaching a beach, develop the habit of shutting down the powerhead in water less than six inches below the hull. If the water under the hull becomes less than three inches, loose particles on the floor will be sucked into the intake grille. The manufacturer recommends keeping the powerhead speed within the idle range when moving through water less than twelve inches under the hull. Once the water depth increases, with more than a foot of water under the hull, the boat speed may be increased. As boat speed increases, the danger of sucking up rocks is greatly reduced. The faster the boat moves, the ability to "suck" up rocks becomes negligible because the short span of time the grille is over the rock.

Powerhead Stall

If a rock should be sucked up and pass through the grille and jam between the jet impeller and the intake wall, the engine will stall, or attempt to stall. If the powerhead

A jet drive unit mounted, adjusted, and ready to serve the owner with a fun day on the water.

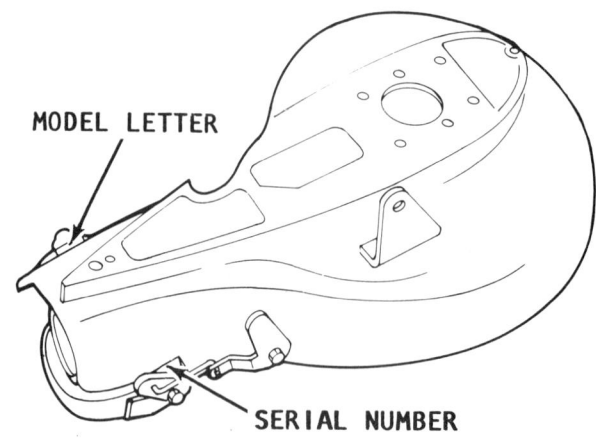

The model letter is embossed on the port side and the serial number on the starboard side, as indicated.

JET DRIVE 1-9

sounds as if it is attempting to stall, shut the powerhead down **IMMEDIATELY**.

The shear pin will most likely prevent any damage to the powerhead. An indication the shear pin has actually done its job and has interrupted the drive train, is a sudden increase in powerhead rpm. Again, the unit should be shut down **AT ONCE**. If the shear pin has "sheared" little or no cooling water will be available to the powerhead. Remove and replace the shear pin. See Chapter 11 for detailed instructions.

Grille Blockage

If a sudden loss of power to move the boat is experienced, the cause is probably an object blocking the intake grille. A paper or plastic bag, large section of kelp or seaweed, or even a large flat rock may be preventing the intake and exhaust of sufficient water to propel the boat.

If the boat if left unattended in shallow water with the jet drive in the water and there is considerable wave action, the cav-

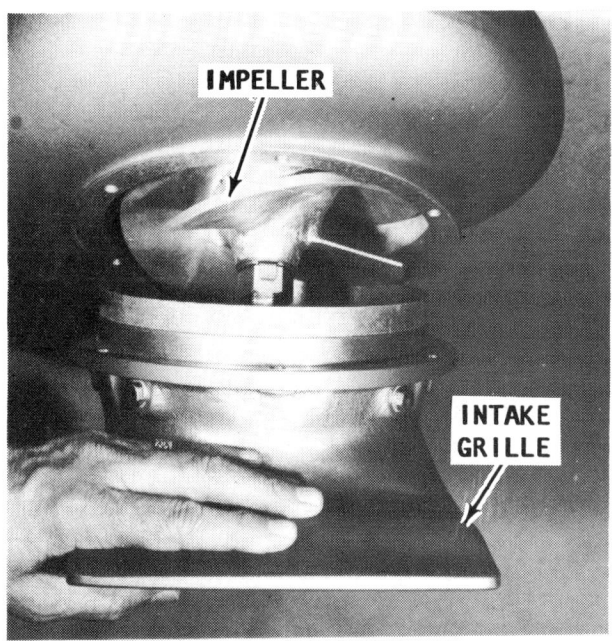

The intake grille must be removed to clear an obstruction lodged between the impeller and the housing.

ity around the impeller could fill with sand in about three hours, or less. Therefore, if the boat is to be left for any period of time under the conditions just described, raise the outboard unit to the full **UP** and locked position.

If a large quantity of sand should be deposited in the jet drive cavity, any attempt to start the powerhead would fail. The outboard unit must be raised, and the sand flushed out with buckets of water before returning the unit to operational status.

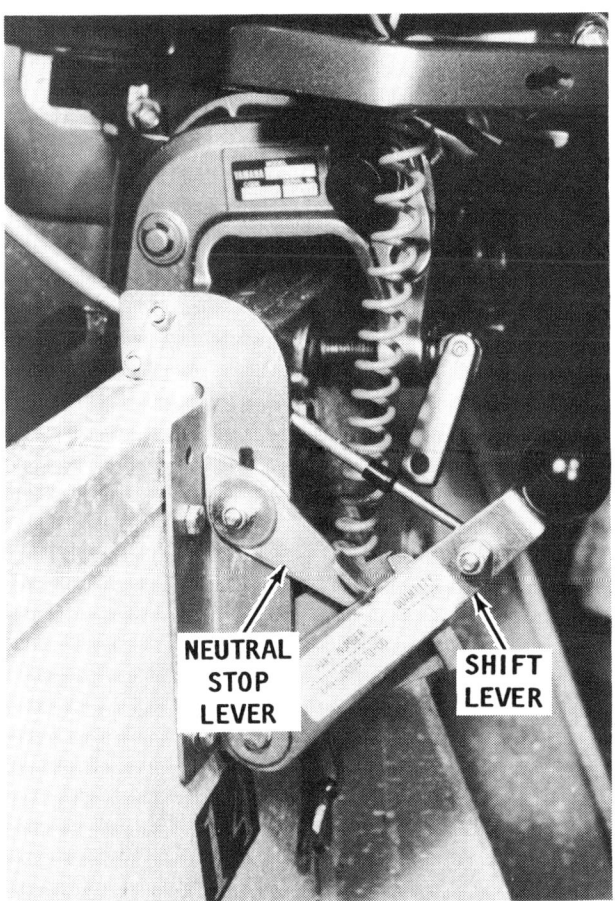

*The **NEUTRAL** stop lever ensures the shift lever will not creep into reverse gear through vibration or by accident.*

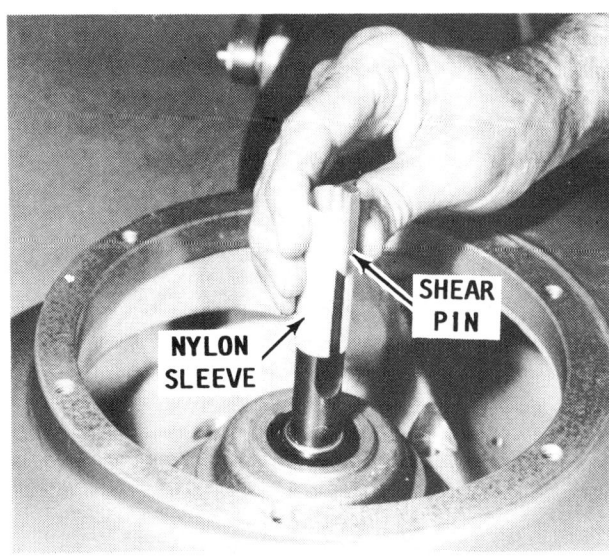

Installing a new shear pin inside the nylon sleeve. The shear pin prevents serious injury to the outboard unit.

1-6 FUEL SYSTEM

With Built-in Fuel Tank

All parts of the fuel system should be selected and installed to provide maximum service and protection against leakage. Reinforced flexible sections should be installed in fuel lines where there is a lot of motion, such as at the engine connection. The flaring of copper tubing should be annealed after it is formed as a protection against hardening.

CAUTION: Compression fittings should **NOT** be used because they are so easily overtightened, which places them under a strain and subjects them to fatigue. Such conditions will cause the fitting to leak after it is connected a second time.

The capacity of the fuel filter must be large enough to handle the demands of the engine as specified by the engine manufacturer.

A manually-operated valve should be installed if anti-siphon protection is not provided. This valve should be installed in the fuel line as close to the gas tank as possible. Such a valve will maintain anti-siphon protection between the tank and the engine.

The supporting surfaces and hold-downs must fasten the tank firmly and they should be insulated from the tank surfaces. This insulation material should be non-abrasive and nonabsorbent material. Fuel tanks installed in the forward portion of the boat should be especially well secured and protected because shock loads in this area can be as high as 20 to 25 g's ("g" equals force of gravity).

Taking On Fuel

The fuel tank of the boat should be kept full to prevent water from entering the system through condensation caused by temperature changes. Water droplets forming is one of the greatest enemies of the fuel system. By keeping the tank full, the air space in the tank is kept to an absolute minimum and there is no room for moisture to form. It is a good practice not to store fuel in the tank over an extended period, say for six months. Today, fuels contain ingredients that change into gums when stored for any length of time. These gums and varnish products will cause carburetor problems and poor spark plug performance. An additive (Yamaha Fuel Conditioner and Stabilizer) is available and can be used to prevent gums and varnish from forming.

Static Electricity

In very simple terms, static electricity is called frictional electricity. It is generated by two dissimilar materials moving over

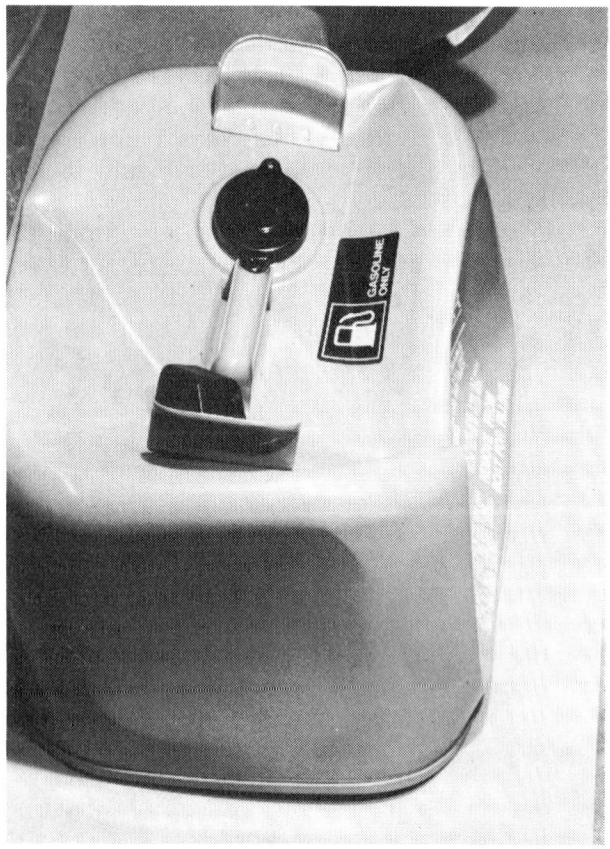

The owner of this portable fuel tank allowed the tank to sit exposed to the hot sun, clamped the cap on tightly, and then moved the tank into a cool area. As the fuel cooled, the volume decreased, causing the inside pressure to also decrease. When the inside pressure became lower than the outside atmospheric pressure, the side of the tank buckled inward. Lesson learned: Leave the cap loose when a "hot" tank is moved to a cool area.

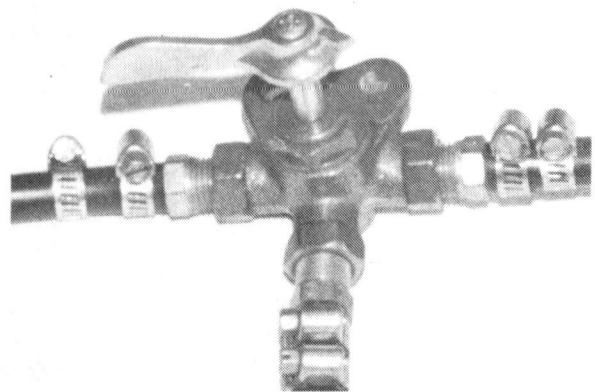

A three-position valve permits fuel to be drawn from either tank or to be shut off completely. Such an arrangement prevents accidental siphoning of fuel from the tank.

each other. One form is gasoline flowing through a pipe or into the air. Another form is when you brush your hair or walk across a synthetic carpet and then touch a metal object. All of these actions cause an electrical charge. In most cases, static electricity is generated during very dry weather conditions, but when you are filling the fuel tank on a boat it can happen at any time.

Fuel Tank Grounding

One area of protection against the buildup of static electricity is to have the fuel tank properly grounded (also known as bonding). A direct metal-to-metal contact from the fuel hose nozzle to the water in which the boat is floating. If the fill pipe is made of metal, and the fuel nozzle makes a good contact with the deck plate, then a good ground is made.

As an economy measure, some boats use rubber or plastic filler pipes because of compound bends in the pipe. Such a fill line does not give any kind of ground and if your boat has this type of installation and you do not want to replace the filler pipe with a metal one, then it is possible to connect the deck fitting to the tank with a copper wire. The wire should be 8 gauge or larger.

The fuel line from the tank to the engine should provide a continuous metal-to-metal contact for proper grounding. If any part of this line is plastic or other non-metallic material, then a copper wire must be connected to bridge the non-metal material. The power train provides a ground through the engine and drive shaft, to the propeller in the water.

Fiberglass fuel tanks pose problems of their own. Fortunately, this material has almost totally disappeared as a suitable substance for fuel tanks. If, however, the boat you are servicing, does have a fiberglass tank, or one is being installed, or repaired, it is almost mandatory that you check with the Coast Guard Recreational Boating Standards Office in your district before proceeding with any work. The new standards are very specific and the Coast Guard is extremely rigid about enforcing the regulations.

Anything you can feel as a "shock" is enough to set off an explosion. Did you know that under certain atmospheric conditions you can cause a static explosion yourself, particularly if you are wearing synthetic clothing. It is almost a certainty you could cause a static spark if you are **NOT** wearing insulated rubber-soled shoes.

As soon as the deck fitting is opened, fumes are released to the air. Therefore, to be safe you should ground yourself before opening the fill pipe deck fitting. One way to ground yourself is to dip your hand in the water overside to discharge the electricity in your body before opening the filler cap. Another method is to touch the engine block or any metal fitting on the dock which goes down into the water.

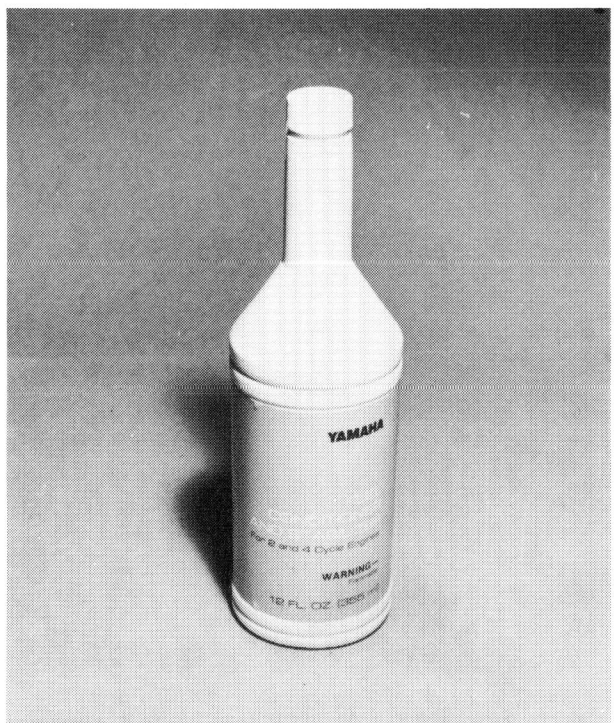

An approved fuel additive will prevent fuel from "souring" for up to twelve full months.

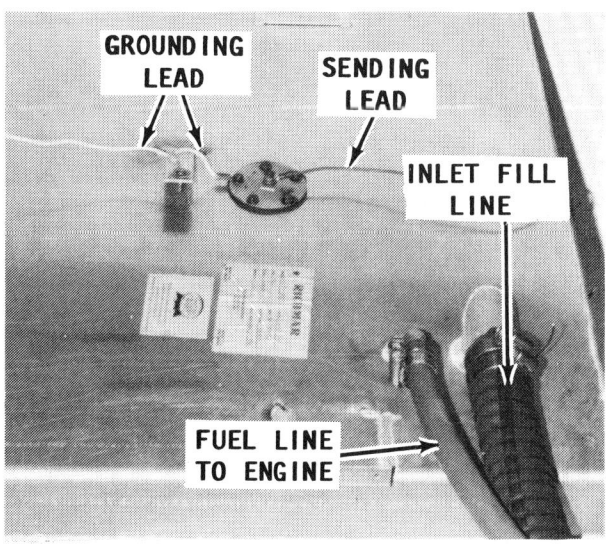

A fuel tank properly grounded to prevent static electricity. Static electricity could be extremely dangerous when taking on fuel.

1-7 LOADING

A plate attached to the hull indicates the U.S. Coast Guard capacity information in pounds for persons and gear. If the plate states the maximum person capacity to be 750 pounds and the assumption is made each person weighs an average of 150 lbs., then the boat could carry five persons safely. If another 250 lbs. is added for motor and gear, and the maximum weight capacity for persons and gear is 1,000 lbs. or more, then the five persons and gear would be within the limit.

Clarification

Much confusion arises from the terms, certification, requirements, approval, regulations, etc. Perhaps the following may clarify a couple of these points.

1- The Coast Guard does not approve boats in the same manner as they "Approve" life jackets. The Coast Guard applies a formula to inform the public of what is safe for a particular craft.

2- If a boat has to meet a particular regulation, it must have a Coast Guard certification plate. The public has been led to believe this indicates approval of the Coast Guard. Not so.

3- The certification plate means a willingness of the manufacturer to meet the Coast Guard regulations for that particular craft. The manufacturer may recall a boat if it fails to meet the Coast Guard requirements.

4- The Coast Guard certification plate, see accompanying illustration, may or may not be metal. The plate is a regulation for the manufacturer. It is only a warning plate and the public does not have to adhere to the restrictions set forth on it. Again, the plate sets forth information as to the Coast Guard's opinion for safety on that particular boat.

5- Coast Guard Approved equipment is equipment which has been approved by the Commandant of the U.S. Coast Guard and has been determined to be in compliance with Coast Guard specifications and regulations relating to the materials, construction, and performance of such equipment.

1-8 HORSEPOWER

The maximum horsepower engine for each individual boat should not be increased by any significant amount without checking requirements from the Coast Guard in the local area. The Coast Guard determines horsepower requirements based on the length, beam, and depth of the hull. **TAKE CARE NOT** to exceed the maximum horsepower listed on the plate or the warranty, and possibly the insurance, on the boat may become void.

1-9 FLOTATION

If the boat is less than 20 ft. overall, a Coast Guard or BIA (Boating Industry of America), now changed to NMMA (National Marine Manufacturers Association) requirement is that the boat must have buoyant

U.S. Coast Guard plate affixed to all new boats. When the blanks are filled in, the plate will indicate the Coast Guard's recommendations for persons, gear, and horsepower to ensure safe operation of the boat. These recommendations should not be exceeded, as explained in the text.

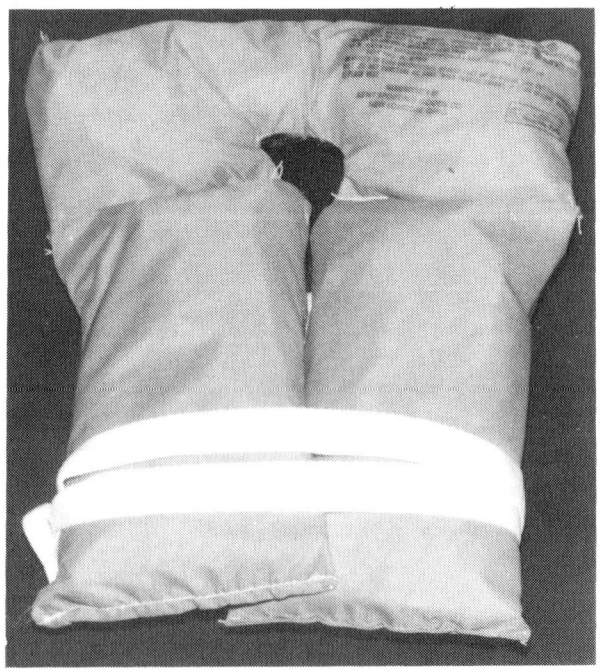

*Type I PFD Coast Guard approved life jacket. This type flotation device provides the greatest amount of buoyancy. **NEVER** use them for cushions or other purposes.*

FLOTATION 1-13

material built into the hull (usually foam) to keep it from sinking if it should become swamped. Coast Guard requirements are mandatory but the NMMA is voluntary.

"Kept from sinking" is defined as the ability of the flotation material to keep the boat from sinking when filled with water and with passengers clinging to the hull. One restriction is that the total weight of the motor, passengers, and equipment aboard does not exceed the maximum load capacity listed on the plate.

Life Preservers — Personal Flotation Devices (PFDs)

The Coast Guard requires at least one Coast Guard approved life-saving device be carried on board all motorboats for each person on board. Devices approved are identified by a tag indicating Coast Guard approval. Such devices may be life preservers, buoyant vests, ring buoys, or buoyant cushions. Cushions used for seating are serviceable if air cannot be squeezed out of it. Once air is released when the cushion is squeezed, it is no longer fit as a flotation device. New foam cushions dipped in a rubberized material are almost indestructable.

Life preservers have been classified by the Coast Guard into five type categories. All PFDs presently acceptable on recreational boats fall into one of these five designations. All PFDs **MUST** be U.S. Coast Guard approved, in good and serviceable condition, and of an appropriate size for the persons who intend to wear them. Wearable PFDs **MUST** be readily accessible and throwable devices **MUST** be immediately available for use.

Type I PFD has the greatest required buoyancy and is designed to turn most **UNCONSCIOUS** persons in the water from a face down position to a vertical or slightly backward position. The adult size device provides a minimum buoyancy of 22 pounds and the child size provides a minimum buoyancy of 11 pounds. The Type I PFD provides the greatest protection to its wearer and is most effective for all waters and conditions.

Type II PFD is designed to turn its wearer in a vertical or slightly backward position in the water. The turning action is not as pronounced as with a Type I. The device will not turn as many different type persons under the same conditions as the Type I. An adult size device provides a minimum buoyancy of 15½ pounds, the medium child size provides a minimum of 11 pounds, and the infant and small child sizes provide a minimum buoyancy of 7 pounds.

Type III PFD is designed to permit the wearer to place himself (herself) in a vertical or slightly backward position. The Type III device has the same buoyancy as the Type II PFD but it has little or no turning ability. Many of the Type III PFD are

A Type IV PFD cushion device intended to be thrown to a person in the water. If air can be squeezed out of the cushion, it is no longer fit for service as a PFD.

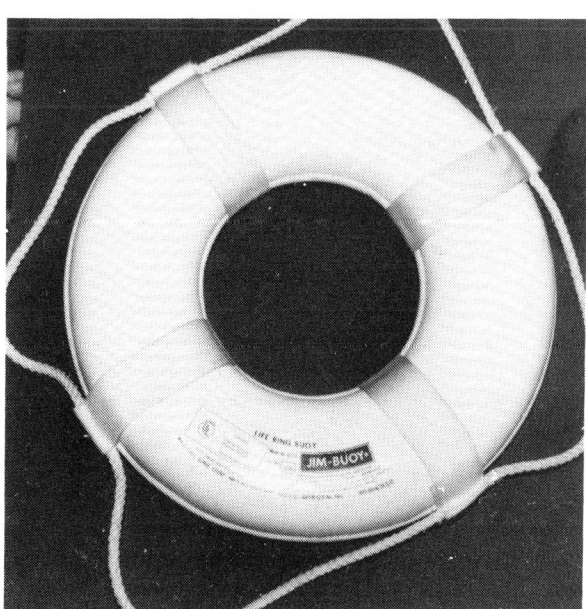

Type IV ring buoy also designed to be thrown to a person in the water. On ocean cruisers, this type device usually has a weighted pole with flag and light attached to the buoy.

designed to be particularly useful when water skiing, sailing, hunting, fishing, or engaging in other water sports. Several of this type will also provide increased hypothermia protection.

Type IV PFD is designed to be thrown to a person in the water and grasped and held by the user until rescued. It is **NOT** designed to be worn. The most common Type IV PFD is a ring buoy or a buoyant cushion.

Type V PFD is any PFD approved for restricted use.

Coast Guard regulations state, in general terms: All boats less than 16 ft. overall, one Type I, II, III, or IV device shall be carried on board for each person in the boat. On boats over 26 ft., one Type I, II, or III device shall be carried on board for each person in the boat **plus** one Type IV device.

It is an accepted fact, most boating people own life preservers, but too few actually wear them. There is little or no excuse for not wearing one because the modern comfortable designs available today do not subtract from an individual's boating pleasure. Make a life jacket available to the crew and advise each member to wear it. If you are a crew member, ask the skipper to issue one, especially when boating in rough weather, cold water, or when running at high speed. Naturally, a life jacket should be a must for non-swimmers any time they are out on the water in a boat.

1-10 EMERGENCY EQUIPMENT

Visual Distress Signals
The Regulation

Since January 1, 1981, Coast Guard Regulations require all recreation boats when used on coastal waters, which includes the Great Lakes, the territorial seas and those waters directly connected to the Great Lakes and the territorial seas, up to a point where the waters are less than two miles wide, and boats owned in the United States, when operating on the high seas, to be equipped with visual distress signals.

The only exceptions are during daytime (sunrise to sunset) for:

Recreational boats less than 16 ft. (5 meters) in length.

Boats participating in organized events such as races, regattas or marine parades.

Open sailboats not equipped with propulsion machinery and less than 26 ft. (8 meters) in length.

Manually propelled boats.

The above listed boats need to carry night signals when used on these waters at night.

Pyrotechnic visual distress signaling devices **MUST** be Coast Guard Approved, in serviceable condition and stowed to be readily accessible. If they are marked with a date showing the serviceable life, this date must not have passed. Launchers, produced before Jan. 1, 1981, intended for use with approved signals are not required to be Coast Guard Approved.

USCG Approved pyrotechnic visual distress signals and associated devices include:

Pyrotechnic red flares, hand held or aerial.

Pyrotechnic orange smoke, hand held or floating.

Launchers for aerial red meteors or parachute flares.

Non-pyrotechnic visual distress signaling devices must carry the manufacturer's certification that they meet Coast Guard requirements. They must be in serviceable

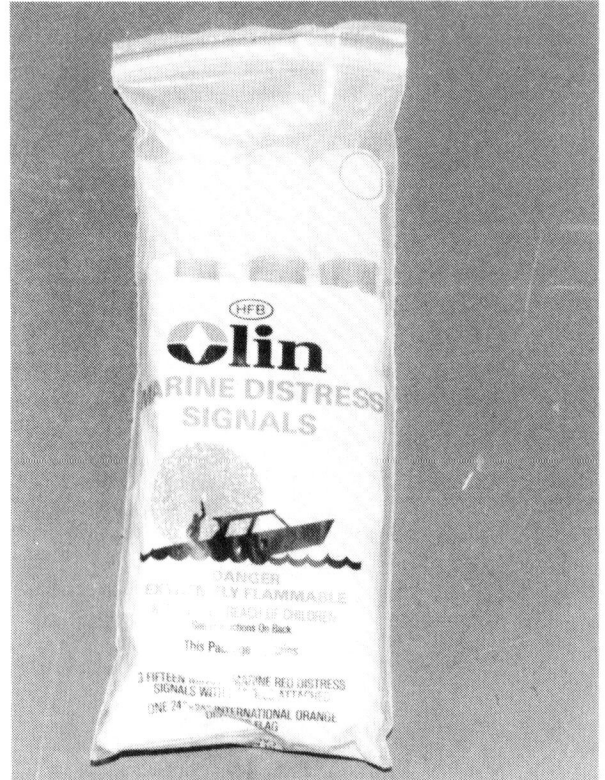

Moisture-protected flares should be carried on board for use as a distress signal.

condition and stowed so as to be readily accessible.

This group includes:

Orange distress flag at least 3 x 3 feet with a black square and ball on an orange background.

Electric distress light -- not a flashlight but an approved electric distress light which **MUST** automatically flash the international **SOS** distress signal (. . . - - . . .) four to six times each minute.

Types and Quantities

The following variety and combination of devices may be carried in order to meet the requirements.

1- Three hand-held red flares (day and night).

2- One electric distress light (night only).

3- One hand-held red flare and two parachute flares (day and night).

4- One hand-held orange smoke signal, two floating orange smoke signals (day) and one electric distress light (day and night).

If young children are frequently aboard your boat, careful selection and proper stowage of visual distress signals becomes especially important. If you elect to carry pyrotechnic devices, you should select those in tough packaging and not easy to ignite should the devices fall into the hands of children.

Coast Guard Approved pyrotechnic devices carry an expiration date. This date can **NOT** exceed 42 months from the date of manufacture and at such time the device can no longer be counted toward the minimum requirements.

SPECIAL WORDS

In some states the launchers for meteors and parachute flares may be considered a firearm. Therefore, check with your state authorities before acquiring such a launcher.

First Aid Kits

The first-aid kit is similar to an insurance policy or life jacket. You hope you don't have to use it but if needed, you want it there. It is only natural to overlook this essential item because, let's face it, who likes to think of unpleasantness when planning to have only a good time. However, the prudent skipper is prepared ahead of time, and is thus able to handle the emergency without a lot of fuss.

Good commercial first-aid kits are available such as the Johnson and Johnson "Marine First-Aid Kit". With a very modest expenditure, a well-stocked and adequate kit can be prepared at home.

Any kit should include instruments, supplies, and a set of instructions for their use. Instruments should be protected in a watertight case and should include: scissors, tweezers, tourniquet, thermometer, safety pins, eye-washing cup, and a hot water bottle. The supplies in the kit should include: assorted bandages in addition to the various sizes of "band-aids", adhesive tape, absorbent cotton, applicators, petroleum jelly, antiseptic (liquid and ointment), local oint-

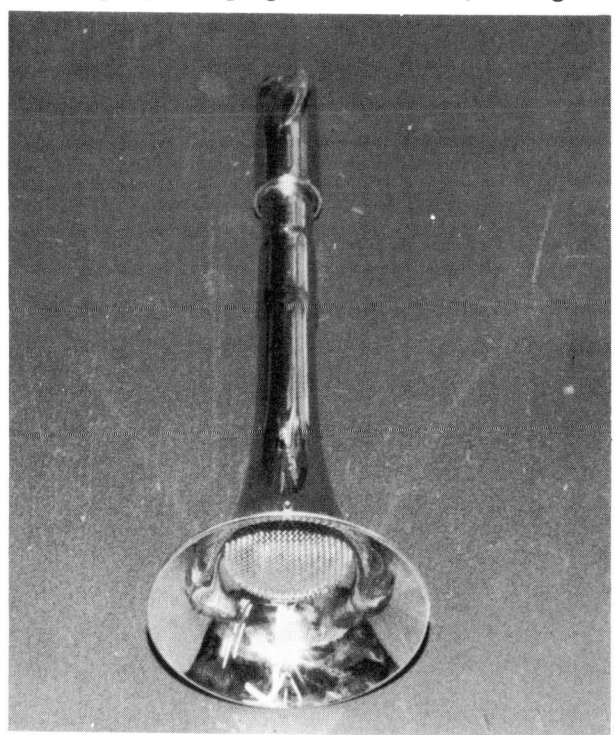

A sounding device should be mounted close to the helmsman for use in sounding an emergency alarm.

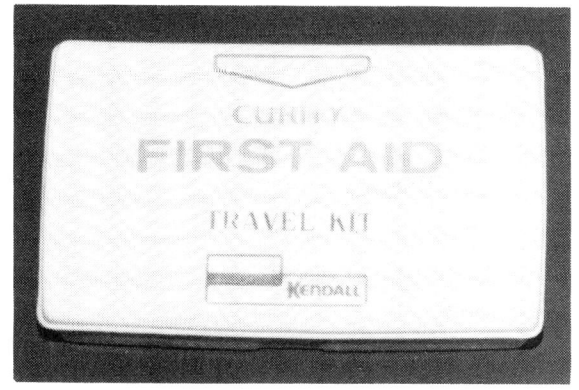

An adequately stocked first aid kit should be on board for the safety of crew and guests.

ment, aspirin, eye ointment, antihistamine, ammonia inhalent, sea-sickness pills, antacid pills, and a laxative. You may want to consult your family physician about including antibiotics. Be sure your kit contains a first-aid manual because even though you have taken the Red Cross course, you may be the patient and have to rely on an untrained crew for care.

Fire Extinguishers

All fire extinguishers must bear Underwriters Laboratory (UL) "Marine Type" approved labels. With the UL certification, the extinguisher does not have to have a Coast Guard approval number. The Coast Guard classifies fire extinguishers according to their size and type.

Type B-I or B-II Designed for extinguishing flammable liquids. Required on all motorboats.

The Coast Guard considers a boat having one or more of the following conditions as a "boat of closed construction" subject to fire extinguisher regulations.

1- Inboard engine or engines.

2- Closed compartments under thwarts and seats wherein portable fuel tanks may be stored.

3- Double bottoms not sealed to the hull or which are not completely filled with flotation materials.

4- Closed living spaces.

5- Closed stowage compartments in which combustible or flammable material is stored.

6- Permanently installed fuel tanks.

Detailed classification of fire extinguishers is by agent and size:

B-I contains 1-1/4 gallons foam, 4 pounds carbon dioxide, 2 pounds dry chemical, and 2-1/2 pounds freon.

B-II contains 2-1/2 gallons foam, 15 pounds carbon dioxide, and 10 pounds dry chemical.

The class of motorboat dictates how many fire extinguishers are required on board. One B-II unit can be substituted for two B-I extinguishers.

Dry chemical fire extinguishers without gauges or indicating devices must be weighed and tagged every 6 months. If the gross weight of a carbon dioxide (CO_2) fire extinguisher is reduced by more than 10% of the net weight, the extinguisher is not acceptable and must be recharged.

READ labels on fire extinguishers. If the extinguisher is U.L. listed, it is approved for marine use.

DOUBLE the number of fire extinguishers recommended by the Coast Guard, because their requirements are a bare **MINIMUM** for safe operation. Your boat, family, and crew, must certainly be worth much more than "bare minimum".

1-11 COMPASS

Selection

The safety of the boat and her crew may depend on her compass. In many areas weather conditions can change so rapidly that within minutes a skipper may find himself "socked-in" by a fog bank, a rain squall, or just poor visibility. Under these conditions, he may have no other means of keeping to his desired course except with the compass. When crossing an open body of water, his compass may be the only means of making an accurate landfall.

During thick weather when you can neither see nor hear the expected aids to navigation, attempting to run out the time on a given course can disrupt the pleasure of the cruise. The skipper gains little comfort in a chain of soundings that does not match those given on the chart for the expected

A suitable fire extinguisher should be mounted close to the helmsperson for emergency use.

area. Any stranding, even for a short time, can be an unnerving experience.

A pilot will not knowingly accept a cheap parachute. A good boater should not accept a bargain in lifejackets, fire extinguishers, or compass. Take the time and spend the few extra dollars to purchase a compass to fit your expected needs. Regardless of what the salesman may tell you, postpone buying until you have had the chance to check more than one make and model.

Lift each compass, tilt and turn it, simulating expected motions of the boat. The compass card should have a smooth and stable reaction.

The card of a good quality compass will come to rest without oscillations about the lubber's line. Reasonable movement in your hand, comparable to the rolling and pitching of the boat, should not materially affect the reading.

Installation

Proper installation of the compass does not happen by accident. Make a critical check of the proposed location to be sure compass placement will permit the helmsman to use it with comfort and accuracy. First, the compass should be placed directly in front of the helmsman and in such a position that it can be viewed without body stress as he sits or stands in a posture of relaxed alertness. The compass should be in the helmsman's zone of comfort. If the compass is too far away, he may have to bend forward to watch it; too close and he must rear backward for relief.

Second, give some thought to comfort in heavy weather and poor visibilty conditions during the day and night. In some cases, the compass position may be partially determined by the location of the wheel, shift lever, and throttle handle.

Third, inspect the compass site to be sure the instrument will be at least two feet from any engine indicators, bilge vapor detectors, magnetic instruments, or any steel or iron objects. If the compass cannot be placed at least two feet (six feet would be better) from one of these influences, then either the compass or the other object must be moved, if first order accuracy is to be expected.

Once the compass location appears to be satisfactory, give the compass a test before installation. Hidden influences may be concealed under the cabin top, forward of the cabin aft bulkhead, within the cockpit ceiling, or in a wood-covered stanchion.

Move the compass around in the area of the proposed location. Keep an eye on the card. A magnetic influence is the only thing that will make the card turn. You can

Do not hesitate to spend a few extra dollars for a good, reliable compass. If in doubt, seek advice from fellow boaters.

The compass is a delicate instrument and deserves respect. It should be mounted securely and in position where it can be easily observed by the helmsperson.

quickly find any such influence with the compass. If the influence can not be moved away or replaced by one of non-magnetic material, test to determine whether it is merely magnetic, a small piece of iron or steel, or some magnetized steel. Bring the north pole of the compass near the object, then shift and bring the south pole near it. Both the north and south poles will be attracted if the compass is demagnetized. If the object attracts one pole and repels the other, then the compass is magnetized. If your compass needs to be demagnetized,

"Innocent" objects close to the compass, such as diet coke in an aluminum can, may cause serious problems and lead to disaster, as these three photos and the accompanying text prove.

take it to a shop equipped to do the job **PROPERLY**.

After you have moved the compass around in the proposed mounting area, hold it down or tape it in position. Test everything you feel might affect the compass and cause a deviation from a true reading. Rotate the wheel from hard over to hard over. Switch on and off all the lights, radios, radio direction finder, radio telephone, depth finder and the shipboard intercom, if one is installed. Sound the electric whistle, turn on the windshield wipers, start the engine (with water circulating through the engine), work the throttle, and move the gear shift lever. If the boat has an auxiliary generator, start it.

If the card moves during any one of these tests, the compass should be relocated. Naturally, if something like the windshield wipers causes a slight deviation, it may be necessary for you to make a different deviation table to use only when certain pieces of equipment are operating. Bear in mind, following a course that is only off a degree or two for several hours can make considerable difference at the end, putting you on a reef, rock, or shoal.

Check to be sure the intended compass site is solid. Vibration will increase pivot wear.

Now, you are ready to mount the compass. To prevent an error on all courses, the line through the lubber line and the compass card pivot must be exactly parallel to the keel of the boat. You can establish the fore-and-aft line of the boat with a stout cord or string. Use care to transfer this line to the compass site. If necessary, shim the base of the compass until the stile-type lubber line (the one affixed to the case and not gimbaled) is vertical when the boat is on an even keel. Drill the holes and mount the compass.

Magnetic Items After Installation

Many times an owner will install an expensive stereo system in the cabin of his boat. It is not uncommon for the speakers to be mounted on the aft bulkhead up against the overhead (ceiling). In almost every case, this position places one of the speakers in very close proximity to the compass, mounted above the ceiling.

As we all know, a magnet is used in the operation of the speaker. Therefore, it is very likely that the speaker, mounted almost under the compass in the cabin will have a very pronounced affect on the compass accuracy.

Consider the following test and the accompanying photographs as prove of the statements made.

First, the compass was read as 190 degrees while the boat was secure in her slip.

Next a full can of diet coke in an **aluminum** can was placed on one side and the compass read as 204 degrees, a good 14 degrees off.

Next, the full can was moved to the opposite side of the compass and again a reading was observed. This time as 189 degrees, 11 degrees off from the original reading.

Finally the contents of the can were consumed, the can placed on both sides of the compass with **NO** affect on the compass reading.

Two very important conclusions can be drawn from these tests.

1- Something must have been in the contents of the can to affect the compass so drastically.

2- Keep even "innocent" things clear of the compass to avoid any possible error in the boat's heading.

REMEMBER, a boat moving through the water at 10 knots on a compass error of just 5 degrees will be almost 1.5 miles off course in only **ONE** hour. At night, or in thick weather, this could very possibly put the boat on a reef, rock, or shoal, with disastrous results.

1-12 ANCHORS

One of the most important pieces of equipment in the boat next to the power plant is the ground tackle carried. The engine makes the boat go and the anchor and its line are what hold it in place when the boat is not secured to a dock or on the beach.

The anchor must be of suitable size, type, and weight to give the skipper "peace of mind" when the boat is at anchor. Under certain conditions, a second, smaller, lighter anchor may help to keep the boat in a favorable position during a non-emergency daytime situation.

In order for the anchor to hold properly, a piece of chain must be attached to the anchor and then the nylon anchor line attached to the chain. The amount of chain should equal or exceed the length of the boat. Such a piece of chain will ensure that the anchor stock will lay in an approximate horizontal position and permit the flutes to dig into the bottom and hold.

1-13 MISCELLANEOUS EQUIPMENT

In addition to the equipment you are legally required to carry in the boat and those previously mentioned, some extra items will add to your boating pleasure and safety. Practical suggestions would include: a bailing device (bucket, pump, etc.), boat hook, fenders, spare propeller, spare engine parts, tools, an auxiliary means of propulsion (paddle or oars), spare can of gasoline, flashlight, and extra warm clothing. The area of your boating activity, weather conditions, length of stay aboard your boat, and the specific purpose will all contribute to the kind and amount of stores you put aboard. When it comes to personal gear, heed the advice of veteran boaters who say, "Decide on how little you think you can get by with, then cut it in half".

Bilge Pumps

Automatic bilge pumps should be equipped with an overriding manual switch. They should also have an indicator in the operator's position to advise the helmsman when

*The weight of the anchor **MUST** be adequate to secure the boat without dragging.*

the pump is operating. Select a pump that will stabilize its temperature within the manufacturer's specified limits when it is operated continuously. The pump motor should be a sealed or arcless type, suitable for a marine atmosphere. Place the bilge pump inlets so excess bilge water can be removed at all normal boat trims. The intakes should be properly screened to prevent the pump from sucking up debris from the bilge. Intake tubing should be of a high quality and stiff enough to resist kinking and not collapse under maximum pump suction condition if the intake becomes blocked.

To test operation of the bilge pump, operate the pump switch. If the motor does not run, disconnect the leads to the motor. Connect a voltmeter to the leads and see if voltage is indicated. If voltage is not indicated, then the problem must be in a blown fuse, defective switch, or some other area of the electrical system.

If the meter indicates voltage is present at the leads, then remove, disassemble, and inspect the bilge pump. Clean it, reassemble, connect the leads, and operate the switch again. If the motor still fails to run, the pump must be replaced.

To test the bilge pump switch, first disconnect the leads from the pump and connect them to a test light.

Next, hold the switch firmly against the mounting location in order to make a good ground. Now, tilt the opposite end of the switch upward until it is activated as indicated by the test light coming on or the ohmmeter showing continuity. Finally, lower the switch slowly toward the mounting position until it is deactivated. Measure the distance between the point the switch was activated and the point it was deactivated. For proper service, the switch should deactivate between 1/2-inch and 1/4-inch from the planned mounting position. **CAUTION: The switch must never be mounted lower than the bilge pump pickup.**

1-14 BOATING ACCIDENT REPORTS

In the United States, new federal and state regulations require an accident report to be filed with the nearest state boating authority within 48 hours, if a person is lost, disappears, or is injured. "Injured" is defined as requiring medical attention beyond "First Aid".

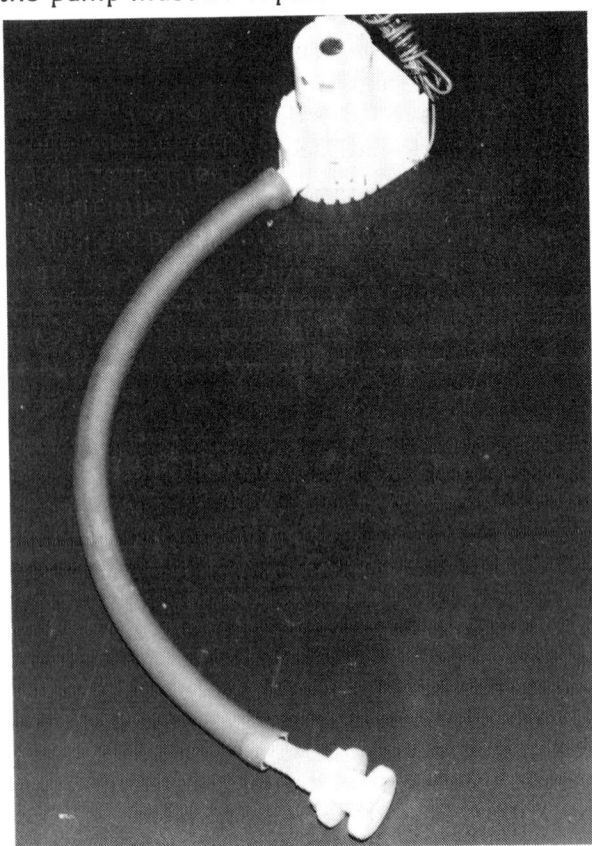

The bilge pump line must be cleaned frequently to ensure the bilge pump will be able to do its job in an emergency.

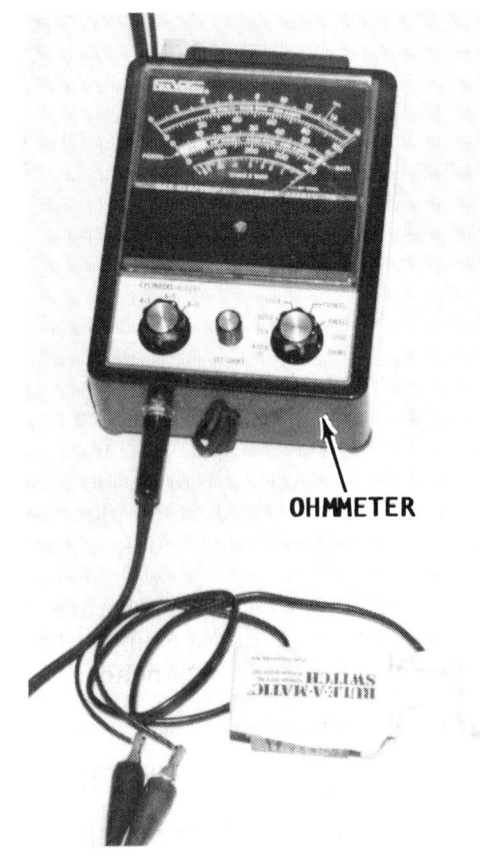

Hookup for testing an automatic bilge pump switch.

NAVIGATION 1-21

Accidents involving only property or equipment damage **MUST** be reported within 10 days if the damage is in excess of $200. Some states are more stringent and require reporting of accidents with property damage less than $200.

A **$500 PENALTY** may be assessed for failure to submit the report.

WORD OF ADVICE

Take time to make a copy of the report to keep for your records or for the insurance company. Once the report is filed, the Coast Guard will not give out a copy, even to the person who filed the report.

The report must give details of the accident and include:

1- The date, time, and exact location of the occurrence.
2- The name of each person who died, was lost, or injured.
3- The number and name of the vessel.
4- The names and addresses of the owner and operator.

If the operator cannot file the report for any reason, each person on board **MUST** notify the authorities, or determine that the report has been filed.

1-15 NAVIGATION

Buoys

In the United States, a buoyage system is used as an assist to all boaters of all size craft to navigate our coastal waters and our navigable rivers in safety. When properly read and understood, these buoys and markers will permit the boater to cruise with comparative confidence that he will be able to avoid reefs, rocks, shoals, and other hazards.

In the spring of 1983, the Coast Guard began making modifications to U.S. aids to navigation in support of an agreement sponsored by the International Associaiton of Lighthouse Authorities (IALA) and signed by representatives from most of the maritime nations of the world. The primary purpose of the modifications is to improve safety by making buoyage systems around the world more alike and less confusing.

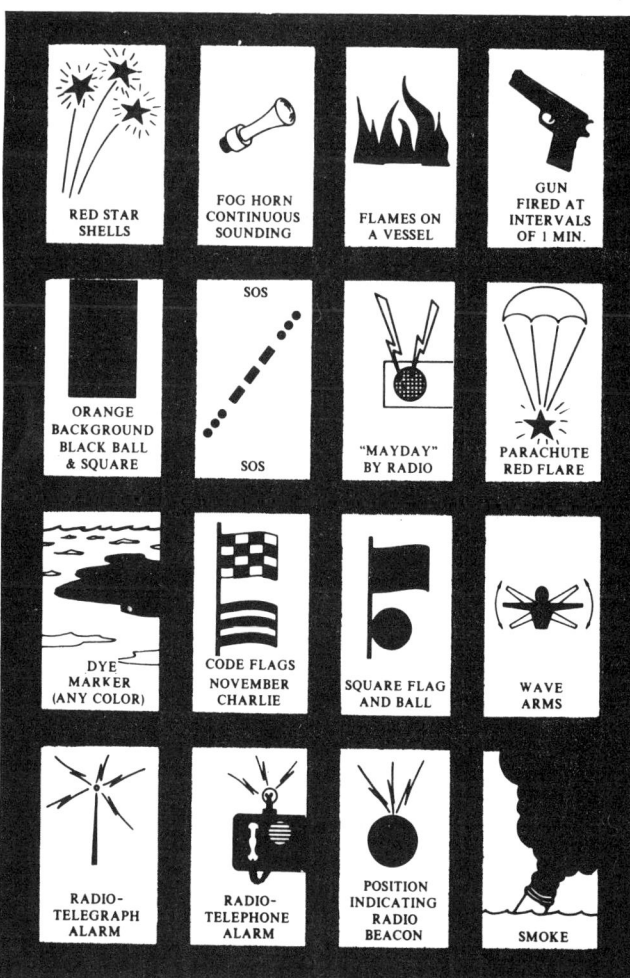

Internationally accepted distress signals.

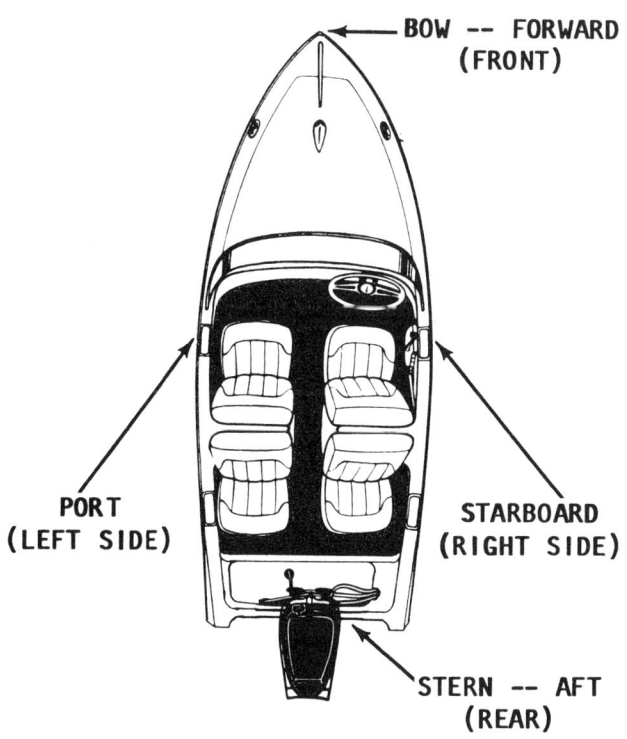

Common terminology used throughout the world for reference designation on boats of all sizes. These are the terms used in this book.

I-22 SAFETY

In nautical terms, the front of the boat is the **bow**; the rear is the **stern.**

The terms **"PORT"** and **"STARBOARD"** are used to refer to the left and right side of the boat, when looking forward. One easy way to remember this basic fundamental is to consider the words "port" and "left" both have four letters and go together.

Waterway Rules

On the water, certain basic safe-operating practices must be followed. You should learn and practice them, for to **know,** is to be able to handle your boat with confidence and safety. Knowledge of what to do, and not do, will add a great deal to the enjoyment you will receive from your boating investment.

Rules of the Road

The best advice possible and a Coast Guard requirement for boats over 39' 4" (12 meters) since 1981, is to obtain an official copy of the "Rules of the Road", which includes Inland Waterways, Western Rivers, and the Great Lakes for study and ready reference.

The following two paragraphs give a **VERY** brief, condensed, and abbreviated -- synopsis of the rules. They should not be considered in any way as covering the entire subject.

Powered boats must yield the right-of-way to all boats without motors, except when being overtaken. When meeting another boat head-on, keep to starboard, unless you are too far to port to make this practical. When overtaking another boat, the right-of-way belongs to the boat being overtaken. If your boat is being passed, you must maintain course and speed.

When two boats approach at an angle and there is danger of collision, the boat to port must give way to the boat to starboard. Always keep to starboard in a narrow channel or canal. Boats underway must stay clear of vessels fishing with nets, lines, or trawls. (Fishing boats are not allowed to fish in channels or to obstruct navigation.)

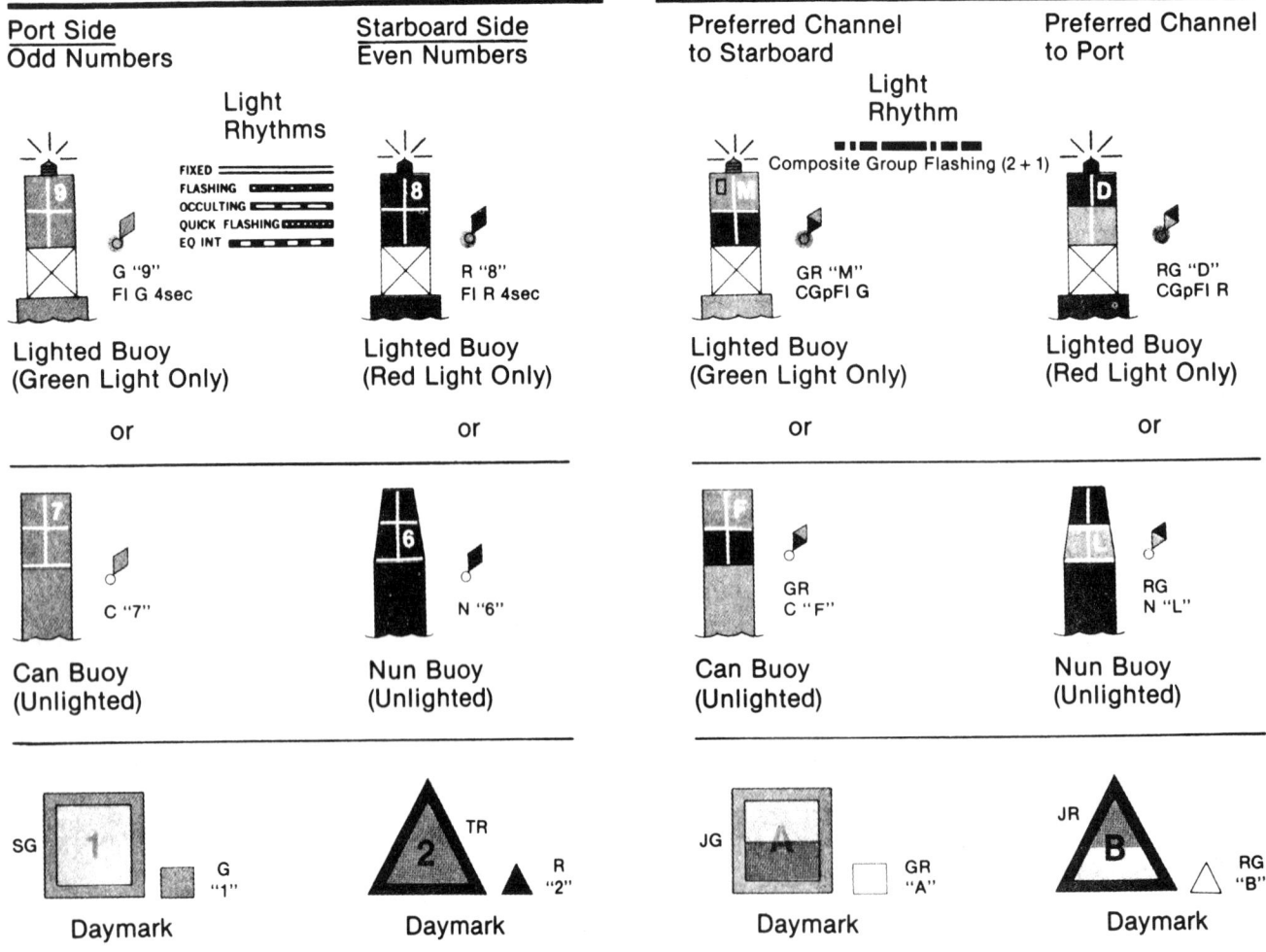

2
TUNING

2-1 INTRODUCTION

The efficiency, reliability, fuel economy and enjoyment available from engine performance are all directly dependent on having it tuned properly. The importance of performing service work in the sequence detailed in this chapter cannot be over emphasized. Before making any adjustments, check the specifications in the Appendix. **NEVER** rely on memory when making critical adjustments.

Before beginning to tune any engine, check to be sure the engine has satisfactory compression. An engine with worn or broken piston rings, burned pistons, or scored cylinder walls, cannot be made to perform properly no matter how much time and expense is spent on the tune-up. Poor compression must be corrected or the tune-up will not give the desired results.

A practical maintenance program that is followed throughout the year, is one of the best methods of ensuring the engine will give satisfactory performance at any time.

The extent of the engine tune-up is usually dependent on the time lapse since the last service. A complete tune-up of the entire engine would entail almost all of the work outlined in this manual. A logical sequence of steps will be presented in general terms. If additional information or detailed service work is required, the chapter containing the instructions will be referenced.

Each year higher compression ratios are built into modern outboard engines and the electrical systems become more complex, especially with electronic (capacitor discharge) units and those controlled by microcomputers. Therefore, the need for reliable, authoritative, and detailed instructions becomes more critical. The information in this chapter and the referenced chapters fulfill that requirement.

2-2 TUNE-UP SEQUENCE

During a major tune-up, a definite sequence of service work should be followed to return the engine to the maximum performance desired. This type of work should not be confused with attempting to locate problem areas of "why" the engine is not performing satisfactorily. This work is classified as "trouble shooting". In many cases, these two areas will overlap, because many times a minor or major tune-up will correct the malfunction and return the system to normal operation.

Outboard mounted on the boat ready to be backed into the water or have a flush attachment connected in preparation to making fine tuning adjustments.

2-2 TUNING

The following list is a suggested sequence of tasks to perform during the tune-up service work. The tasks are merely listed here. Generally procedures are given in subsequent sections of this chapter. For more detailed instructions, see the referenced chapter.

1- Perform a compression check of each cylinder. See Chapter 8.
2- Inspect the spark plugs to determine their condition. Test for adequate spark at the plug. See Chapter 6.
3- Start the engine in a body of water and check the water flow through the engine. See Chapter 11.
4- Check the gear oil in the lower unit. See Chapter 11.
5- Check the carburetor adjustments and the need for an overhaul. See Chapter 4.
6- Check the fuel pump for adequate performance and delivery. See Chapter 4.
7- Make a general inspection of the ignition system. See Chapter 6.
8- Test the starter motor and the solenoid, if so equipped. See Chapter 9.
9- Check the internal wiring.
10- Check the timing and synchronization. See Chapter 7.

2-3 COMPRESSION CHECK

A compression check is extremely important, because an engine with low or uneven compression between cylinders **CANNOT** be tuned to operate satisfactorily. Therefore, it is essential that any compression problem be corrected before proceeding with the tune-up procedure. See Chapter 8.

If the powerhead shows any indication of overheating, such as discolored or scorched paint, inspect the cylinders visually thru the transfer ports for possible scoring. It is possible for a cylinder with satisfactory compression to be scored slightly. Also, check the water pump. The overheating condition may be caused by a faulty water pump.

Checking Compression

Remove the spark plug wires. **ALWAYS** grasp the molded cap and pull it loose with a twisting motion to prevent damage to the connection. Remove the spark plugs and keep them in **ORDER** by cylinder for evaluation later. Ground the spark plug leads to the engine to render the ignition system inoperative while performing the compression check.

Removing the spark plug high tension lead. Always use a pulling and twisting motion on only the cap, not the wire, to prevent damage to the cap or the boot.

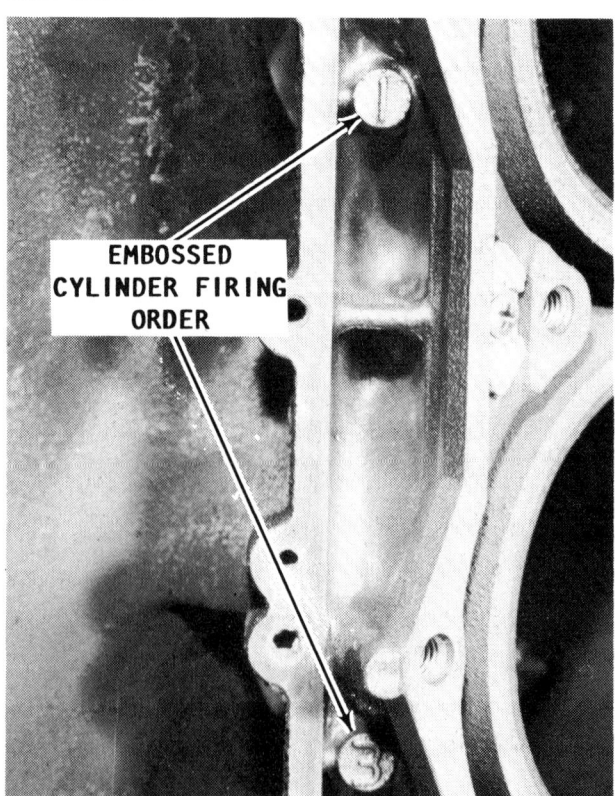

The firing order, by cylinder number, is embossed on V4 and V6 powerheads.

COMPRESSION CHECK 2-3

Insert a compression gauge into the No. 1, spark plug opening. On all V4 and V6 powerheads covered in this manual, cylinder firing takes place in numerical order. The firing order numbers are therefore the same as the cylinder numbers and are embossed on the aft side of each cylinder head. No. 1 is the uppermost cylinder on the starboard side.

Crank the engine with the cranking motor thru at least four complete strokes with the throttle at the wide open position, to obtain the highest possible reading. Record the reading. Repeat the test and record the compression for each cylinder. A variation between cylinders is far more important than the actual readings. A variation of more than 15 psi (103 kPa), between cylinders indicates the lower compression cylinder is defective. The problem may be worn, broken, or sticking piston rings, scored pistons, or worn cylinders.

Special Words for V6 Models

In recent years, the manufacturer has modified the cylinder head design on some models in an attempt to keep the temperature of each head the same. This action has changed the shape of the combustion chamber, and therefore the volume and compression pressure of each cylinder.

As a general rule, the pressure between pairs of cylinders which share the same crankshaft throw, should be approximately the same. Cylinder No. 1 should be the same as cylinder No. 2; cylinder No. 3 should be the same as cylinder No. 4; and so on.

Normally, on a V6 powerhead, cylinder No. 1 and No. 2 will have the highest compression pressure. Cylinder No. 5 and No. 6 will have the lowest compression pressure. The design modification has brought about one exception to this rule, and this is for the 175hp and 200hp in 1986 only. Cylinder No. 3 and No. 4 will have an even higher compression pressure than cylinder No. 1 and No. 2. Cylinder No. 5 and No. 6 will still have the lowest compression pressure.

Use of an engine cleaner, available at any automotive parts house, will help to free stuck rings and to dissolve accumulated carbon. Follow the directions on the container.

2-4 SPARK PLUG INSPECTION

Inspect each spark plug for badly worn electrodes, glazed, broken, blistered, or lead fouled insulators. Replace all of the plugs, if one shows signs of excessive wear.

Make an evaluation of the cylinder performance by comparing the spark condition with those shown in Chapter 6. Check each spark plug to be sure they are all by the same manufacturer and have the same heat range rating.

Inspect the threads in the spark plug opening of the block, and clean the threads before installing the plug.

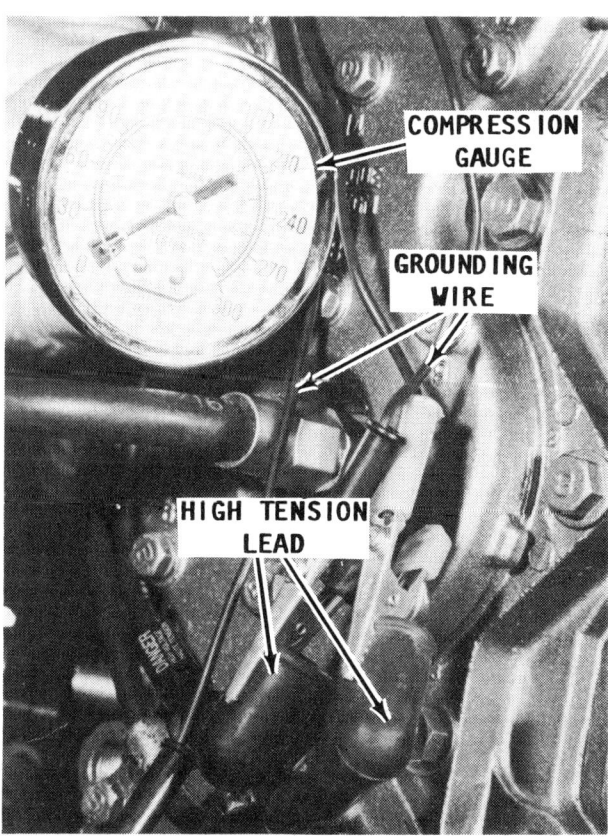

All spark plugs should be "grounded" while making compression tests. This action will prevent placing an extra load on the ignition coil.

Damaged spark plugs. Notice the broken electrode on the left plug. The broken part MUST be found and removed before returning the powerhead to service.

2-4 TUNING

When purchasing new spark plugs, **ALWAYS** ask the marine dealer if there has been a spark plug change for the engine being serviced.

Crank the engine through several revolutions to blow out any material which might have become dislodged during cleaning.

ALWAYS use a new gasket and wipe the seats in the block clean. The gasket must be fully compressed on clean seats to complete the heat transfer process and to provide a gas tight seal in the cylinder.

Install the spark plugs and tighten them to the torque value given in Chapter 8 for the powerhead being serviced. Overtightening the spark plugs may cause the porcelain insulator to crack, undertightening the spark plugs may cause them to unseat due to constant vibration.

Broken Reed

A broken reed is usually caused by metal fatigue over a long period of time. The failure may also be due to the reed flexing too far because the reed stop has not been adjusted properly or the stop has become distorted.

If the reed is broken, the loose piece **MUST** be located and removed, before the powerhead is returned to service. The piece of reed may have found its way into the crankcase, behind the intake manifold. If the broken piece cannot be located, the powerhead must be completely disassembled until it is located and removed.

An excellent check for a broken reed on an operating powerhead is to hold an ordinary business card in front of the carburetor. Under normal operating conditions, a very small amount of fine mist will be noticeable, but if fuel begins to appear rapidly on the card from the carburetor, one of the reeds is broken and causing the backflow through the carburetor onto the card.

A broken reed will cause the powerhead to operate roughly and with a "pop" back through the carburetor.

The reeds must **NEVER** be turned over in an attempt to correct a problem. Such action would cause the reed to flex in the opposite direction and the reed would break in a very short time.

2-5 IGNITION SYSTEM

Only one ignition system is used on the powerheads covered in this manual. The 220hp and 225hp is equipped with the Yamaha Microcomputer Ignition System, YMIS,

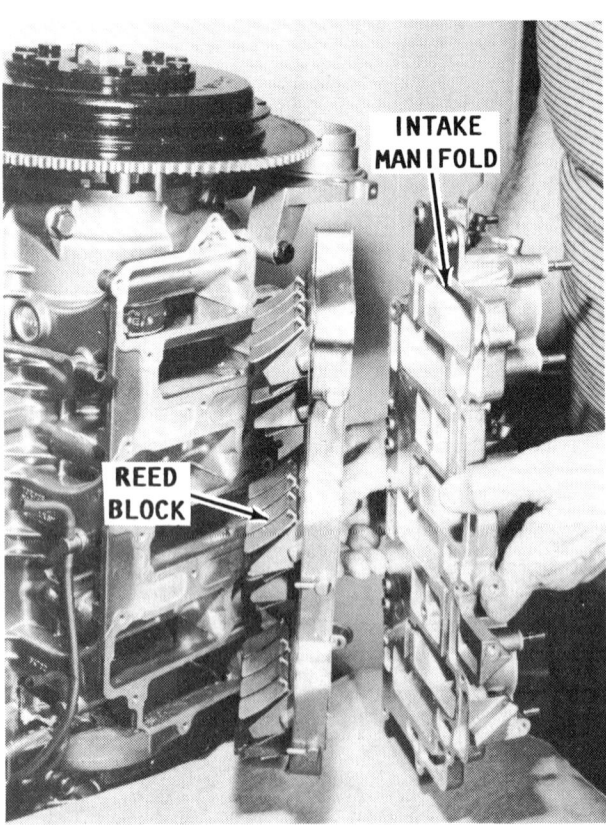

The reed block on the V4 and V6 powerheads covered in this manual may be serviced without disassembling the powerhead.

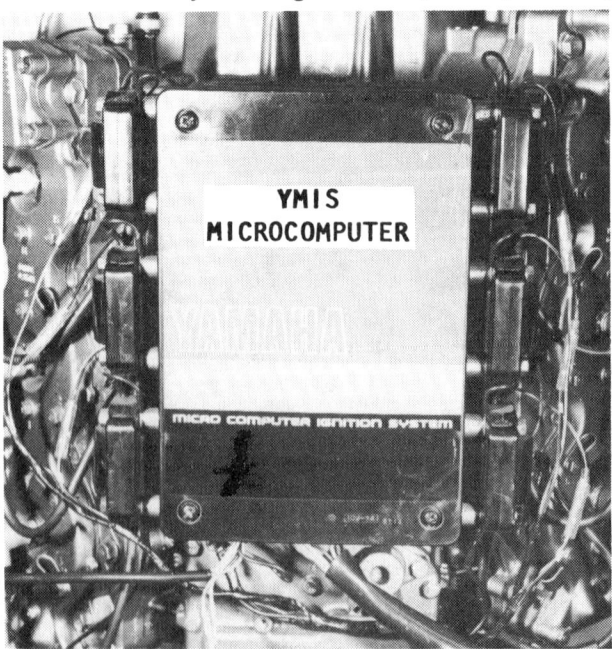

The unmistakable "black box" mounted at the rear of the powerhead houses the microcomputer for models equipped with YMIS (Yamaha Microcomputer Ignition System).

BATTERY CHECK 2-5

but essentially still has a standard CDI type ignition, computer "assisted". If the powerhead performance is less than expected, and the ignition is diagnosed as the problem area, refer to Chapter 6 for detailed service procedures. To properly time and synchronize the ignition system with the fuel system, see Chapter 7.

2-6 TIMING AND SYNCHRONIZING

Correct timing and synchronization are essential to efficient engine operation. An engine may be in apparent excellent mechanical condition, but perform poorly, unless the timing and synchronization have been adjusted precisely. To time and synchronize the powerhead, see the Table of Contents -- Chapter 7.

2-7 BATTERY CHECK

Inspect and service the battery, cables and connections. Check for signs of corrosion. Inspect the battery case for cracks or bulges, dirt, acid, and electrolyte leakage. Check the electrolyte level in each cell.

Fill each cell to the proper level with distilled water or water passed thru a demineralizer.

Clean the top of the battery. The top of a 12-volt battery should be kept especially clean of acid film and dirt, because of the high voltage between the battery terminals.

For best results, first wash the battery with a diluted ammonia or baking soda solution to neutralize any acid present. Flush the solution off the battery with clean water. Keep the vent plugs tight to prevent the neutralizing solution or water from entering the cells.

Check to be sure the battery is fastened securely in position. The hold-down device should be tight enough to prevent any movement of the battery in the holder, but not so tight as to place a strain on the battery case.

If the battery posts or cable terminals are corroded, the cables should be cleaned separately with a baking soda solution and a wire brush. Apply a thin coating of Multipurpose Lubricant to the posts and cable clamps before making the connections. The lubricant will help to prevent corrosion.

If the battery has remained under-charged, check for high resistance in the charging circuit. If the battery appears to be using too much water, the battery may be defective, or it may be too small for the job.

A series of link rods and levers connect the carburetor throttle plate opening to the stator plate to advance and retard the timing.

A check of the electrolyte in the battery should be a regular task on the maintenance schedule on any boat.

Jumper Cables

If booster batteries are used for starting an engine the jumper cables must be connected correctly and in the proper sequence to prevent damage to either battery, or diodes in the circuit.

Jumper cables may be used on the 220hp and 225hp models equipped with YMIS (Yamaha Microcomputer Ignition System). No damage will occur to the microcomputer **IF** the following procedure is closely followed. A sudden power surge may "fry" the electronic components and lead to a costly replacement of the "black box".

Be sure the main switch is in the **OFF** position and the booster battery is fully charged.

ALWAYS connect a cable from the positive terminal of the "dead" battery to the positive terminal of the good battery **FIRST**. **NEXT**, connect one end of the other cable to the negative terminal of the good battery and the other end to a good ground on the powerhead to be cranked. By making the ground connection to the powerhead, if an arc is created while making the connection, the arc will not be near the battery. An arc near the battery could cause an explosion, destroying the battery and causing serious personal injury.

After all the connections are secure, crank and attempt to start the powerhead using the key switch.

CRITICAL WORDS

Keep the powerhead running at about 3,000 rpm for at least three minutes with the jumper cables connected. Shut down the powerhead and disconnect the jumper cables.

DO NOT attempt to disconnect the jumper cables while the powerhead is running, especially on the 220hp and 225hp models with YMIS. Such action could lead to a power variance, damaging the circuits in the microcomputer.

DISCONNECT the battery ground cable before replacing any part of the ignition or cranking system, or before connecting any type of meter to the ignition system.

If it is necessary to use a fast charger on a dead battery, **ALWAYS** disconnect one of the boat cables from the battery first, to prevent burning out the diodes in the circuit.

NEVER use a fast charger as a booster to start the engine because the diodes will be **DAMAGED**.

Connect a voltmeter across the battery posts, and measure the unloaded battery voltage. Without a "full" charge, the battery will probably fail to provide sufficient energy to the cranking motor and the microcomputer to start the powerhead. The battery must be recharged.

The 220hp and 225hp models equipped with YMIS (Yamaha Microcomputer Ignition System), must have a fully charged battery at all times to ensure proper function of the microcomputer. If the operating voltage falls to below 7 volts or a connection at the battery becomes loose or corroded, the output of the microcomputer may become severely impaired.

A 20 amp fuse is installed between the starter relay and the rectifier/voltage regulator. The purpose of this fuse is to protect the rectifier/voltage regulator from damage should the battery be accidentally connected in reverse. This fuse is easily identifiable, because it has a decal on the outer casing with the words "FUSE FUSIBLE" on it. (Something lost here in the translation, but this is the wording on the decal.)

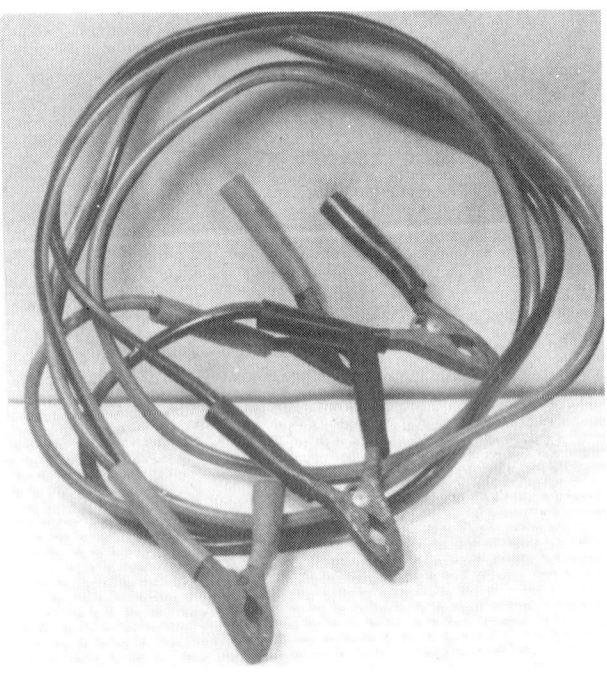

Common set of jumper cables for use with a second battery to crank and start the engine. ***EXTREME*** *care should be exercised when using a second battery, as explained in the text.*

If the key switch is turned on and the powerhead fails to start, this fuse would be one of the first areas to check. especially if the battery had been removed or disconnected for any reason.

A second area to check is all terminals for corrosion. Disconnect, clean, and reconnect, as required, especially the gound leads and terminals.

2-8 FUEL AND FUEL TANKS

Take time to check the fuel tank and all of the fuel lines, fittings, couplings, valves, flexible tank fill and vent. Turn on the fuel supply valve at the tank. If the gas was not drained at the end of the previous season, make a careful inspection for gum formation. When gasoline is allowed to stand for long periods of time, particularly in the presence of copper, gummy deposits form.

This gum can clog the filters, lines, and passageway in the carburetor. Chapter 4 has good information regarding "sour" fuel, unleaded fuel, and gasohol.

Check the fuel tank for the following:
1- Adequate air vent in the fuel cap.
2- Fuel line of sufficient size, should be 5/16" to 3/8" (8mm to 9.5mm).
3- Filter on the end of the pickup is too small or is clogged.
4- Fuel pickup tube is too small.

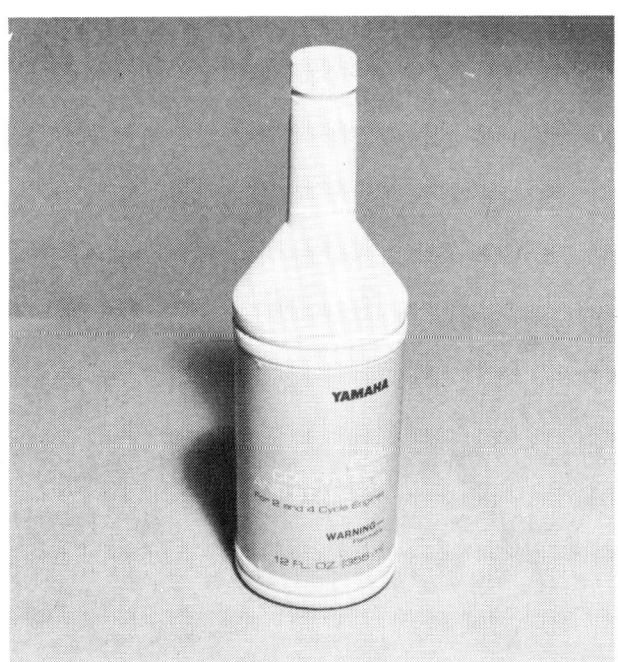

Yamaha fuel conditioner and stabilizer may be used to prevent fuel from "souring" for up to twelve full months.

2-9 CARBURETOR ADJUSTMENTS

SPECIAL WORDS ON TACHOMETERS AND CONNECTIONS

A tachometer is installed as standard equipment on all powerheads covered in this manual.

Due to local conditions, it may be necessary to adjust the carburetor while the outboard unit is running in a test tank or with the boat in a body of water. For maximum performance, the idle rpm should be adjusted under actual operating conditions.

Under such conditions it might be necessary to attach a tachometer closer to the powerhead than the one installed on the control panel.

Open the remote control box. Disconnect the Black and Green leads. Connect the Black lead to the ground terminal of the auxiliary tachometer and the Green lead to the "input" or "hot" terminal of the auxiliary tachometer.

If a tachometer is purchased from the manufacturer for installation, it should be calibrated for the model matching the particular model powerhead.

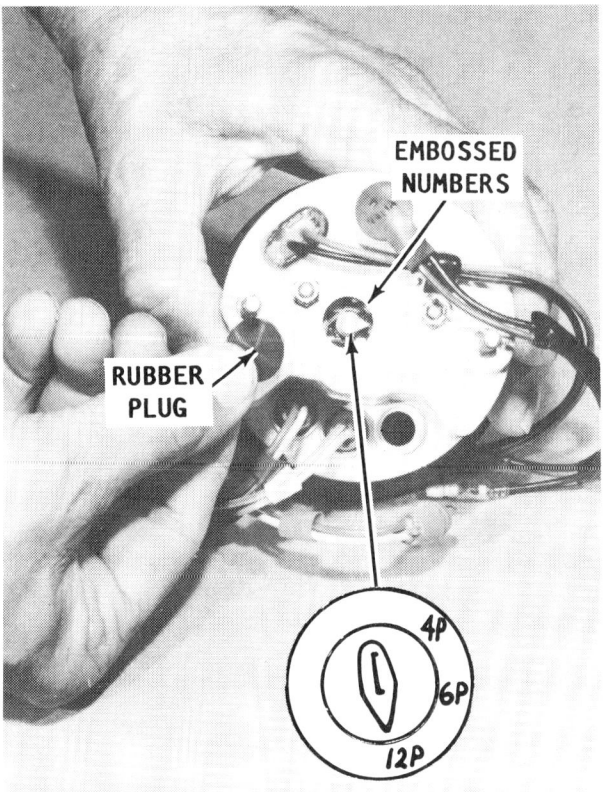

Checking a tachometer for correct calibration, as explained in the text.

2-8 TUNING

Three different tachometer designs are installed on V4 and V6 powerheads. Two are the gauge type and the third is a digital type. Of the two gauge types, one has a small rubber plug and the other has a terminal on the back of the meter.

Calibration of digital tachometers, the units having the rubber plug, is as follows: Remove the rubber plug from the back of the meter. Observe the ring with 4P, 6P, and 12P embossed around the ring. These numbers indicate the number of possible poles used on the Yamaha flywheel magnetos. Using a slotted screwdriver, move the arrow until it points toward the 12 pole setting. This setting is to be used for all powerheads covered in this manual.

Calibration of the tachometer with terminals is as follows: Identify the two Black leads connected to terminals on the back of the meter. Make sure these two leads are connected to the "12" and the "B" terminals.

Start the engine and allow it to warm to operating temperature.

REMEMBER, the powerhead will **NOT** start without the emergency tether in place behind the "kill" switch knob.

CAUTION
Water must circulate through the lower unit to the powerhead anytime the powerhead is operating to prevent damage to the water pump in the lower unit. Just five seconds without water will damage the water pump impeller.

NEVER, AGAIN, NEVER operate the engine at high speed with a flush device attached. The engine, operating at high speed with such a device attached, would **RUNAWAY** from lack of a load on the propeller, causing extensive damage.

Air/Fuel Mixture Adjustment

The pilot screw regulates the air/fuel mixture. The setting for this screw varies for each unit. The setting is given in Chapter 4, for each model covered in this manual. Check the Table of Contents, Chapter 4 for the model being serviced. Actually, this setting is **NOT** an adjustment, it is a specification.

Idle Speed Adjustment

The idle speed is regulated by the throttle stop screw which "sets" the position of the throttle plate inside the carburetor throat. Idle speed recommendations are given in Chapter 4 and also in the Appendix for each unit. Rotating the throttle stop **CLOCKWISE** increases powerhead speed, and rotating the screw **COUNTERCLOCKWISE** decreases powerhead speed.

The idle rpm is adjusted under actual operation to 800 rpm for all models covered in this manual, **EXCEPT** the 115hp, the 130hp, the 220hp, and the 225hp models. These models are adjusted for 750 rpm.

If the condition of the fuel is in doubt, drain, clean, and fill the tank with fresh fuel.

Check the line between the fuel pump and the carburetor while the powerhead is operating and the line between the fuel tank and the pump when the powerhead is not operating. A leak between the tank and the pump many times will not appear when the powerhead is operating, because the suction

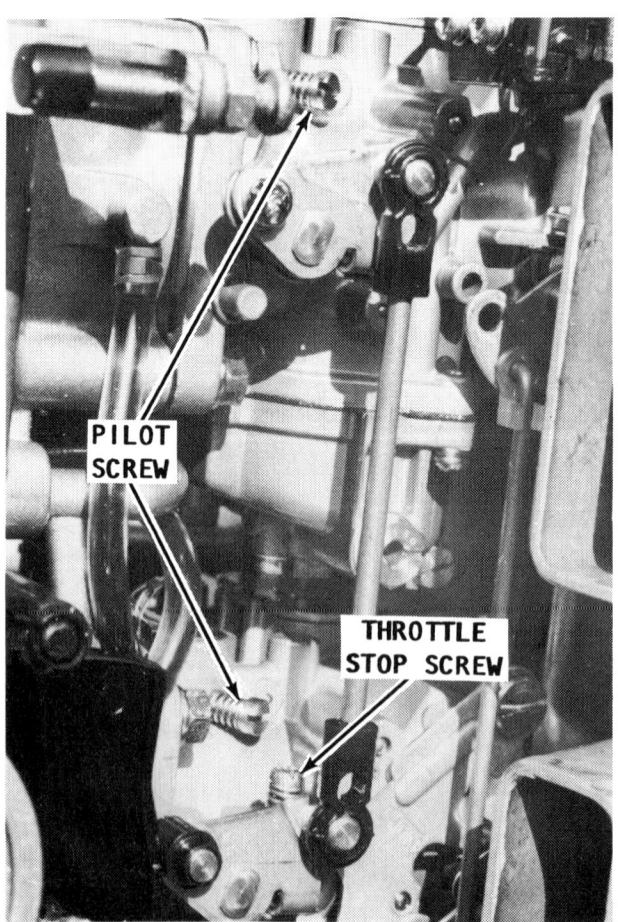

Location of the throttle stop screw and the pilot screws. The function of these screws is explained in the text.

created by the pump drawing fuel will not allow the fuel to leak. Once the powerhead is shut down and the suction no longer exists, fuel may begin to leak.

Repairs and Adjustments

Chapter 4 contains detailed, comprehensive procedures to disassemble, clean, assemble, and adjust the single type carburetor used on the powerheads covered in this manual.

2-10 FUEL PUMPS

If the powerhead operates as if the load on the boat is being constantly increased and decreased, even though an attempt is being made to hold a constant powerhead speed, the problem can most likely be attributed to the fuel pump.

Many times, a defective fuel pump diaphragm is mistakenly diagnosed as a problem in the ignition system. The most common problem is a tiny pin-hole in the diaphragm or a bent check valve inside the fuel pump. Such a small hole will permit gas to enter the crankcase and wet foul the spark plug at idle-speed. During high-speed operation, gas quantity is limited, the plug is not foul and will therefore fire in a satisfactory manner.

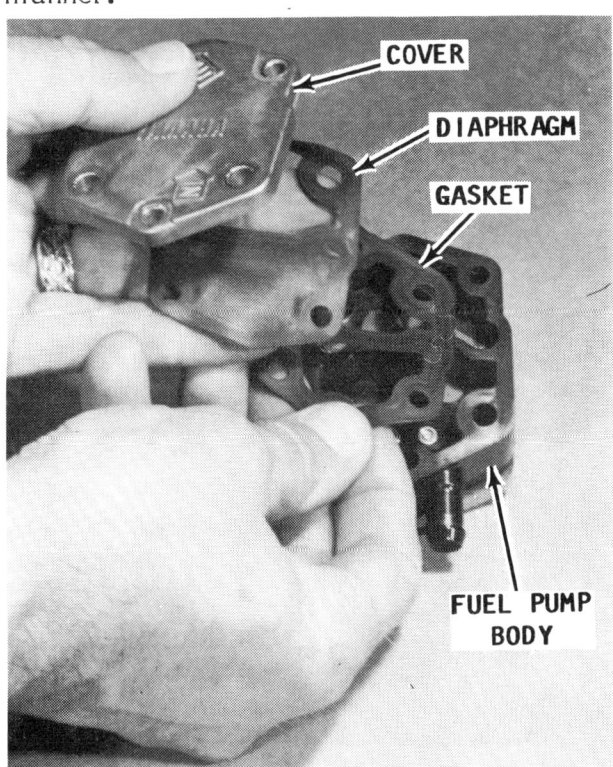

Arrangement of vacuum operated fuel pump parts. A tiny hole in the diaphragm can affect performance.

If the fuel pump fails to perform properly, an insufficient fuel supply will be delivered to the carburetor. This lack of fuel will cause the engine to run lean, lose rpm or cause piston scoring.

Tune-up Task

Remove the fuel filter on the carburetor. Wash the parts in solvent and then dry them with compressed air. Install the clean element. A fuel pump pressure test should be made any time the engine fails to perform satisfactorily at high speed.

NEVER use liquid Neoprene on fuel line fittings. Always use Permatex when making fuel line connections. Permatex is available at almost all marine and hardware stores.

To service the fuel pump, see Chapter 4.

2-11 CRANKING MOTOR

Cranking Motor Test

Check to be sure the battery has a 70-ampere rating and is fully charged. Would you believe, many cranking motors are needlessly disassembled, when the battery is actually the culprit.

Lubricate the pinion gear and screw shaft with No. 10 oil.

Connect one lead of a voltmeter to the positive terminal of the cranking motor. Connect the other meter lead to a good ground on the engine. Check the battery voltage under load by turning the ignition switch to the START position and observing the voltmeter reading.

If the reading is 9-1/2 volts or greater, and the cranking motor fails to operate, repair or replace the cranking motor. See Chapter 9.

2-12 INTERNAL WIRING HARNESS

Check the internal wiring harness if problems have been encountered with any of

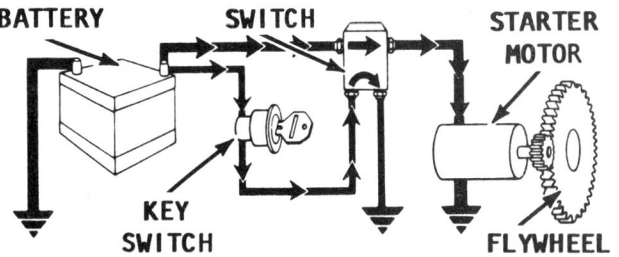

Functional diagram of a typical cranking circuit.

2-10 TUNING

Harness connector, with the prongs properly cleaned, ready for continued service.

the electrical components. Check for frayed or chafed insulation and/or loose connections between wires and terminal connections.

Check the harness connector for signs of corrosion. Inspect the electrical "prongs" to be sure they are not bent or broken. If the harness shows any evidence of the foregoing problems, the problem must be corrected before proceeding with any harness testing.

Verify the "prongs" of the harness connector are clean and free of corrosion. Convince yourself a good electrical connection is being made between the harness connector and the remote control harness.

2-13 WATER PUMP CHECK

FIRST, SOME GOOD WORDS

The water pump **MUST** be in very good condition for the engine to deliver satisfactory service. The pump performs an extremely important function by supplying enough water to properly cool the engine.

Worn water pump impeller, unfit for service.

Therefore, in most cases, it is advisable to replace the complete water pump assembly at least once a year, or anytime the lower unit is disassembled for service.

Sometimes during adjustment procedures, it is necessary to run the engine with a flush device attached to the lower unit. **NEVER** operate the engine over 1000 rpm with a flush device attached, because the engine may **"RUNAWAY"** due to the no-load condition on the propeller. A "runaway" engine could be severely damaged.

As the name implies, the flush device is primarily used to flush the engine after use in salt water or contaminated fresh water. Regular use of the flush device will prevent salt or silt deposits from accumulating in the water passageway. During and immediately after flushing, keep the outboard unit in an upright position until all of the water has drained from the intermediate housing. This will prevent water from entering the power head by way of the intermediate housing and the exhaust ports, during the flush. It will also prevent residual water from being trapped in the intermediate housing and other passageways.

All powerheads covered in this manual have water exhaust ports which deliver a tattle-tale stream of water, if the water pump is functioning properly during engine operation. Water pressure at the cylinder

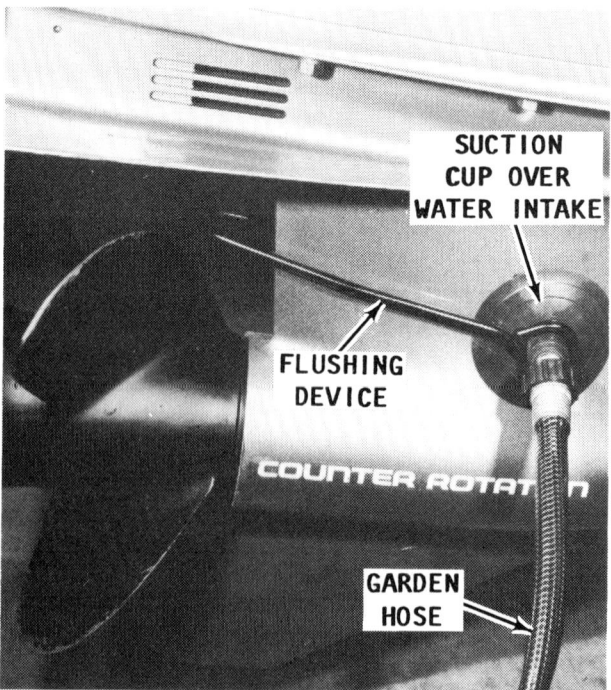

This popular and inexpensive flushing device may be purchased at almost any marine store. This item should be included in every boat owner's service "kit".

block should be checked if an overheating condition is detected or suspected.

To test the water pump, the lower unit, or jet drive, **MUST** be placed in a test tank or the boat moved into a body of water. The pump must now work to supply a volume of water to the powerhead. A tattle-tale stream of water should be visible from the pilot hole beneath the cover cowling.

Lack of adequate water supply from the water pump thru the engine will cause any number of powerhead failures, such as stuck rings, scored cylinder walls, burned pistons, etc.

For water pump service, see Chapter 11.

2-14 PROPELLER

Check the propeller blades for nicks, cracks, or bent condition. If the propeller is damaged, the local marine dealer can make repairs or send it out to a shop specializing in such work.

Remove the propeller and the thrust hub. Check the propeller shaft seal to be sure it is not leaking. Check the area just forward of the seal to be sure a fish line is not wrapped around the shaft.

Operation At Recommended RPM

Check with the local marine dealer, or a propeller shop for the recommended size and pitch for a particular size engine, boat, and intended operation. The correct propeller should be installed on the engine to enable operation at the upper end of the factory recommended rpm.

2-15 LOWER UNIT

NEVER remove the vent or filler plugs when the lower unit is hot. Expanded lubricant would be released through the plug hole. Check the lubricant level after the unit has been allowed to cool. Add only Yamaha Gear Case Lubricant. **NEVER** use regular automotive-type grease in the lower unit, because it expands and foams too much. Outboard lower units do not have provisions to accommodate such expansion.

If the lubricant appears milky brown, or if large amounts of lubricant must be added to bring the lubricant up to the full mark, a thorough check should be made to determine the cause of the loss.

Propeller properly installed with anti-seizing compound applied to the propeller shaft.

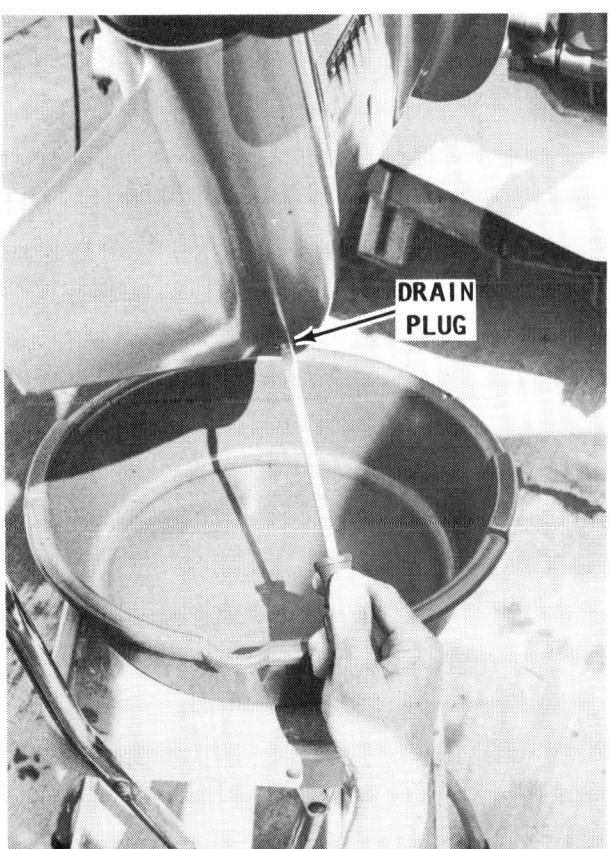

A suitable container positioned under a fully tilted lower unit, ready to receive lubricant as it drains.

2-12 TUNING

Special Words for Counterrotating Models

The first few times gearcase lubricant is drained, from a counterrotating lower unit, a discoloration may be noticed. This condition is quite normal and is no cause for alarm unless accompanied by the presence of metal chips. The discoloration is due to the use of molybdenum-disulfide assembly grease at the factory.

Draining Lower Unit

Tilt the lower unit upward until the drain plug is at the lowest point to enable all the fluid to drain.

Remove the **FILL** plug from the lower end of the gear housing on the port side and the **OIL LEVEL** plug just above the anti-cavitation plate.

Filling Lower Unit

Position the drive unit approximately vertical and without a list to either port or starboard. Insert the lubricant tube into the **OIL FILL** hole at the bottom plug hole, and inject lubricant until the excess begins to come out the **OIL LEVEL** hole. Install the **OIL LEVEL** plug first then replace the **OIL FILL** plug with **NEW** gaskets. Check to be sure the gaskets are properly positioned to prevent water from entering the housing.

For detailed lower unit service procedures, **AND** lower unit lubrication capacities, see Chapter 9.

2-16 JET DRIVE UNITS

The gate of a jet drive unit **MUST** be properly adjusted to obtain maximum performance from the outboard unit. When properly adjusted, the gate will permit the pump to deliver its full potential of thrust with no drag.

Gate Position and Shift Lever

The shift lever adjustment should be checked from time to time to ensure the gate is firmly against the rubber pad beneath the pump housing, when the unit is in the **FORWARD** position (wide open).

The lower unit is filled with lubricant as directed in the text. Capacity for various units is presented in chart form in Chapter 11.

A jet drive ready for a day on the water. The gate is in the NEUTRAL position.

JET DRIVE 2-13

The leaf spring "snaps" onto the top surface of the shift lever to hold the FORWARD position.

With the gate against the rubber pad, any rattle noises will be avoided as the boat moves through the water. Proper positioning of the gate in forward gear will prevent wave action from accidentally shifting the gate into reverse as the boat is operated through violent maneuvers.

When the shift lever is in the **FORWARD** position, a leaf spring prevents the lever from returning to the **NEUTRAL** position and allowing the gate to rise. Shifting may be accomplished with one hand by moving the leaf spring to one side with a thumb, and then moving the lever.

A simple "tuning" task with the jet drive is to make a thorough check of the linkage and gate movement to ensure maximum thrust and use of the horsepower developed by the powerhead.

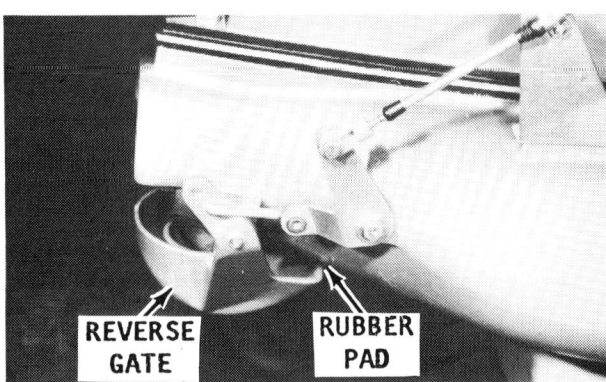

When the gate is in the FORWARD position, the gate should make firm contact with the rubber pad on the underside of the housing, to prevent "rattling".

Impeller

Excessively rounded jet impeller edges will reduce the efficiency of the jet drive. See Chapter 3, Section 3-12. for impeller care.

A proper scoop angle for the water intake grille will enhance the drive's performance. The vertical axis of the outboard drive shaft must be completely vertical or slightly inclined under and toward the boat to provide the best scoop angle, as indicated in the accompanying illustration.

Trimming the outboard unit to position the jet drive away from the boat reduces the scoop angle and may cause jet impeller slippage and even cavitation "burns" on the impeller blades.

The term cavitation "burn" is a common expression used throughout the world among people working with pumps, impeller blades, and forceful water movement.

"Burns" on the impeller blades are caused by cavitation air bubbles exploding with considerable force against the impeller blades. The edges of the blades may develop small "dime size" areas resembling a porus sponge, as the aluminum is actually "eaten" by the condition just described.

2-17 BOAT TESTING

Operation of the outboard unit, mounted on a boat with some type of load, is the ultimate test. Failure of the power unit or the boat under actual movement through the water may be detected much more quickly than operating the power unit in a test tank.

Hook and Rocker

Before testing the boat, check the boat bottom carefully for marine growth or evidence of a "hook" or a "rocker" in the

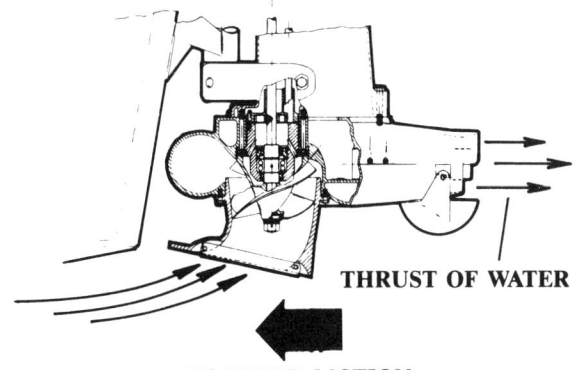

The best angle for the inlet grille is obtained when the surface of the grille is at a slight incline under and toward the boat, as indicated.

2-14 TUNING

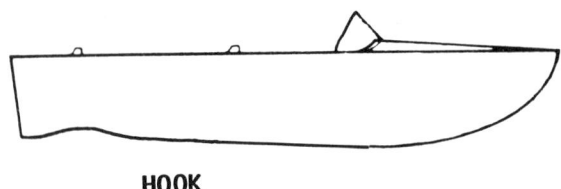
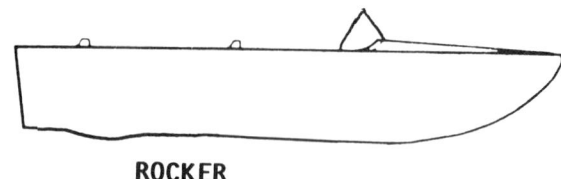

HOOK **ROCKER**

Boat performance will be drastically impaired, if the bottom is damaged by a dent (hook), or by a bulge (rocker).

bottom. Either one of these conditions will greatly reduce performance.

Performance

Mount the motor on the boat. Install the remote control cables (if used), and check for proper adjustment.

Make an effort to test the boat with what might be considered an average gross load. The boat should ride on an even keel, without a list to port or starboard. Adjust the motor tilt angle, if necessary, to permit the boat to ride slightly higher than the stern. If heavy supplies are stowed aft of the center, the bow will be light and the boat will "plane" more efficiently. For this test the boat must be operated in a body of water.

If the motor is equipped with an adjustable trim tab, the tab should be adjusted to permit boat steerage in either direction with equal ease.

Check the engine rpm at full throttle. The rpm should be within the Specifications in the Appendix. If the rpm is not within specified range, a propeller change may be in order. A higher pitch propeller will decrease rpm, and a lower pitch propeller will increase rpm.

High Performance Propellers

The accompanying illustration of a high performance propeller clearly shows the unique design of the long blades and other features. This propeller is the weedless type, having extra sharp blades.

Installation of a high performance propeller requires raising the transom height, trim out after planing, and installation of a special design trim tab. Installation of a water pressure gauge is highly recommended, because raising the transom height will affect the amount of water entering the lower unit through the intake holes. An inadequate amount of water taken in will certainly cause powerhead cooling problems.

The high performance propeller has standard attaching hardware with the exception of the inner spacer which is a special three-pronged design, as shown in the accompanying illustration.

If the outboard unit is equipped with a jet drive, check to be sure the vertical axis of the outboard drive shaft is vertical or slightly inclined under and toward the boat to provide the best scoop angle.

For maximum low speed engine performance, the idle mixture and the idle rpm should be readjusted under actual operating conditions.

Example of a high performance propeller. Notice how the blades are more narrow with more pitch than a standard propeller of the same size.

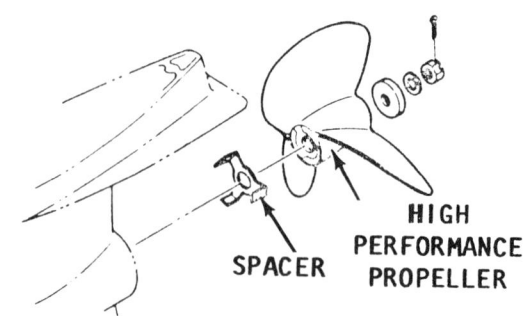

Arrangement of attaching hardware for a high performance propeller.

3
MAINTENANCE

3-1 INTRODUCTION

GOOD WORDS

The authors estimate 75% of engine repair work can be directly or indirectly attributed to lack of proper care for the engine. This is especially true of care during the off-season period. There is no way on this green earth for a mechanical engine, particularly an outboard motor, to be left sitting idle for an extended period of time, say for six months, and then be ready for instant satisfactory service.

Imagine, if you will, leaving your automobile for six months, and then expecting to turn the key, have it roar to life, and be able to drive off in the same manner as a daily occurrence.

It is critical for an outboard engine to be run at least once a month, preferably, in the water, but if this is not possible, then a flush attachment **MUST** be connected to the lower unit.

CAUTION

Water must circulate through the lower unit to the powerhead anytime the powerhead is operating to prevent damage to the water pump in the lower unit. Just five seconds without water will damage the water pump impeller.

NEVER, AGAIN NEVER, operate the engine at high speed with a flush device attached. The engine, operating at high speed with such a device attached, would **RUNAWAY** from lack of load on the propeller, causing extensive damage.

At the same time, the shift mechanism should be operated through the full range several times and the steering operated from hard-over to hard-over.

Only through a regular maintenance program can the owner expect to receive long life and satisfactory performance at minimum cost.

The material presented in this chapter is divided into four general areas.

1- General information every boat owner should know.

2- Maintenance tasks that should be performed periodically to keep the boat operating at minimum cost.

3- Care necessary to maintain the appearance of the boat and to give the owner that "Pride of Ownership" look.

4- Winter storage practices to minimize damage during the off-season when the boat is not in use.

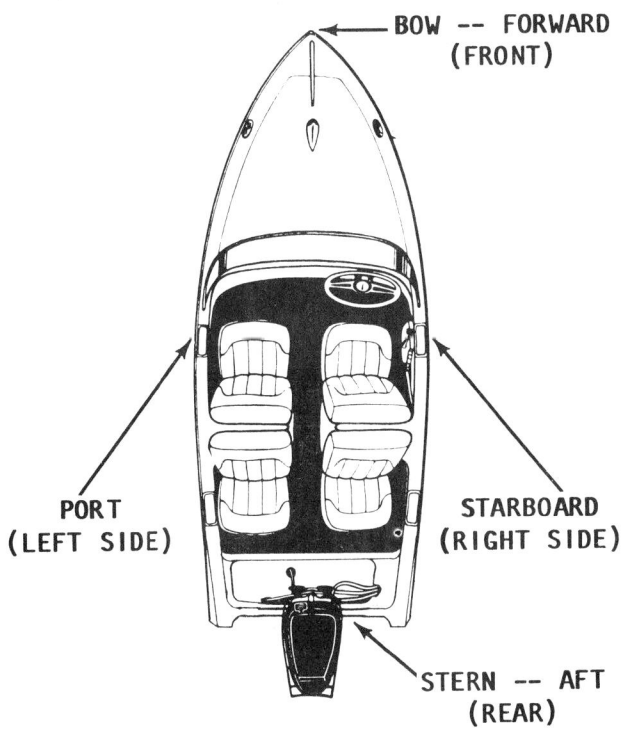

*Common terminology used throughout the world for reference designation on boats of **ALL** sizes. "Port", "Starboard", "Forward", and "Aft", never change, even if standing on your head.*

3-2 MAINTENANCE

In nautical terms, the front of the boat is the **bow**; the rear is the **stern**; the right side, when facing forward, is the **starboard** side; and the left side is the **port** side. All directional references in this manual use this terminology. Therefore, the direction from which an item is viewed is of no consequence, because **starboard** and **port** **NEVER** change no matter where the individual is located or his position -- even standing on his/her head.

3-2 OUTBOARD SERIAL NUMBERS

The outboard serial number and the engine serial number are the manufacturer's key to engine changes. These numbers identify the year of manufacture, the qualified horsepower rating, and the parts book identification. If any correspondence or parts are required, the engine serial number **MUST** be used or proper identification is not possible. The accompanying illustration will be very helpful in locating the engine identification tag for the various models.

The outboard number is stamped on the plate usually attached to the port side of the clamp bracket.

The powerhead serial number is usually stamped on the port side of the cylinder block.

ONE MORE WORD

As a theft prevention measure, a special label with the outboard serial number is bonded to the starboard side of the clamp bracket. Any attempt to remove this label will result in cracks across the serial number.

Serial numbers on the identification plate of a late model outboard unit covered in this manual. This is a standard location for the plate, although some units may have it elsewhere.

3-3 EMERGENCY TETHER

All outboard units covered in this manual are equipped with an emergency tether by the manufacturer. This tether must be in place behind the "kill" switch or the powerhead cannot be started. If the powerhead is operating and the tether is removed, the unit will immediately shut down.

Explanation

The plastic tether acts as a spacer, moving the "kill" button out slightly and allowing an internal contact to be closed, permitting the ignition circuit to be completed and the powerhead to be started.

The "kill" button may now be depressed to shut the powerhead down with the tether in place. If the tether is pulled free from behind the "kill" button, the button pops inward and the ignition circuit is opened. With the "kill" switch in this position, no amount of cranking will result in powerhead startup.

The purpose of this tether is two fold.

As a Safety Feature

When this tether is used as intended by the manufacturer, the boat operater attaches the belt hook onto his/her clothing at any convenient location. Should the operator be thrown overboard or knocked forward away from the outboard unit, the tether will be pulled free of the "kill" button, and the powerhead will be shut down.

As a Security Feature

If the boat is moored and will be left unattended, the owner may take the entire emergency tether with him/her. Without the tether in place behind the "kill" button, the powerhead cannot be started. Any attempt to start the powerhead and steal the boat will be unsuccessful -- unless the thief is familiar with this particular security device.

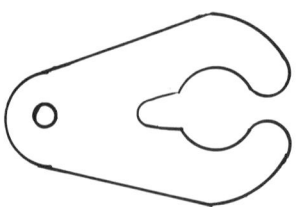

Pattern to be used to fabricate a "homemade" emergency tether, as explained in the text.

EMERGENCY TETHER 3-3

Temporary Replacement

If the boat owner loses the emergency tether and is unable to obtain one immediately from the local Yamaha dealer, an emergency substitute tether may be made using only a common knife and a couple pieces of plastic.

First, obtain a piece of plastic from the cover of a container of margarine, whipped topping, or similar product.

Next, using the pattern shown on this page cut out about four shapes, as shown. Stack the four cutouts together, secure them with a paper clip, or similar object, and then insert them behind the "kill" switch.

SPECIAL WORDS

If the material described is not available, obtain some other pliable material and cut the shape indicated. The thickness of the substitute tether device should be approximately 1/8" (3mm). About 100 pages (50 sheets) of this manual is approximately the proper thickness.

REMEMBER, use this device only in an emergency situation and purchase the proper tether from the local Yamaha dealer at the first opportunity. By substituting this "home made" tether, both the safety and the security features intended by the manufacturer have been lost.

3-4 LUBRICATION - COMPLETE UNIT

As with every type mechanical invention with moving parts, lubrication plays a prominent role in operation, enjoyment, and longevity of the unit.

If an outboard unit is operated in salt water the frequency of applying lubricant to fittings is usually cut in half for the same fitting if the unit is used in fresh water. The few minutes involved in moving around the outboard applying lubricant and at the same time making a visual inspection of its general condition will pay in rich rewards with years of continued service.

It is not uncommon to see outboard units well over 20 years of age moving a boat through the water as if the unit had recently been purchased from the current line of models. An inquiry with the proud owner will undoubtedly reveal his main credit for its performance to be regular periodic maintenance.

The accompanying chart can be used as a guide to periodic maintenance while the outboard is being used during the season.

In addition to the normal lubrication listed in the lubrication chart, the prudent owner will inspect and make checks on a regular basis as listed in the accompanying chart.

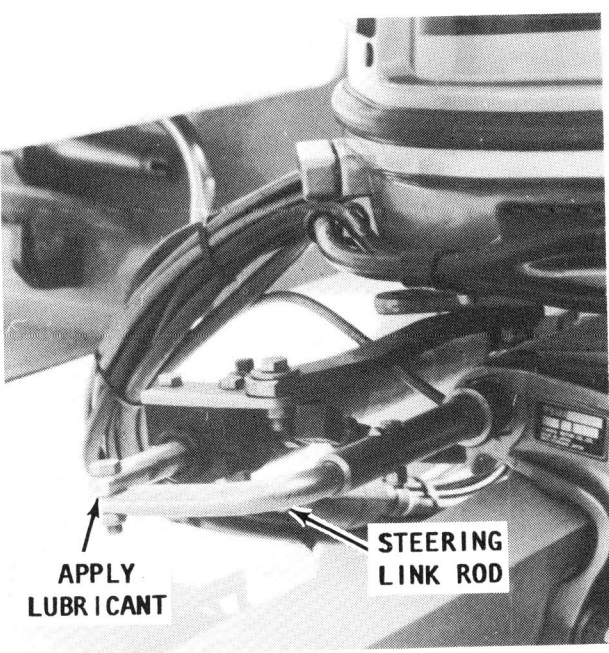

The steering cable should be lubricated with Yamaha All Purpose Grease.

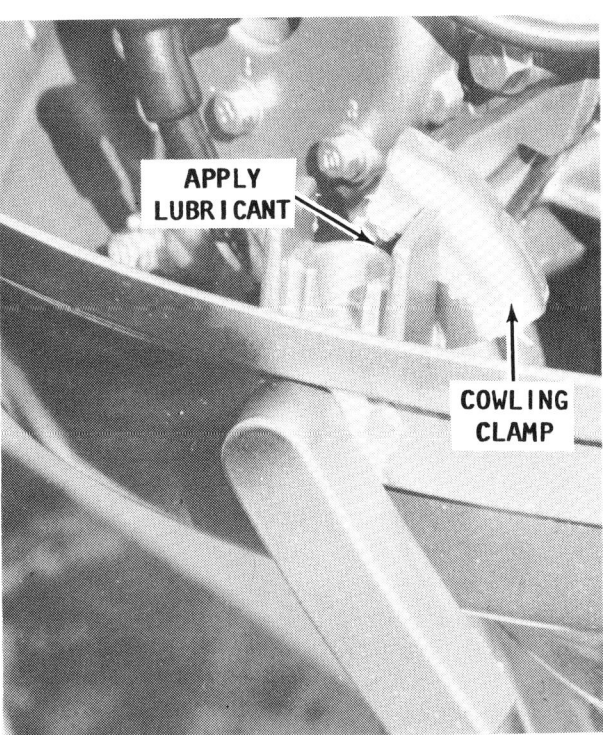

The cowling clamp should also be lubricated with Yamaha All Purpose Grease.

LUBRICATION POINT/FREQUENCY CHART

DESCRIPTION	LUBRICANT	FREQUENCY FRESH WATER	FREQUENCY SALT WATER
Throttle Linkage	Yamalube All-Purpose Lubricant	Every 100 hrs. or 60 days	Every 50 hrs. or 30 days
Throttle Control Lever			
Throttle Grip Housing			
Throttle Link Journal			
Shift Lever Journal			
Shift Mechanism			
Steering Pivot Shaft			
Top or Bottom Cowling Clamp Lever Journal			
Choke Lever			
Swivel Bracket			
Clamp Bolt			
Tilt Mechanism			
Propeller Shaft			
Gear Oil	Yamaha Gear Case Lubricant	Every 100 hrs. or 60 days	Every 50 hrs. or 30 days
Jet Drive	Yamalube All Purpose Lubricant	Every 10 hours	Every 5 hours

3-5 PRE-SEASON PREPARATION

Satisfactory performance and maximum enjoyment can be realized if a little time is spent in preparing the outboard unit for service at the beginning of the season. Assuming the unit has been properly stored, as outlined in Section 3-14, a minimum amount of work is required to prepare the unit for use.

The following steps outline an adequate and logical sequence of tasks to be performed before using the outboard the first time in a new season.

1- Lubricate the outboard according to the manufacturer's recommendations. Refer to the lubrication chart. Remove, clean, inspect, adjust, and install the spark plugs with new gaskets (if they require gaskets). Make a thorough check of the ignition sys-

PRE-SEASON PREP 3-5

tem. This check should include: the ignition coils, stator assembly, condition of the wiring, and the battery electrolyte level and charge.

2- If a built-in fuel tank is installed, take time to check the gasoline tank and all fuel lines, fittings, couplings, valves, including the flexible tank fill and vent. Turn on the fuel supply valve at the tank. If the fuel was not drained at the end of the previous season, make a careful inspection for gum formation. If a six-gallon fuel tank is used, take the same action. When gasoline is allowed to stand for long periods of time, particularly in the presence of copper, gummy deposits form. This gum can clog the filters, lines, and passageways in the carburetor. See Chapter 4, Fuel System Service.

3- Check the oil level in the lower unit by first removing the vent screw on the port side just above the anti-cavitation plate. Insert a short piece of wire into the hole and check the level. Fill the lower unit according to procedures outlined in Section 3-11.

GOOD WORDS

The manufacturer recommends the fuel filter be replaced at the start of each season or at least once a year. The manufacturer also recommends oil be added to the fuel tank at the ratio of 25:1 for the first ten hours of operation after the unit is brought out of storage. This ratio will **ENSURE** adequate lubrication of moving parts which have been drained of oil during the storage period.

After the first ten hours of operation, the normal 100:1 oil/fuel mixture may be used. Use only outboard marine oil in the mixture, never automotive oils. Four stroke automotive engine oil is not designed to burn completely, only to lubricate. Therefore, automotive oil, if used in a two stroke powerhead, will leave an undesirable residue.

ALL UNITS

4- Close all water drains. Check and replace any defective water hoses. Check to be sure the connections do not leak. Replace any spring-type hose clamps, if they have lost their tension, or if they have distorted the water hose, with band-type clamps.

5- The engine can be run with the lower unit in water to flush it. If this is not practical, a flush attachment may be used. This unit is attached to the water pick-up in the lower unit. Attach a garden hose, turn on the water, allow the water to flow into the engine for awhile, and then run the engine.

Make a pre-season check of the fuel line coupling at the fuel joint to ensure a proper and clean connection.

This popular and inexpensive flushing device should be included in every boat owner's maintenance "kit".

3-6 MAINTENANCE

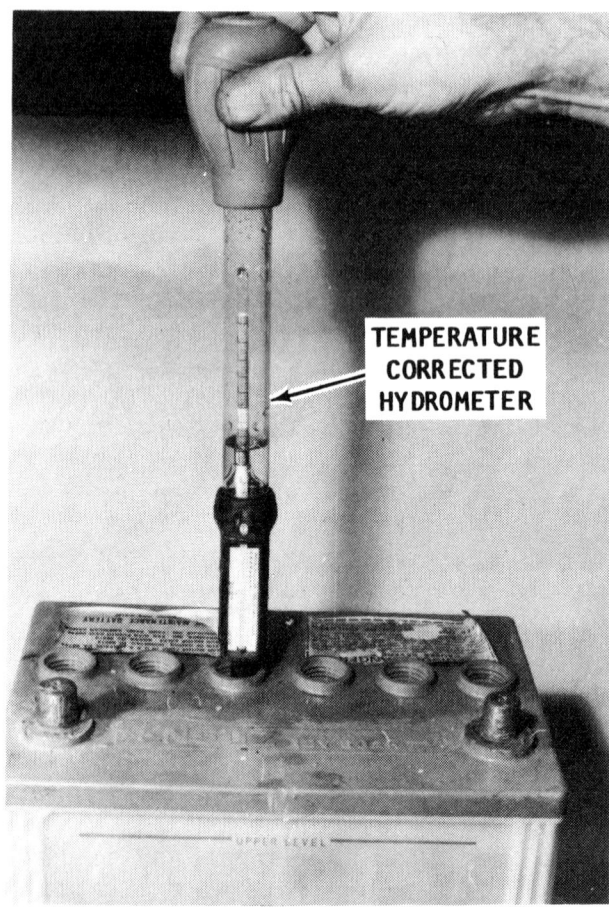

Checking the condition of the battery electrolyte using a temperature corrected hydrometer.

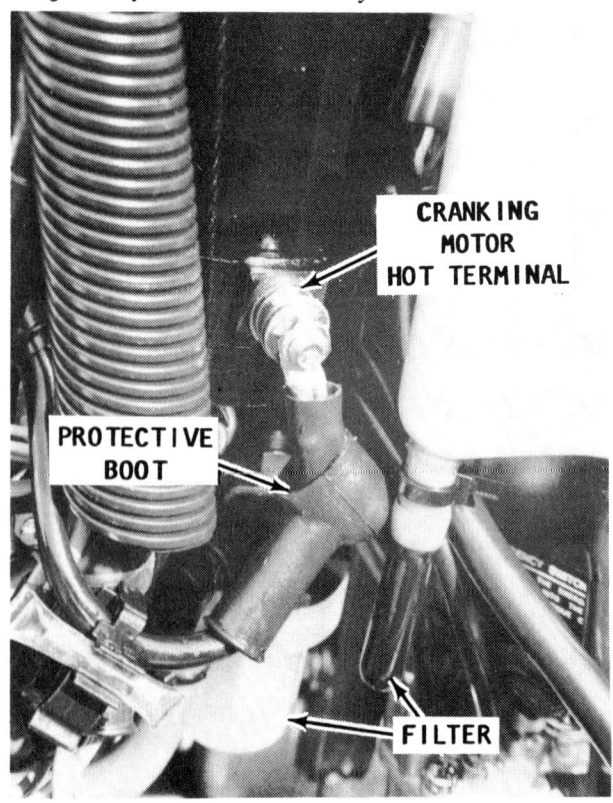

Electrical and fuel system components should be checked on a regular basis.

CAUTION

Water must circulate through the lower unit to the powerhead anytime the powerhead is operating to prevent damage to the water pump in the lower unit. Just five seconds without water will damage the water pump impeller.

Check the exhaust outlet for water discharge. Check for leaks. Check operation of the thermostat.

6- Check the electrolyte level in the battery and the voltage for a full charge. Clean and inspect the battery terminals and cable connections. **TAKE TIME** to check the polarity, if a new battery is being installed. Cover the cable connections with grease or special protective compound as a prevention to corrosion formation. Check all electrical wiring and grounding circuits.

7- Check all electrical parts on the engine and lower portions of the hull to be sure they are not of a type that could cause ignition of an explosive atmosphere. Rubber boots help keep electrical connections clean and reduce the possibility of arcing.

8- If a water separating filter is installed between the fuel tank and the powerhead fuel filter, replace the element at least once each season. This filter removes water and fuel system contaminants such as

A water separating fuel filter installed inside the boat on the transom.

PRE-SEASON PREP 3-7

dirt, rust, and other solids, thus reducing potential problems.

Electric cranking motors and high tension wiring harnesses should be of a marine type that cannot cause an explosive mixture to ignite.

ONE FINAL WORD

Before putting the boat in the water, **TAKE TIME** to **VERIFY** the drain plugs are installed. Countless number of boating excursions have had a very sad beginning because the boat was eased into the water only to have the boat begin to fill with the "wet stuff" from the river, lake, reservoir, etc.

3-6 SEALANTS, ADHESIVES, LUBRICANTS, HYDRAULIC FLUID & FUEL STABILIZER

The manufacturer has made a wide range of sealants, adhesives, and lubricants, available for use with Yamaha outboard units. Throughout this manual, the authors recommend the application of the manufacturer's products as a first choice. All listed are alternatives of equal value, if the use of the manufacturer's line is not possible for any number of reasons.

Sealants

Four sealants are recommended and they are **NOT** interchangeable. Each is designed to perform under a different set of conditions.

VHT^r (very high temperature) Quick Gasket is a silicone rubber based adhesive recommended for use where rubber gaskets are used next to a metal surface at high temperature. This sealant is used on the exhaust manifold cover gaskets. The substance has a milky white color.

Adhesives

Permatex #27 is a high temperature sealant and is a product of the Loctite Corporation. This material is usually applied to gaskets.

Yamabond #4 is a non hardening material and is recommended for metal to metal joints such as the crankcase halves and the lower crankshaft oil seal housing. This substance is highly resistant to oil and gasoline.

Loctite or Lock N' Seal is a non hardening adhesive. This material is recommended for application to the threads of load bearing fasteners. Loctite helps prevent loosening of the bolt due to vibration, thread wear, and corrosion. Loctite **MUST** be applied **ONLY** to clean and dry parts.

Lubricants

Four different lubricants are recommended in the lubrication procedures presented in this manual. These lubricants are **NOT** interchangeable, each is designed to perform under varying conditions.

Yamaha or Yamalube All-Purpose Marine Grease is a general outboard lubricant, chemically formulated to resist salt water. This lubricant is recommended for application to bearings, bushings, and oil seals.

Yamalube Lubricant is a two-stroke engine oil. It is a petroleum based, clean burning lubricant. Yamalube reduces carbon deposits and ensures maximum protection against engine wear. No oil additives are recommended by the manufacturer. Yam-

Yamaha recommended lubricants and additives will not only keep the unit within the limits of the warranty, but will be a major contributing factor to dependable performance and reduced maintenance costs.

Yamaha products available from the local dealer will do much to keep the outboard unit looking "sharp" and promote that "pride of ownership" feeling.

3-8 MAINTENANCE

alube contains ashless detergent to minimize piston rings from sticking.

Yamaha or Yamalube Gearcase Lubricant contains high viscosity additives to protect the lower unit gears at high speed operation. The lubricant will extend gear life, reduce gear noise, minimize friction, and has a cooling affect on the lower unit moving parts.

Hydraulic Fluid

Yamaha or Yamalube Power Trim and Tilt Fluid is a highly refined hydraulic fluid. This product has a high detergent content and additives to keep seals pliable.

A high grade automatic transmission fluid, Dexron or Type F, may be substituted if the Yamalube fluid is not available. If a substitute fluid is used to "top off" the reservoir, the system should be drained and filled with a single product at the first opportunity. Brand names should **NEVER** be mixed, if its possible to avoid.

Fuel Stabilizer

Yamaha Fuel Conditioner and Stabilizer is recommended for use at all times, during operation of the powerhead and during the storage period. The material absorbs water in the fuel system and protects against corrosion.

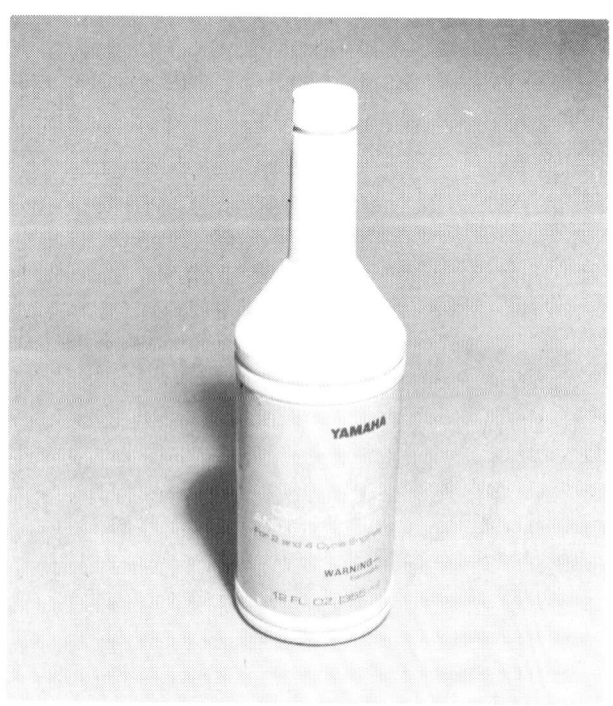

Yamaha Gasoline Stabilizer and Conditioner may be used to prevent the fuel from "souring" for up to twelve full months.

If used during operation, this fuel additive will prevent the formation of gum and varnish deposits and greatly extend the period between required carburetor overhaul.

When added to the fuel during storage, the additive will prevent the fuel from "souring" for up to twelve full months.

3-7 FIBERGLASS HULLS

Fiberglass reinforced plastic hulls are tough, durable, and highly resistant to impact. However, like any other material they can be damaged. One of the advantages of this type of construction is the relative ease with which it may be repaired. Because of its break characteristics, and the simple techniques used in restoration, these hulls have gained popularity throughout the world. From the most congested urban marina, to isolated lakes in wilderness areas, to the severe cold of far off northern seas, and in sunny tropic remote rivers of primitive islands or continents, fiberglass boats can be found performing their daily task with a minimum of maintenance.

A fiberglass hull has almost no internal stresses. Therefore, when the hull is broken or stove-in, it retains its true form. It will not dent to take an out-of-shape set. When the hull sustains a severe blow, the impact will be either absorbed by deflection of the laminated panel or the blow will result in a definite, localized break. In addition to hull damage, bulkheads, stringers, and other stiffening structures attached to the hull may also be affected and therefore, should be checked. Repairs are usually confined to the general area of the rupture.

3-8 BELOW WATERLINE SERVICE

A foul bottom can seriously affect boat performance. This is one reason why racers, large and small, both powerboat and sail, are constantly giving attention to the condition of the hull below the waterline.

In areas where marine growth is prevalent, a coating of vinyl, anti-fouling bottom paint should be applied. If growth has developed on the bottom, it can be removed with a solution of muriatic acid applied with a brush or swab and then rinsed with clear water. **ALWAYS** use rubber gloves when working with muriatic acid and **TAKE EXTRA CARE** to keep it away from your face and hands. The **FUMES ARE TOXIC.**

Therefore, work in a well-ventilated area, or if outside, keep your face on the windward side of the work.

Barnacles have a nasty habit of making their home on the bottom of boats which have not been treated with anti-fouling paint. Actually they will not harm the fiberglass hull, but can develop into a major nuisance.

If barnacles or other crustaceans have attached themselves to the hull, extra work will be required to bring the bottom back to a satisfactory condition. First, if practical, put the boat into a body of fresh water and allow it to remain for a few days. A large percentage of the growth can be removed in this manner. If this remedy is not possible, wash the bottom thoroughly with a high-pressure fresh water source and use a scraper. Small particles of hard shell may still hold fast. These can be removed with sandpaper.

3-9 INSIDE THE BOAT

The following points may be lubricated with Yamalube All-Purpose lubricant:

a- Remote control cable ends next to the hand nut. **DO NOT** over-lubricate the cable.

b- Steering arm pivot socket.

c- Exposed shaft of the cable passing through the cable guide tube.

d- Steering link rod to steering cable.

3-10 SUBMERGED ENGINE SERVICE

A submerged engine is always the result of an unforeseen accident. Once the engine is recovered, special care and service procedures **MUST** be closely followed in order to return the unit to satisfactory performance.

NEVER, again we say **NEVER** allow an engine that has been submerged to stand more than a couple hours before following the procedures outlined in this section and making every effort to get it running. Such delay will result in serious internal damage. If all efforts fail and the engine cannot be started after the following procedures have been performed, the engine should be disassembled, cleaned, assembled, using new gaskets, seals, and O-rings, and then started as soon as possible.

Submerged engine treatment is divided into three unique problem areas: Submersion in salt water; submerged while powerhead was running; and a submerged unit in fresh water.

The most critical of these three circumstances is the engine submerged in salt water, with submersion while running, a close second.

Salt Water Submersion

NEVER attempt to start the engine after it has been recovered. This action will only result in additional parts being damaged and the cost of restoring the engine increased considerably. If the engine was submerged in salt water the complete unit **MUST** be disassembled, cleaned, and assembled with new gaskets, O-rings, and seals. The corrosive effect of salt water can only be eliminated by the complete job being properly performed.

Submerged While Running
Special Instructions

If the engine was running when it was submerged, the chances of internal engine damage is greatly increased. After the engine has been recovered, remove the spark plugs to prevent compression in the cylinders. Make an attempt to rotate the crankshaft with the flywheel. On larger horsepower engines, use a socket wrench on

Easy removal of the exhaust cover will provide access to inspect the condition of the cylinder bores, the pistons, and the rings. Such inspection may reveal the cause of "strange" noises in the powerhead.

the flywheel nut to rotate the crankshaft. If the attempt fails, the chances of serious internal damage, such as: bent connecting rod, bent crankshaft, or damaged cylinder, is greatly increased. If the crankshaft cannot be rotated, the powerhead must be completely disassembled.

CRITICAL WORDS

Never attempt to start a powerhead that has been submerged. If there is water in the cylinder, the piston will not be able to compress the liquid. The result will most likely be a bent connecting rod.

Submerged Engine — Fresh Water
SPECIAL WORDS

As an aid to performing the restoration work, the following steps are numbered and should be followed in sequence. However, illustrations are not included with the procedural steps because the work involved is general in nature.

1- Recover the engine as quickly as possible.

2- Remove the cowling and the spark plugs.

3- Remove the carburetor float bowl cover, or the bowl.

4- Flush the outside of the engine with fresh water to remove silt, mud, sand, weeds, and other debris. **DO NOT** attempt to start the engine if sand has entered the powerhead. Such action will only result in serious damage to powerhead components. Sand in the powerhead means the unit must be disassembled.

CRITICAL WORDS

Never attempt to start a powerhead that has been submerged. If there is water in the cylinder, the piston will not be able to compress the liquid. The result will most likely be a bent connecting rod.

5- Remove as much water as possible from the powerhead. Most of the water can be eliminated by first holding the engine in a horizontal position with the spark plug holes **DOWN,** and then cranking the powerhead with a socket wrench on the flywheel nut. Rotate the crankshaft through at least 10 complete revolutions. If you are satisfied there is no water in the cylinders, proceed with Step 6 to remove moisture.

6- Alcohol will absorb moisture. Therefore, pour alcohol into the carburetor throat and again crank the powerhead.

7- Rotate the outboard in the horizontal position until the spark plug openings are facing **UPWARD.** Pour alcohol into the spark plug openings and again rotate the crankshaft.

8- Rotate the outboard in the horizontal position until the spark plug openings are again facing **DOWN.** Pour engine oil into the carburetor throat and, at the same time, rotate the crankshaft to distribute oil throughout the crankcase.

9- Rotate the outboard in the horizontal position until the spark plug holes are again facing **UPWARD.** Pour approximately one teaspoon of engine oil into each spark plug opening. Rotate the crankshaft to distribute the oil in the cylinders.

10- Install and connect the spark plugs.

11- Install the carburetor float bowl cover, or the bowl.

12- Obtain **FRESH** fuel and attempt to start the engine. If the powerhead will start, allow it to run for approximately an hour to eliminate any unwanted moisture remaining in the powerhead.

CAUTION

Water must circulate through the lower unit to the powerhead anytime the powerhead is operating to prevent damage to the water pump in the lower unit. Just five seconds without water will damage the water pump impeller.

If an outboard unit is submerged, the unit may be restored to top performance through **IMMEDIATE** *and* **PROPER** *service, as explained in the text.*

LOWER UNIT   3-11

13- If the powerhead fails to start, determine the cause, electrical or fuel, correct the problem, and again attempt to get it running. **NEVER** allow a powerhead to remain unstarted for more than a couple hours without following the procedures in this section and attempting to start it. If attempts to start the powerhead fail, the unit should be disassembled, cleaned, assembled, using new gaskets, seals, and O-rings, as **SOON** as possible.

3-11 LOWER UNIT
STANDARD PROPELLER DRIVE

Draining Lower Unit

Tilt the lower unit upward until the drain plug is at the lowest point to enable all the fluid to drain.

Remove the **FILL** plug from the lower end of the gear housing on the port side and the **VENT** plug just above the anti-cavitation plate.

CAUTION WORDS

Do not remove the plugs if the outboard unit has been operated recently, or if the unit has been sitting exposed to the hot sun. If one of the plugs should be removed when the lubricant is hot, the material will squirt out under considerable pressure.

Add only Yamaha gear case lubricant. **NEVER** use regular automotive-type grease in the lower unit because it expands and foams too much. Lower units do not have provisions to accommodate such expansion.

If the lubricant appears milky brown, or if large amounts of lubricant must be added to bring the lubricant up to the full mark, a thorough check should be made to determine the cause of the loss.

Special Words For Counterrotating Models

The first few times gearcase lubricant is drained from the lower unit, a discoloration may be noticed. This condition is quite normal and is no cause for alarm unless accompanied by the presence of metal chips. The discoloration is due to the use of molybdenum-disulfide assembly grease at the factory.

Water in the Lower Unit

Water in the lower unit is usually caused by fish line becoming entangled around the propeller shaft behind the propeller and damaging the propeller seal. If the line is not removed, it will cut the propeller shaft seal and allow water to enter the lower unit. Fish line has also been known to cut a groove in the propeller shaft.

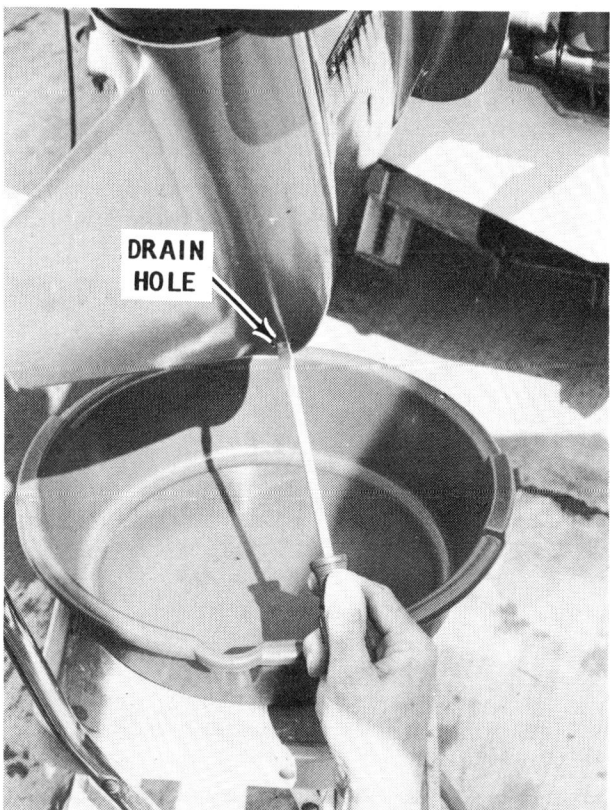

A suitable container positioned under a fully tilted lower unit, ready to receive lubricant as it drains.

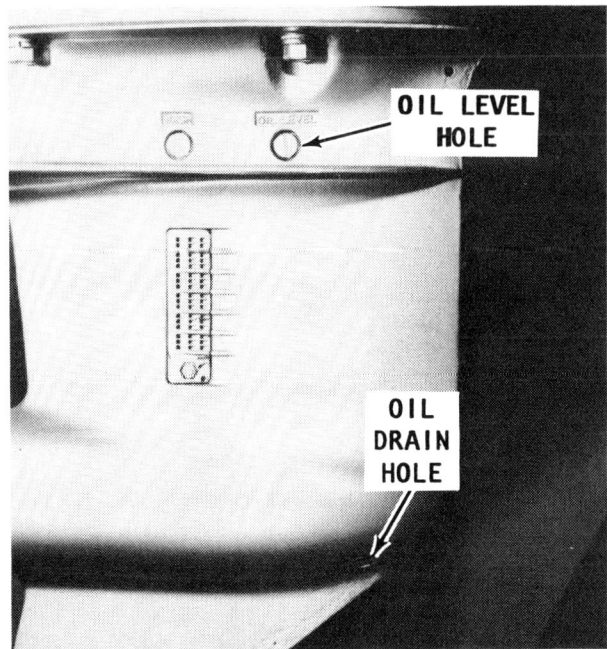

The amount of lubricant in the lower unit should be checked on a daily basis during the operating season. A fish line wrapped around the propeller shaft could cut through a seal and oil be lost without warning.

The propeller should be removed each time the boat is hauled from the water at the end of an outing and any material entangled behind the propeller removed before it can cause expensive damage. The small amount of time and effort involved in pulling the propeller is repaid many times by reduced maintenance and service work, including the replacement of expensive parts.

Filling Lower Unit

Position the drive unit approximately vertical and without a list to either port or starboard. Insert the lubricant tube into the **FILL/DRAIN** hole at the bottom plug hole, and inject lubricant until the excess begins to come out the **VENT** hole. Install the **VENT** and **FILL** plugs with **NEW** gaskets. Check to be sure the gaskets are properly positioned to prevent water from entering the housing.

See Chapter 11 for lower unit capacities.

3-12 LOWER UNIT JET DRIVE

Lubrication

The bearing and seal housing installed on the jet unit drive shaft is lubricated through an externally mounted zerk fitting. This fitting is located at the top of the jet drive casing on the port side.

To lubricate through the fitting, first push the coupling aside to release the coupling and the hose from the fitting. **DO NOT** attempt to pull on the coupling or the hose. Pushing the coupling aside is the answer.

Pump Yamalube All Purpose Marine Grease into the fitting until the old grease emerges from the hose coupler. Internal passageways are provided through the outer jet drive housing and the bearing and the seal housing to route the lubricant. When the old grease emerges from the hose coupling, there is no doubt the system has been properly lubricated. Wipe off the excess grease, and then snap the coupler back into place over the fitting.

The fitting should be lubricated every ten hours of jet drive operation.

Every 50 hours of jet drive operation, pump enough new grease into the fitting to replace the old grease. A distinct change in color between the old grease and the new grease will be noted, indicating the unit is filled with the new lubricant.

When the unit is new, a slight discoloration may be expected, as the new seals are "broken in".

If the old grease contains tiny beads of water, the seals are beginning to break down. If the old grease emerging from the hose coupling is a dark dirty grey color, the seals have already broken down and water is attacking the bearings. See Chapter 11 for service of the Jet Drive.

Lubricate the gate control linkage pivot points at regular intervals. These points are indicated in the accompanying illustration.

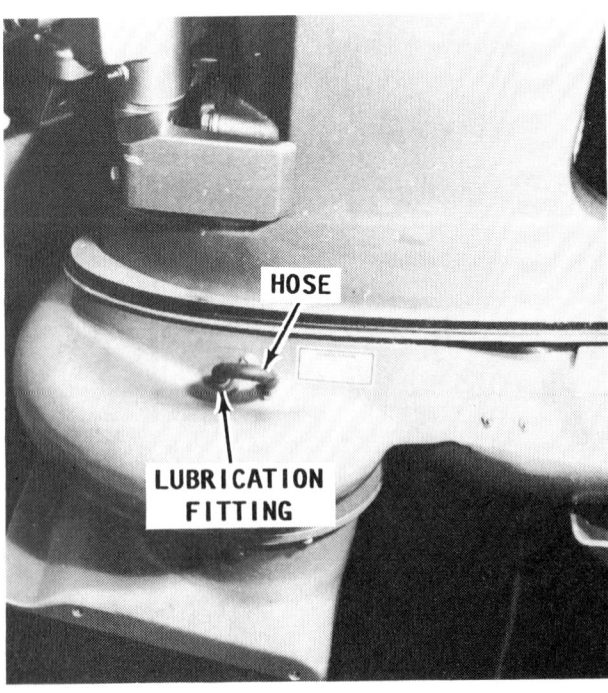

Removing the hose is accomplished by deflecting the hose to one side to snap it free of the fitting. Do NOT attempt to "pull" the hose off the fitting.

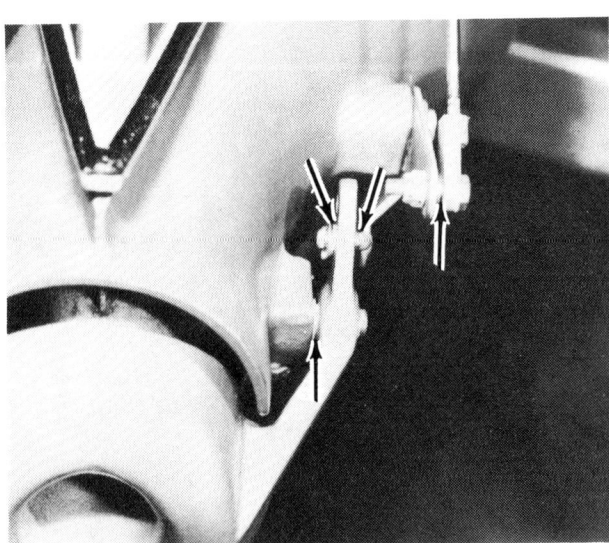

The arrows indicate pivot points to be lubricated with Yamaha All Purpose Grease.

Gate Adjustment

The position of the gate determines the direction of thrust of the expelled water.

Improper adjustment of the gate can cause the gate to be unintentionally thrown into reverse, tilting the jet drive, and the boat, up out of the water.

Detailed adjustment procedures are outlined in Chapter 11.

Impeller Clearance

The clearance between the outer edge of the jet drive impeller blade and the water intake casing cone wall should be maintained at approximately 1/32" (0.8mm). This distance can be checked visually by shining a flashlight up through the intake grille and estimating the distance between the impeller and the casing cone, as indicated in the accompanying illustrations. It is not humanly possible to accurately measure this clearance, but by observing closely and estimating the clearance, the results should be fairly accurate.

After continued use, the clearance will increase. The jet drive need not be removed from the intermediate housing to perform the procedural steps to bring the jet impeller closer to the casing cone wall.

See the appropriate section in Chapter 11 and perform the steps as if the jet drive were separated from the intermediate housing.

Adjust the spacers to bring the jet impeller further into the tapered (the cone) part of the housing and reduce the clearance. Remove spacers from the "nut" end of the jet impeller and place them at the "intermediate housing" end of the impeller.

Jet Impeller

The jet impeller is a precisely machined and dynamically balanced aluminum spiral. Observe the drilled recesses at exact locations to achieve the delicate balancing.

Excessive vibration of the jet drive may be attributed to an out-of-balance condition caused by the impeller being struck excessively by rocks, gravel or cavitation "burn".

The term cavitation "burn" is a common expression used throughout the world among people working with pumps, impeller blades, and forceful water movement.

"Burns" on the impeller blades are caused by cavitation air bubbles exploding with considerable force against the impeller blades. The edges of the blades may develop small "dime size" areas resembling a porous sponge, as the aluminum is actually "eaten" by the condition just described.

Excessive rounding of the impeller edges will reduce efficiency and performance. Therefore, the impeller should be inspected at regular intervals. If rounding is detected, the impeller should be removed from the

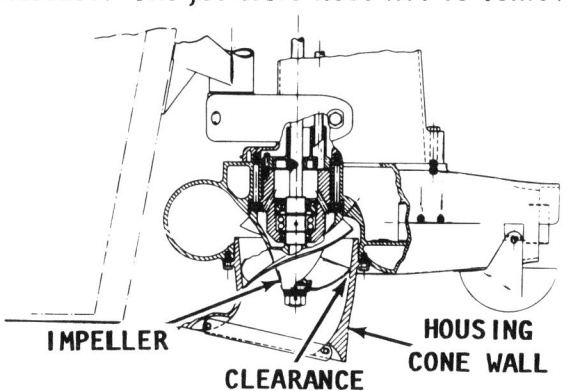

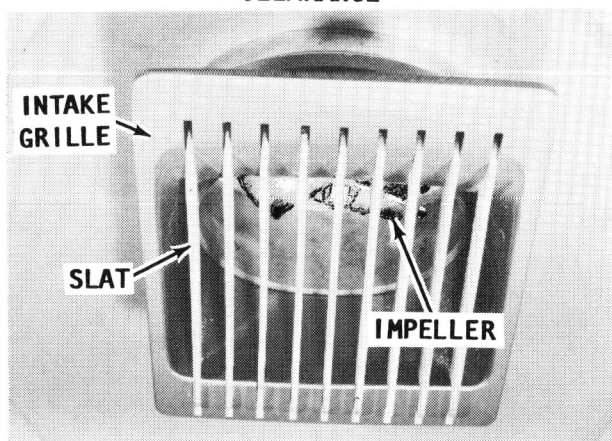

The clearance between the jet impeller and the casing cone, as indicated in the cross-section line drawing above, can be fairly well estimated by shining a flashlight up through the grille and visually checking the distance between the impeller and the cone.

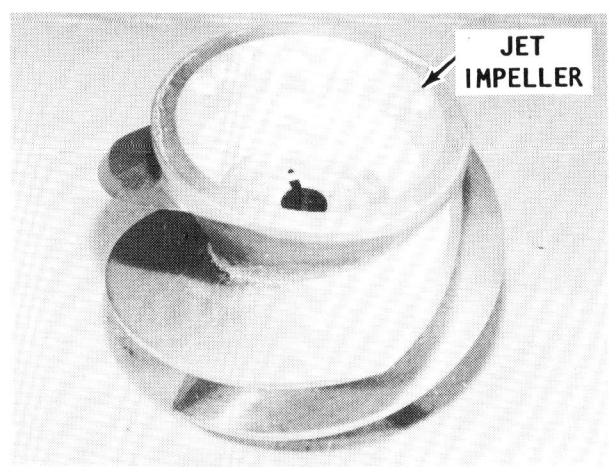

The edges of the impeller should be kept sharp for maximum efficiency and performance.

3-14 MAINTENANCE

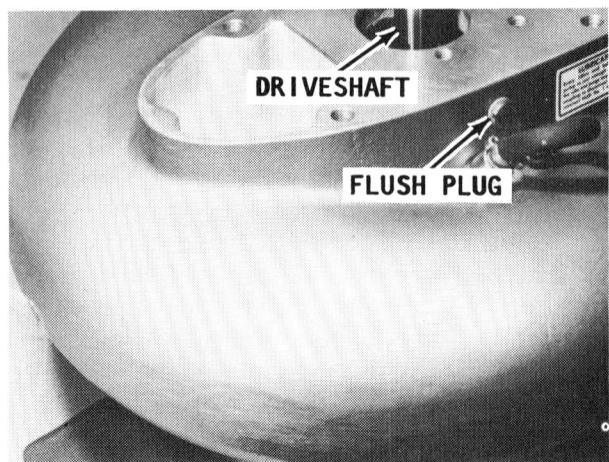

All jet drive units, 1987 and on, are provided with a flush plug on the port side of the housing.

housing, placed on a work bench and the edges restored to as sharp a condition as possible using a file.

Jet Drive and Flushing Device

Regular flushing of the jet drive will prolong the life of the powerhead, by clearing the cooling system of possible obstructions.

Some jet drives manufactured after January, 1987, are equipped with a plug on the port side just above the lubrication hose.

Remove the plug and gasket, install the flush adaptor, connect the garden hose and turn on the water supply.

Start and operate the powerhead at a fast idle for about 15 minutes. Disconnect the flushing adaptor and replace the plug.

SPECIAL WORDS

The procedure just described will only flush the powerhead cooling system, not the jet drive. To flush the jet drive unit, direct a stream of high pressure water through the intake grille.

Models Without A Flushing Plug

Units manufactured prior to January, 1987 must have a small hole drilled and tapped in the port side of the jet drive unit to accept a flush adaptor. Proceed as follows:

First, the jet drive must be "dropped" from the intermediate housing and disassembled according to the procedures outlined in Chapter 11.

Deflect the lubrication hose coupler sideways to remove it from the grease fitting and secure the hose out of the way.

Drill a 15/64" hole above the grease fitting, as shown in the accompanying illustration. Using a 8mm x 1.25 tap thread the hole. Blow away all metal chips from the interior and exterior of the housing, using compressed air.

Install the plug and gasket (available at the local Yamaha dealer). Assemble and install the jet drive to the intermediate

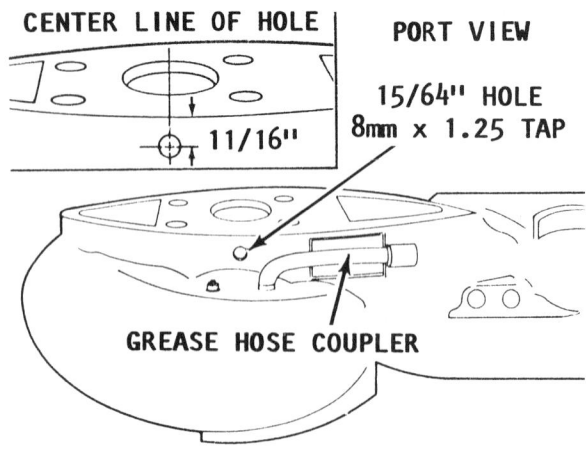

On jet drive units prior to 1987, a hole should be drilled and tapped to receive a flush plug. The dimensions indicate exactly where the hole is to be drilled.

An application of Perfect Seal on the propeller shaft splines will prevent the propeller from "freezing" to the shaft and facilitate easier removal for the next service.

housing according to the procedures outlined in Chapter 11.

The powerhead may now be flushed as described in the previous paragraphs for units with the flush plug.

3-13 PROPELLER SERVICE

The propeller should be checked regularly to be sure all the blades are in good condition. If any of the blades become bent or nicked, this condition will set up vibrations in the motor. Remove and inspect the propeller. Use a file to trim nicks and burrs. **TAKE CARE** not to remove any more material than is absolutely necessary. For a complete check, take the propeller to your marine dealer where the proper equipment and knowledgeable mechanics are available to perform a proper job at modest cost.

Inspect the propeller shaft to be sure it is still true and not bent. If the shaft is not perfectly true, it should be replaced.

Install the thrust hub. Coat the propeller shaft splines with Perfect Seal No. 4, and the rest of the shaft with a good grade of anti-corrosion lubricant. Install the front spacer, the propeller, the washer and the propeller nut.

Position a block of wood between the propeller and the anti-cavitation tab to keep the propeller from turning. Tighten the propeller nut to the torque specifications given in Chapter 11. Adjust the nut to enable the cotter pin to be threaded through the propeller nut and shaft. Bend the two ends of the cotter pin in opposite directions around the nut. This action will prevent the nut from backing off the shaft.

Trim Tabs and Lead Wires

Check the trim tab and the anode (zinc) heads. Replace them, if necessary. The trim tab must make a good ground inside the lower unit. Therefore, the trim tab and the cavity **MUST NOT** be painted. In addition to trimming the boat, the trim tab acts as a zinc electrode to prevent electrolysis from acting on more expensive parts. It is normal for the tab to show signs of erosion. The tabs are inexpensive and should be replaced frequently.

A block of wood, inserted between the propeller and the anti-cavitation plate will prevent the propeller from turning while the nut is being removed or installed.

Excellent view of rope and fish line entangled behind the propeller. Entangled fish line can actually cut through the seals allowing water to enter and oil escape from lower unit. Check this area constantly.

In addition to trimming the boat for easier steering, the trim tab is also an anode to prevent electrolysis from acting on more expensive parts.

Such extensive erosion of a trim tab compared with a new tab, suggests an electrolysis problem, or complete disregard for periodic maintenance.

Clean the exterior surface of the unit thoroughly. Inspect the finish for damage or corrosion. Clean any damaged or corroded areas, and then apply primer and matching paint.

Check the entire unit for loose, damaged, or missing parts.

An anode is attached to the base of one of the clamp brackets on all models up to 1988. Since 1988 this anode has been redesigned. It is now much larger and is mounted across both clamp brackets. It also serves as protection for the coil of hydraulic hoses beneath the trim/tilt unit between the brackets.

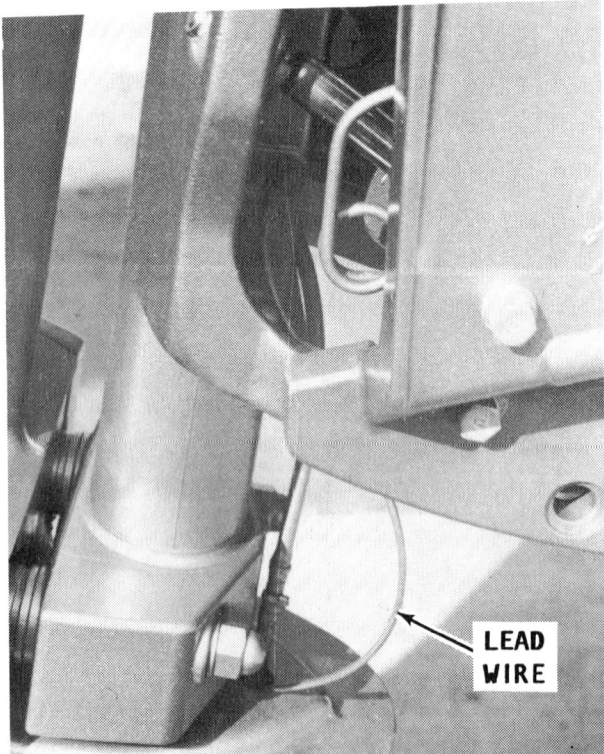

One of many lead wires used to connect bracket parts, as an assist in reducing corrosion.

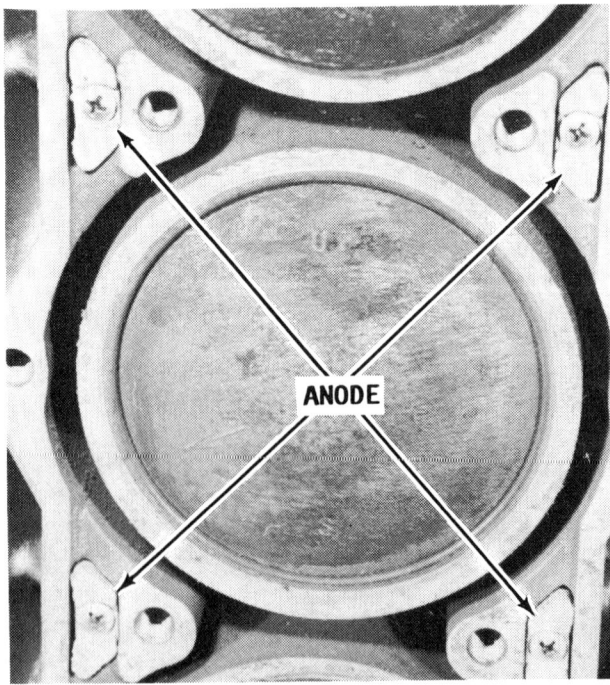

Anodes are installed in each cylinder bank as a corrosion prevention measure. The V4 powerheads have a total of four anodes and the V6 powerheads have eight. These anodes are only accessible during powerhead disassembling.

Lead wires provide good electrical continuity between various brackets which might be isolated from the trim tab by a coating of lubricant between moving parts.

3-14 WINTER STORAGE

Taking extra time to store the boat properly at the end of each season, will increase the chances of satisfactory service at the next season. **REMEMBER**, idleness is the greatest enemy of an outboard motor. The unit should be run on a monthly basis. The boat steering and shifting mechanism should also be worked through complete cycles several times each month. The owner who spends a small amount of time involved in such maintenance will be rewarded by satisfactory performance, and greatly reduced maintenance expense for parts and labor.

Proper storage involves adequate protection of the unit from physical damage, rust, corrosion, and dirt.

The following steps provide an adequate maintenance program for storing the unit at the end of a season.

A Kleen-Klip fuel hose protector, available at marine dealers and used on the end of a disconnected fuel line, will keep the fitting free of almost any type contamination and most damage.

1- Empty all fuel from the carburetor.

For many years there has been the widespread belief simply shutting off the fuel at the tank and then running the powerhead until it stops is the proper procedure before storing the engine for any length of time. Right? **WRONG!**

First, it is **NOT** possible to remove all fuel in the carburetor by operating the powerhead until it stops. Considerable fuel is trapped in the float chamber and other passages and in the line leading to the

This lower unit was destroyed because the bearing carrier was "frozen" -- possibly due to negligent maintenance. The drastic action used did not damage other expensive parts, which were saved for further service.

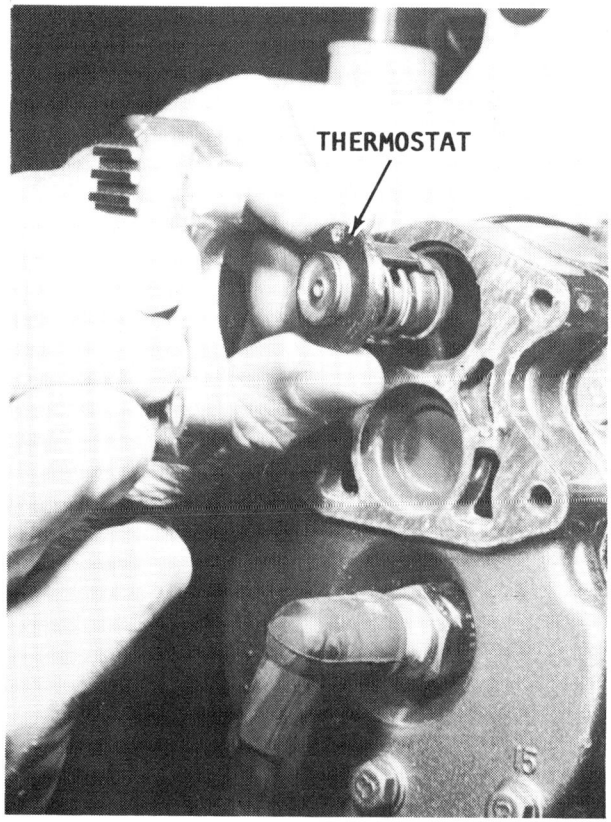

The thermostat is usually located in an accessible place for easy maintenance and replacement.

carburetor. The **ONLY** guaranteed method of removing **ALL** fuel is to take the time to remove the carburetor, and drain the fuel.

Secondly, if the powerhead is operated with the fuel supply shut off until it stops, the fuel and oil mixture inside the powerhead is removed, leaving bearings, pistons, rings, and other parts without any protective lubricant.

2- Drain the fuel tank and the fuel lines. Pour approximately one quart (0.96 liters) of benzol (benzine) into the fuel tank, and then rinse the tank and pickup filter with the benzol. Drain the tank. Store the fuel tank in a cool dry area with the vent **OPEN** to allow air to circulate through the tank. **DO NOT** store the fuel tank on bare concrete. Place the tank to allow air to circulate around it.

3- Clean the carburetor fuel filter with benzol, see Chapter 4, Carburetor Repair Section.

4- Drain, and then fill the lower unit with Yamaha Gear Case Lubricant, as outlined in Section 3-11.

5- Lubricate the throttle and shift linkage. Lubricate the steering pivot shaft with Yamalube or equivalent.

Clean the outboard unit thoroughly. Coat the powerhead with a commercial corrosion and rust preventative spray. Install the cowling, and then apply a thin film of fresh engine oil to all painted surfaces.

Remove the propeller. Apply Perfect Seal or a waterproof sealer to the propeller shaft splines, and then install the propeller back in position.

FINAL WORDS

Be sure all drain holes in the gear housing are open and free of obstruction. Check to be sure the **FLUSH** plug has been removed to allow all water to drain. Trapped water could freeze, expand, and cause expensive castings to crack.

ALWAYS store the outboard unit off the boat with the lower unit below the powerhead to prevent any water from being trapped inside.

BATTERY STORAGE

Remove the batteries from the boat and keep them charged during the storage period. Clean the batteries thoroughly of any dirt or corrosion, and then charge them to full specific gravity reading. After they are fully charged, store them in a clean cool dry place where they will not be damaged or knocked over.

NEVER store the battery with anything on top of it or cover the battery in such a manner as to prevent air from circulating around the fillercaps. All batteries, both new and old, will discharge during periods of storage, more so if they are hot than if they remain cool. Therefore, the electrolyte level and the specific gravity should be checked at regular intervals. A drop in the specific gravity reading is cause to charge them back to a full reading.

In cold climates, **EXERCISE CARE** in selecting the battery storage area. A fully-charged battery will freeze at about 60 degrees below zero. A discharged battery, almost dead, will have ice forming at about 19 degrees above zero.

ALWAYS remove the drain plug and position the boat with the bow higher than the stern. This will allow any rain water and melted snow to drain from the boat and prevent "trailer sinking". This term is used to describe a boat that has filled with rain water and ruined the interior, because the plug was not removed or the bow was not high enough to allow the water to drain properly.

4
FUEL

4-1 INTRODUCTION

The carburetion and ignition principles of two-cycle engine operation **MUST** be understood in order to perform a proper tune-up on an outboard motor.

If you have any doubts concerning your understanding of two-cycle engine operation, it would be best to study the Introduction section in the first portion of Chapter 8, before tackling any work on the fuel system.

The fuel system includes the fuel tank, fuel pump, fuel filters, carburetors, a squeeze bulb, and the associated parts to connect it all together. Regular maintenance of the fuel system to obtain maximum performance, is limited to changing the fuel filter at regular intervals and using fresh fuel.

If a sudden increase in gas consumption is noticed, or if the engine does not perform properly, a carburetor overhaul, including boil-out, or replacement of the fuel pump may be required.

4-2 GENERAL CARBURETION INFORMATION

The carburetor is merely a metering device for mixing fuel and air in the proper proportions for efficient engine operation. At idle speed, an outboard engine requires a mixture of about 8 parts air to 1 part fuel. At high speed or under heavy duty service, the mixture may change to as much as 12 parts air to 1 part fuel.

Float Systems

A small chamber in the carburetor serves as a fuel reservoir. A float valve admits fuel into the reservoir to replace the fuel consumed by the engine. If the carburetor has more than one reservoir, the fuel level in each reservoir (chamber) is controlled by identical float systems.

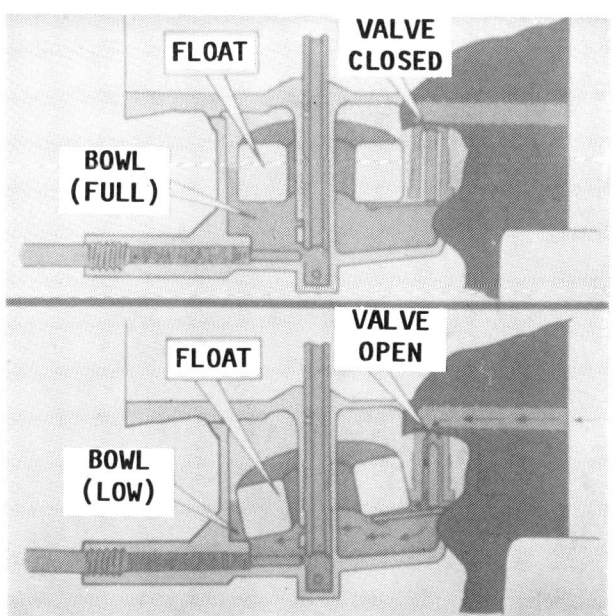

Fuel flow principle of a modern carburetor.

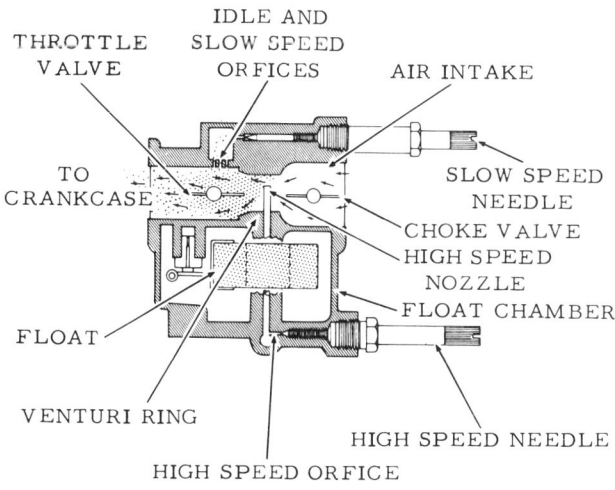

Fuel flow through the venturi, showing principle and related parts controlling intake and outflow.

4-2 FUEL

Fuel level in each chamber is extremely critical and must be maintained accurately. Accuracy is obtained through proper adjustment of the float. This adjustment will provide a balanced metering of fuel to each cylinder at all speeds.

Following the fuel through its course, from the fuel tank to the combustion chamber of the cylinder, will provide an appreciation of exactly what is taking place. In order to start the engine, the fuel must be moved from the tank to the carburetor by a squeeze bulb installed in the fuel line. This action is necessary because the fuel pump does not have sufficient pressure to draw fuel from the tank during cranking before the engine starts.

After the engine starts, the fuel passes from the tank through the fuel filter, on to the fuel pump and finally to the carburetors. At the carburetor, the fuel passes through the inlet passage to the needle and seat, and then into the float chamber (reservoir). A float in the chamber rides up and down on the surface of the fuel. After fuel enters the chamber and the level rises to a predetermined point, a tang on the float closes the inlet needle and the flow entering the chamber is cutoff. When fuel leaves the chamber as the engine operates, the fuel level drops and the float tang allows the inlet needle to move off its seat and fuel once again enters the chamber. In this manner a constant reservoir of fuel is maintained in the chamber to satisfy the demands of the engine at all speeds.

A fuel chamber vent hole is located near the top of the carburetor body to permit atmospheric pressure to act against the fuel in each chamber. This pressure assures an adequate fuel supply to the various operating systems of the powerhead.

Air/Fuel Mixture

A suction effect is created each time the piston moves upward in the cylinder. This suction draws air through the throat of the carburetor. A restriction in the throat, called a venturi, controls air velocity and has the effect of reducing air pressure at this point.

The difference in air pressures at the throat and in the fuel chamber, causes the fuel to be pushed out of metering jets extending down into the fuel chamber. When the fuel leaves the jets, it mixes with the air passing through the venturi. This fuel/air mixture should then be in the proper proportion for burning in the cylinders for maximum engine performance.

In order to obtain the proper air/fuel mixture for all engine speeds, some models have high and low speed jets. These jets have adjustable needle valves which are used to compensate for changing atmospheric conditions. In almost all cases, the high-speed circuit has fixed high-speed jets and are not adjustable.

Air flow principle of a modern carburetor.

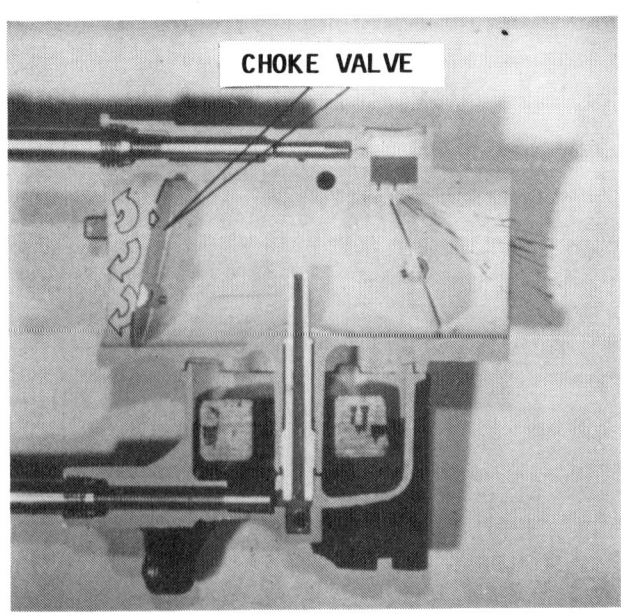

Choke valve location in the carburetor venturi. The choke valve on most carburetors covered in this manual is located in front of the venturi.

CARBURETOR MODELS 4-3

A throttle valve controls the flow of air/fuel mixture drawn into the combustion chambers. A cold powerhead requires a richer fuel mixture to start and during the brief period it is warming to normal operating temperature. A choke valve is placed ahead of the metering jets and venturi. As this valve begins to close, the volume of air intake is reduced, thus enriching the mixture entering the cylinders.

When this choke valve is fully closed, a very rich fuel mixture is drawn into the cylinders.

The throat of the carburetor is usually referred to as the "barrel". Carburetors with single, double, or four barrels have individual metering jets, needle valves, throttle and choke plates for each barrel. Single and two barrel carburetors are fed by a single float and chamber.

4-3 CARBURETOR MODELS
V4 AND V6 MODELS

All V4 and V6 models covered in this manual have the same design carburetor. However, this carburetor is not interchangeable between models due to variations in choke, venturi, and main bore diameters. Main nozzle, main jet, pilot jet and pilot air jet are also different for each model.

The V4 models have two carburetors. Each carburetor has two barrels. The V6 models have three carburetors. Again, each carburetor has two barrels.

The carburetors are mounted on the intake manifold, and each barrel supplies one cylinder with air/fuel mixture. Service procedures for each carburetor are identical. Therefore, tasks to recondition the carburetors are listed only once. Simply repeat the work for the second and third carburetor, as required.

Pilot Screw

Each carburetor has two pilot screws. Pilot screw adjustments are the same for the carburetors installed on the same powerhead, but the adjustment does vary from year to year. Some models have different settings for the port and starboard pilot screws.

Float Level

Float level settings are the same for the carburetors installed on the same powerhead, but they can vary from year to year.

Correct positioning of the electric choke solenoid is essential to ensure proper movement of the choke plate.

Starboard view of the carburetor installation on a V4 powerhead. Each two barrel carburetor has two pilot screws, one on the carburetor port side and the other on the starboard side.

Check the listing in the Appendix for pilot screw settings. Float levels for each model are given in the procedural steps.

Carburetor Vendors

In 1986 Yamaha changed carburetor vendors from Teikei to Nikki. The service procedures for both manufactured carburetors remain identical, except for the float arrangement. Units prior to 1986, the Teikei carburetor installed on all V4 and V6 models, have a figure "8" shaped float in an elongated float bowl. This carburetor also has a single needle and seat assembly.

After the vendor change, in 1986, and until press time for this manual, all V4 and V6 models are equipped with Nikki carburetors. This carburetor has two individual float bowls, two floats -- one for each bowl -- with a separate needle and seat assembly for each float.

Carburetor Identification

The carburetors are easily identified. The shape of the bowl is a "dead" giveaway. The single float bowl identifies the Teikei and the double float bowl identifies the Nikki.

The float level and pilot screw settings are different for each vendor carburetor. Because the change occurred mid-year -- 1986 -- check to be sure the proper specifications are being referenced in the Appendix for the unit being serviced.

4-4 FUEL SYSTEM

The fuel system includes the fuel tank, fuel pump, fuel filter, carburetor, connecting lines with a squeeze bulb, and the associated parts to connect it all together. Regular maintenance of the fuel system to obtain maximum performance, is limited to changing the fuel filter at regular intervals and using fresh fuel. Even with the high price of fuel, removing gasoline that has been standing unused over a long period of time, is still the easiest and least expensive preventive maintenance possible. In most cases this old gas, even with some oil mixed with it, can be used without harmful effects in an automobile using regular gasoline.

If a sudden increase in gas consumption is noticed, or if the engine does not perform properly, a carburetor overhaul, including boil-out, or replacement of the fuel pump may be required.

LEADED GASOLINE AND GASOHOL

The manufacturer of the units covered in this manual recommends the powerheads be operated using either regular unleaded or regular leaded gasoline having a minimum octane rating of 86 or higher for all models with standard CDI ignition. However, the manufacturer recommends the use of premium unleaded or regular leaded gasoline with a minimum octane rating of 89 or higher for the models equipped with YMIS (Yamaha Microcomputer Ignition System).

In the United States, the Environmental Protection Agency (EPA) has slated a proposed national phase-out of leaded fuel, "Regular" gasoline, by 1988. Lead in gasoline boosts the octane rating (energy). Therefore, if the lead is removed, it must be replaced with another agent. Unknown to the general public, many refineries are adding alcohol in an effort to hold the octane rating.

Alcohol in gasoline can have a deteriorating effect on certain fuel system parts. Seals can swell, pump check valves can swell, diaphragms distort, and other rubber or neoprene composition parts in the fuel system can be affected.

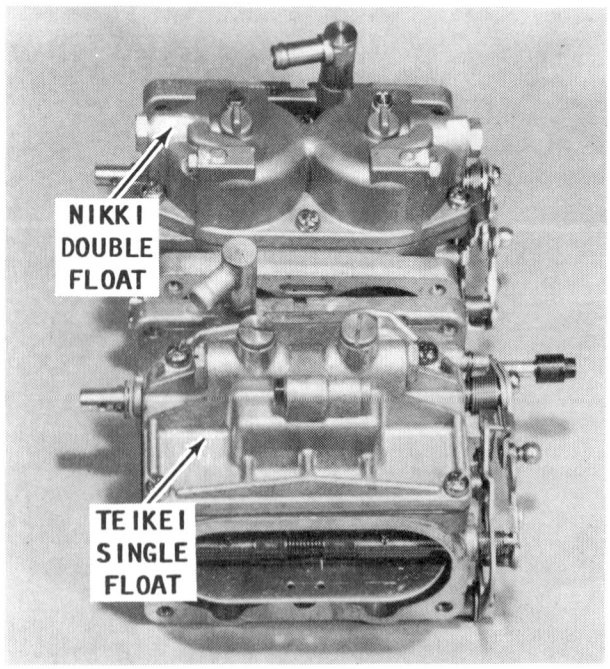

The powerheads covered in this manual have either Nikki or Teikei carburetors. Basically, both units are identical, except for the float arrangement. Other differences are noted in the text.

Since about 1981, all manufacturers have made every effort to use materials that will resist the alcohol being added to fuels.

Fuels containing alcohol will slowly absorb moisture from the air. Once the moisture content in the fuel exceeds about 1%, it will separate from the fuel taking the alcohol with it. This water/alcohol mixture will settle to the bottom of the fuel tank. The engine will fail to operate. Therefore, storage of this type of gasoline for use in marine engines is not recommended for more than just a few days.

One temporary, but aggravating, solution to increase the octane of "unleaded" fuel is to purchase some aviation fuel from the local airport. Add about 10 to 15 percent of the tank's capacity to the unleaded fuel.

REMOVING FUEL FROM THE SYSTEM

For many years there has been the widespread belief that simply shutting off the fuel at the tank and then running the engine until it stops is the proper procedure before storing the engine for any length of time. Right? **WRONG.**

It is **NOT** possible to remove all of the fuel in the carburetor by operating the engine until it stops. Some fuel is trapped in the float chamber and other passages and in the line leading to the carburetor. The **ONLY** guaranteed method of removing **ALL** of the fuel is to take the time to remove the carburetor, and drain the fuel.

If the engine is operated with the fuel supply shut off until it stops, the fuel and oil mixture inside the engine is removed, leaving bearings, pistons, rings, and other parts with little protective lubricant, during long periods of storage.

Proper procedure involves: Shutting off the fuel supply at the tank; disconnecting the fuel line at the tank; operating the engine until it begins to run **ROUGH;** then stopping the engine, which will leave some fuel/oil mixture inside; and finally removing and draining the carburetor. By disconnecting the fuel supply, all **SMALL** passages are cleared of fuel even though some fuel is left in the carburetor. A light oil should be put in the combustion chamber as instructed in the Owner's Manual. On the carburetor covered in this manual, a drain plug on the base of the fuel bowl can be removed to drain the fuel from the carburetor.

For short periods of storage, simply running the carburetor dry may help prevent severe gum and varnish from forming in the carburetor. This is especially true during hot weather.

4-5 TROUBLESHOOTING

The following paragraphs provide an orderly sequence of tests to pinpoint problems in the system. It is very rare for the carburetor by itself to cause failure of the engine to start.

FUEL PROBLEMS

Many times fuel system troubles are caused by a plugged fuel filter, a defective fuel pump, or by a leak in the line from the fuel tank to the fuel pump. A defective choke may also cause problems. **WOULD YOU BELIEVE,** a majority of starting troubles which are traced to the fuel system are the result of an empty fuel tank or aged "sour" fuel.

"SOUR" FUEL

Under average conditions (temperate climates), fuel will begin to breakdown in about four months. A gummy substance forms in the bottom of the fuel tank and in

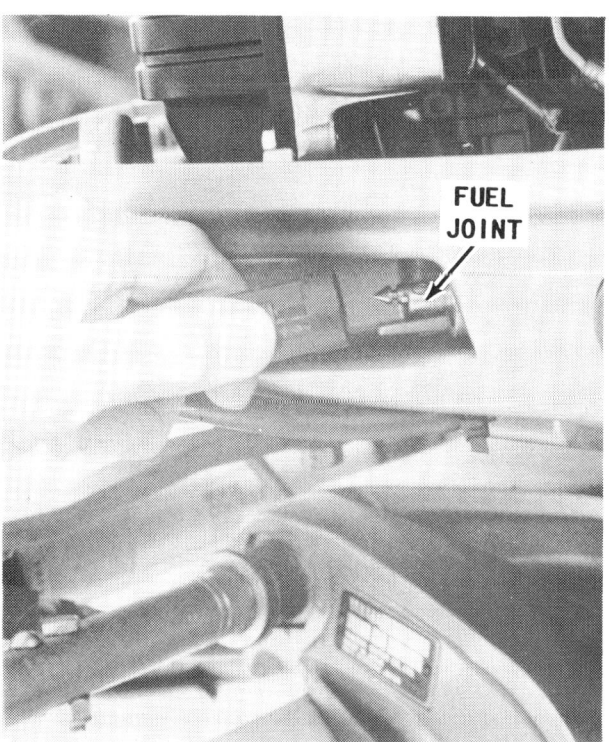

Typical fuel line quick disconnect fitting.

other areas. The filter screen between the tank and the carburetor and small passages in the carburetor will become clogged. The gasoline will begin to give off an odor similar to rotten eggs. Such a condition can cause the owner much frustration, time in cleaning components, and the expense of replacement or overhaul parts for the carburetor.

Even with the high price of fuel, removing gasoline that has been standing unused over a long period of time is still the easiest and least expensive preventive maintenance possible. In most cases, this old gas can be used without harmful effects in an automobile using regular gasoline.

The gasoline preservative additive Yamaha Fuel Conditioner and Stabilizer for 2 and 4 cycle engines, will keep the fuel "fresh" for up to twelve months. If this particular product is not available in your area, other similar additives are produced under various trade names.

Choke Problems

When the engine is hot, the fuel system can cause starting problems. After a hot engine is shut down, the temperature inside the fuel bowl may rise to 200°F and cause the fuel to actually boil. All carburetors are vented to allow this pressure to escape to the atmosphere. However, some of the fuel may percolate over the main nozzle.

If the choke should stick in the open position, the engine will be hard to start. If the choke should stick in the closed position, the engine will flood making it very difficult to start.

In order for this raw fuel to vaporize enough to burn, considerable air must be added to lean out the mixture. Therefore, the only remedy is to remove the spark plugs; ground the leads; crank the powerhead through about ten revolutions; clean the plugs; install the plugs again; and start the engine.

If the needle valve and seat assembly is leaking, an excessive amount of fuel may enter the reed housing in the following manner: After the powerhead is shut down, the pressure left in the fuel line will force fuel past the leaking needle valve. This extra fuel will raise the level in the fuel bowl and cause fuel to overflow into the reed housing.

A continuous overflow of fuel into the reed housing may be due to a sticking inlet needle or to a defective float which would cause an extra high level of fuel in the bowl and overflow into the reed housing.

FUEL PUMP TEST

CAUTION Gasoline will be flowing in the engine area during this test. Therefore, guard against fire by grounding the high-tension wire to prevent it from sparking.

A good safety procedure is to ground each spark plug lead. Disconnect the fuel line at the carburetor. Place a suitable container over the end of the fuel line to catch the fuel discharged. Open the valve at the fuel tank. Squeeze the primer bulb and observe if there is satisfactory flow of fuel from the line.

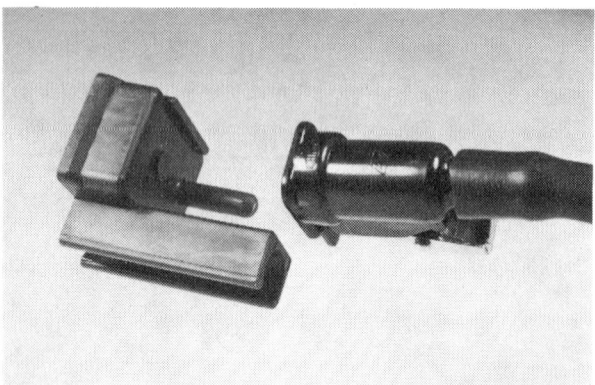

An excellent method of protecting fuel hose connectors against damage and contamination is an end cap by "Kleen Klip". More information on this device in Chapter 3.

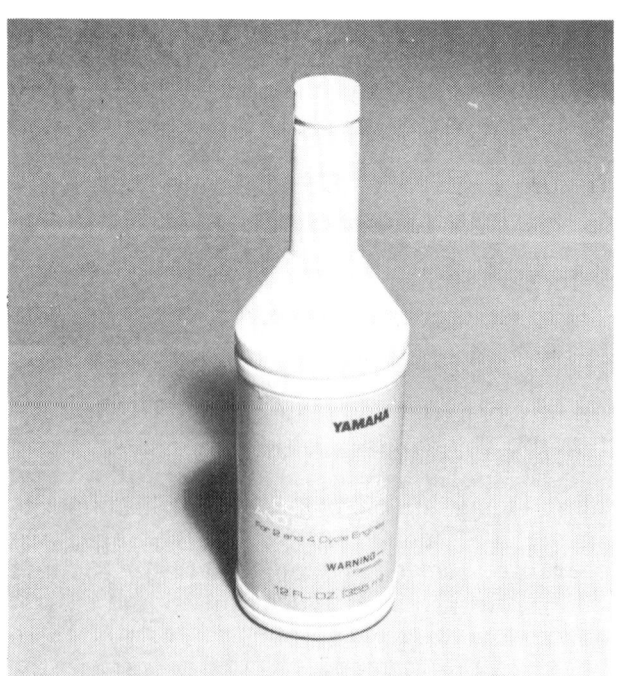

Yamaha Gasoline Stabilizer and Conditioner may be used to prevent the fuel from "souring" for up to twelve full months.

SYSTEM TESTS 4-7

If there is no fuel discharged from the line, the check valve in the squeeze bulb may be defective, or there may be a break or obstruction in the fuel line.

If there is a good fuel flow, then crank the powerhead. If the fuel pump is operating properly, a healthy stream of fuel should pulse out of the line.

Continue cranking the engine and catching the fuel for about 15 pulses to determine if the amount of fuel decreases with each pulse or maintains a constant amount. A decrease in the discharge indicates a restriction in the line. If the fuel line is plugged, the fuel stream may stop. If there is fuel in the fuel tank but no fuel flows out of the fuel line while the engine is being cranked, the problem may be in one of four areas:

1- The line from the fuel pump to the carburetor may be plugged as already mentioned.

2- The fuel pump may be defective.

3- The line from the fuel tank to the fuel pump may be plugged; the line may be leaking air; or the squeeze bulb may be defective.

4- If the engine does not start even though there is adequate fuel flow from the fuel line, the fuel filter in the carburetor inlet may be plugged or the fuel inlet needle valve and the seat may be gummed together and prevent adequate fuel flow.

FUEL LINE TEST

On most installations, the fuel line is provided with quick-disconnect fittings at the tank and at the engine. If there is reason to believe the problem is at the quick-disconnects, the hose ends should be replaced as an assembly. For a small additional expense, the entire fuel line can be replaced and thus eliminate this entire area as a problem source for many future seasons.

The primer squeeze bulb can be replaced in a short time. First, cut the hose line as close to the old bulb as possible. Slide a small clamp over the end of the fuel line from the tank. Next, install the **SMALL** end of the check valve assembly into this side of the fuel line. The check valve always goes towards the fuel tank. Place a large clamp over the end of the check valve assembly. Use Primer Bulb Adhesive when the connections are made. Tighten the clamps. Repeat the procedure with the other side of the bulb assembly and the line leading to the engine.

Location of the two fuel pumps installed on V6 powerheads. The V4 powerhead has only one fuel pump.

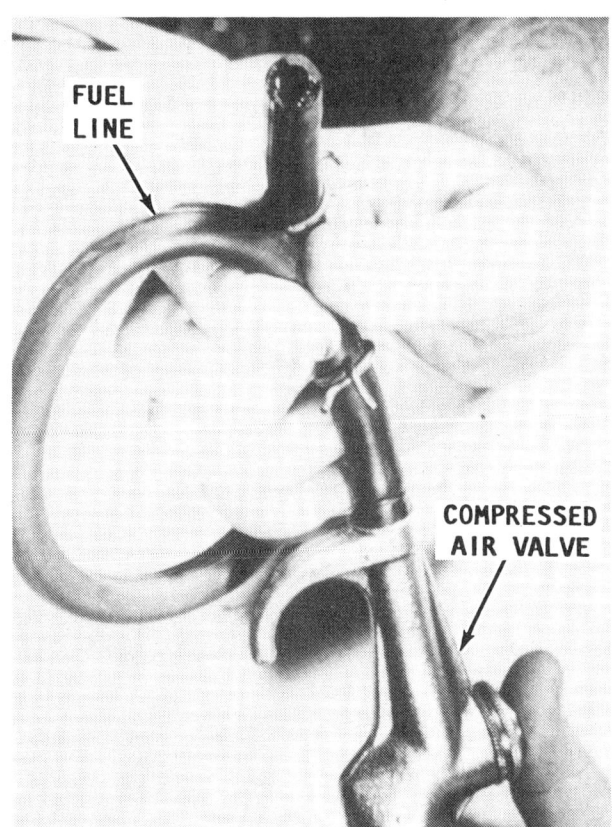

*Many times, restrictions such as foreign material may be cleared from the fuel line using compressed air. Use **CARE** to be sure the open end of the hose is pointing clear to avoid personal injury to the eyes.*

4-8 FUEL

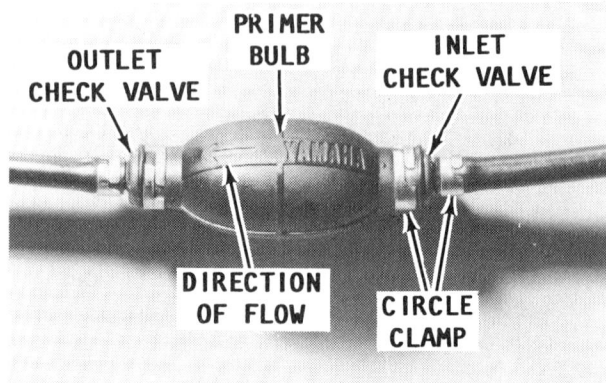

Major parts of a typical fuel line squeeze bulb, used to prime the system and deliver fuel to the carburetor until the powerhead is operating and the pump can deliver the required amount of fuel on its own.

ROUGH ENGINE IDLE

If an engine does not idle smoothly, the most reasonable approach to the problem is to perform a tune-up to eliminate such areas as: defective points; faulty spark plugs; and timing out of adjustment.

Other problems that can prevent an engine from running smoothly include: An air leak in the intake manifold; uneven compression between the cylinders; and sticky or broken reeds.

Of course any problem in the carburetor affecting the air/fuel mixture will also prevent the engine from operating smoothly at idle speed. These problems usually include: Too high a fuel level in the bowl; a heavy float; leaking needle valve and seat; defective automatic choke; and improper adjustments for idle mixture or idle speed.

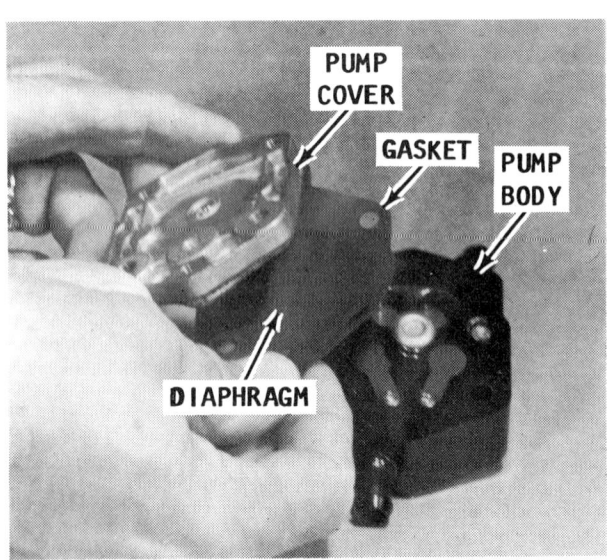

Disassembly of the fuel pump used on the powerheads covered in this manual.

Lack of adequate fuel, possibly a defective fuel pump, caused the burn condition damage to this piston.

EXCESSIVE FUEL CONSUMPTION

Excessive fuel consumption can be the result of any one of three conditions, or a combination of all three.

1- Inefficient engine operation.
2- Faulty condition of the hull, including excessive marine growth.
3- Poor boating habits of the operator.

If the fuel consumption suddenly increases over what could be considered normal, then the cause can probably be attributed to the engine or boat and not the operator.

Marine growth on the hull can have a very marked effect on boat performance. This is why sail boats always try to have a haul-out as close to race time as possible. While you are checking the bottom, take note of the propeller condition. A bent blade or other damage will definitely cause poor boat performance.

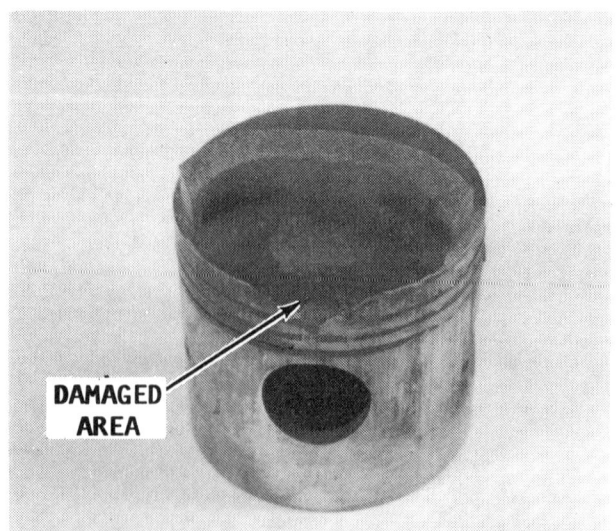

Damaged piston, possibly caused by insufficient oil mixed with the fuel; using too-low an octane fuel; or using fuel that has "soured" (stood too long without a preservative added).

Fouled spark plug, possibly caused by the operator's habit of overchoking or a malfunction holding the choke closed. Either of these conditions delivered a too-rich fuel mixture to the cylinder.

If the hull and propeller are in good shape, then check the fuel system for possible leaks. Check the line between the fuel pump and the carburetor while the engine is running and the line between the fuel tank and the pump when the engine is not running. A leak between the tank and the pump many times will not appear when the engine is operating, because the suction created by the pump drawing fuel will not allow the fuel to leak. Once the engine is turned off and the suction no longer exists, fuel may begin to leak.

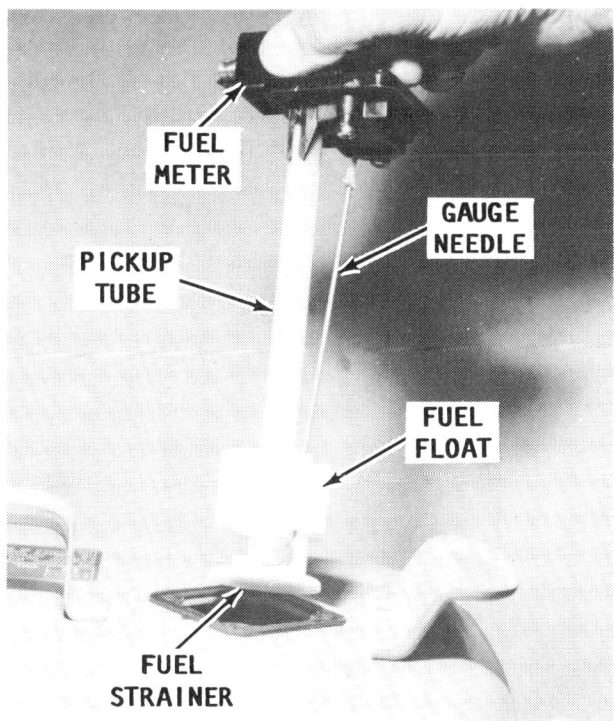

The fuel pump pickup tube removed from a Yamaha fuel tank. As the fuel level drops in the tank, a float slides down the tube and deflects a needle. The upper end of the needle registers the fuel level on the fuel meter.

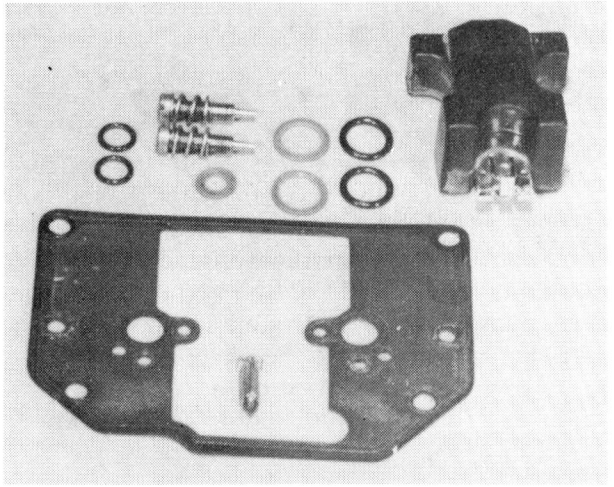

Parts included (at press time), in the Teikei carburetor repair kit.

If a minor tune-up has been performed and the spark plugs, points, and timing are properly adjusted, then the problem most likely is in the carburetor and an overhaul is in order. Check the needle valve and seat for leaking. Use extra care when making any adjustments affecting the fuel consump-

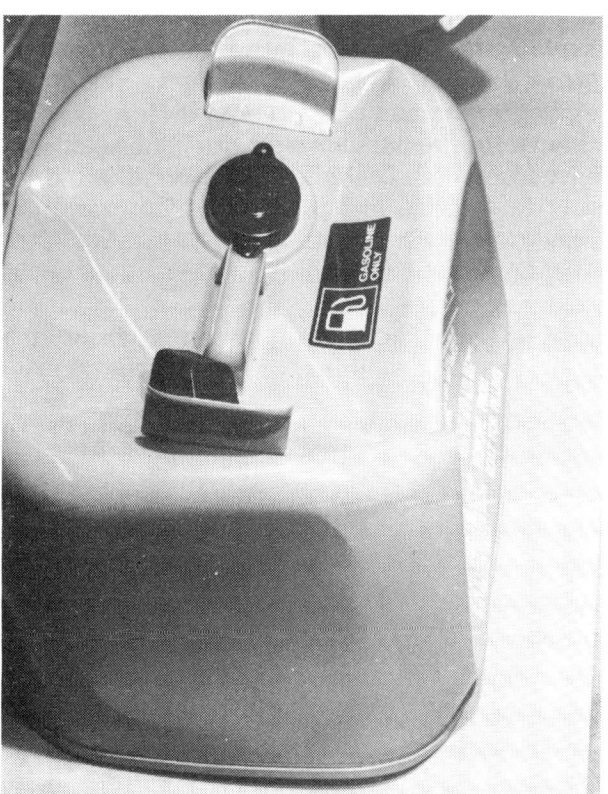

The owner of this portable fuel tank allowed the tank to sit exposed to the hot sun, clamped the cap on tightly, and then moved the tank into a cool area. As the fuel cooled, the volume decreased, causing the inside pressure to also decrease. When the inside pressure became lower than the outside atmospheric pressure, the side of the tank buckled inward. Lesson learned: Leave the cap loose when a "hot" tank is moved to a cool area.

tion, such as the float level or automatic choke.

ENGINE SURGE

If the engine operates as if the load on the boat is being constantly increased and decreased, even though an attempt is being made to hold a constant engine speed, the problem can most likely be attributed to the fuel pump, or a restriction in the fuel line between the tank and the carburetor.

Operational description and service procedures for the fuel pump are given in Section 4-13.

4-6 CARBURETOR SERVICE

Two carburetors for the V4 powerheads and three carburetors for the V6 powerheads are used on the outboard units covered in this manual. Complete, detailed, illustrated, procedures to remove, service, and install the carburetors are presented in this chapter. Removal procedures may vary slightly due to differences in linkage. As explained in the description of the carburetors, Section 4-3, this chapter, the float arrangement and adjustments differ due to a change in carburetor vendor. These differences are noted whenever they make a substantial change in a procedure.

REMOVAL

1- Disconnect the fuel line from the tank at the fuel joint. Remove the screws

securing the outer silencer cover to the inner silencer cover. V4 models have eight attaching bolts, V6 models have ten attaching bolts.

V4 Models Only

2- Pry off the little O-ring which serves as a retainer for the choke solenoid pull wire. Take care not to lose this small part. Slip off the end of the pull wire from the carburetor linkage. Remove the choke pull wire and plunger from the solenoid. Disconnect the single Blue lead at its quick disconnect fitting. The solenoid may remain mounted to the inner silencer cover. The cover is removed in a later step.

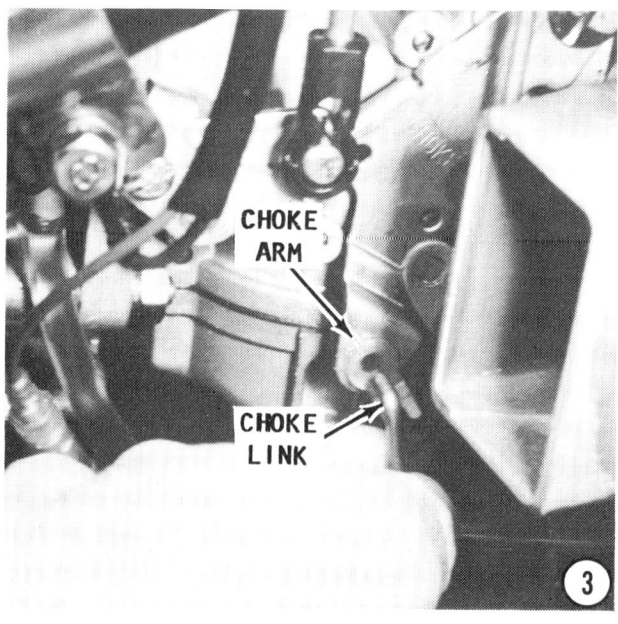

CARBURETOR REMOVAL 4-11

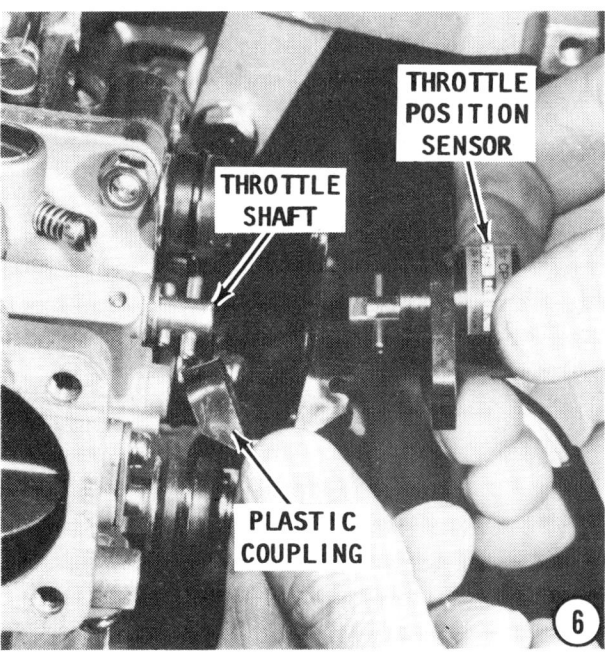

All Models

3- Pry the choke link from its retainer on the portside of the bottom carburetor.

4- Snip the three plastic hose retainers and gently pull off the fuel supply hoses from the carburetors.

5- Remove the bolts securing the inner silencer cover to the carburetors. Separate the cover from the carburetors a short distance, and then gently pull off the oil tank breather hose (top) and the fuel recirculation hose (bottom) from the aftside of the inner silencer cover.

Very Important Words for Models Equipped with YMIS

Mark the original location of the throttle position sensor **BEFORE** removal. If this sensor is misaligned during installation, a **DIGITAL** type voltmeter is needed to correctly reset the sensor on its mounting bracket, because voltages in the 1/10 range must be accurately read. A misaligned sensor could send misleading signals to the microcomputer and consequently effect the ignition timimg.

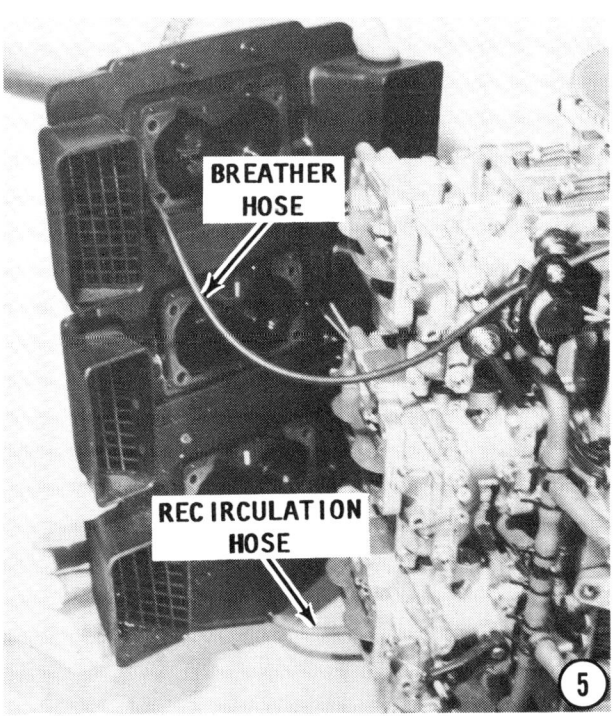

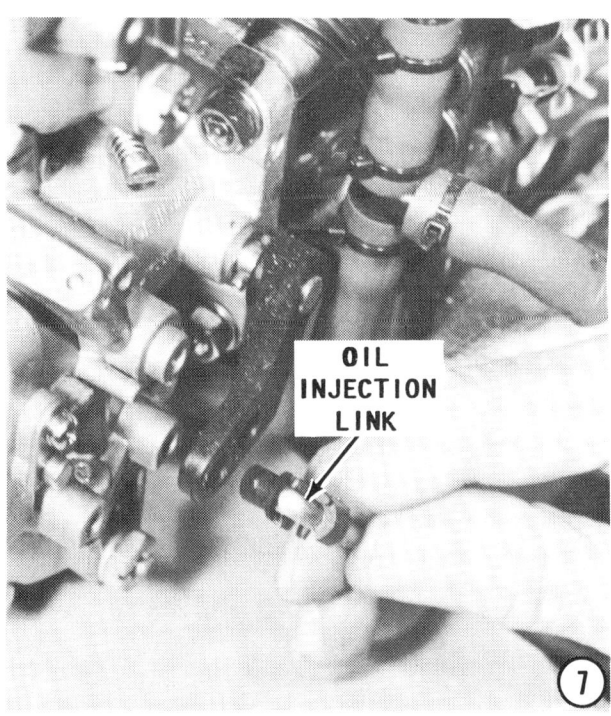

4-12 FUEL

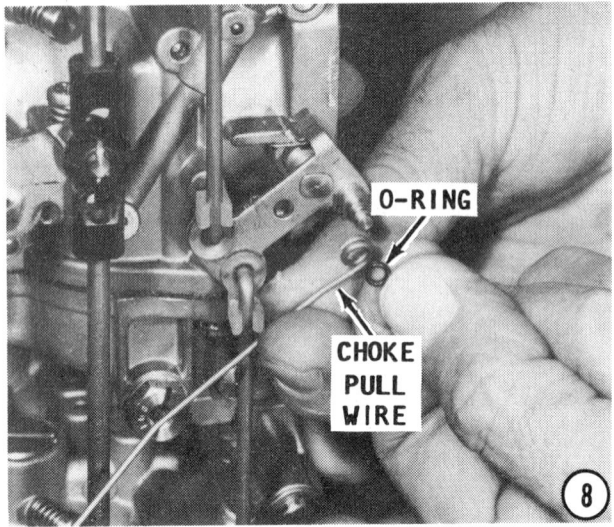

YMIS Models Only

6- After marking the position of the throttle sensor, remove the two securing screws and lift out the sensor from its plastic coupling. Slide the plastic coupling from the throttle shaft.

All Models

7- On the starboard side of the bottom carburetor: pry the oil injection link from the linkage.

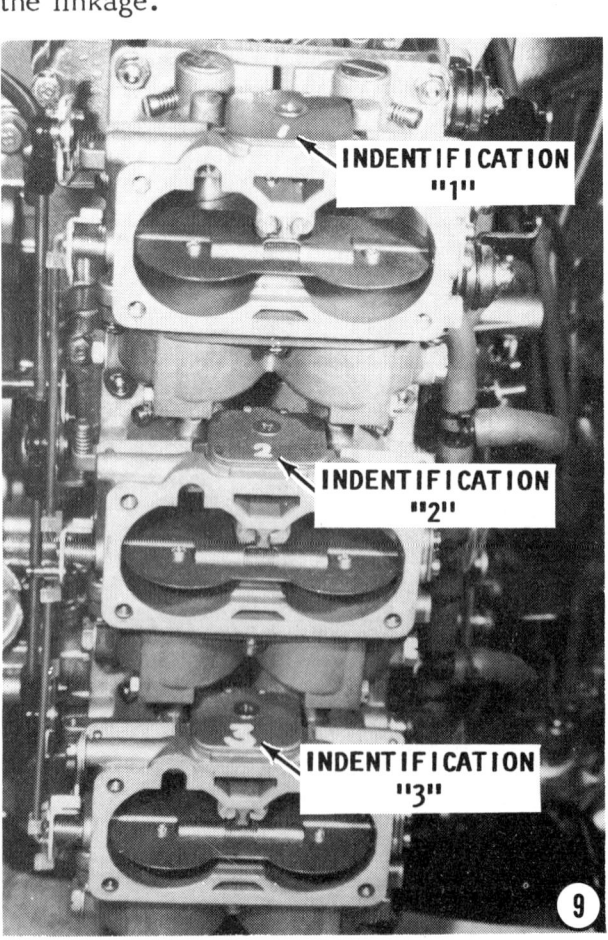

V6 Models Only

8- Pry off the little O-ring which serves as a retainer for the choke solenoid pull wire. Take care not to lose this small part. Slip off the end of the pull wire from the carburetor linkage. The choke solenoid need not be removed on these models.

9- Identify each carburetor by inscribing or painting a 1, 2 and 3 (if applicable) on the mixing chamber cover to ensure each carburetor will be installed back into the same position from which it was removed.

10- Remove the mounting nuts, four on each carburetor, and then remove the carburetors as an assembly. Each carburetor has a separate gasket which may either come away with the carburetor, or remain on the intake manifold. Remove and discard these gaskets.

Place the carburetor assembly on the workbench and remove each piece of linkage one at a time. Arrange the linkage on the workbench as it was installed on the carburetors, as an assist during assembling.

4-7 DISASSEMBLING

The following procedures pick up the work after the carburetors have been re-

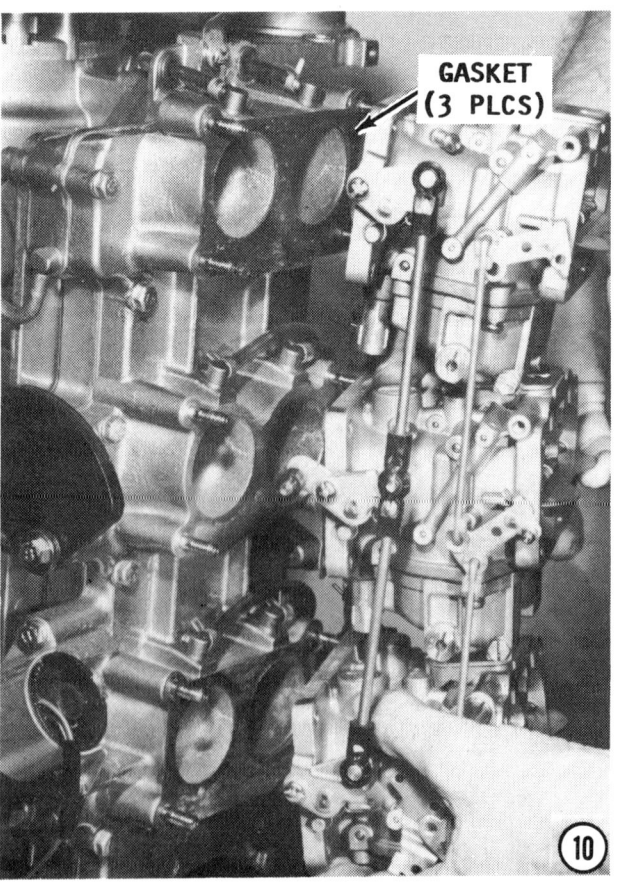

CARBURETOR DISASSEMBLING 4-13

moved from the powerhead, as outlined in the previous steps. The procedures for each of the two or three carburetors is identical, except for the powerheads equipped with YMIS. The top carburetor on a powerhead equipped with YMIS has an extension on the throttle shaft for the throttle position sensor.

Where differences occur in the single (Teikei) and dual (Nikki) float designs, both sets of procedures are included.

Therefore, perform the following procedures for each of the three carburetors.

SPECIAL WORDS

Any differences in float drop measurement, pilot screw turns, jets, or any other adjustments will be clearly identified for the carburetor including location and powerhead hp model.

Teikei Carburetor

1- Use a small screwdriver and remove the pilot air jets and the main air jets. On some Teikei carburetors, the main air jets are not removeable, because the jet does not have a screwdriver slot. The jet is flush with the carburetor body.

Keep the pilot air jets and the main air jets separated **AND** identified. They look identical and may easily be confused, one for the other. The main air jets have the higher numerical value embossed on them.

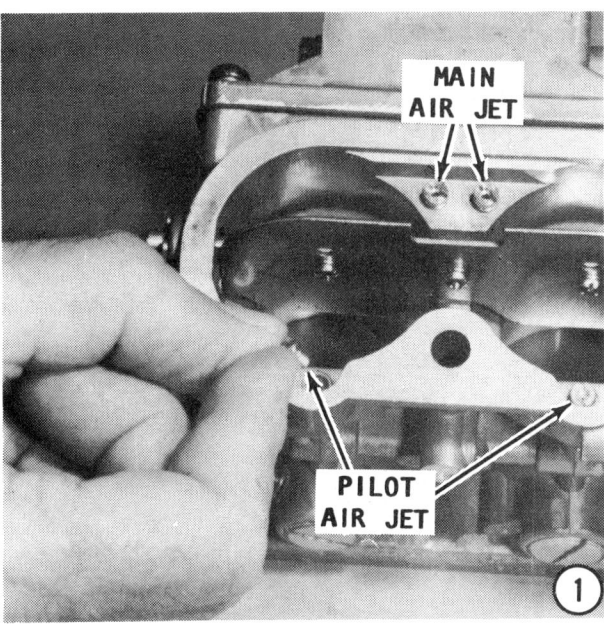

The Teikei carburetor has non-removeable main air jets. Notice: jets are flush with carburetor body and do not have a screwdriver slot.

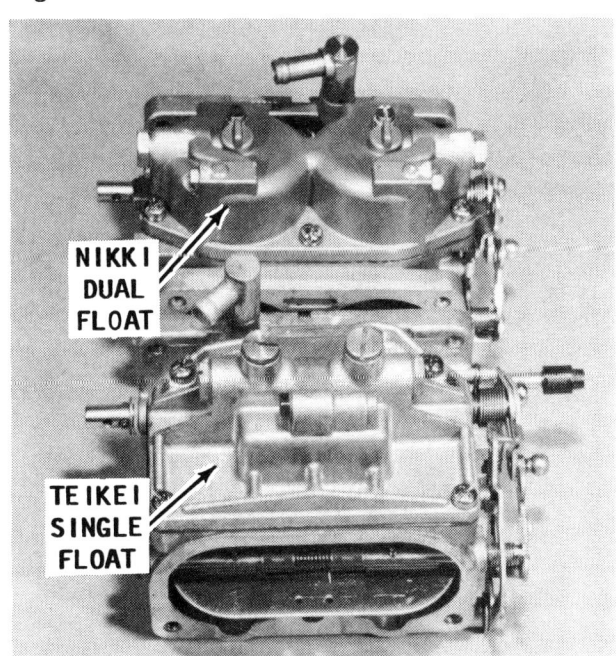

Procedures are included in the text for the Teikei single float carburetor (bottom), and for the Nikki dual float carburetor (top). Any other differences are clearly indicated.

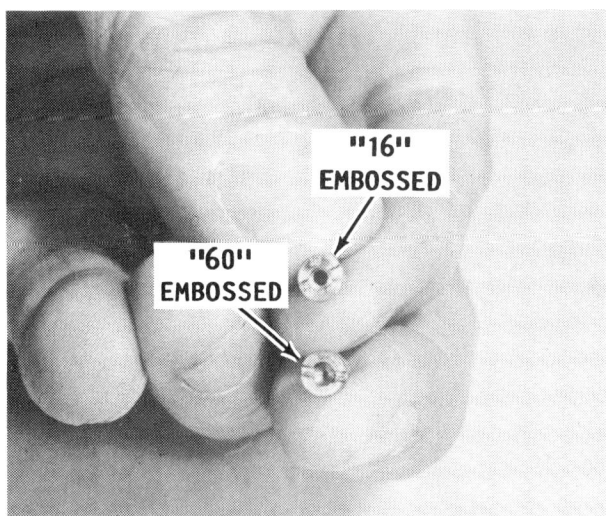

*The pilot jet (above), and the main jet (bottom), look similar, but are **NOT** interchangeable. The main jet has a greater numercial number embossed on its face, as indicated.*

4-14 FUEL

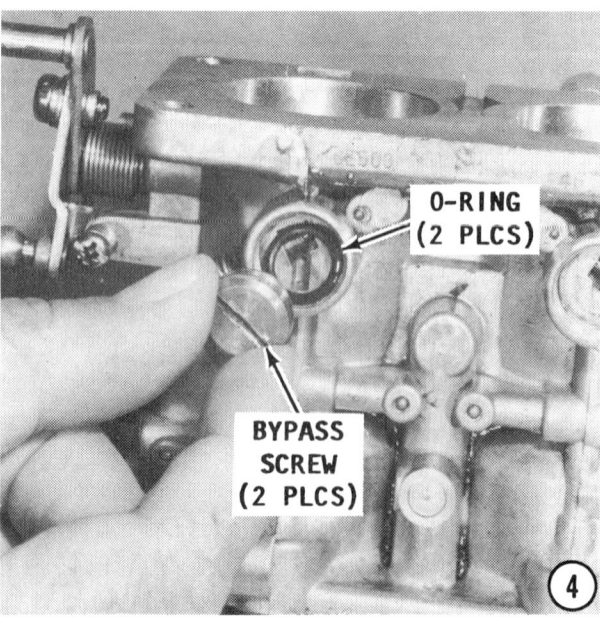

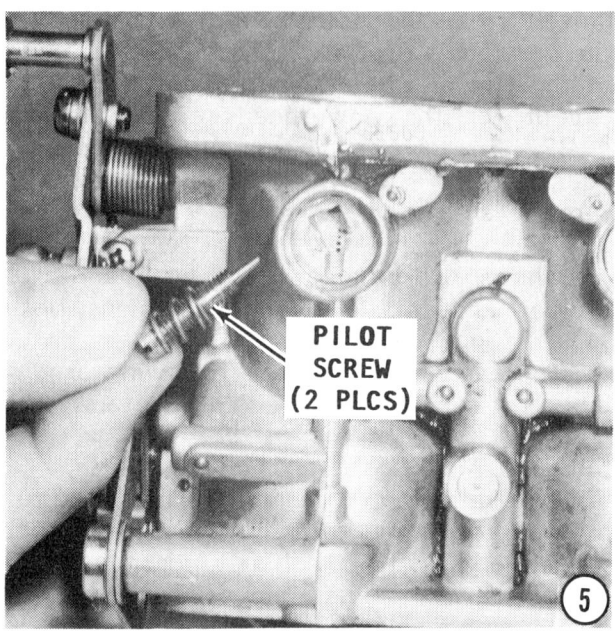

Nikki Carburetor

2- Remove the Phillips screw securing the jet access cover to the carburetor. Remove the access cover and the gasket beneath the cover. Discard the gasket.

3- Use a small screw driver and remove the main air jets located under the access cover and the pilot air jets located at the front of the carburetor.

Keep the pilot air jets and the main air jets separated **AND** identified. They look identical and may easily be confused, one for the other. The main air jets have the higher numerical value embossed on them.

All Carburetors

4- Remove the two bypass screws from the mixing chamber. Remove and discard the two O-rings.

5- Remove both pilot screws and springs. It is not necessary to count the

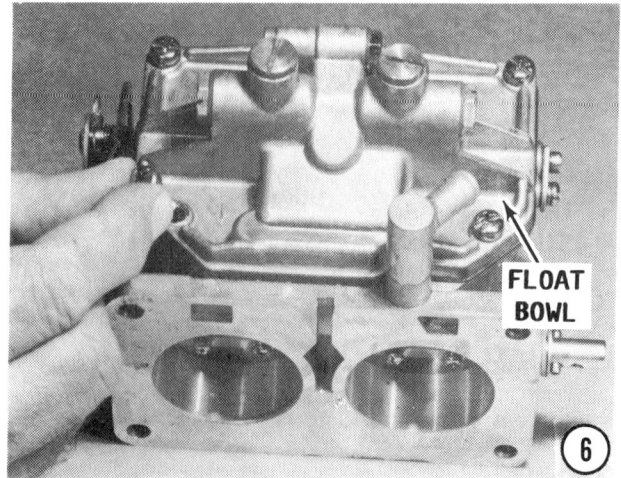

CARBURETOR DISASSEMBLING 4-15

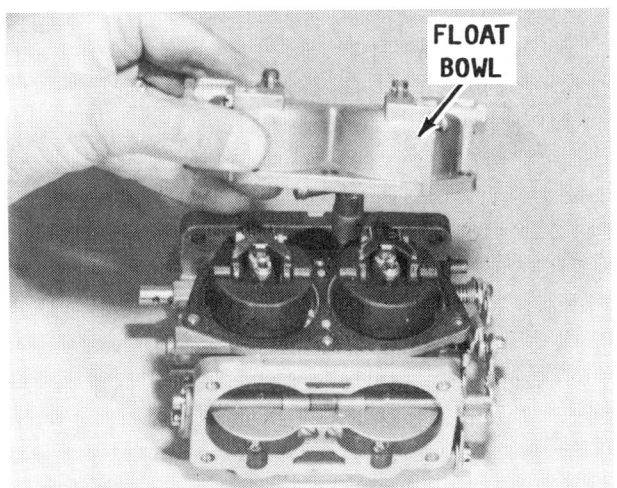

Removing the dual float bowl from a Nikki carburetor.

number of "turns in to a lightly seated position", as a guide for installation. The number of turns will be specified in the installation procedures.

6- Remove the securing screws and lift off the float bowl. Do **NOT** attempt to remove the gasket at this time.

7- Remove the two main jets and drain screws from the underneath side of the float bowl. Remove and discard the gaskets and O-rings.

Teikei Carburetor

8- Remove the small Phillips head screw retaining the hinge pin.

9- Grasp the float lightly. Use a small pair of needle nose pliers and draw out the hinge pin from its mounting posts.

10- Gently lift up the float. The needle valve, attached to the tang on the float, will also slide out of the needle seat.

11- Unhook the wire loop and needle valve from the tang on the float.

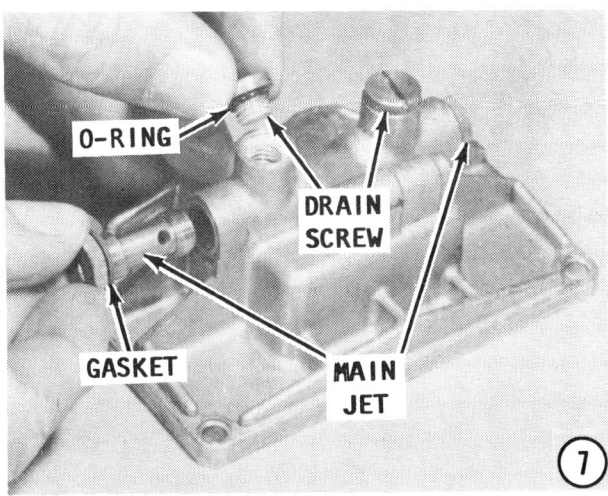

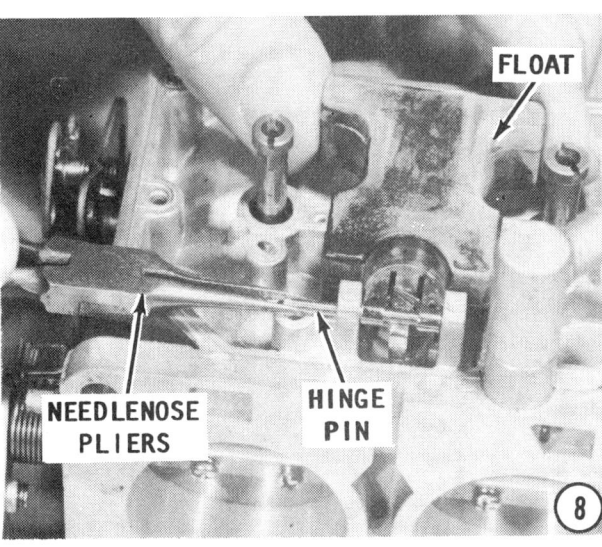

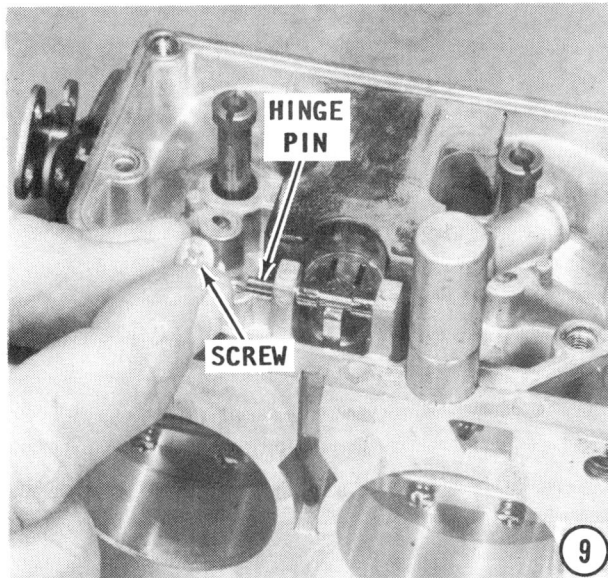

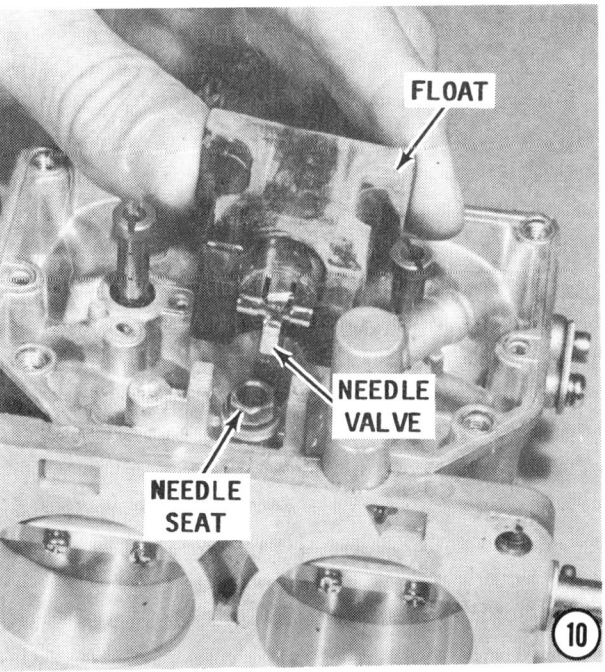

4-16 FUEL

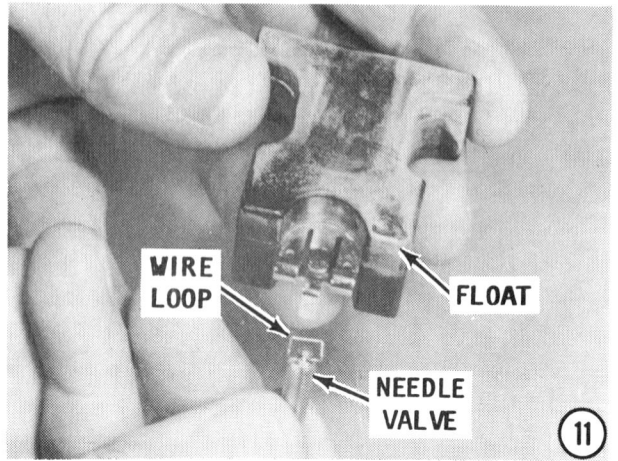

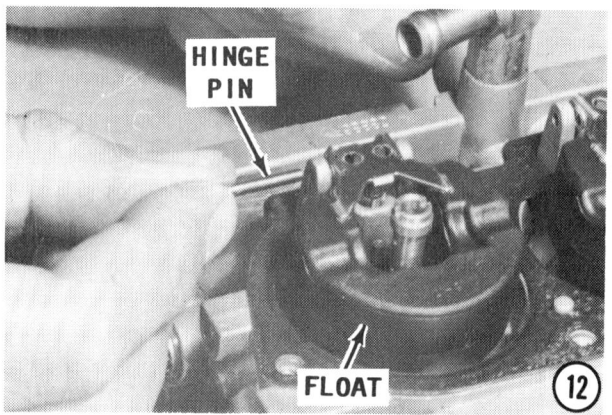

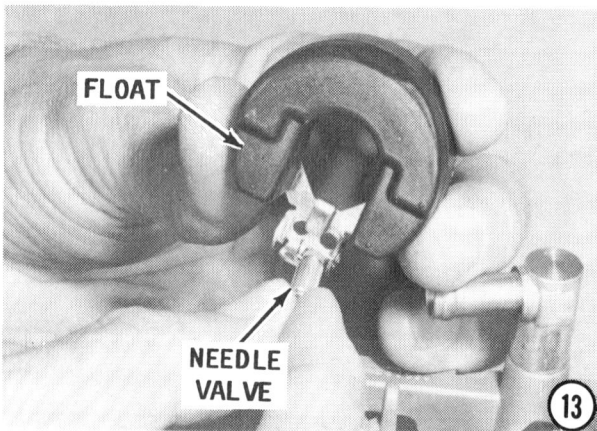

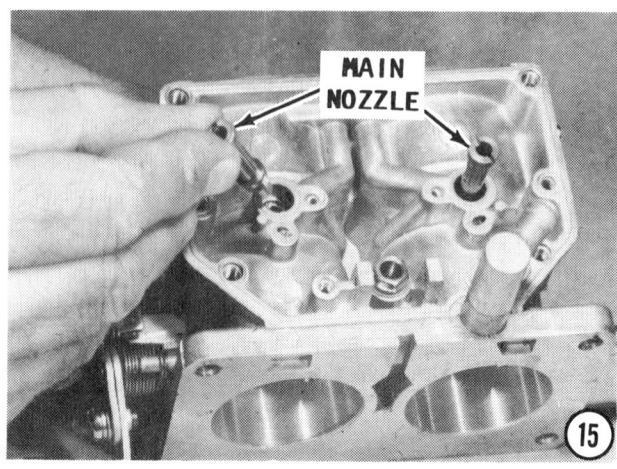

Nikki Carburetor

12- Slide the hinge pin outward to release the float.

13- Gently lift up the float. The needle valve, attached to the tang on the float, will also slide out of the needle seat. Unhook the wire loop and needle valve from the tang on the float. Remove the other float in a similar manner.

All Carburetors

14- Remove and discard the float bowl gasket from the mixing chamber.

15- Remove the two main nozzles from the mixing chamber.

16- Obtain the correct size thin walled socket and remove the needle seat.

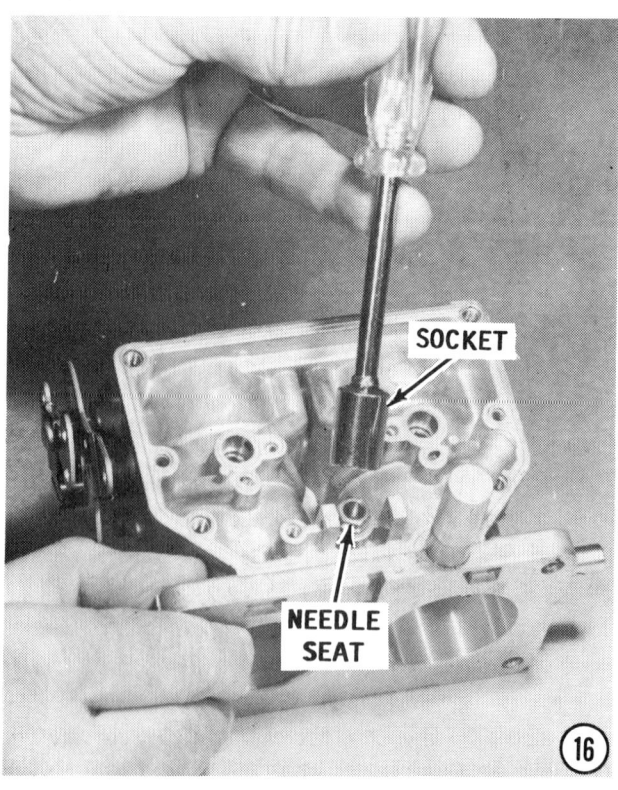

CLEANING & INSPECTING 4-17

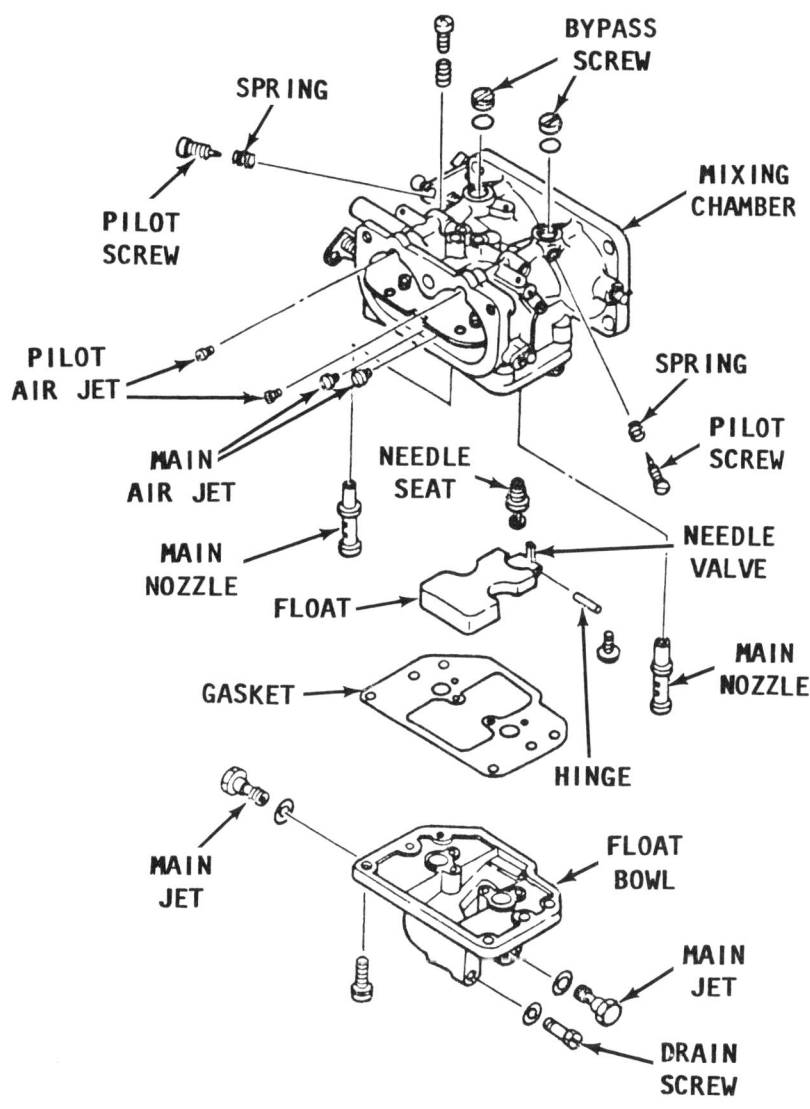

Exploded drawing of a Teikei single float carburetor. The Nikki carburetor incorporates a double float arrangement. This and other differences are clearly indicated in the text and with supporting illustrations throughout this chapter.

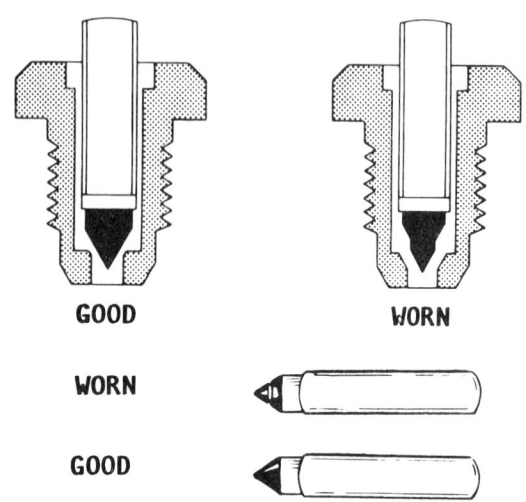

Needle and seat arrangement on the carburetor covered in this section, showing a worn and new needle for comparison.

4-8 CLEANING AND INSPECTING

NEVER dip rubber or plastic parts in carburetor cleaner. These parts should be cleaned **ONLY** in solvent, and then blown dry with compressed air.

Place all metal parts in a screen type tray and dip them in carburetor cleaner until they appear completely clean, then blow them dry with compressed air.

Blow out all passages in the castings with compressed air. Check all parts and passages to be sure they are not clogged or contain any deposits. **NEVER** use a piece of wire or any type of pointed instrument to clean drilled passages or calibrated holes in a carburetor.

Move the throttle and choke shafts back and forth to check for wear. If the shaft appears to be too loose, replace the complete mixing chamber because individual replacement parts are **NOT** available.

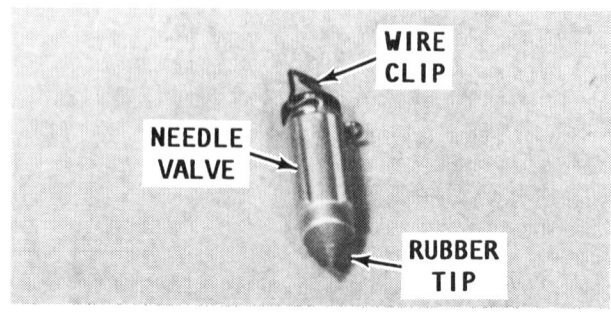

A wire clip secures the needle valve to the float. If this wire should break or slip free of the float, the fuel supply will be cut off.

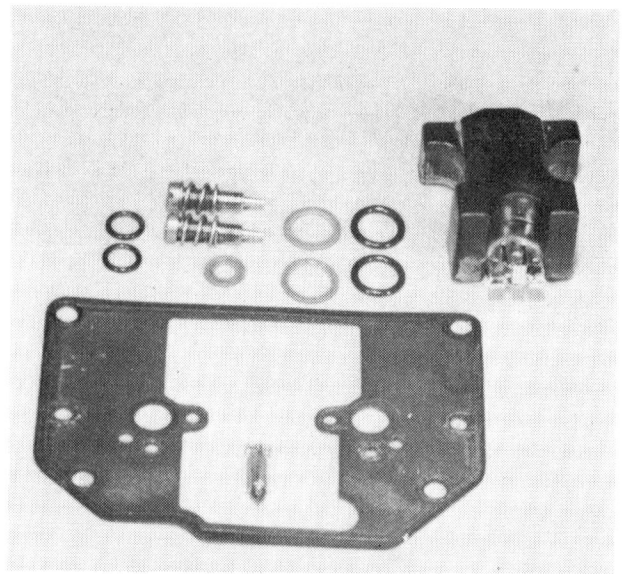

Parts included (at press time), in the Teikei carburetor repair kit.

Inspect the mixing chamber, and fuel bowl gasket surfaces for cracks and burrs which might cause a leak. Check the float/s for deterioration. Check to be sure the needle valve loop has not been stretched. If any part of the float is damaged, the float must be replaced. Check the needle valve tip contacting surface and replace the needle valve if this surface has a groove worn in it.

Inspect the tapered section of the pilot screw and replace the screw if it has developed a groove.

As previously mentioned, most of the parts which should be replaced during a carburetor overhaul are included in an overhaul kit available from your local marine dealer. One of these kits will contain a matched fuel inlet needle and seat. This combination should be replaced each time the carburetor is disassembled as a precaution against leakage.

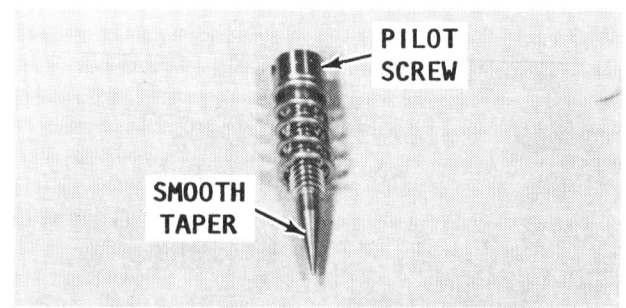

Inspect the taper on the end of the pilot screw for ridges or signs of roughness. Good shop practice dictates a new pilot screw be installed each time the carburetor is overhauled.

CARBURETOR ASSEMBLING 4-19

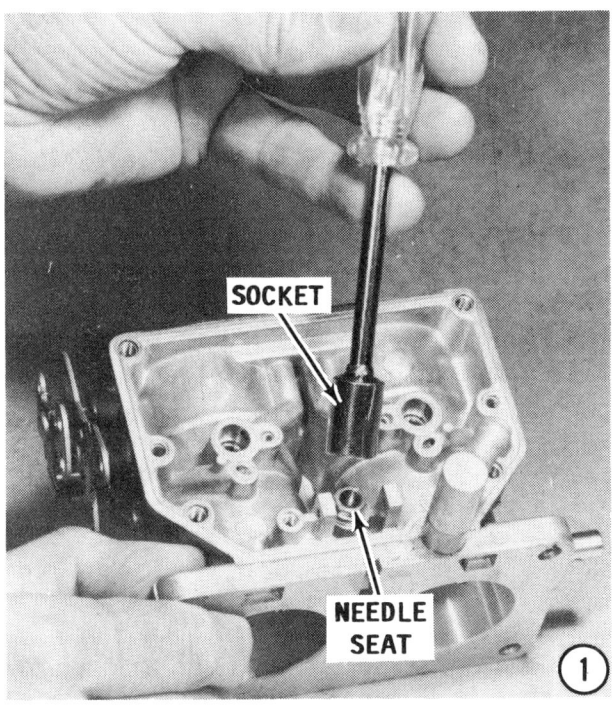

4-9 ASSEMBLING

All Carburetors

1- Install and tighten the needle seat snugly, using a thin walled socket.

2- Install the two main nozzles into the mixing chamber. Tighten them securely.

3- Place a new float bowl gasket in position over the mixing chamber.

Teikei Carburetor

4- Check to be sure the wire loop is securely in position around the needle valve. Slide the loop over the tang on the float, and then check to ensure the needle valve can be moved freely.

5- Lower the float assembly into the mixing chamber guiding the needle valve into the needle seat.

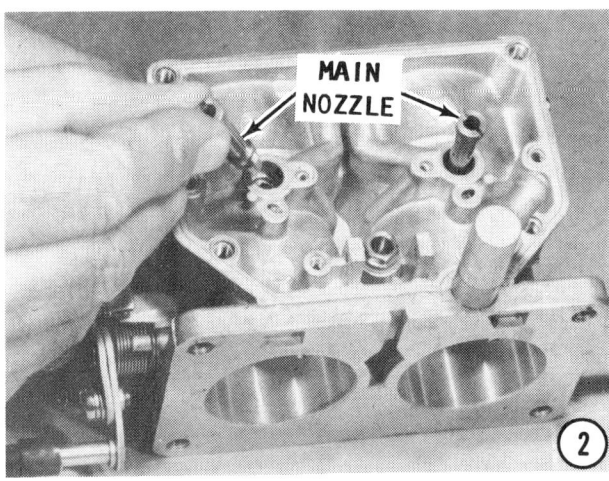

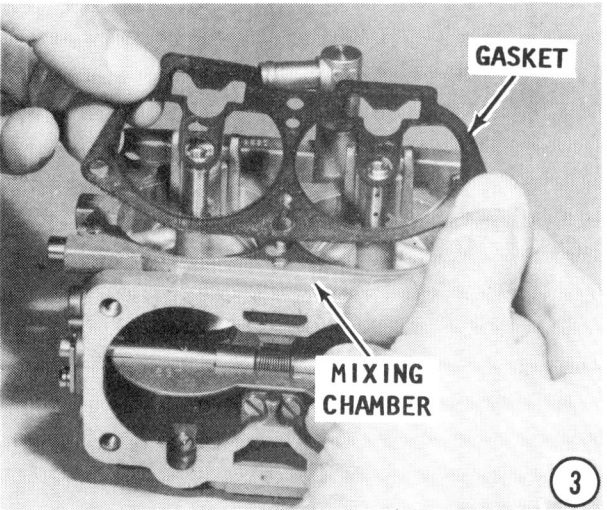

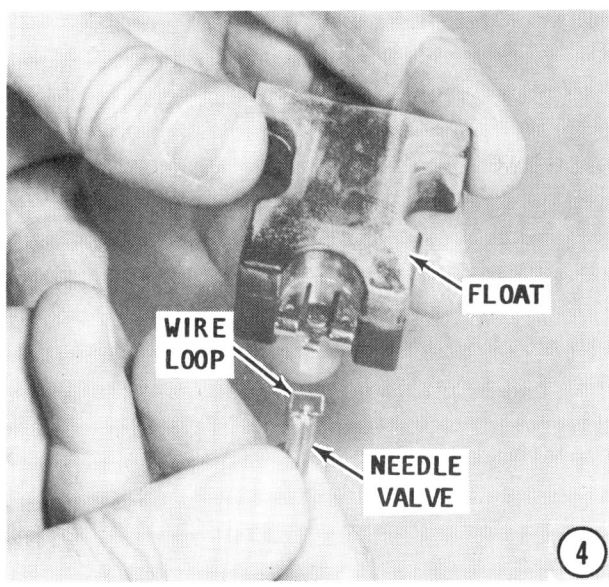

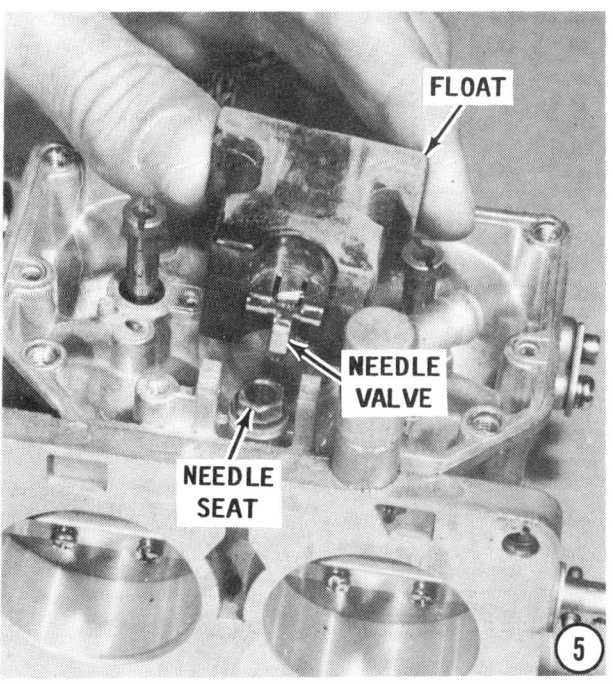

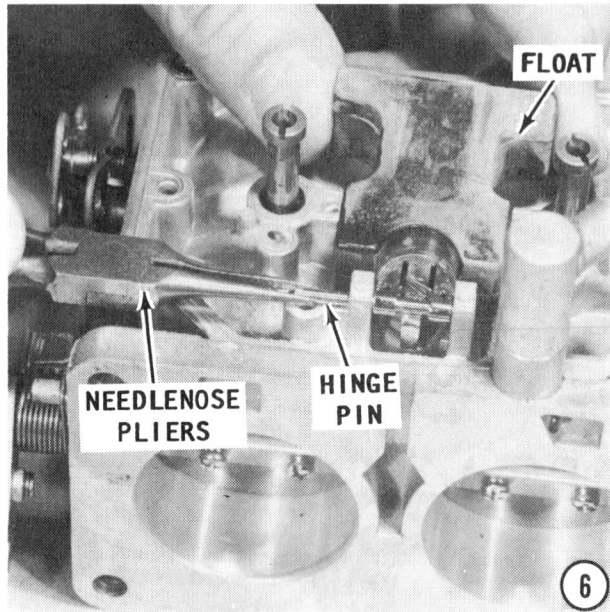

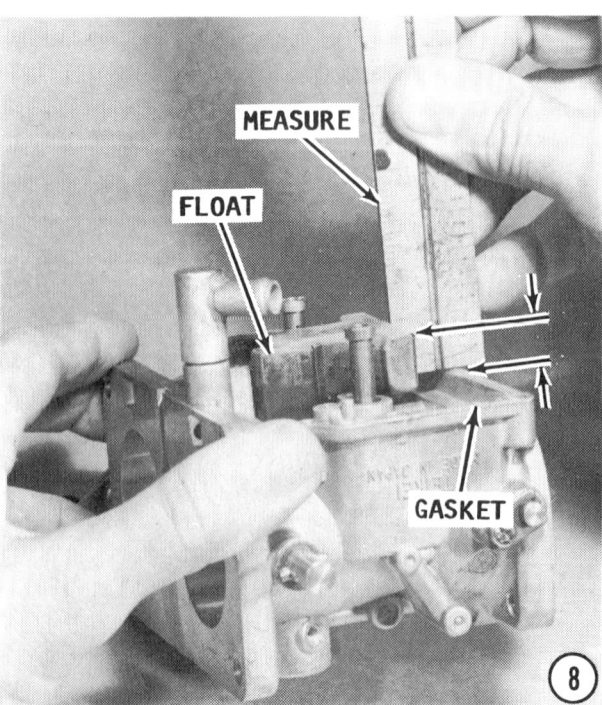

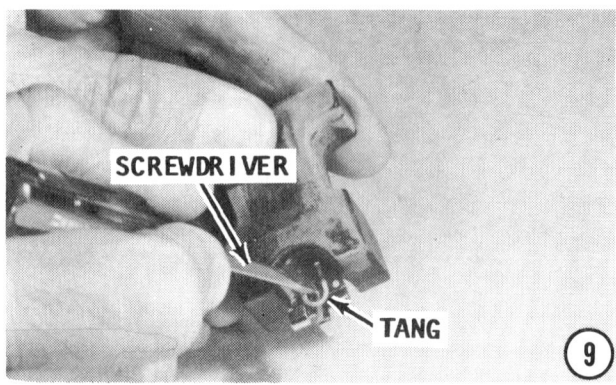

6- Use a small pair of needle nose pliers and slide the hinge pin through the mounting post and the float hinge. Position the end of the hinge pin to clear its retaining screw.

7- Install and tighten the retaining screw.

8- Hold the mixing chamber in the inverted position, (as it has been held during the past few steps). Measure the distance between the top of the float and the gasket. This distance should be 1/2" (12.5 mm).

9- If the distance is not as specified, remove the float and needle valve. Gently bend the tab on the float using a small screwdriver to correct the float level measurement.

Nikki Carburetor

10- Check to be sure the wire loop is securely in position around the needle valve. Slide the loop over the tang on the float, and then check to ensure the needle valve can be moved freely. Lower the float assembly into the mixing chamber guiding the needle valve into the needle seat. Install the other float in the same manner.

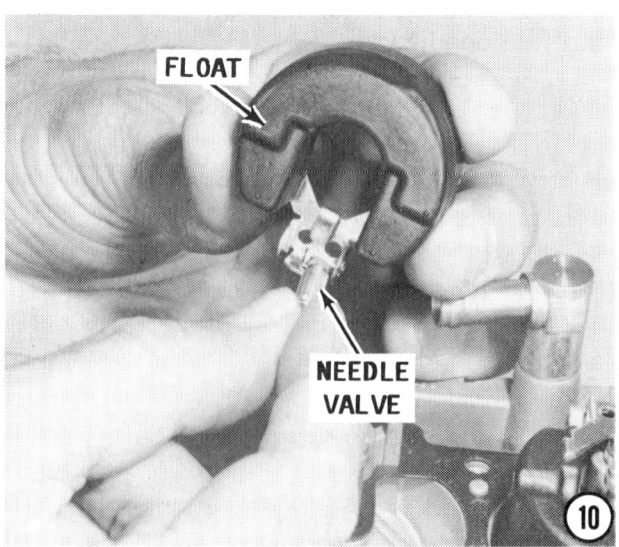

CARBURETOR ASSEMBLING   4-21

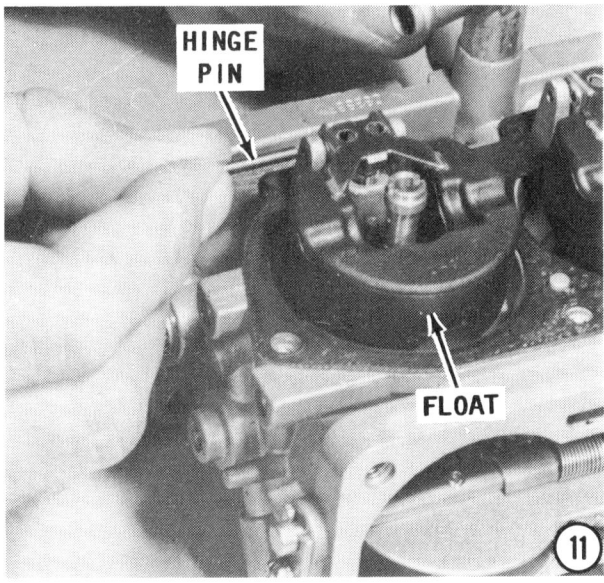

11- Slide the hinge pin through the mounting posts and float hinge, until the pin ends are flush with the mounting posts.

12- Hold the mixing chamber in the inverted position, (as it has been held during the past few steps). Measure the distance between the top of the float and the gasket. This distance should be 5/8" (16 mm).

If the distance is not as specified, remove the float and needle valve. Gently bend the tab on the float using a small screwdriver to correct the float level measurement. Repeat the float level measurement for the other float.

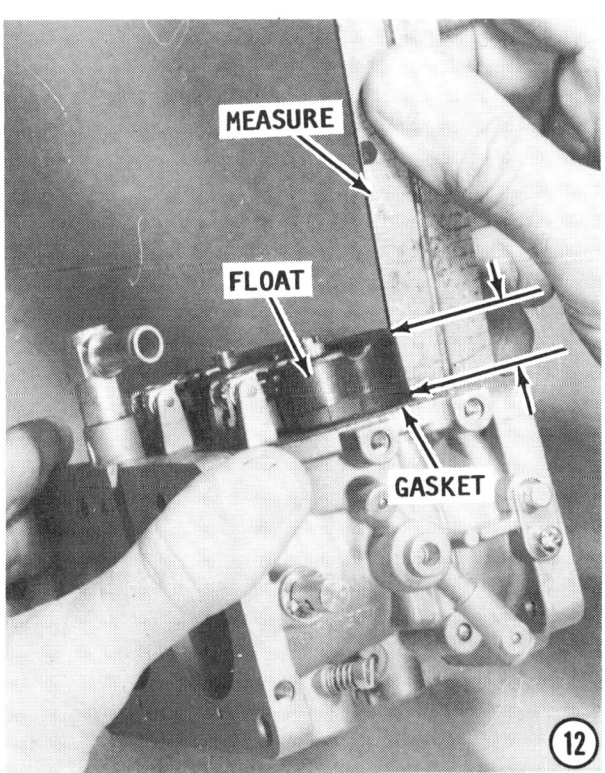

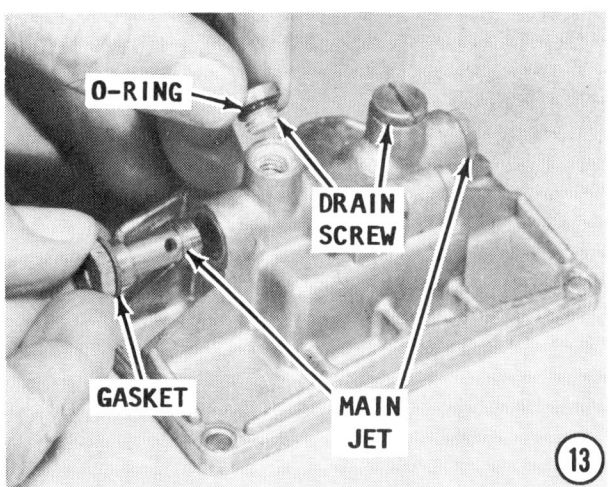

All Carburetors

13- Install new gaskets on the main jets. Install the jets into the float bowl and tighten them securely. Install new O-rings around the two drain screws. Install the screws into the float bowl and tighten them securely.

14- Lower the float bowl over the float/s. Take care not to disturb the float level adjustment. Install and tighten the attaching hardware.

15- Slide new springs over the pilot screws. Install the pilot screws into the carburetor. Tighten each screw until it **BARELY** seats. From this position, back out the screw the specified number of turns as given in the Appendix.

SPECIAL WORDS

Take notice, each year of manufacture could have a different pilot screw setting. Furthermore, on certain models, the port screw has a different setting from the starboard screw.

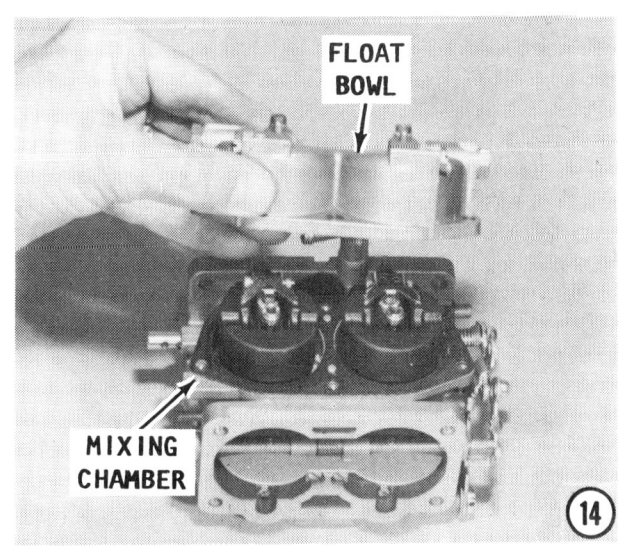

4-22 FUEL

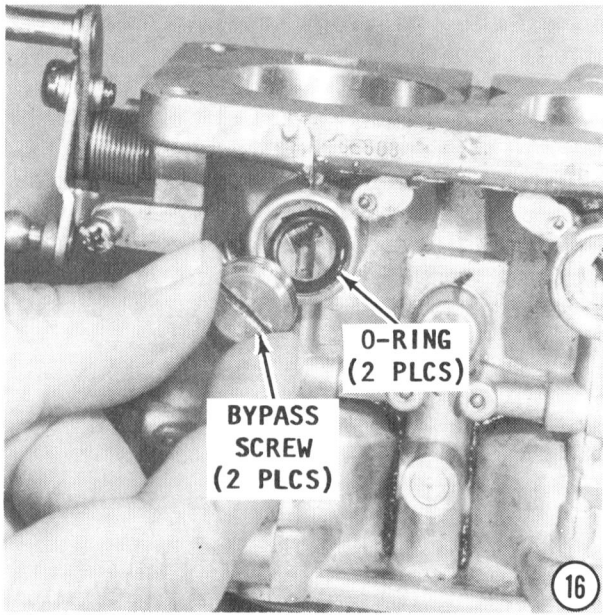

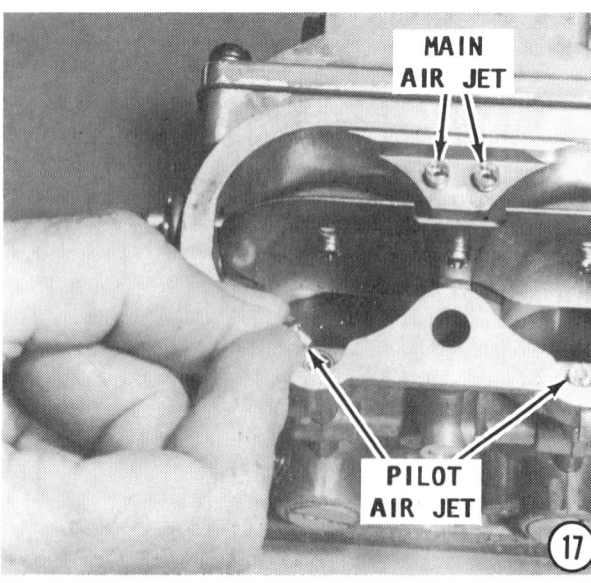

16- Install new O-rings around the two bypass screws, and then install them into the mixing chamber.

Teikei Carburetor

17- Identify the pilot air jets and the main air jets. The main air jets have a higher numerical value embossed on them. Install the two sets of jets in the locations indicated in the accompanying illustration. On some Teikei carburetors, the main air jets are not replacable.

Nikki Carburetors

18- Identify the pilot air jets and the main air jets. The main air jets have a larger number embossed on them. Install the main air jets under the access cover and the pilot air jets located at the front of the carburetor.

19- Position a new gasket over the mixing chamber and place the jet access cover over the gasket. Install and tighten the Phillips head screw securing the cover.

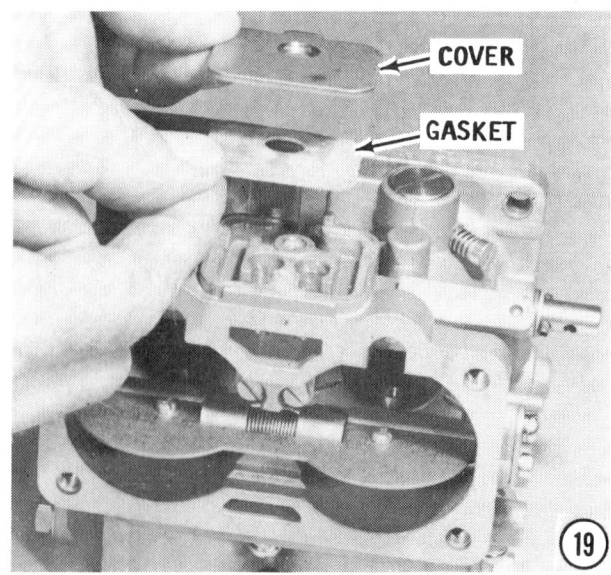

CARBURETOR INSTALLATION

4-10 INSTALLATION

1- Identify each carburetor by the mark scribed on the mixing chamber during removal. Check to be sure the small pieces of linkage were installed onto the correct carburetor. On V4 models, the throttle roller is used on the lower carburetor. On V6 models, the throttle roller is used on the center carburetor.

The choke and oil injection link retainers are located on the bottom carburetor. Place the carburetors in line on the work bench. Install the throttle and choke linkage. Place a **NEW** gasket onto the studs of the intake manifold. The manufacturer recommends **NO** sealant at this location. Install and tighten the four mounting nuts for each carburetor to a torque value of 5.8 ft lbs (8Nm).

V6 Models Only

2- Hook the choke solenoid pull wire loop into the linkage between the two choke rods. Slide the little O-ring over the linkage ball to retain the wire loop. The choke plunger adjustment is performed following Step 9.

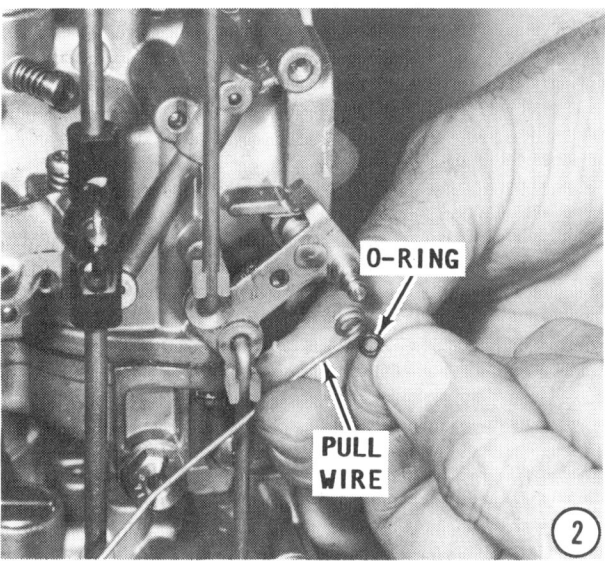

All Models

3- On the starboard side of the bottom carburetor: snap the oil injection link rod into the linkage. If the length of this rod was accidentally changed, see Chapter 5, Section 5-5 to adjust the length to specifications.

YMIS Models Only

4- Slide one end of the plastic coupling over the throttle shaft extension and the other end of the coupling over the throttle position sensor shaft. Engage the pins of both shafts into the slots of the plastic coupling. Install the sensor onto the bracket on the top carburetor, using the attaching hardware, in **EXACTLY** the same location, matching the marks made during removal.

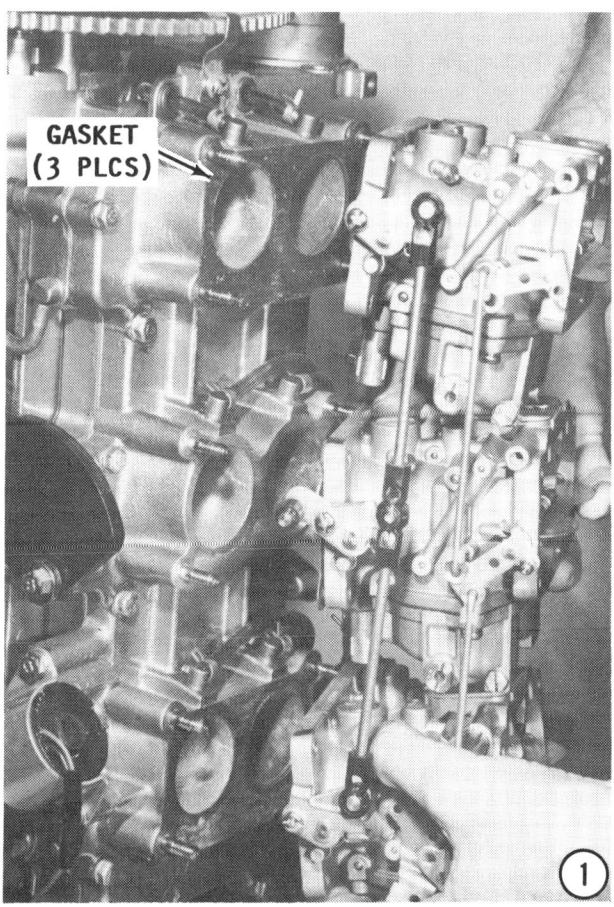

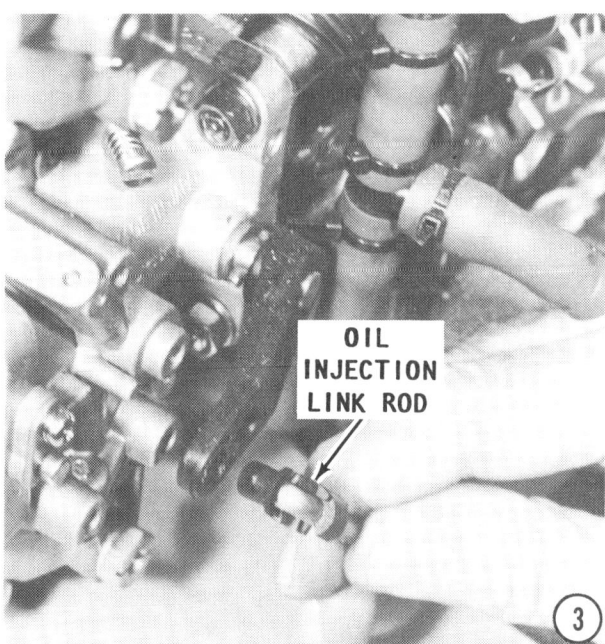

4-24 FUEL

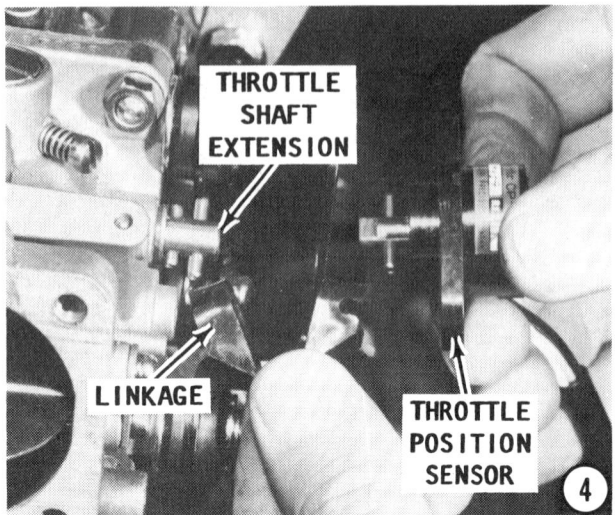

Very Important Words for Models Equipped with YMIS

If this sensor is misaligned during installation, a **DIGITAL** type voltmeter is needed to correctly reset the sensor on its mounting bracket because voltages in the 1/10 range must be accurately read. A misaligned sensor will send misleading signals to the microcomputer and consequently affect the ignition timimg.

If no locating marks were made during removal, carefully examine the bracket for traces of an outline made by the sensor and try to mount the sensor as near to the original location as possible. A difference of 0.04" (1mm) corresponds to 2° of throttle angle.

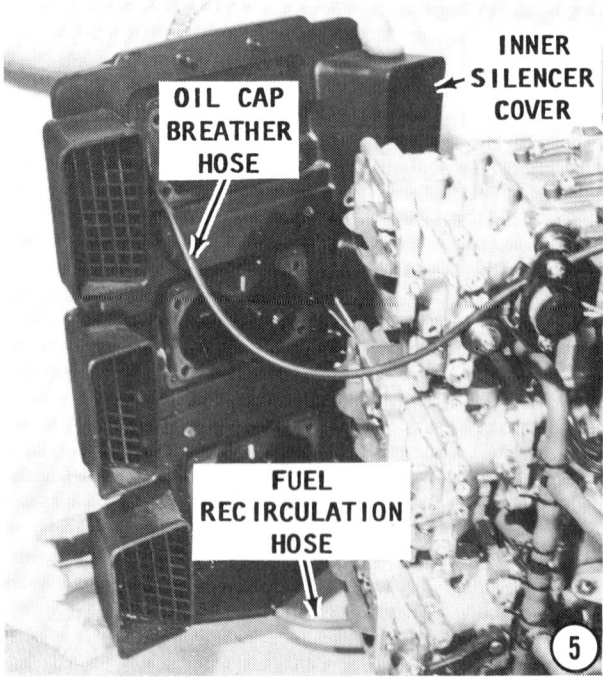

Throttle position sensor adjustment is covered in Chapter 6, Section 6-16, using a **DIGITAL** type voltmeter.

All Models

5- Bring the inner silencer cover up to the installed carburetors. Install the hose leading from the oil tank to the uppermost fitting on the cover. Install the fuel recirculation hose onto the lower fitting on the cover.

6- Place new hose clamps over the three fuel supply hoses. Slide the hoses over the inlet fittings, and then secure them with the hose clamps.

7- On the portside of the bottom carburetor, snap the choke link onto the carburetor choke arm. Check the action of the choke linkage on the side of the carburetors by moving the choke lever in and out.

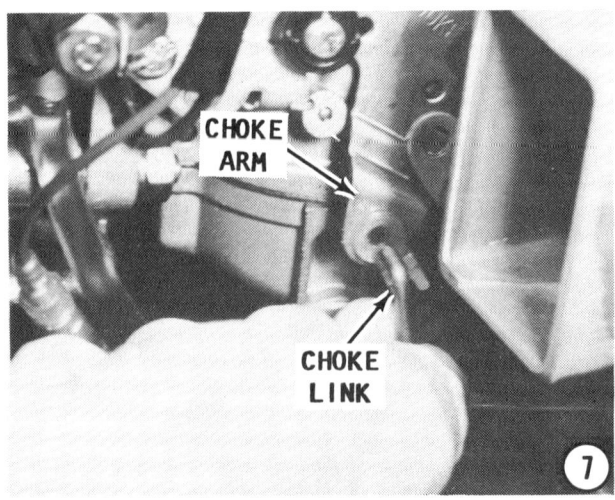

CARBURETOR INSTALLATION 4-25

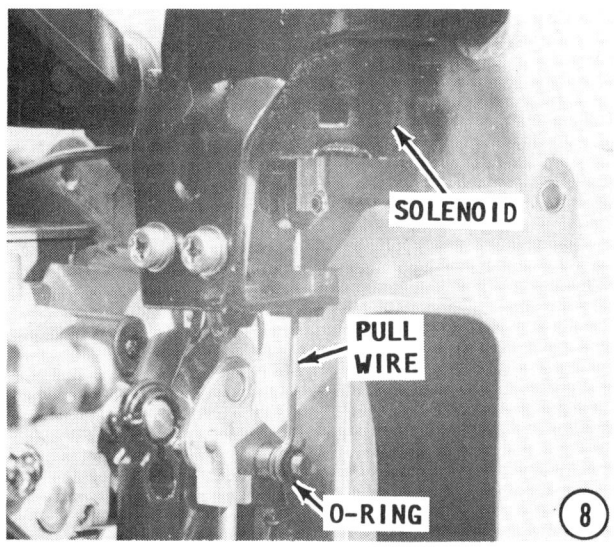

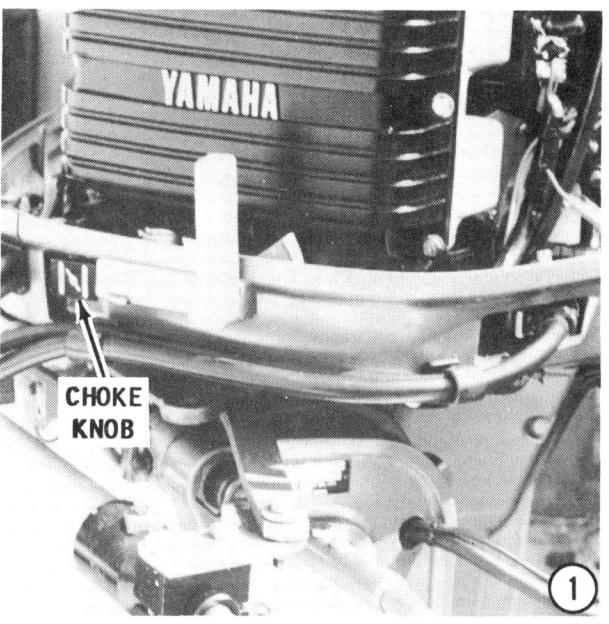

V4 Models Only

8- Hook the choke solenoid pull wire loop into the linkage between the two choke rods. Slide the little O-ring over the linkage ball to retain the wire loop. The choke plunger adjustment is performed following Step 9. Connect the single Blue lead at the quick disconnect fitting.

All Models

9- Install the outer silencer cover to the inner silencer cover. V4 Models have eight attaching screws; V6 models have ten attaching screws. Connect the fuel line at the fuel joint.

Choke Solenoid Adjustment

1- Pull out the choke knob on the lower cowling, to fully close the choke butterflies.

2- Check to ensure the mark on the plunger is flush with the surface of the solenoid. Illustration No. 2 shows a V4

model choke solenoid, and illustration No. 2A shows a V6 model choke solenoid.

If the mark is not aligned, adjustment of the solenoid mounting position is required. Loosen the attaching hardware which secures the solenoid in its bracket and move the solenoid until the mark on the plunger is flush with the surface of the solenoid. Tighten the bolts to hold this newly adjusted position.

CLOSING TASKS

Mount the outboard unit in a test tank, on a boat in a body of water, or connect a flush attachment and hose to the lower unit. Connect a tachometer to the powerhead.

SPECIAL WORDS ON TACHOMETERS AND CONNECTIONS

A tachometer is installed as standard equipment on all powerheads covered in this manual.

Due to local conditions, it may be necessary to adjust the carburetor while the outboard unit is running in a test tank or with the boat in a body of water. For maximum performance, the idle rpm should be adjusted under actual operating conditions.

Under such conditions it might be necessary to attach a tachometer closer to the powerhead than the one installed on the control panel.

Open the remote control box. Disconnect the Black and Green leads. Connect the Black lead to the ground terminal of the auxiliary tachometer and the Green lead to the "input" or "hot" terminal of the auxiliary tachometer.

NEVER, AGAIN, NEVER operate the engine at high speed with a flush device attached. The engine, operating at high speed with such a device attached, would **RUNAWAY** from lack of a load on the propeller, causing extensive damage.

Start the engine and check the completed work.

CAUTION
Water must circulate through the lower unit to the powerhead anytime the powerhead is operating to prevent damage to the water pump in the lower unit. Just five seconds without water will damage the water pump impeller.

3- Allow the powerhead to warm to normal operating temperature. Adjust the throttle stop screw until the powerhead idles at the speed listed in the Appendix. Rotating the throttle stop screw **CLOCKWISE** increases powerhead speed. Rotating the screw **COUNTERCLOCKWISE** decreases powerhead speed.

To time and synchronize the fuel system with the ignition system, see Chapter 7.

4-11 FUEL PUMP

DESCRIPTION AND OPERATION

This short section briefly describes operation of the fuel pump used on powerheads covered in this manual. One fuel pump is installed on V4 powerheads and two fuel pumps are installed on V6 powerheads. This description section is followed by detailed procedures to test fuel pump pressure, test pump output volume, and the necessary steps to remove, service and install the fuel pump.

The pump is a diaphragm displacement type. The pump is attached to the crankcase and is operated by crankcase impulses. A hand-operated squeeze bulb is installed in the fuel line to fill the fuel pump and carburetor with fuel prior to powerhead start. After the powerhead is operating, the pump is able to supply an adequate fuel supply to the carburetor to meet engine demands under all speeds and conditions.

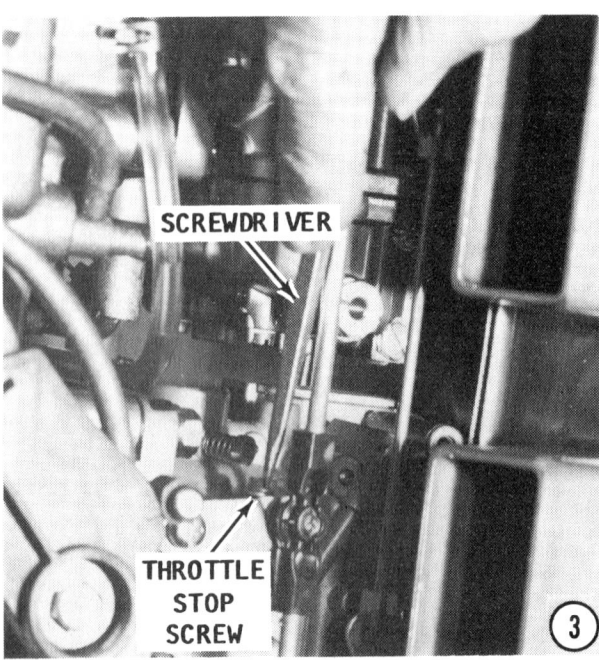

PUMP TESTING 4-27

The pump consists of a spring loaded inner diaphragm, a spring loaded outer diaphragm, two valves, one for inlet (suction) and the other for outlet (discharge), and a small opening leading directly into the crankcase. The suction and compression created as the piston travels up and down in the cylinder, causes the diaphragms to flex.

The pump is a diaphragm displacement type. The pump is attached to the crankcase and is operated by crankcase impulses. A hand-operated squeeze bulb is installed in the fuel line to fill the fuel pump and carburetor with fuel prior to powerhead start. After the powerhead is operating, the pump is able to supply an adequate fuel supply to the carburetor to meet engine demands under all speeds and condition.

As the piston moves upward, the inner diaphragm will flex inward displacing volume on its opposite side to create suction. This suction will draw fuel in through the inlet valve.

When the piston moves downward, compression is created in the crankcase. This compression causes the inner diaphragm to flex in the opposite direction. This action causes the discharge valve to lift off its seat. Fuel is then forced through the discharge valve into the carburetor.

The function of the outer diaphragm is to absorb the pulsations of the fuel and allow a smooth uninterrupted fuel flow.

This design fuel pump has the capacity to lift fuel two feet and deliver approximately five gallons per hour at four pounds pressure psi.

Problems with the fuel pump are limited to possible leaks in the flexible neoprene suction lines; a punctured diaphragm; air leaks between sections of the pump assembly; or possibly from the valves becoming distorted or not seating properly.

4-12 FUEL PUMP PRESSURE CHECK

FIRST, THESE WORDS

Lack of an adequate fuel supply will cause the powerhead to run lean, lose rpm, or cause piston scoring.

Fuel pressure should be checked if a fuel tank, other than the one supplied by the outboard unit's manufacturer, is being used. When the tank is checked, be sure the fuel cap has an adequate air vent. Verify the size of the fuel line from the tank to be sure it is of adequate size to accommodate powerhead demands.

An adequate size line would be one measuring from 5/16" to 3/8" (7.94 to 9.52mm) ID (inside diameter). Check the fuel strainer on the end of the pickup in the fuel tank to be sure it is not too small and is not clogged. Check the fuel pickup tube. The tube must be large enough to accommodate the powerhead fuel demands under all conditions. Be sure to check the filter at the carburetor. Sufficient quantities of fuel cannot pass through into the carburetor to meet powerhead demands if this screen becomes clogged.

TO TEST

Obtain about a foot (30cm) of fuel hose the same size as used on the powerhead being serviced. Obtain a "T" fitting compatable with the fuel hose and a pressure gauge.

Vacuum Check

If servicing the pump on a V4 powerhead, remove the hose clamps and fuel hose between the fuel filter and the fuel pump. If servicing the pump on a V6 powerhead, remove the fuel filter and existing "T" fitting.

Replace this hose with the "temporary" fuel hose cut into workable lengths, with a "T" fitting installed at the mid-point of the new hose.

Connect the pressure gauge to the open "T" fitting.

Mount the outboard unit in a test tank, connect a flush attachment to the lower unit, or move the boat into a body of water.

Start the powerhead.

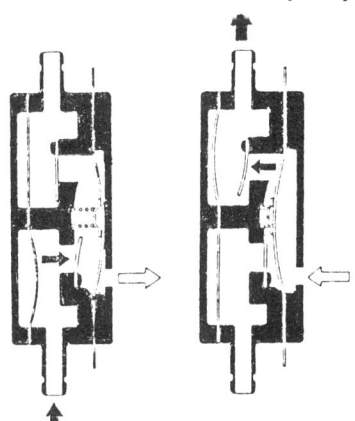

Cross section drawing to depict the diaphragm displacement type fuel pump installed on V4 and V6 powerheads.

4-28 FUEL

REMEMBER, the powerhead will **NOT** start without the emergency tether in place behind the "kill" switch knob.

NEVER, AGAIN, NEVER operate the engine at high speed with a flush device attached. The engine, operating at high speed with such a device attached, would **RUNAWAY** from lack of a load on the propeller, causing extensive damage.

CAUTION

Water must circulate through the lower unit to the powerhead anytime the powerhead is operating to prevent damage to the water pump in the lower unit. Just five seconds without water will damage the water pump impeller.

Observe the vacuum reading on the pressure gauge. When the powerhead is operating between 4700 and 5500 rpm, the gauge should register 2.13 lb/in^2 vacuum.

If the fuel pump/s do not pull this amount of vacuum, insufficient fuel is reaching the cylinders. This may cause a surge, or a loss of power. Check the fuel hoses for a restriction before removing and replacing the fuel pump/s.

Remove the "temporary" fuel hoses and install the original hoses. Secure the hoses to the inlet fittings on the carburetors with new hose clamps. If a decision has been reached to replace the fuel pump/s, see Section 4-13, this chapter.

Pressure Check

If servicing a pump on a V4 powerhead, remove the hose clamps and fuel hose between the fuel pump and existing "T" fitting. If servicing a pump on a V6 powerhead, remove the hose between the existing "T" fitting at the top carburetor and the existing 4-way "T" fitting at the middle carburetor. Replace this hose with a "temporary" hose cut into workable lengths, with the new "T" fitting installed at the midpoint of the new hose.

Connect the pressure gauge to the open "T" fitting.

Mount the outboard unit in a test tank, connect a flush attachment to the lower unit, or move the boat into a body of water.

Start the powerhead.

REMEMBER, the powerhead will **NOT** start without the emergency tether in place behind the "kill" switch knob.

CAUTION

Water must circulate through the lower unit to the powerhead anytime the powerhead is operating to prevent damage to the water pump in the lower unit. Just five seconds without water will damage the water pump impeller.

Observe the pressure reading on the gauge. When the powerhead is operating between 4700 and 5500 rpm, the gauge should register 5.7 lb/in^2 positive pressure.

If the fuel pump/s do not produce this amount of pressure, insufficient fuel is reaching the cylinders. This may cause a surge, or a loss of power. Check the fuel hoses for a restriction before removing and replacing the fuel pump/s. Remove the "temporary" fuel hoses and install the original hoses. Secure the hoses to the inlet fittings on the carburetors with new hose clamps. If a decision has been reached to replace the fuel pump/s, see Section 4-13, this chapter.

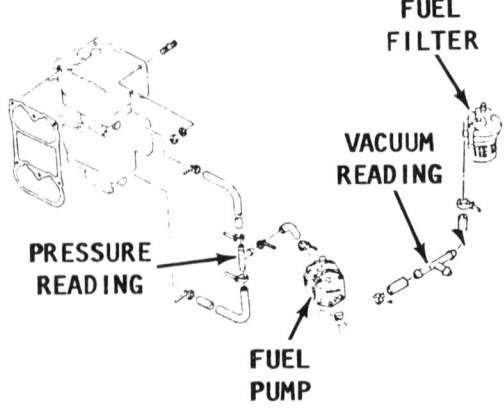

Gauge connections required to perform vacuum and pressure tests on V4 powerhead fuel pump.

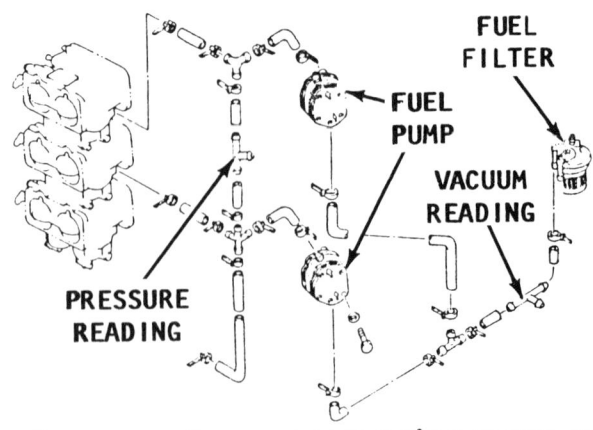

Gauge connections required to perform vacuum and pressure tests on V6 powerhead fuel pumps.

PUMP SERVICE 4-29

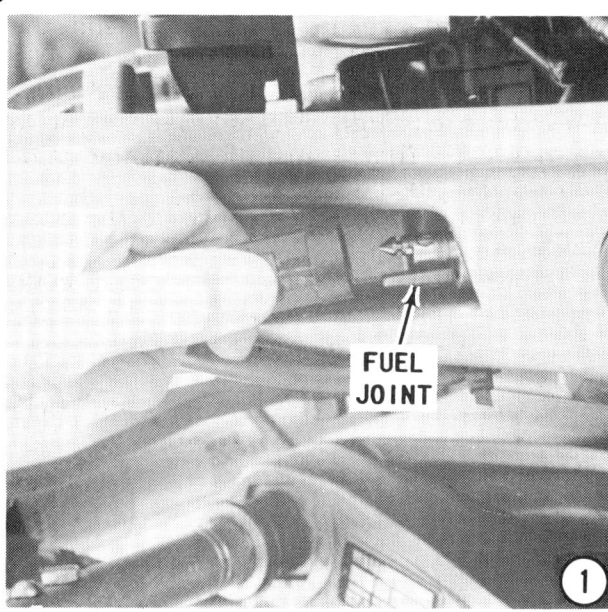

4-13 SERVICING THE FUEL PUMP

Disassembly and assembling should be performed on a clean work surface. Make every effort to prevent foreign material from entering the fuel pump or adhering to the diaphragms.

REMOVAL AND DISASSEMBLING

1- Disconnect the fuel line fuel joint.
2- Disconnect the inlet and outlet hose from the fuel pump/s. Remove the two bolts securing the pump/s to the crankcase.

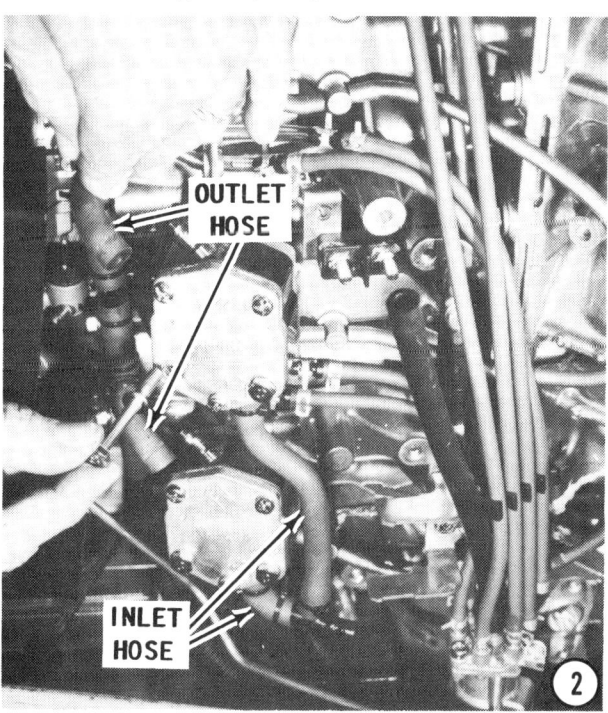

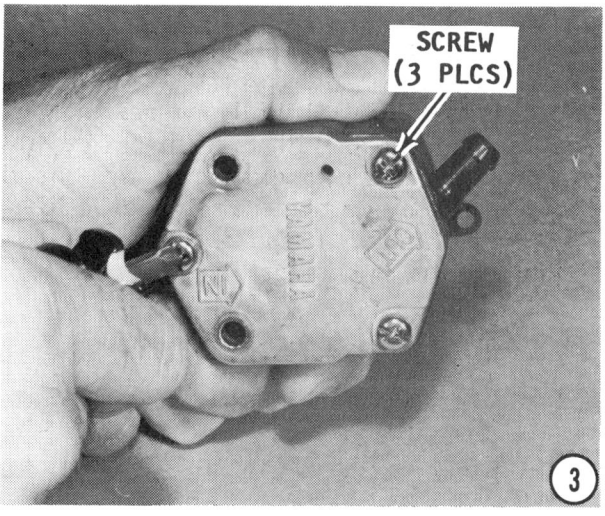

3- Remove the pump/s and move it to a suitable clean work surface. Remove the three screws securing the pump together. **TAKE CARE** not to let the spring fly out or to lose the cup.

4- Separate the back cover from the pump body. If the gasket and diaphragm are to be used again, take **GREAT CARE** in peeling them away from the surface of the cover. Remove the spring and cup. Separate the parts and keep them in **ORDER** as an assist in assembling.

5- Remove the front cover from the pump body. If the diaphragm and gasket are to be used again, take **GREAT CARE** in peeling them away from the surface of the cover. Separate the parts and keep them in **ORDER** as an assist in assembling.

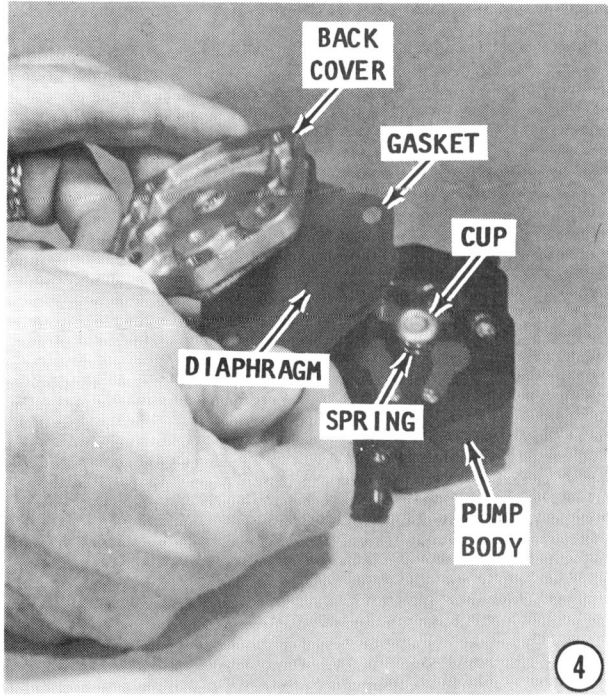

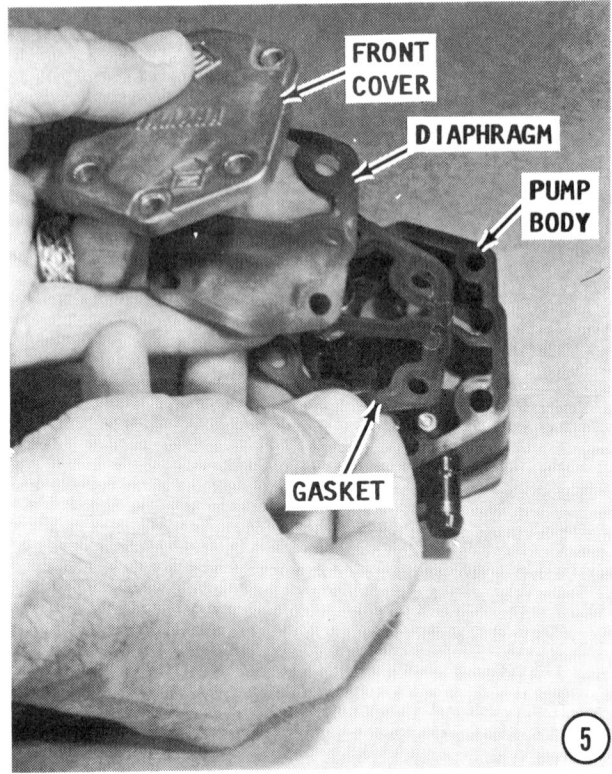

6- Remove the check valves and **TAKE TIME** to **OBSERVE** and **REMEMBER** how each valve faces, because it **MUST** be installed in exactly the same manner, or the pump will not function.

CLEANING AND INSPECTING

Wash all metal parts thoroughly in solvent, and then blow them dry with compressed air. **USE CARE** when using compressed air on the check valves. **DO NOT** hold the nozzle too close because the check valve can be damaged from an excessive blast of air.

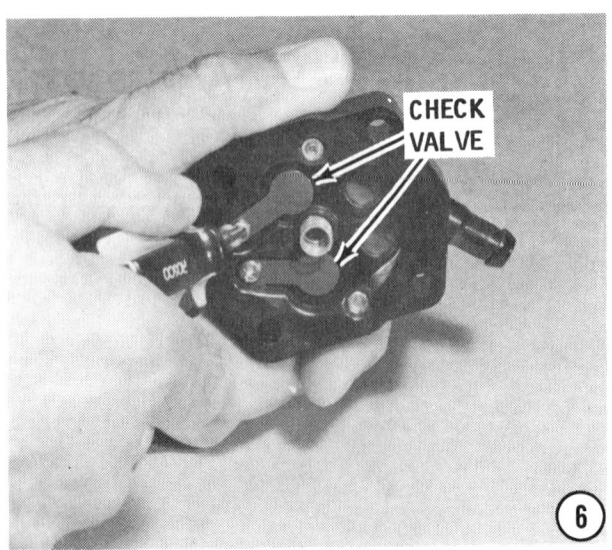

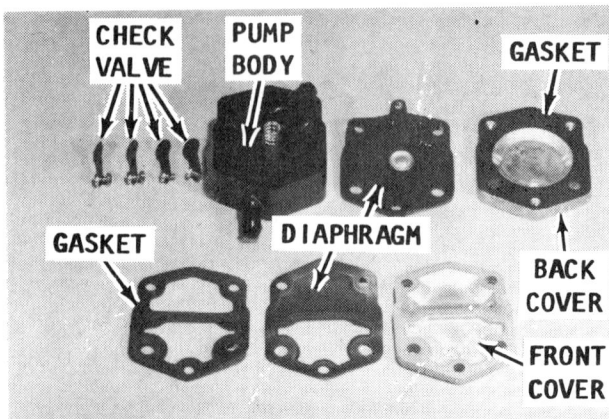

Arrangement showing parts of the fuel pump used on the V4 and V6 powerheads covered in this manual.

Inspect each part for wear and damage. Verify that the valve seats provide a flat contact area for the valve. Tighten all check valve connections firmly as they are replaced.

Test each check valve by blowing through it with your mouth. In one direction the valve should allow air to pass through. In the other direction, air should not pass through.

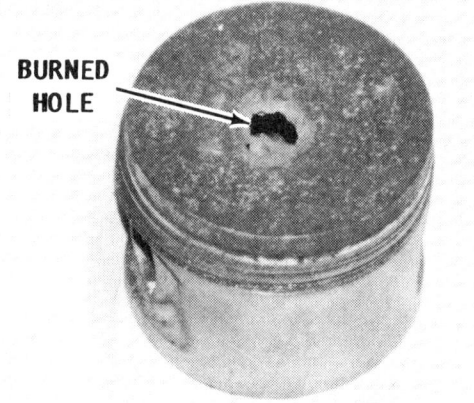

Lack of adequate fuel, possibly a defective fuel pump, caused the burn condition damage to this piston.

If the diaphragm in the fuel pump should rupture, an excessive amount of fuel would enter the cylinder and foul the spark plug, as shown.

PUMP SERVICE 4-31

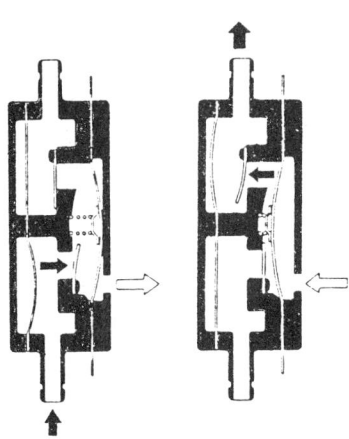

Cross section drawing of the diaphragm displacement type fuel pump installed on the powerheads covered in this manual.

Check the diaphragms for pin holes by holding it up to the light. If pin holes are detected or if the diaphragm is not pliable, it **MUST** be replaced.

ASSEMBLING AND INSTALLATION

Proper operation of the fuel pump is essential for maximum powerhead performance. Therefore, always use **NEW** gaskets.

NEVER use any type of sealer on fuel pump gaskets.

1- Place the appropriate check valves on the appropriate sides of the pump body with the "fold" in the valve facing **UP**. **TAKE CARE** not to damage the very fragile and flat surface of the valve. Secure each check valve in place with a Phillips head screw. Tighten the screw securely.

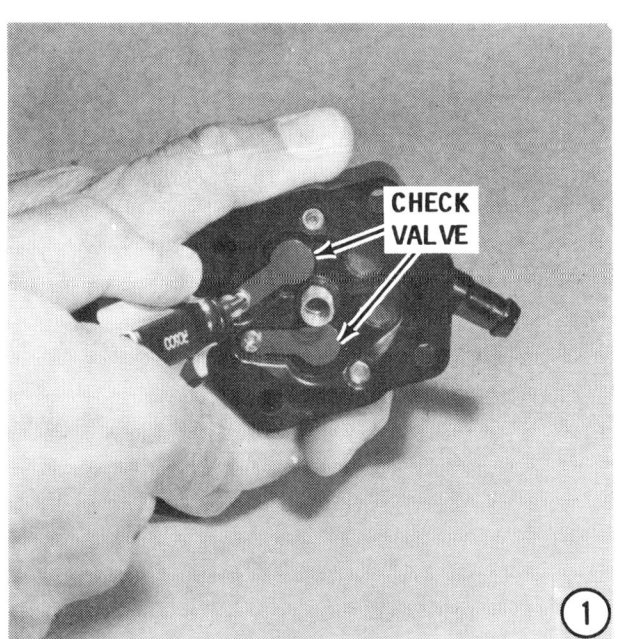

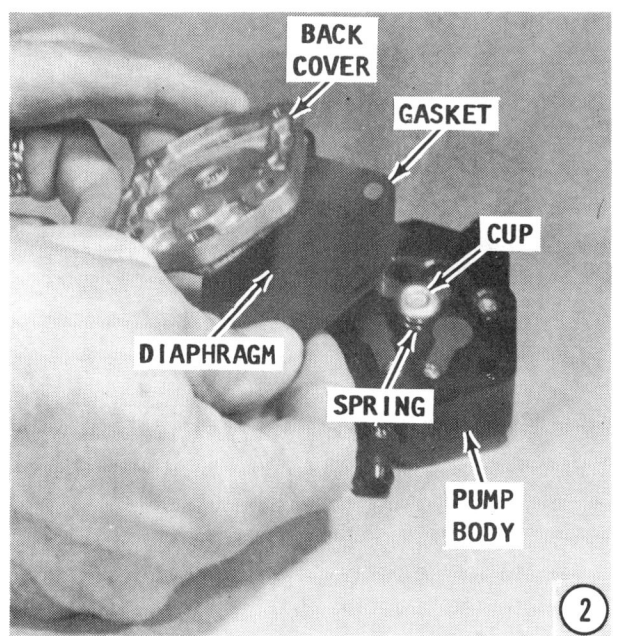

2- Place the following parts in order on the pump body: the spring, the cup (on top of the spring), the diaphragm, the gasket, and finally the inner cover. Hold these parts together and turn the pump over.

3- Install the gasket and then the diaphragm onto the pump body. Install the front cover.

4- Check to be sure the holes for the screws are all aligned through the cover, diaphragms, and gaskets. If the diaphragms are not properly aligned, a tear would surely develop when the screws are installed.

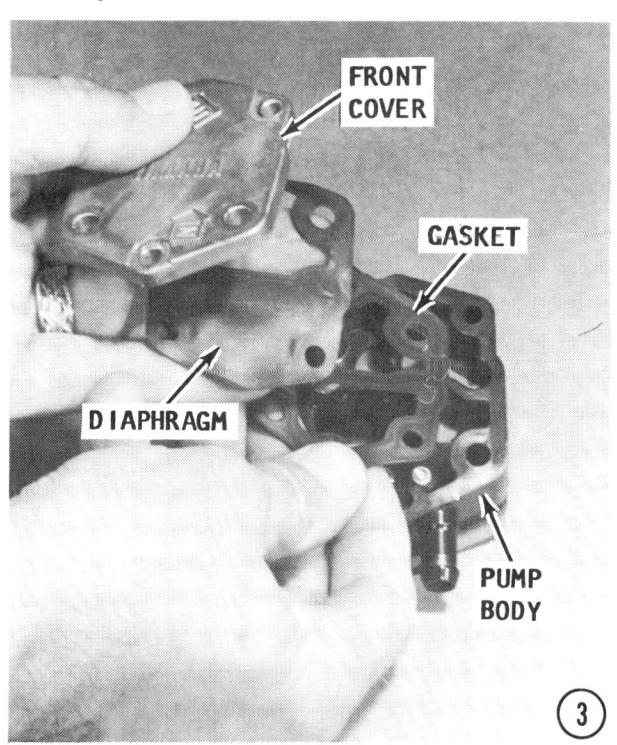

4-32 FUEL

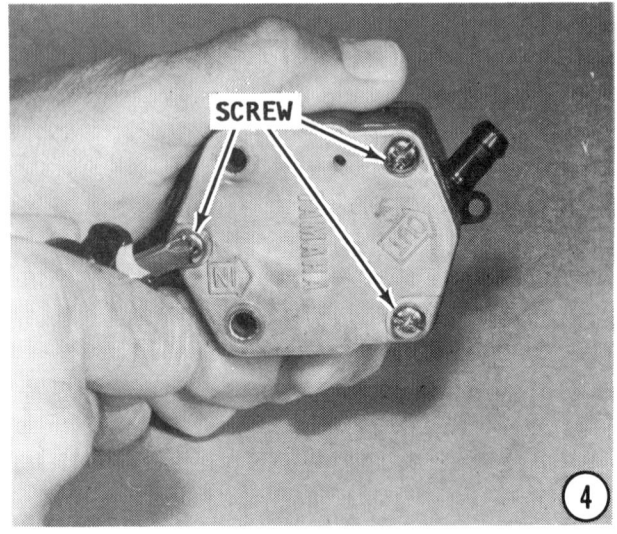

Install the three Phillips head screws through the various parts and tighten the screws securely.

5- Position the mounting gasket onto the crankcase and the fuel pump against the gasket.

Secure the pump/s to the powerhead with the two bolts.

Connect the fuel line from the filter to the fitting embossed with the word **IN**. Connect the fuel line to the carburetor onto the fitting embossed with the word **OUT**.

Secure the fuel lines with the wire type clamps.

6- Connect the fuel supply at the fuel joint.

Mount the outboard unit in a test tank, connect a flush attachment to the lower unit, or move the boat into a body of water.

Start the powerhead and check the completed work.

CAUTION

Water must circulate through the lower unit to the powerhead anytime the powerhead is operating to prevent damage to the water pump in the lower unit. Just five seconds without water will damage the water pump impeller.

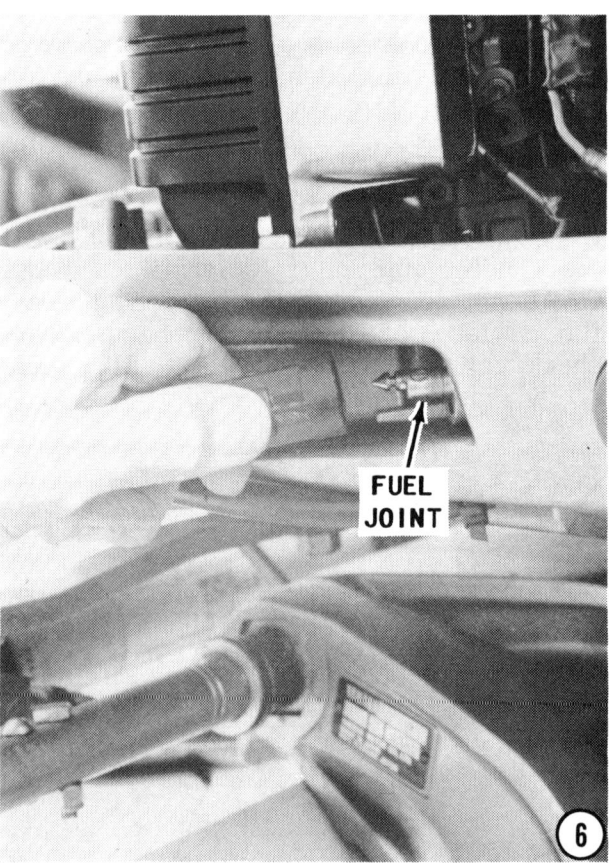

5
OIL INJECTION

5-1 DESCRIPTION AND OPERATION

Since outboard units have grown in number of cylinders with accompanying increases in horsepower, and because the size of the fuel tanks also grew to handle the increased demand of these larger powerheads, the requirement for a more sophisticated method of mixing oil with the fuel for internal lubrication became a primary design objective.

Almost all outboard manufacturers have now developed their own method to provide adequate oil delivery to the cylinders under all demands of the powerhead. Each system has its own trade name.

Precision Blend

Yamaha engineers designed and developed their oil injection system as an independent unit, separate from the fuel system until the lubricant enters the cylinder. The trade name used is "Precision Blend".

This oil injection system replaces the age old method of manually mixing oil with the fuel for lubrication of internal moving parts in the powerhead. The Precision Blend system is standard equipment -- factory installed -- on all models covered in this manual.

Unlike most other marine oil injection systems, which premix oil and fuel before carburetion, the Precision Blend system injects oil directly into the intake manifold between the carburetor and the reed valve for each cylinder. This arrangement provides immediate oil delivery response to changes in rpm and load demands. Because the oil bypasses the carburetor, gummy deposits in the carburetor passages are avoided.

Oil Mixture

The rate of oil injection is controlled by the throttle position and engine rpm. After the initial break-in period, oil is injected at a rate giving approximately a 200:1 mixture at low and idle speed. The mixture increases to 50:1 at advanced WOT (wide open throttle).

In order to obtain extra lubrication during the first 10 hours of break-in, the manufacturer recommends a premix of 50:1 mixture directly into the fuel tank for all units. The rod between the carburetor and the oil injection pump lever is to remain connected during the break-in period. This arrangement will increase the oil mixture at idle rpm to approximately 100:1 during the first 10 hours of powerhead operation.

SYSTEM COMPONENTS

The oil injection system installed on powerheads covered in this manual is identical whether or not the powerhead is equipped with YMIS (Yamaha Microcomputer Ignition System), with one exception. On a powerhead equipped with YMIS, there is an extra lead from the oil injection control unit to the microcomputer.

The oil injection components on a Model 130hp powerhead are easily accessible.

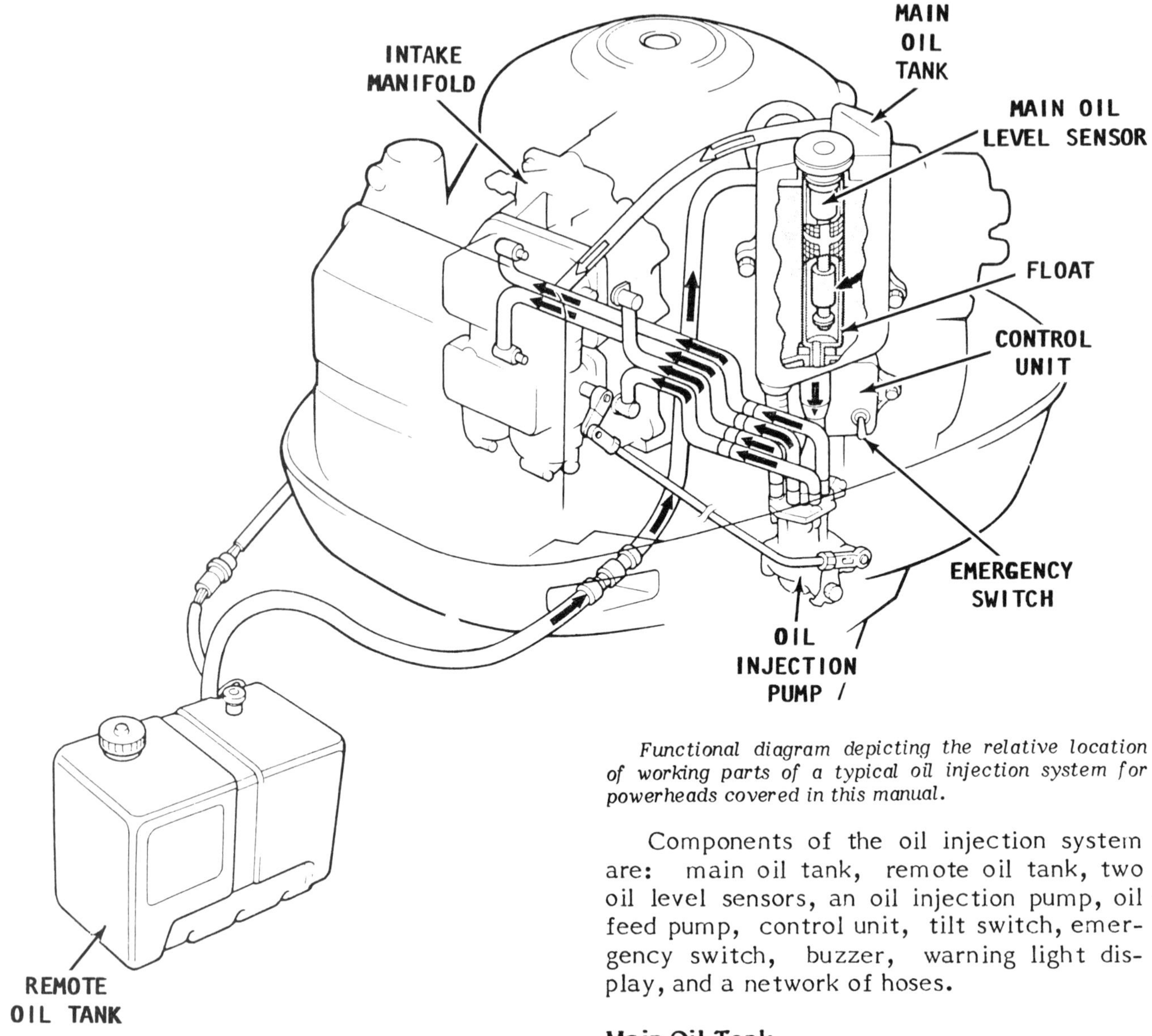

5-2 OIL INJECTION

Functional diagram depicting the relative location of working parts of a typical oil injection system for powerheads covered in this manual.

Components of the oil injection system are: main oil tank, remote oil tank, two oil level sensors, an oil injection pump, oil feed pump, control unit, tilt switch, emergency switch, buzzer, warning light display, and a network of hoses.

Main Oil Tank

The main oil tank is mounted on the powerhead. The oil tank filler is easily accessible on top of the cowling. The capacity of this tank is 0.9 U.S. quarts (0.85 litres) for all models. Since 1988, a transparent plastic water/dust trap has been introduced, located at the bottom of the main oil tank.

Oil supply to the oil injection pump is gravity fed from the oil tank. Therefore, a breather for the tank is located next to the filler cap. This breather **MUST** remain open at all times.

Remote Oil Tank

The remote oil tank is installed at a convenient location in the boat. The capacity of this tank is 2.77 U.S. gallons (10.5 litres).

Since 1988, the powerhead has a transparent water-dust trap located at the bottom of the main oil tank.

SYSTEM COMPONENTS 5-3

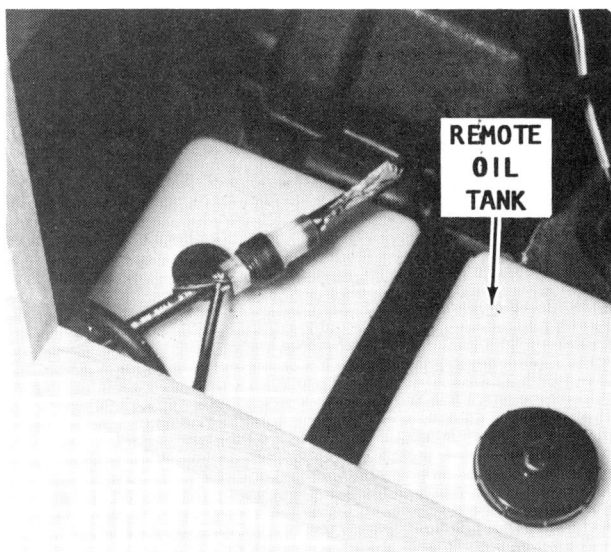

The remote oil tank may be installed in a convenient, but secure, location in the boat close to the outboard unit.

Main Tank Oil Sensor

An oil sensor is mounted in the main oil tank. The oil level sensor consists of a float sliding up and down in the sensor shaft between stops. The float rises and falls with the level of the oil. This sensor monitors the remaining oil level in the tank

The design of the main oil sensor on powerheads equipped with YMIS (Yamaha Microcomputer Ignition System), differs from the sensor installed on powerheads without YMIS. A small oil filter is incorporated inside both sensors.

and sends a signal to a series of lights to indicate oil level and a signal to the control unit. The sensor is also connected to a buzzer in the remote control box.

When the oil level in the main tank falls to 0.53 U.S. quarts (0.5 litres), the sensor signals the feed pump in the remote tank to replenish the main tank. Oil will be pumped up from the remote tank (assuming there is oil in the remote tank) to the main tank, until the level in the main tank reaches 0.9 U.S. quarts (0.85 litres). If there is no oil in the remote tank, the Yellow warning light will come on to inform the operator of the condition of the remote tank. Under this condition, the powerhead will operate strictly on the oil remaining in the main tank.

When the oil level in the main tank falls to 0.3 U.S. quarts (0.3 litres), the oil level sensor will activate the Red warning light, cause the buzzer to sound and send a signal to the CDI unit. The signal received by the CDI unit will automatically cause powerhead rpm to be reduced.

There is no ignition cutout switch. Theoretically, it is possible to operate the powerhead when the main oil tank is dry, leading to overheating and seizure. **HOWEVER**, the manufacturer has incorporated enough visual and audible danger signals to alert the operator well before any internal engine damage can occur. Even an emergency supply is available at this point, which will be covered in the paragraph describing the emergency switch.

Remote Tank Oil Sensor

The oil level sensor in the remote tank consists of a float sliding up and down in the

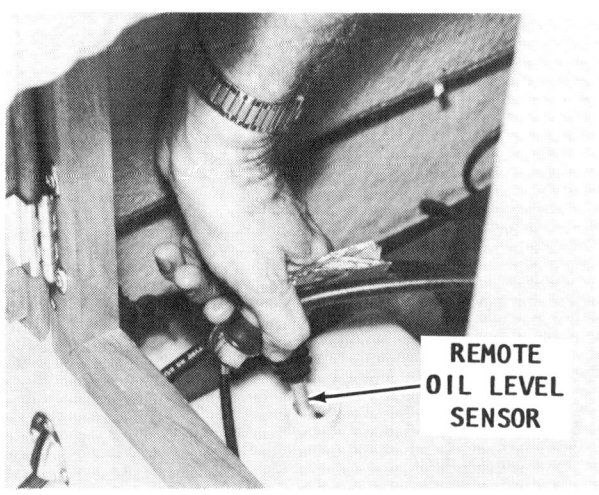

Removing the remote oil level sensor.

sensor shaft between stops. The float rises and falls with the level of the oil. Action of the remote tank sensor is similar to the action of the main tank sensor, but the two sensors are **NOT** interchangeable. The remote tank sensor will send a signal to the warning lamps when the amount of oil in the tank falls to less than 1.6 U.S. quarts (1.5 litres).

Oil Injection Pump

The oil injection pump is a positive displacement type unit driven by a worm gear and two short shafts from the lower end of the crankshaft. A gear pressed onto the lower end of the crankshaft drives the worm gear and a short shaft indexed with a second short shaft to the pump. This second short shaft to the pump drives the plunger cam. With each revolution of the plunger cam, the plunger moves up and down three times pumping oil to the cylinders.

A link from the lower carburetor operates the oil injection pump lever. This lever affects the movement of the plunger cam by limiting the plunger stroke. In this manner, the amount of oil leaving the pump is regulated. As throttle movement is advanced and crankshaft rotation is increased, the amount of oil entering the intake manifold will increase. The mixing ratio depends upon the angle of the pump lever shaft. The pump lever shaft is directly connected to the throttle plate via a link rod. If the lever angle is between 0° and 5°, the mixing ratio is 200:1. If the lever angle is between 5° and 50°, the mixing ratio is between 200:1 and 50:1.

The oil leaves the pump and is delivered to the cylinders through a series of four or six transparent hoses.

Oil Feed Pump

The oil feed pump is located at the rear of the remote oil tank. The unit is an electrically operated gear type pump capable of delivering oil from the remote tank to the main tank.

The pump is activated to start and stop by signals from the oil sensor in the main tank. Under normal operation conditions, when the oil level in the main tank falls to a predetermined "low" level, the sensor sends a signal and the oil feed pump is activated. The pump will deliver oil to the main tank until the oil level in the main tank reaches a predetermined "full" quantity. At this time, another signal is sent to the oil feed pump and pumping from the remote tank ceases.

The oil feed pump will pump oil until only 1.6 U.S. quarts (1.5 litres) remain in the tank. **HOWEVER**, a manual override of the main oil tank sensor will permit the pump to be activated and the remote tank drained of all oil. This override action is accomplished by activating the emergency switch on the control box. A detailed explanation of the emergency switch is given in later paragraphs.

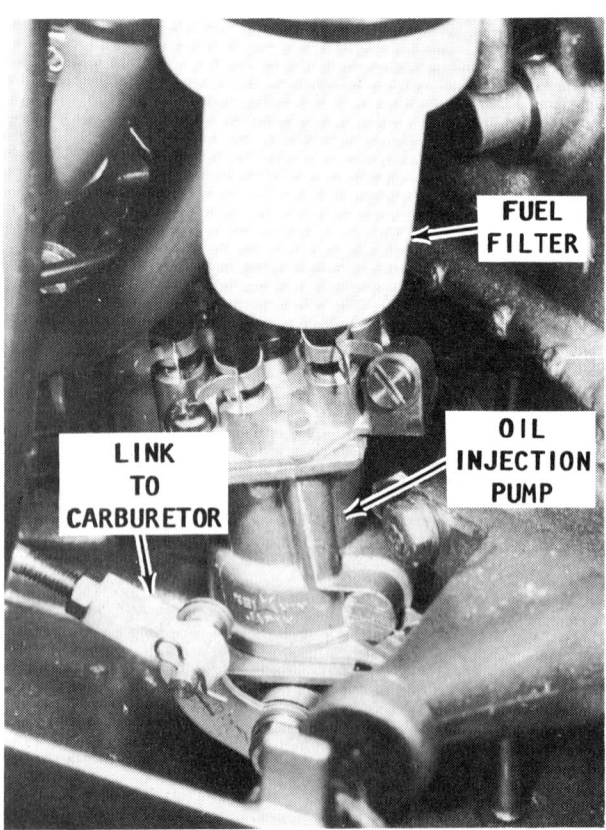

Typical location of the oil injection pump mounted on the powerhead.

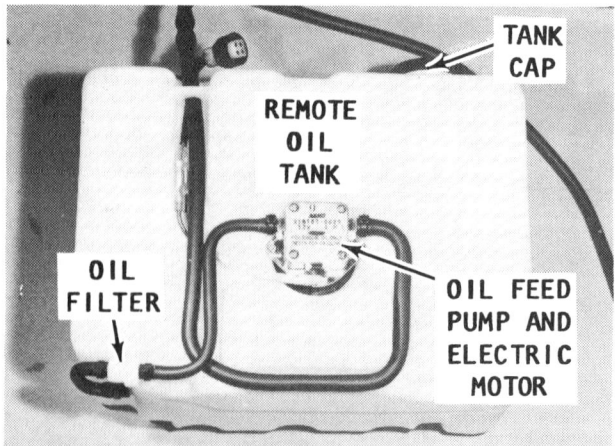

The remote oil tank removed from the boat to show the oil feed pump and electric motor. The oil filter, normally hidden, is also visible.

SYSTEM COMPONENTS 5-5

Control Unit

The control unit is mounted directly beneath the main oil injection tank on the powerhead. The control unit receives input signals from the oil level sensors in both oil tanks and determines which light should illuminate on the display. The control unit also sends signals to the buzzer in the remote control box and to the CDI unit to reduce powerhead speed under specified conditions.

Tilt Switch

The tilt switch is a function of the control unit and is incorporated inside the unit. Should the powerhead be tilted during operation, the sensors in both tanks would register a false reading. This condition could cause unnecessary transfer of oil from the remote tank to the main tank, possibly filling and overflowing the tank. The tilt switch senses the angle of the powerhead at all times. If the unit is tilted beyond $35°$ to the vertical, the tilt switch cuts out the oil feed pump circuit until the powerhead returns to $20°$ or less.

SPECIAL WORDS

Testing procedures are given in Section 5-2 to test this aspect of the control unit.

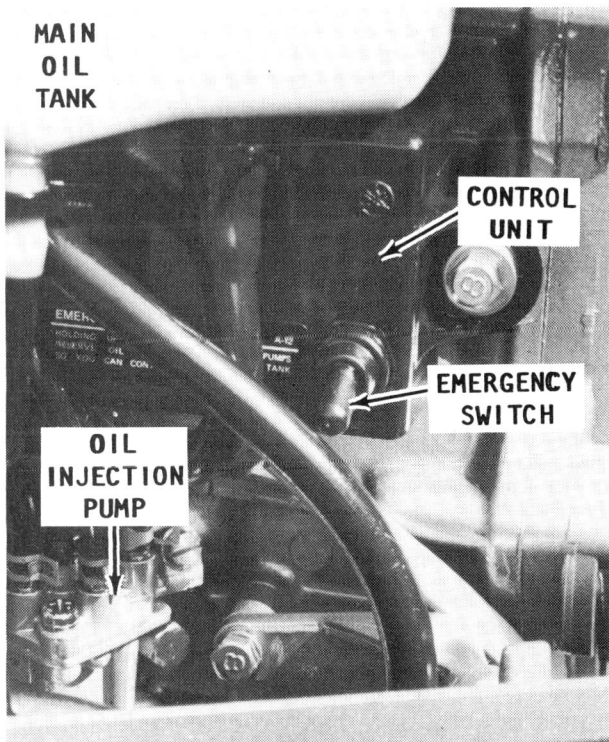

Location of the control unit in relation to other oil injection components. The tilt switch is incorporated inside the control unit.

Emergency Switch

The emergency switch is located on the face of the control unit directly beneath the main oil tank on the powerhead. A decal on the control box descibes the purpose of the switch. Because the decal may no longer be legible, the wording is given here:

"Holding up the emergency switch pumps reserve oil into the main oil tank so you can continue cruising."

When the switch is held in the **UP** position, the operator is able to manually override the signal from the main oil tank sensor and cause the oil feed pump to completely drain the remote tank.

The emergency switch should only be activated if the operator is unable to replenish the remote tank and is forced to use the tank's reserve capacity. If the emergency switch is released before all the oil is pumped to the main tank, the necessity of "bleeding" the oil feed pump may be avoided. If the remote tank is pumped completely dry, the feed pump must be purged of air according to the procedures given in Section 5-4.

Buzzer

The buzzer is installed in the remote control box and warns the operator of a possible problem. The most common cause is low oil level in the remote tank. However, the buzzer can be activated by a signal from three different sensors -- the overrev. sensor, the overheating sensor, and the low oil sensor. If the buzzer sounds due

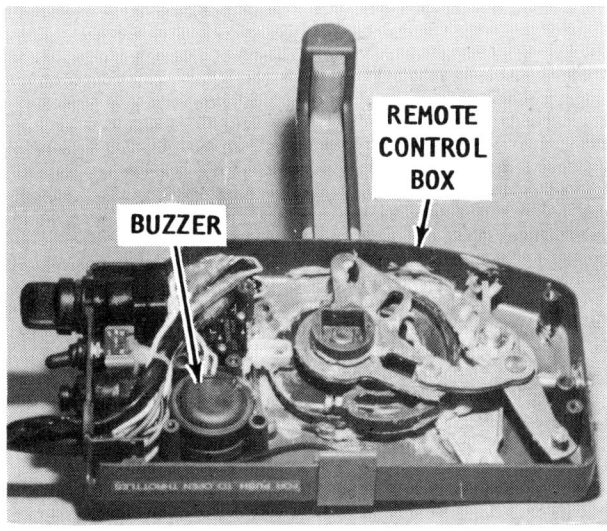

The buzzer and its harness are neatly tucked inside the crowded remote control box.

to a signal from the overrev. or overheating sensor, the signal will be accompanied by a sudden (noticeable) drop in powerhead rpm and the Red warning light will come on.

If the buzzer sounds and the Red light comes on but powerhead rpm is not reduced, then the signal responsible for activating the buzzer and light is from the sensor in the remote oil tank. Adding oil to the tank will correct the problem.

Warning Light Display

The warning light display is built into the tachometer using incandescent lamps. The light colors and their significance are explained in the following paragraphs.

SPECIAL WORDS

If the warning system fails -- no lights come on -- the oil level must be checked **IMMEDIATELY**. Operate the powerhead at reduced speed and with caution until the unit can be properly serviced.

Three colored warning lights, a Green round light, a Yellow "oil can" or round light, and a Red round light are used. These lights warn the operator of a low oil or overheating condition. Different combinations of lights with or without the buzzer can indicate a variety of conditions.

Green Light Comes On
Buzzer Does Not Sound

During normal operating conditions, if the powerhead has an adequate supply of oil, the Green light is always on. This green light informs the operator, more than 0.5 U.S. quarts (0.5 litres) is in the main tank, and at least 1.6 U.S. quarts (1.5 litres) in the remote tank.

Green or Yellow Light Comes On
Buzzer Sounds

This condition should be accompanied with a sudden drop in engine rpm. The oil level in both tanks is acceptable. The problem lies in one or both banks of cylinders because an overheating condition is developing. The powerhead should be shut down **IMMEDIATELY**; the problem isolated; and corrective action taken. Continued operation of the powerhead with one of these lights on and the buzzer sounding could lead to serious and expensive internal damgage and seizure.

Yellow Light Comes On
Buzzer Does Not Sound

If the Yellow light comes on, the operator is advised only a reserve quantity of oil remains in the remote tank. In order to pump this reserve oil, approximately 1.6 U.S. quarts (1.5 litres) to the main tank, the operator must hold the emergency switch in the **UP** position. The Yellow light will remain on until the remote tank is replenished with oil.

Red Light Comes On
Buzzer Does Not Sound

See next paragraph, and check the operation of the buzzer as soon as possible. If the Red light is on, the buzzer should also be sounding.

Red Light Comes On
Buzzer Sounds

If this condition is accompanied with a sudden drop in powerhead rpm, the problem is an overheating condition. One or both banks of cylinders has experienced prolonged excessive temperatures and is in immediate danger of "seizing". Shut the powerhead down **IMMEDIATELY**.

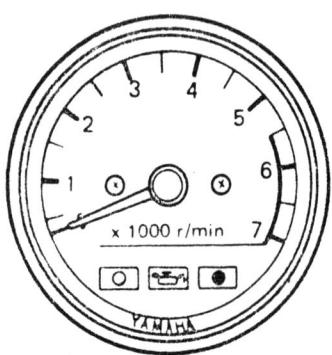

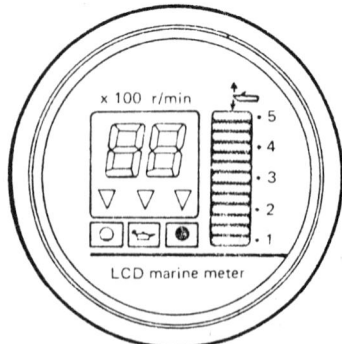

The warning light display is a part of the tachometer face. The upper drawing depicts the older gauge type tachometer. The lower drawing depicts the newer digital type tachometer with trim gauge.

If this signal is not accompanied with a drop in powerhead rpm, then a dangerously low oil level condition exists, which requires the operator's **IMMEDIATE** attention. There is almost **NO** oil left. The main tank has less than 0.3 U.S. quarts (0.3 litres) and the remote tank has less than 1.6 U.S. quarts (1.5 litres). Only the operator knows how much less!!

Green and Red Lights Come On
Buzzer Does Not Sound

See next paragraph, and check the operation of the buzzer as soon as possible. If the Red light is on, the buzzer should also be sounding.

Green and Red Lights Come On
Buzzer Sounds

If the Green and Red lights both come on simultaneously, a problem has arisen in the transfer of oil from the remote tank to the main tank. There is adequate oil in the remote tank but either the oil feed pump has failed or there is a blockage in one of the oil lines. Again, this situation demands **IMMEDIATE** action by the operator. Shut down the powerhead at once. In order to return the boat to its point of origin, the oil in the remote tank must be manually transferred to the main tank. Once the main tank has an adequate supply, the powerhead may be restarted.

Hose Network

Oil from the main oil tank is gravity fed to the oil injection pump. From the injection pump, oil is routed through a series of four or six transparent hoses and delivered to the intake manifold. One of the accompanying illustrations depicts the oil delivery network of hoses for the V4 powerheads. The other illustration shows the network for the V6 powerheads.

5-2 TROUBLESHOOTING

Problems with the oil injection system can be classed in one of two categories: oil delivery or electrical circuitry or sensors.

Insufficient oil delivery to the powerhead, will cause the oil level to drop slower than normal; the powerhead will overheat due to inadequate lubrication. Moving internal parts will wear more quickly.

Excessive oil delivery to the cylinders will cause the oil level to drop at a much faster rate than normal. The engine will smoke -- especially at idle speed and the spark plugs will become fouled causing the powerhead to misfire.

First Checks
Oil Delivery

If any of the oil related problems listed above should occur, any one or more of the following areas may require attention:

Control unit defective.
Oil pump defective.
Oil feed pump defective.
Oil pump link rod adjustment.
Oil pump link rod binding.

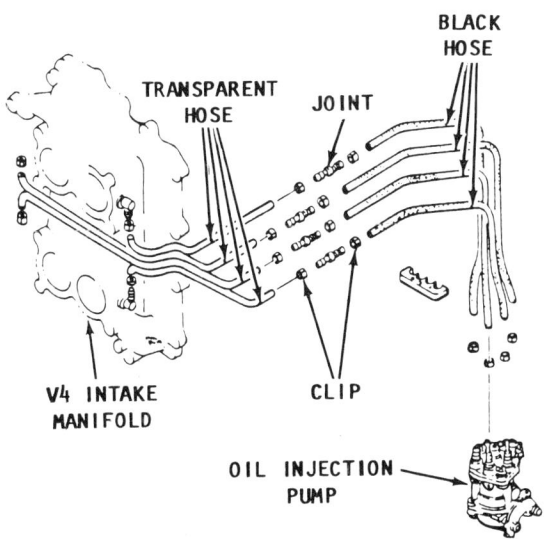

Line drawing to illustrate the oil delivery hose network on a V4 powerhead.

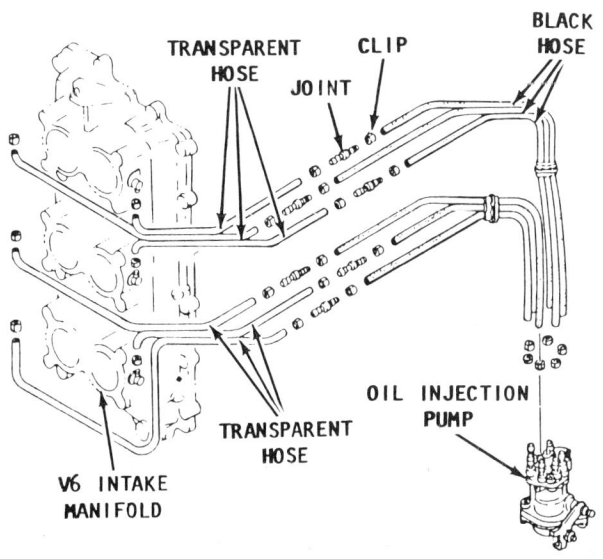

Line drawing to illustrate the oil delivery hose network on a V6 powerhead.

5-8 OIL INJECTION

Vent on the oil tank not clear.
Kinked oil hose.
Oil hose blockage.
Oil filter clogged.
Air in the system.
Leaking oil hose.
Carburetion and synchronization.
Idle fuel adjustment.
Quality of fuel being used.
Quality of two-stroke oil being used.

First Checks
Electrical System

If an electrical problem is encountered, bear in mind, the manufacturer has designed this particular electrical system to use the ground side of any electrical component for switching purposes, where normally the "hot" or power side would be used. This does not affect operation of any part except the part which is "hot" at all times --while the switch is in the **ON** position.

If a wire is shorted to ground, the electrical component will operate continuously instead of blowing a fuse. This condition could weaken the battery over a long period of time. A check may be made to determine if any electrical part is shorted to ground.

With the switch in the **OFF** position, connect an ammeter in series at the battery. The meter should register zero amps. If a reading is obtained, and it is not immediately obvious what is causing the amperage draw, i.e. a light always on, then systematically check the color code of all the wires at all connections with the wiring diagrams.

After checking the connections, check for a short to ground by disconnecting each control wire at the component.

1- Testing Each Warning Light

Disconnect the leads and wire harness connectors from the back of the tachometer. Remove the tachometer from the control panel. Notice there are six wires leading from the three lights. These leads feed into a plastic sleeve, **BUT** only four leads emerge from the sleeve into the harness connector. Inside the sleeve, three Yellow wires become one.

Obtain a 12 volt battery. Make contact with the negative battery terminal lead to the Yellow female terminal in the connector. In turn, one at a time, make contact with the positive battery terminal lead to

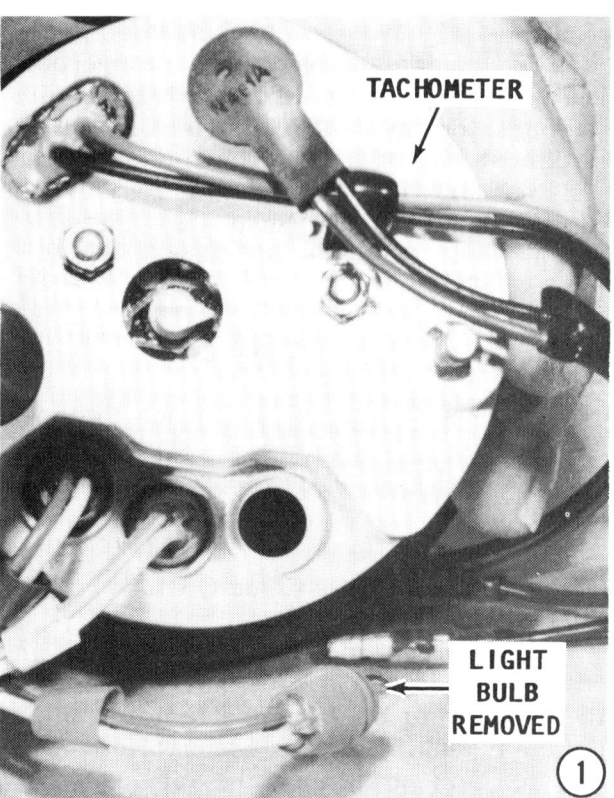

the Green lead from the Green light; the Red/Yellow lead from the Yellow light; and the Green/Red lead from the Red light. Each lamp should light. If a lamp fails to light, replace the lamp.

2- Testing Main Oil Tank Sensor With Ohmmeter

Disconnect the White, Brown, Red, and Black leads. Pull the sensor out of the oil tank. Move the sensor to a suitable clean work area for the following resistance tests.

Remove the circlip and the surrounding tube.

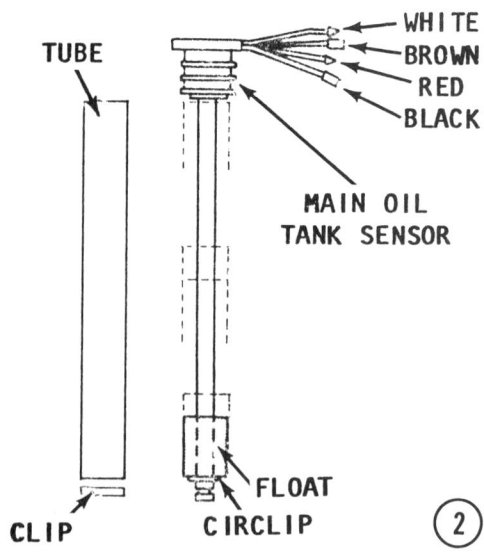

Obtain an ohmmeter and select the Rx1000 ohm scale. Position the float at its highest level and make contact with the meter test leads to the Black and White leads. The meter should register continuity. Make contact with the meter test leads to the Black and Brown leads. With the float still at its highest level, the meter should indicate no continuity.

Slide the float down to the half way point with the meter leads still making contact with the Black and Brown leads. At this point the meter should register continuity.

Continue sliding the float down. The meter should once more register no continuity. Slide the float to its highest level once more. Make contact with the Black meter lead to the Black lead of the sensor and the Red meter lead with the Red sensor lead. The meter should register between 400 and 600 ohms for all positions of the float except for the lowest position, against the circlip when installed. At the lowest position the meter should register continuity.

If the sensor fails to give resistance, as indicated, the sensor should be replaced.

3- Testing Remote Oil Tank Sensor
With Ohmmeter

Disconnect the Black and Black/Red leads. Pull the sensor out of the oil tank and move the sensor to a suitable clean work area for the following resistance tests. Do not remove either of the two clips around the sensor shaft.

Obtain an ohmmeter and select the Rx1000 ohm scale. Make contact with the Red meter lead to the Black/Red sensor lead and the Black meter lead to the Black sensor lead. With the float resting on the lower clip, the meter should register no continuity. Raise the float to make contact with the upper clip. The meter should register continuity.

If the sensor fails to give resistance, as indicated, the sensor should be replaced.

4- Testing Control Unit
With Ohmmeter

Disconnect all eleven leads: Yellow, Brown, Blue, Black, Green/Red, Green, Black/Red, Yellow/Red, White, Brown, and Red at their quick disconnect fittings or harness wire connectors. Remove the control unit from the powerhead through the attaching hardware.

CRITICAL WORDS

The control unit **MUST** be removed from the powerhead and placed on a level surface for testing.

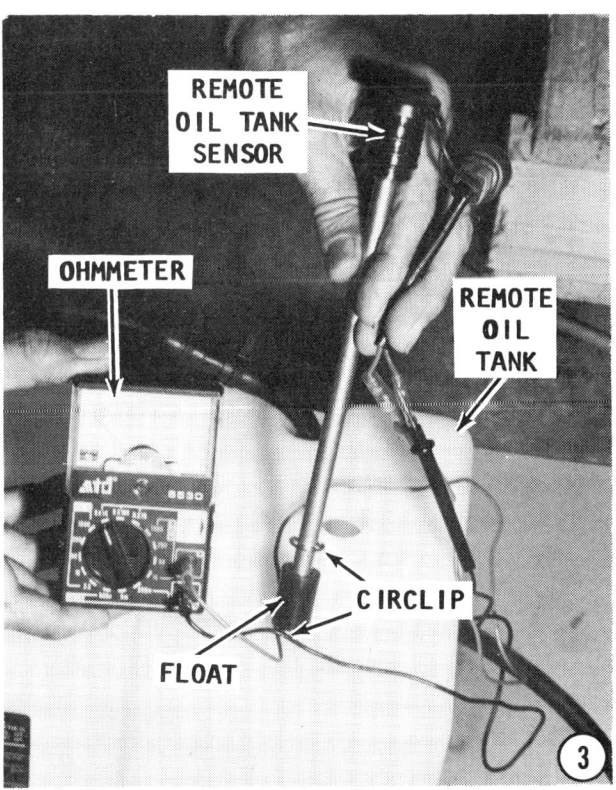

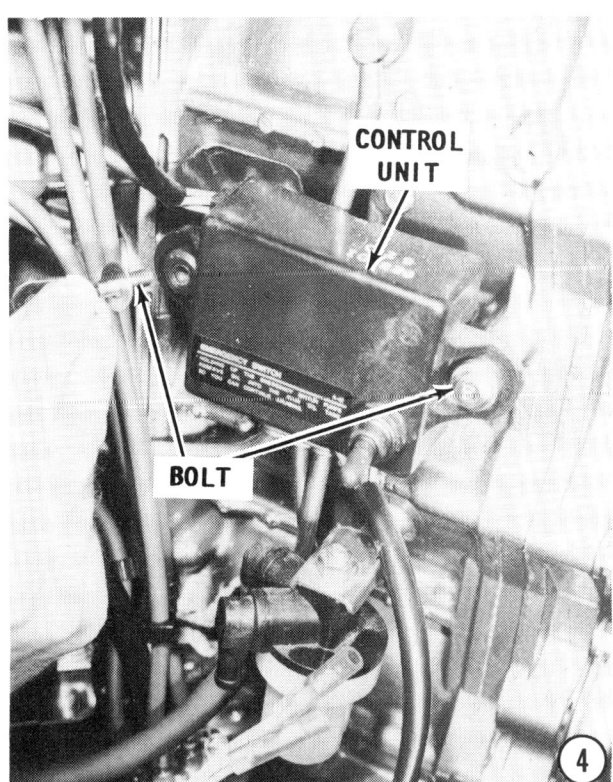

5-10 OIL INJECTION

RESISTANCE TESTS -- OIL INJECTION CONTROL UNIT

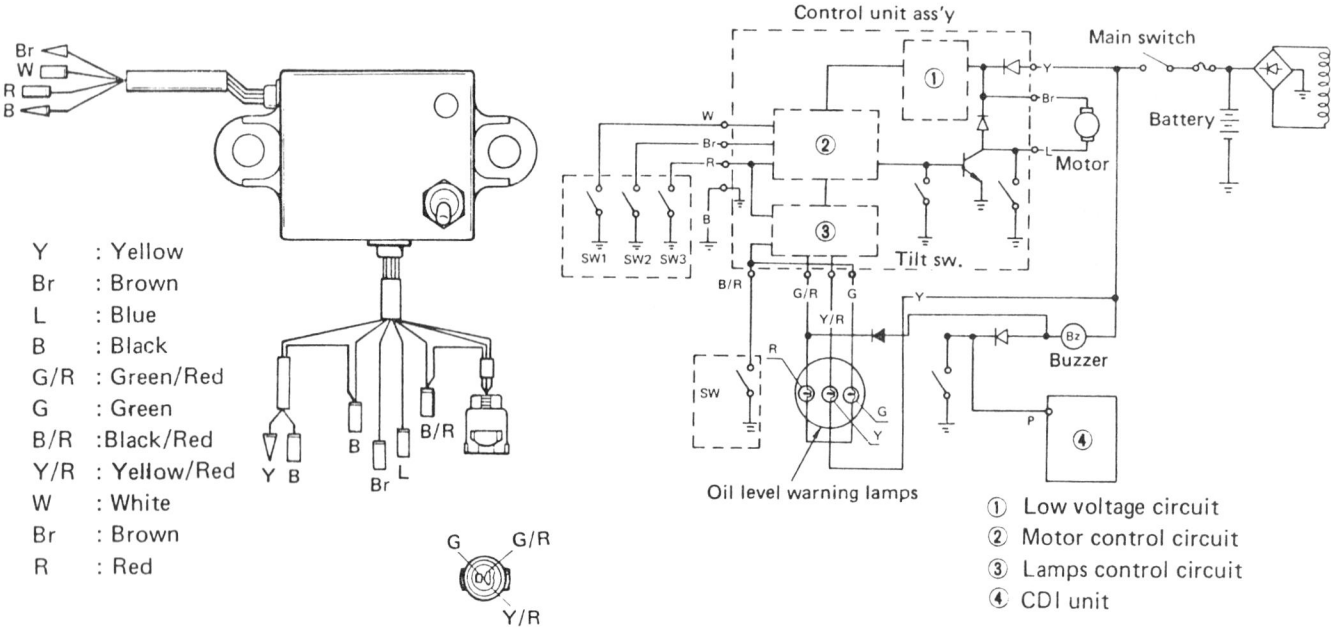

Y : Yellow
Br : Brown
L : Blue
B : Black
G/R : Green/Red
G : Green
B/R : Black/Red
Y/R : Yellow/Red
W : White
Br : Brown
R : Red

① Low voltage circuit
② Motor control circuit
③ Lamps control circuit
④ CDI unit

Unit: kΩ

Tester (+) \ Tester (−)		To Main switch		To Feed pump motor		Ground	To Lamps			To oil level sw.		
		① Y	①*1 Y	② Br	③ L	③*2 L	④ B	⑤ G/R Red	⑥ G B/R Green	⑦ Y/R Yellow	⑧ W SW1	⑨ Br SW2	⑩ R SW3
To Main switch	① Y			3.2~4.8	12~18	4.8~7.2	4.8~7.2	16~24	16~24	16~24	16~24	16~24	16~24
	①*1 Y			3.2~4.8	11.2~16.8	4~6	4~6	16~24	16~24	12~18	16~24	16~24	16~24
To Feed pump motor	② Br	∞	∞		4.8~7.2	1.6~2.4	1.6~2.4	8~12	6.4~9.6	4.8~7.2	8~12	8~12	8~12
	③ L	∞	∞	3.2~4.8			4.8~7.2	16~24	16~24	16~24	16~24	16~24	16~24
	③*2 L	∞	∞	1.6~2.4			0	8~12	8~12	3.2~4.8	8~12	8~12	8~12
Ground	④ B	∞	∞	1.6~2.4	3.2~4.8	0		8~12	8~12	3.2~4.8	8~12	8~12	8~12
To Lamps	⑤ G/R Red	∞	∞	∞	∞	∞	∞		∞	∞	∞	∞	0
	⑥ G B/R Green	∞	∞	∞	∞	∞	∞	∞		∞	∞	∞	∞
	⑦ Y/R Yellow	∞	∞	∞	∞	∞	∞	∞	∞		∞	∞	∞
To oil level sw.	⑧ W SW1	∞	∞	8~12	16~24	8~12	8~12	16~24	16~24	16~24		16~24	16~24
	⑨ Br SW2	∞	∞	8~12	16~24	8~12	8~12	16~24	16~24	16~24	16~24		16~24
	⑩ R SW3	∞	∞	∞	∞	∞	∞	0	∞	∞	∞	∞	

SPECIAL NOTES

*1 Perform resistance tests while tilting control unit, as shown in the accompanying illustration on the next page.

*2 Perform resistance tests while holding the emergency switch in the **ON** position, as shown in the accompanying illustration on the next page.

SERVICING 5-11

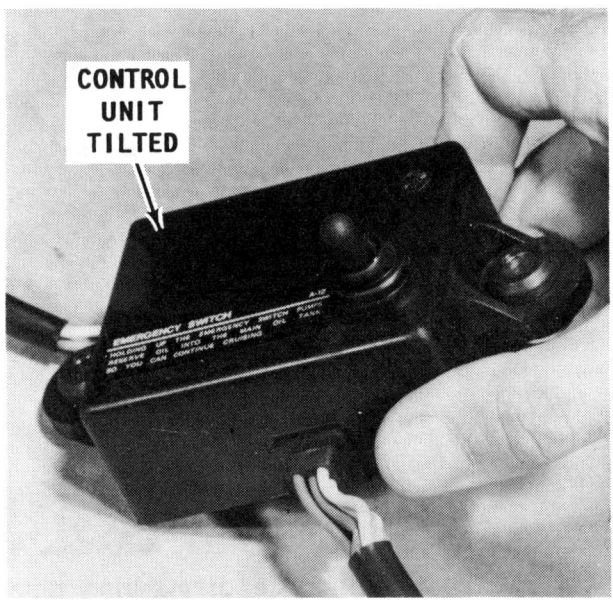

Tilting the oil injection control unit while performing the resistance tests indicated in the chart on the preceding page.

Select the Rx1000 scale on the meter. Systematically perform the following resistance tests according to the accompanying table. If any of the tests fail to give the desired resistance reading the control unit must be replaced. Service or adjustment is not possible.

SPECIAL WORDS

Using the accompanying resistance test chart on the previous page is actually quite simple. The following paragraph gives an example of how the main switch is tested. The other components follow the same procedure.

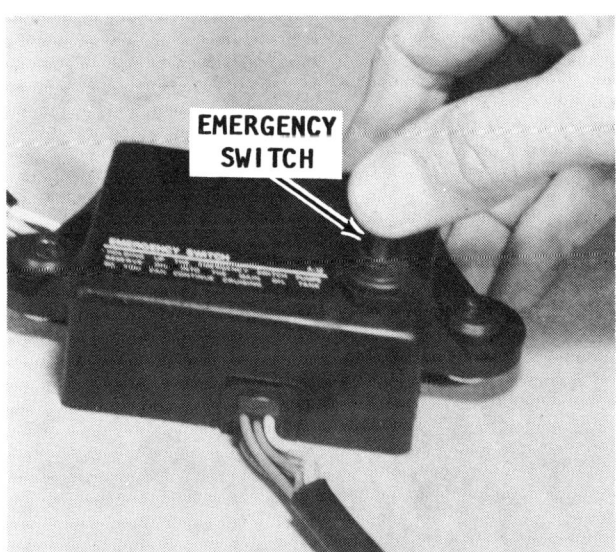

Holding the emergency switch in the UP position while performing the resistance tests outlined in the chart on the preceding page.

Reading across the top of the chart and observing the color code indicated: the negative test lead is connected to the main switch lead. Reading down the left side of the chart, the positive test lead is connected to the feed pump motor. The meter should indicate infinity.

In the vertical column, the main switch leads and the oil feed pump motor leads are tested first with the control unit placed horizontally on the workbench; next, with the unit tilted at an angle as depicted in the accompanying illustration; and finally, with the emergency switch held in the **ON** position during testing.

5-3 SERVICING OIL INJECTION SYSTEM

Main Oil Tank Removal

1- Place some shop cloths in the bottom of the lower cowling to catch any oil which may drain during disconnection of oil supply lines. Disconnect the oil supply line leading from the main oil tank to the oil pump beneath the tank by compressing the wire clamp with a pair of needle nose pliers and pushing the clamp up along the oil line. Squeeze the oil line to restrict the flow of oil while pulling it free of the fitting. Plug the line quickly with a suitable screw to prevent loss of oil.

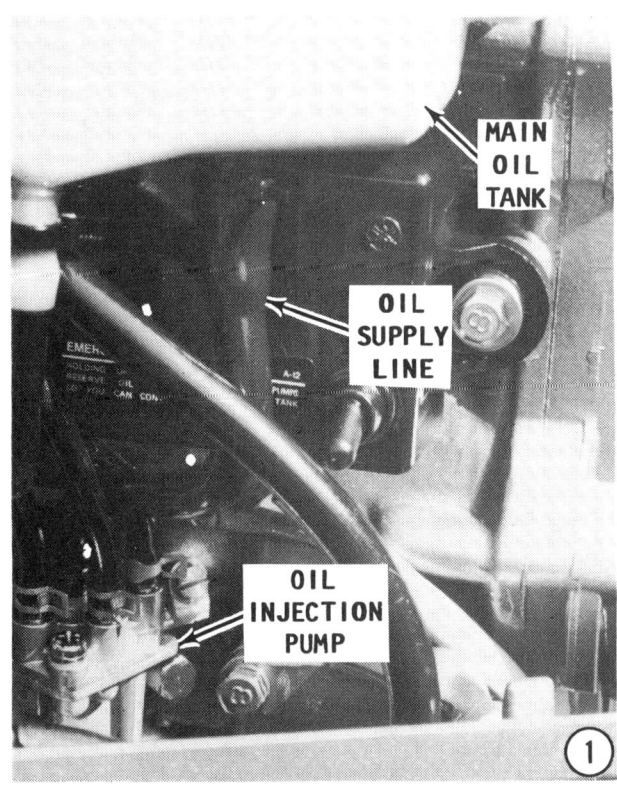

5-12 OIL INJECTION

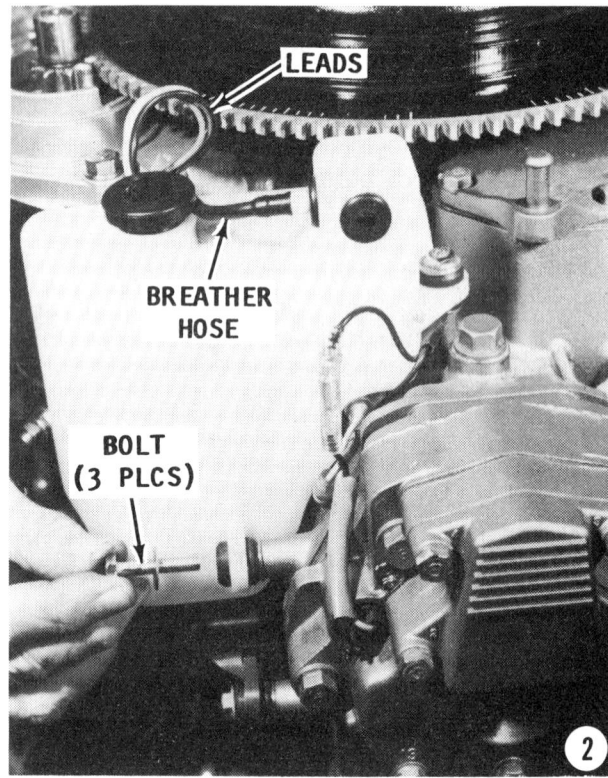

2- Remove the main oil tank breather hose from the air box cover. Disconnect the leads from the oil level sensor cap at their quick disconnect fittings. Loosen and remove the three bolts securing the oil tank. Carefully lift the tank free of the powerhead. Set the tank upright in a safe place.

Oil Pump Removal

3- Pull out the tiny cotter pin from the oil pump shaft. Slide the plastic spacer from the shaft. Pry the oil pump link rod from the ball joint on the pump shaft.

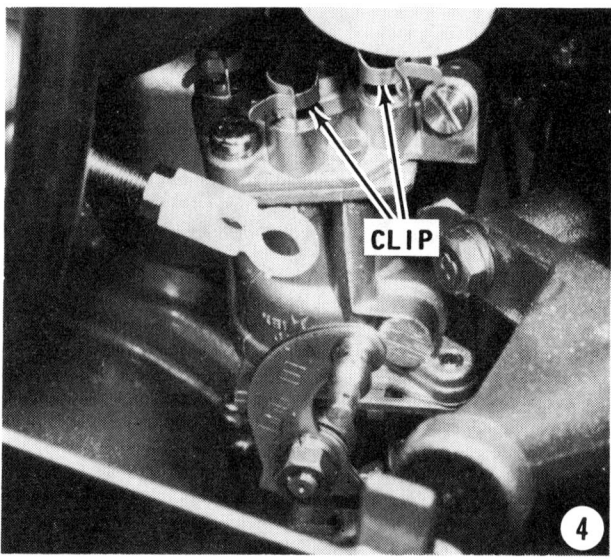

4- Wrap a small piece of masking tape around each of the oil supply lines and write the cylinder number on each piece of tape. Taking time to identify the lines will ensure each will be connected back in its original location.

GOOD WORDS

An illustration showing the numbered discharge fittings will be given later in the assembly procedures as a guide to installing the correct hose onto the correct fitting.

Slide the clips back, one on each oil line. Gently pull each line free of its fitting at the pump cover. Remove the two bolts securing the oil pump. Lift the oil pump from the powerhead. Remove and discard the O-ring.

5- Pull out the driven gear shaft.

DISASSEMBLING OIL PUMP   5-13

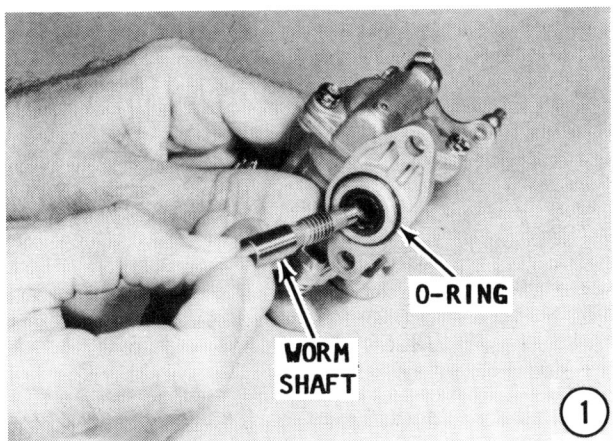

DISASSEMBLING OIL PUMP

1- Remove and discard the O-ring around the worm shaft. Pull out the worm shaft.

2- Remove the two Phillips head screws, and then remove the shaft housing. Remove and discard the O-ring.

3- Normally the shaft housing would remain intact, because individual replacement parts are not available. The only reason the housing would be disassembled is to replace a broken or missing return spring.

To replace the spring, first loosen and remove the nut at the end of the lever shaft. Next, unhook the return spring from the lever and count the number of turns needed to unwind the spring. If the spring is missing or broken, remove what is left. Pull out the lever shaft.

4- Remove the two Phillips head screws, and then lift off the pump cover. Remove the spring and distributor plate. Remove and discard the sealing ring.

5- Push out the plunger cam and distri-

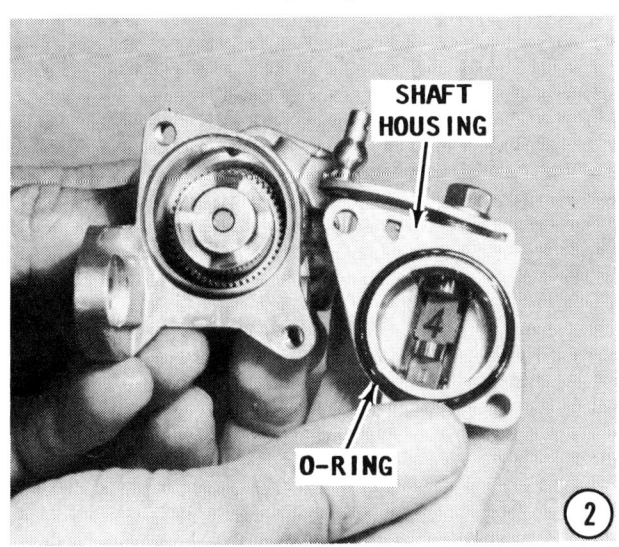

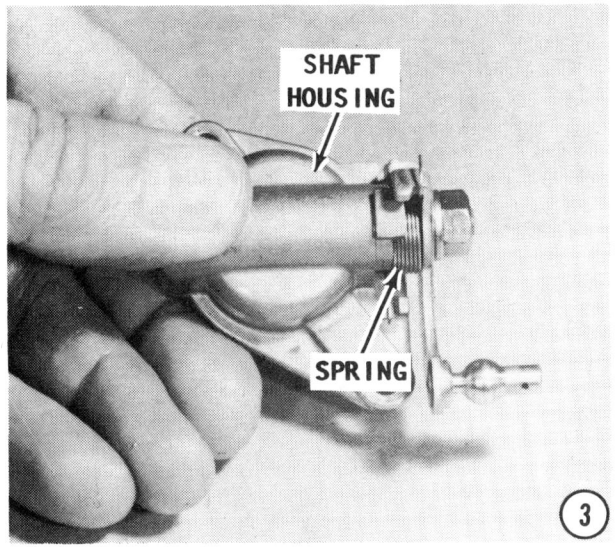

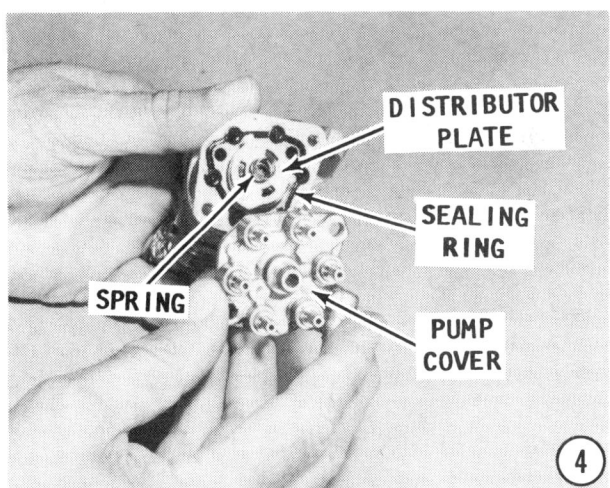

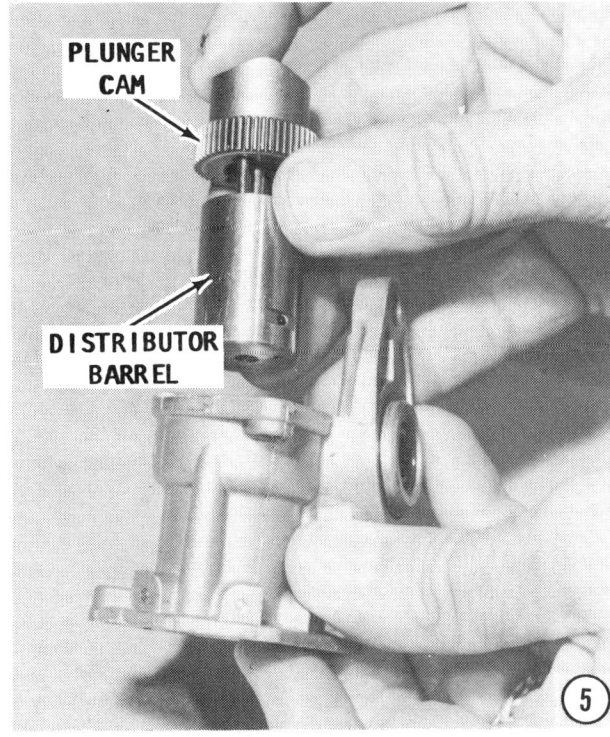

5-14 OIL INJECTION

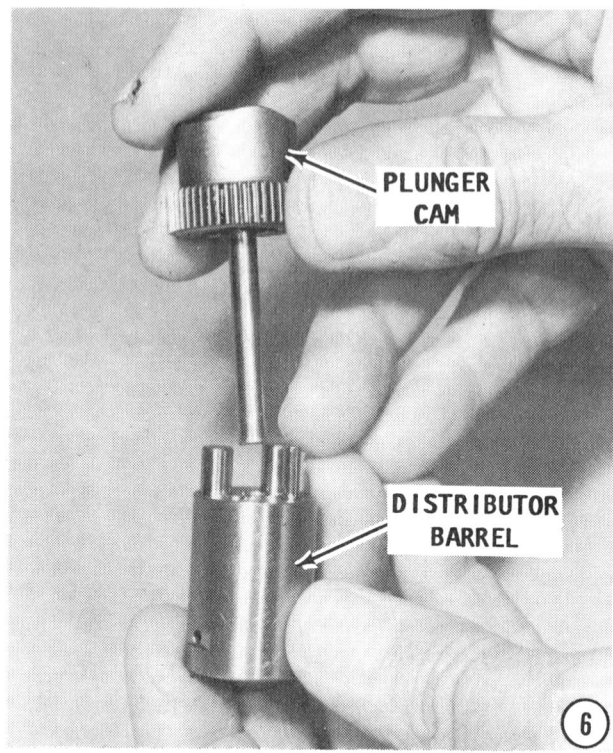

butor barrel as an assembly. Do not allow the assembly to fly apart.

6- Remove the plunger cam from the distributor barrel.

7- Lift out the two stepped shafts and springs from the distributor barrel. Notice the smaller end of the shaft and the spring are located inside the barrel.

CLEANING AND INSPECTING

Rinse the pump body and pump cover in solvent or soak them in contact cleaner for

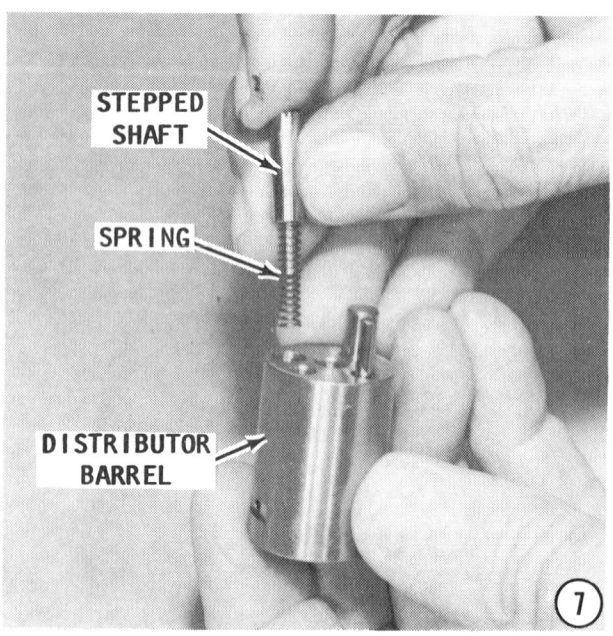

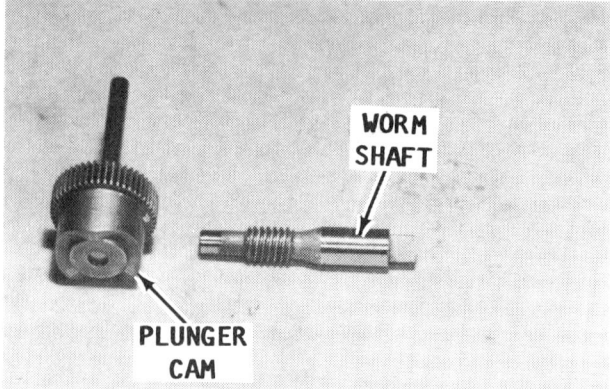

Inspect the teeth/threads on the plunger cam and on the worm shaft.

several hours and then lightly blow dry with compressed air.

Check all parts and passages to be sure they are not clogged or contain any deposits. Do not aim the compressed air directly into the pump cover because such action may damage the small check balls at the discharge fittings by causing them to spin. These check balls are not serviceable, as it is impossible to separate the discharge fittings from the pump cover. If it is determined a check ball is stuck in the open position, a new oil pump must be purchased and installed.

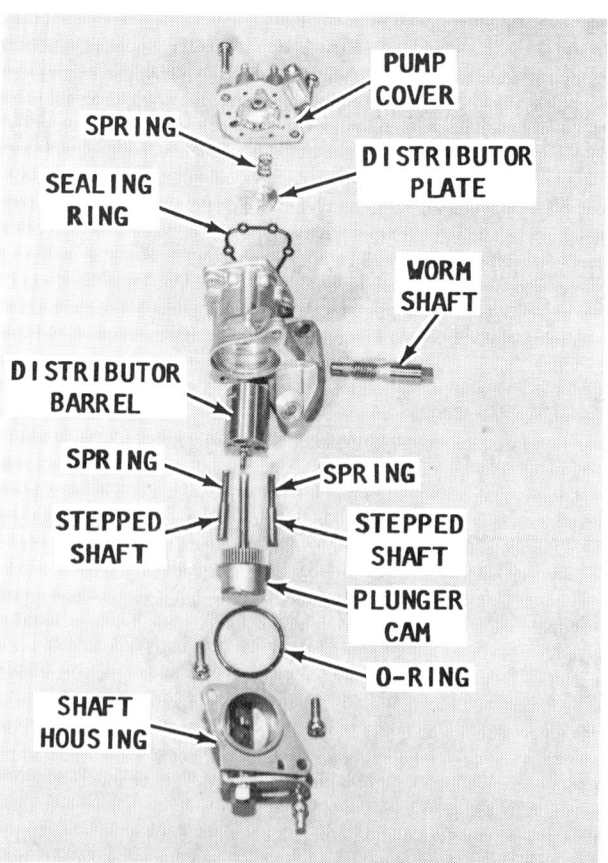

Arrangement of the oil injection pump parts.

ASSEMBLING OIL PUMP 5-15

The discharge ports may be cleaned by spraying contact cleaner past each check ball from the underside of the pump cover. This may cause the check balls to be lifted from their seats. To reseat the balls, place the pump cover on a clean shop cloth with the discharge fittings facing upward. Obtain a 1/16" drill bit and a small hammer. Insert the blunt end of the bit into a discharge fitting and lightly tap on the pointed end with the hammer to seat the ball. Repeat this procedure for all the fittings.

Inspect the condition of the worm shaft threads/teeth and the plunger cam gear teeth for excessive wear and missing or chipped teeth.

If the pump had a tendency to leak oil, only two possible areas may be at fault. The first would be the gasket under the bleed screw. If this gasket is missing, then the pump would definitely leak at this point. The second area to check is the O-ring around the worm shaft. If the O-ring is distorted in any way, the pump will leak profusely at this point.

Inspect the condition of the oil seal around the worm shaft. Remove this seal only if another is available to take its place. If the seal is still in a serviceable condition, leave it in place, because the seal will be destroyed during removal.

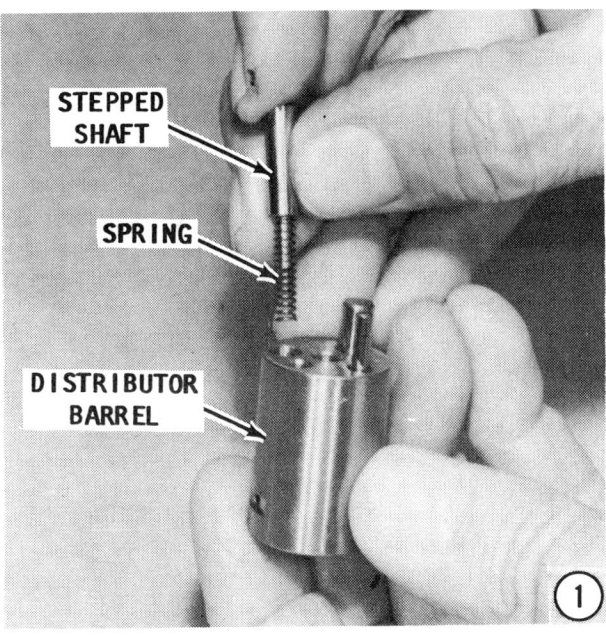

ASSEMBLING OIL PUMP

1- Slide the spring over the smaller end of the stepped shaft and insert this end first into the distributor barrel. Install the other shaft in a similar manner.

2- Slide the shaft of the plunger cam into the distributor barrel. Notice the small stubby shaft on the distributor barrel is the only shaft which will index into one of the three notches on the gear of the cam. The other two stepped shafts rest against the shoulder of the gear. This installation **IS** correct as it provides the pumping action when the plunger cam oscillates.

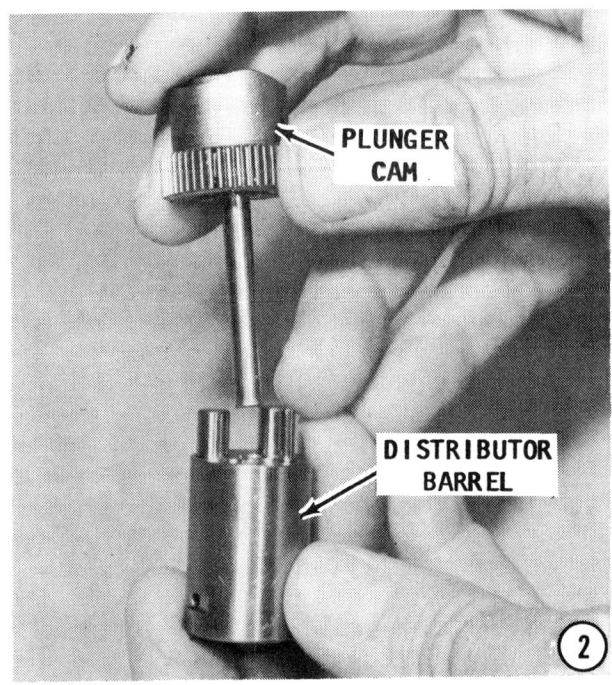

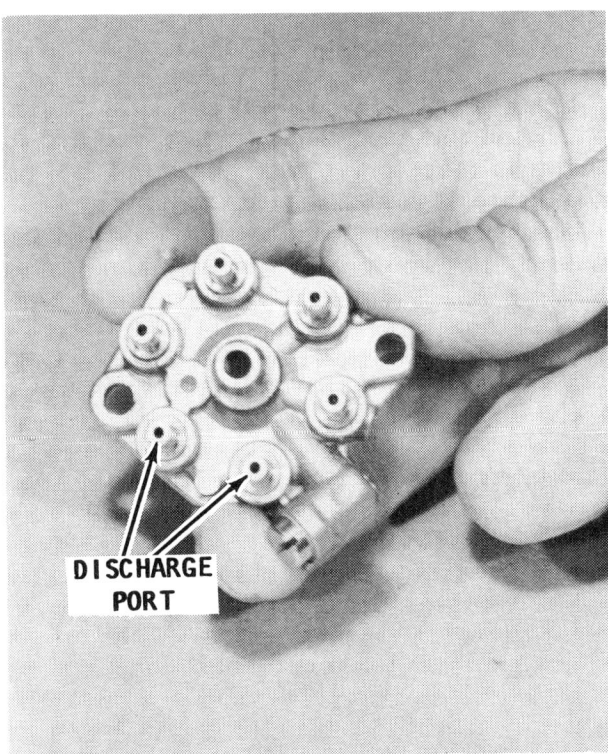

The discharge ports may be cleaned using contact cleaner and a 1/16" drill bit, as explained in the text.

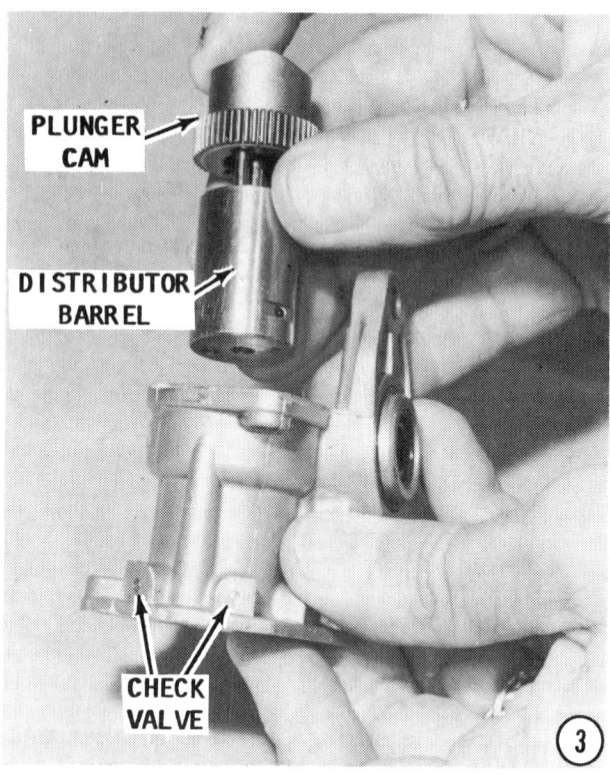

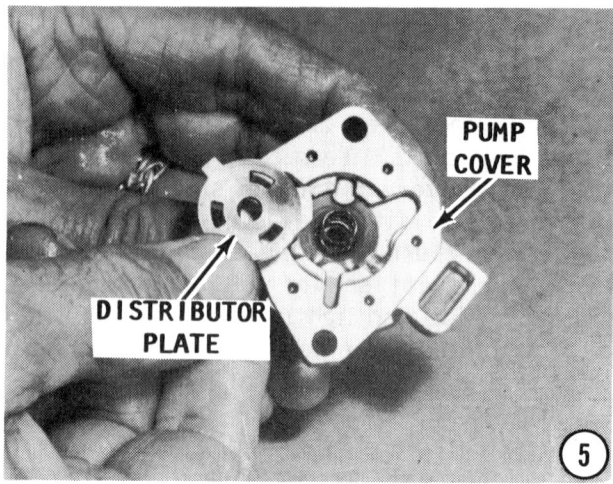

in the cover when it is installed. Center the spring onto the distributor plate.

Bring the pump cover down over the pump body, the spring must center itself inside the cover and the plate must index into the notches of the cover. Hold the cover down against the spring tension while the two Phillips head screws are installed and tightened securely.

3- Hold the assembly together and insert it through the body of the pump with the barrel end closest to the check valves on the pump body after installation.

4- Install a new O-ring around the shaft housing. Install the housing over the pump body and secure it in place with the two attaching Phillips head screws. Tighten the screws securely.

5- Match the pump cover to the pump body. Separate them and notice how the distributor plate aligns with the cover following installation.

6- Install a new sealing ring around the pump body. Place the distributor plate over the distributor barrel to match the notches

7- If the shaft housing was disassembled in Step 3 to replace the return spring, begin to assemble the housing by sliding the lever shaft into the cover. Next, hook the L-shaped end of the spring behind the projection on the housing, and then wind the spring **COUNTERCLOCKWISE** seven full turns. Hold onto the tensioned spring, and at the same time install the pump lever over the shaft. Seat the lever onto the shaft with the ball joint facing **UPWARD** (as installed on the powerhead). Hook the curved end of the tensioned spring over the stop. Install and tighten the lockwasher and nut.

8- If the oil seal was removed from around the worm shaft, install a new one

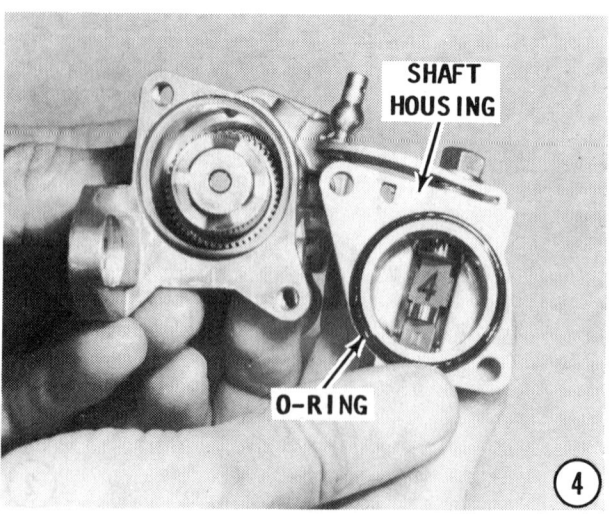

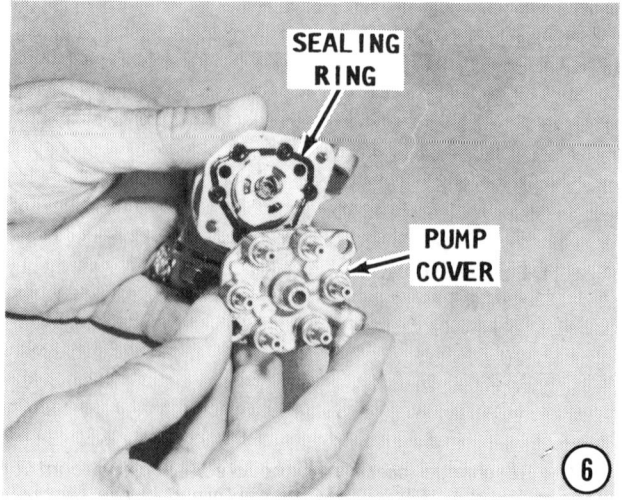

ASSEMBLING OIL PUMP 5-17

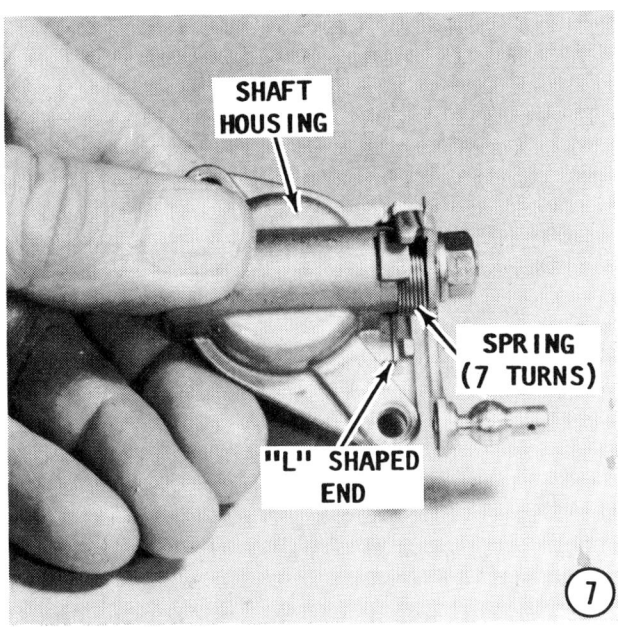

with the lip of the seal facing outward to the exterior of the pump.

Push the worm shaft through the oil seal until it seats inside the pump body. Install a new O-ring around the worm shaft and push it firmly into the groove in the mating surface. An incorrectly installed O-ring will almost certainly cause the pump to leak oil at this location. The pump is now ready for installation onto the powerhead.

INSTALLATION

The following procedures pickup the work after the oil pump has been cleaned, inspected, and assembled as outlined in the previous pages.

Oil Pump Installation

1- Push the driven gear shaft into the powerhead to index with the bronze drive gear around the crankshaft.

2- Install a new O-ring on the oil injection pump. Check to be sure the shaft of the oil injection pump will index into the slot at the center of the crankshaft driven gear. If the two are no longer aligned, rotate the slotted shaft on the pump to match the slot in the driven gear. Install the oil pump with the pump shaft indexed with the slot in the crankshaft driven gear.

Secure the pump with the two attaching bolts. Tighten the bolts securely.

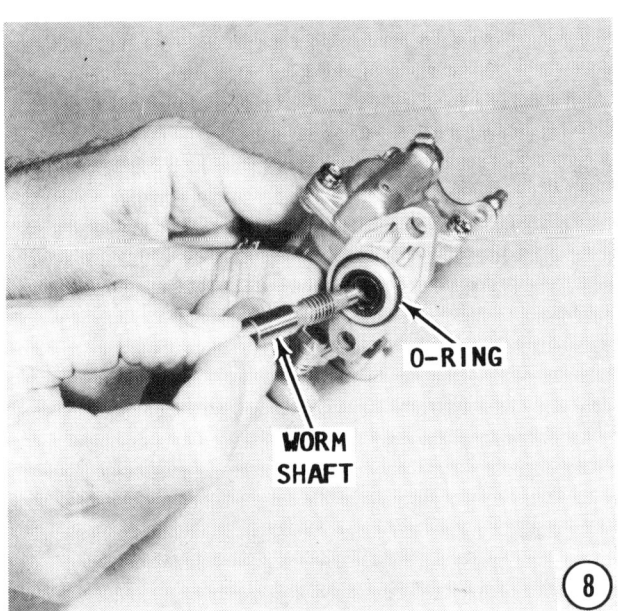

5-18 OIL INJECTION

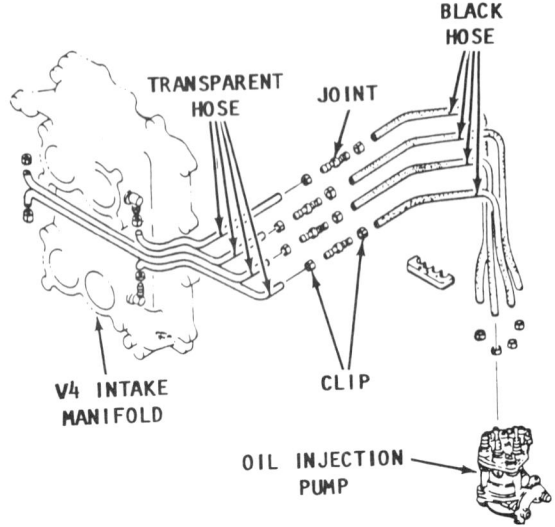

Line drawing to illustrate the oil delivery hose network on a V4 powerhead.

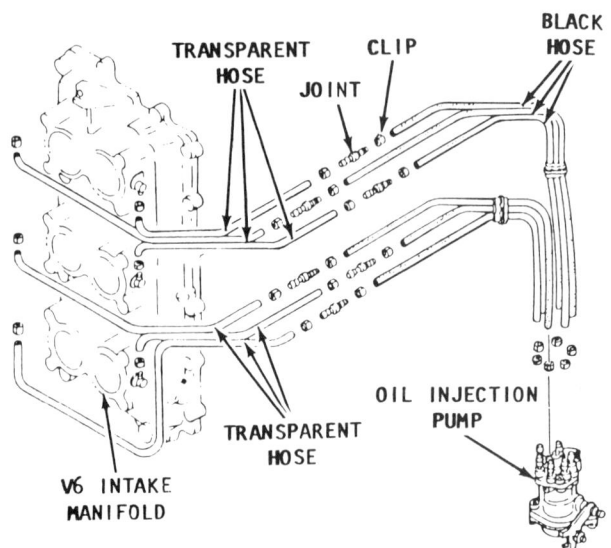

Line drawing to illustrate the oil delivery hose network on a V6 powerhead.

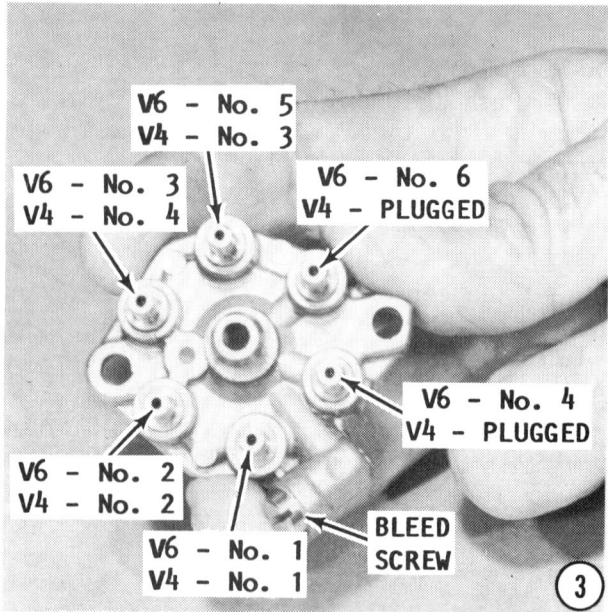

3- Install each of the oil supply lines onto their respective discharge fittings. If the lines were not identified as to which cylinder they originated from, the lines will have to be traced back. On V4 powerheads, there are two long lines leading from cylinders 1 and 3, and two short lines leading from cylinders 2 and 4. On V6 powerheads there is one long line leading from cylinder No. 1; four medium length lines leading from cylinders No. 2, 3, 4, and 5, and one short line leading from cylinder No. 6.

4- Snap the oil injection link rod back onto the ball joint on the pump shaft. Slide the plastic spacer onto the shaft and install the tiny cotter pin. If the length of the rod was accidentally altered, refer to the procedures in Section 5-5 to adjust the rod to the specified length.

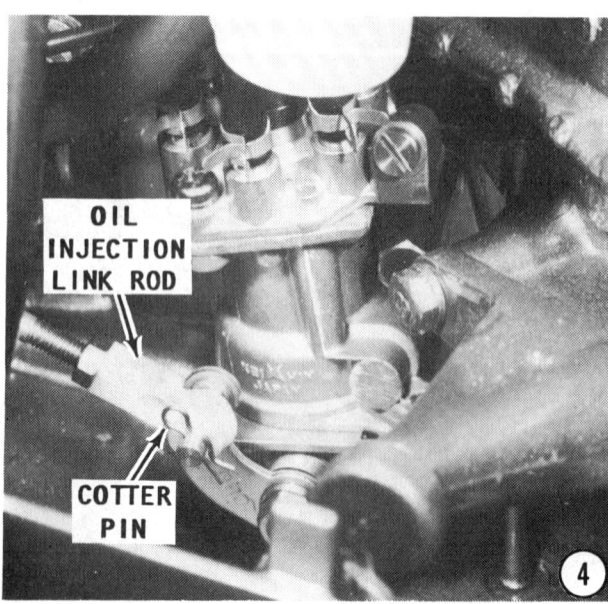

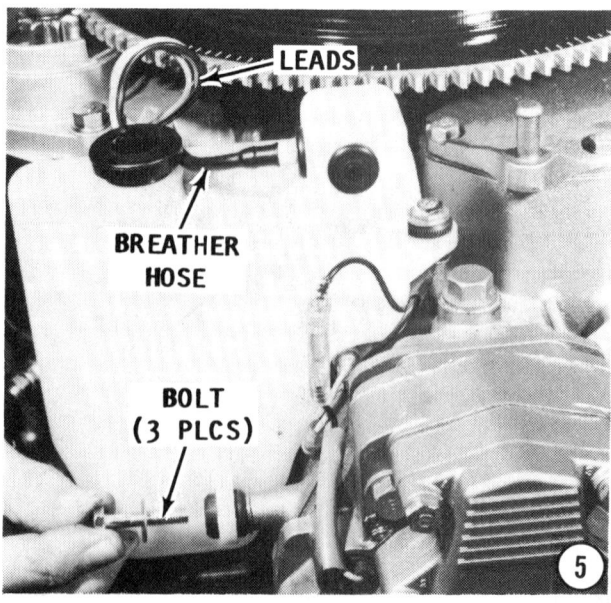

INSTALLATION 5-19

Main Oil Tank Installation

5- Install the oil tank to the powerhead. Secure the tank with the attaching bolts. Install the main oil tank breather hose to the air box cover. Check around the tank to be sure no oil line will be crimped or flattened when the tank is installed. Secure the line with the wire clamp. A free flow of oil through an unrestricted line is **CRITICAL** for adequate powerhead lubrication.

6- Install the oil line from the tank to the pump at the inlet fitting.

Connect the leads from the oil level sensor cap at their quick connect fittings.

7- Fill the oil tank to the top line embossed on the tank. Use only Yamalube two cycle outboard oil (or an equivalent 2-stroke engine oil with a BIA certified rating TC-W). This oil is suitable for use through a temperature range of $14^{\circ}F$ to $140^{\circ}F$ ($-10^{\circ}C$ to $60^{\circ}C$).

CAUTION

Any time the oil tank hose is disconnected, the oil injection pump **MUST** be purged ("bled") of any trapped air. Failure to "bleed" the system could lead to powerhead seizure due to lack of adequate lubrication. Therefore, "bleed" the system per the instructions outlined in Section 5-4.

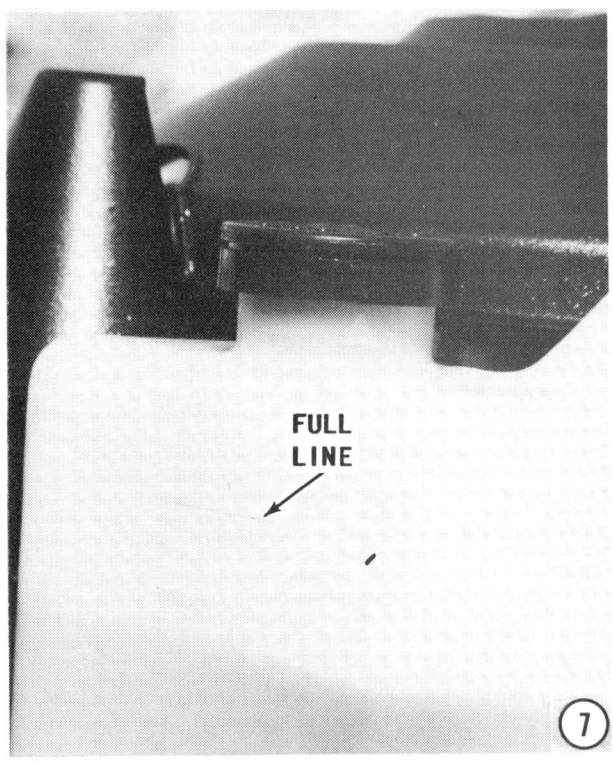

Measuring Oil Flow Through Oil Injection Pump

1- Pull out the tiny cotter pin from the oil pump shaft. Slide the plastic spacer from the shaft. Pry the oil pump link rod from the ball joint on the pump shaft. Turn

5-20 OIL INJECTION

the pump lever all the way **CLOCKWISE** to set the pump stroke to maximum.

Prepare a 50:1 gasoline/oil mixture to operate the powerhead. "Jury rig" a setup to provide this mixture to the fuel pump during this step.

CAUTION
Water must circulate through the lower unit to the powerhead anytime the powerhead is operating to prevent damage to the water pump in the lower unit. Just five seconds without water will damage the water pump impeller.

Start the powerhead. Operate the powerhead **ONLY** at idle speed throughout this procedure.

Allow the powerhead to idle for about ten minutes.

2- Obtain a small container and have a shop towel handy to catch any oil drips. Using a pair of needlenose pliers, squeeze the clamp on the No. 1 cylinder delivery hose. Push the clamp up the hose. Gently pull the hose from the fitting on the intake manifold and immediately hold it over the container. Observe the flow of oil for five full minutes to verify the pump is functioning correctly.

SPECIAL WORDS
The oil/fuel mixture prepared in Step 1 will provide adequate lubrication during this test.

A steady slow pulsing flow with no air bubbles may be expected. The manufacturer gives no specified amount per minute in this case. Reconnect the line. Shut down the powerhead and remove the flushing device.

3- Snap the oil injection link rod back onto the ball joint on the pump shaft. Slide the plastic spacer onto the shaft and install the tiny cotter pin. If the length of the rod was accidentally altered, refer to the procedures in Section 5-5 to adjust the rod to the specified length.

Remote Oil Tank Removal and Oil Feed Pump Service

First, These Words

Each boat will probably have the remote oil tank mounted in a different location using different securing hardware. All tanks are mounted in a holder with a securing strap across the top of the tank. Therefore the following removal procedures begin with this strap.

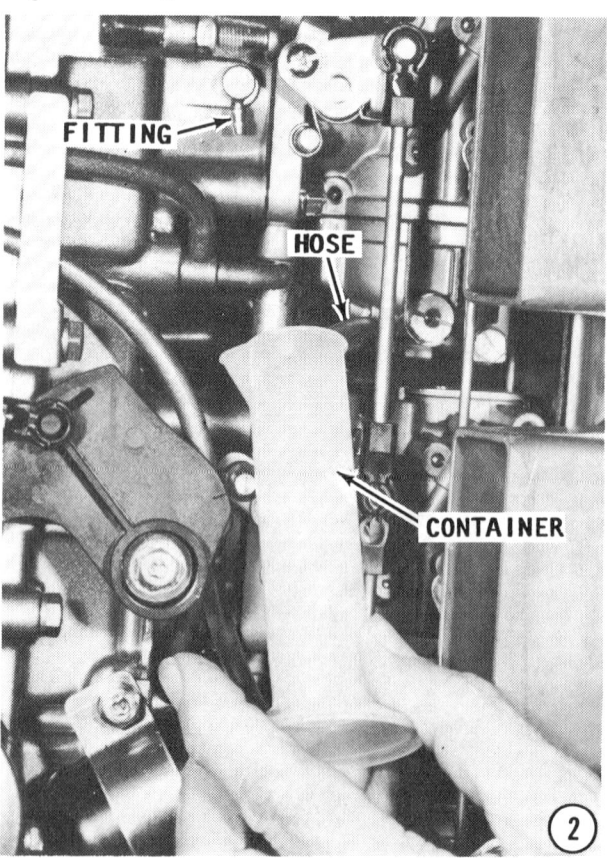

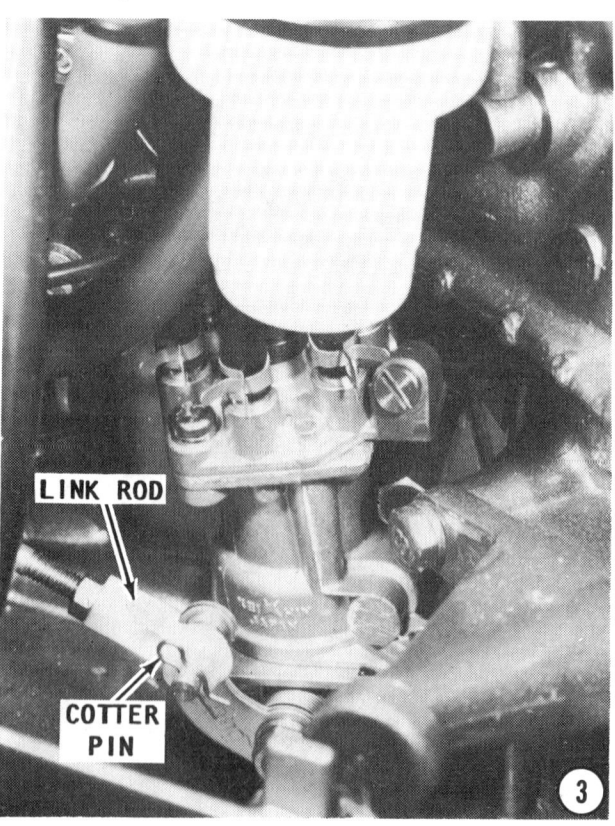

TANK SERVICE 5-21

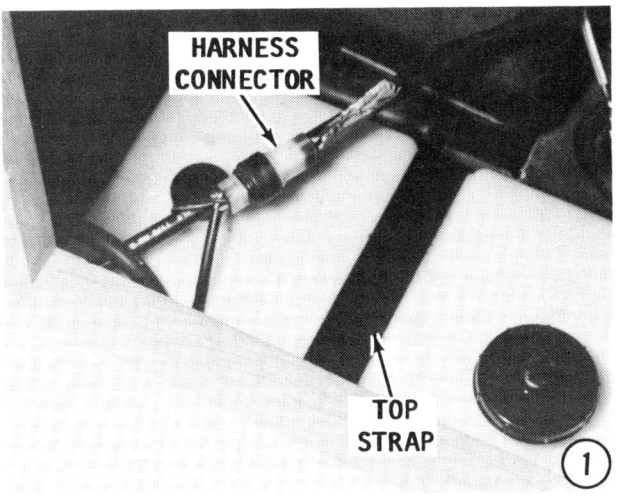

1- Remove the nut and long bolt securing the top strap to the holder. Set aside the strap. Disconnect the wire harness connector from the oil feed pump motor and the oil level sensor.

Place some shop cloths around the oil supply line hose joint to catch any oil which may drain from the hose. Snip the clamp with a pair of dykes. Squeeze the oil line to restrict the flow of oil while pulling it free of the fitting. Plug the line quickly with a suitable screw to prevent loss of oil. Lift out the remote tank.

Measuring Oil Discharge

2- Check to see if the remote tank contains at least 3/4 U.S. quart (0.7 litres).

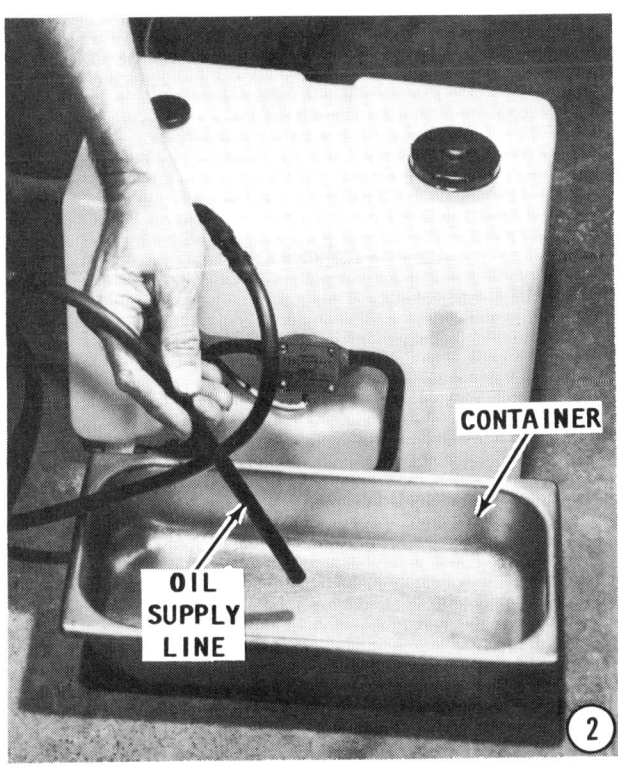

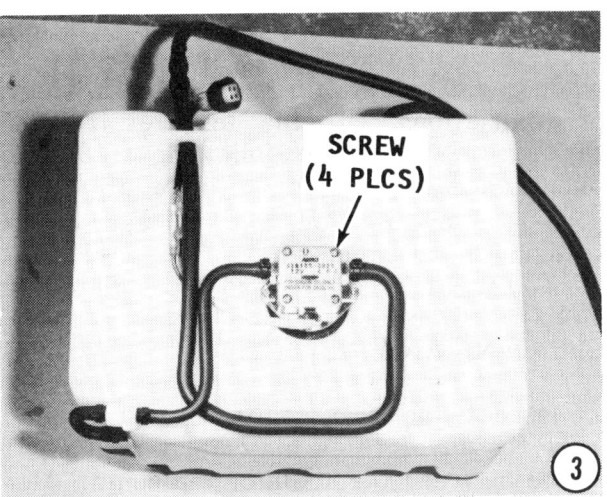

Obtain a 12 volt battery and two leads. Position a suitable container under the oil supply line. Make contact with the lead from the negative battery terminal to the Brown female terminal in the harness connector. Make contact with the lead from the positive battery terminal to the Blue female terminal in the harness connector. The oil line should emit a strong, solid flow of oil.

Obtain a measuring cup or cylinder and measure the rate at which oil is delivered. The flow of oil should be 0.2 U.S. quarts (0.2 litres) per minute. Perform the test over a three minute period and average the reading.

If the pump is allowed to empty the tank of oil and pump air, the feed pump must be purged of air according to the directions given in Section 5-4.

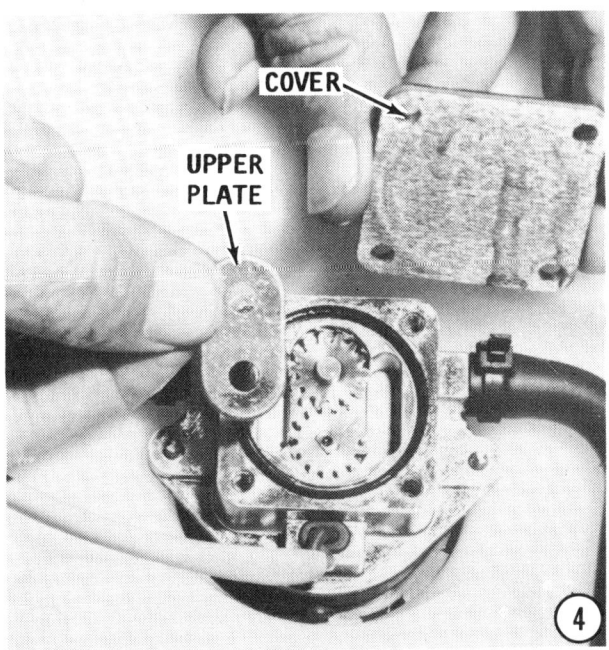

5-22 OIL INJECTION

If the flow of oil is less than specified, proceed with these steps to service the pump and check all oil lines for leaks, kinks, or restrictions. If the test result were satisfactory, skip the following steps and pickup the work in Step 3 of installation.

3- Remove the four Phillips head screws securing the pump cover.

4- Lift off the cover and upper plate.

CLEANING AND INSPECTING

Inspect the two cavities of the pump body for any foreign material which might wedge between the teeth of the pump. Momentarily connect the Brown and Blue leads to the 12 volt battery as described in Step 3. Check to ensure the teeth rotate smoothly. Inspect the condition of the O-ring in the groove of the pump body, replace if necessary.

ASSEMBLING

1- Fill the cavities and coat the gear teeth with oil to prime the pump. Install the upper plate over the two small shafts. Check to be sure the O-ring is correctly seated in the groove of the pump body. Place the pump cover in position over the pump, with the screw holes aligned.

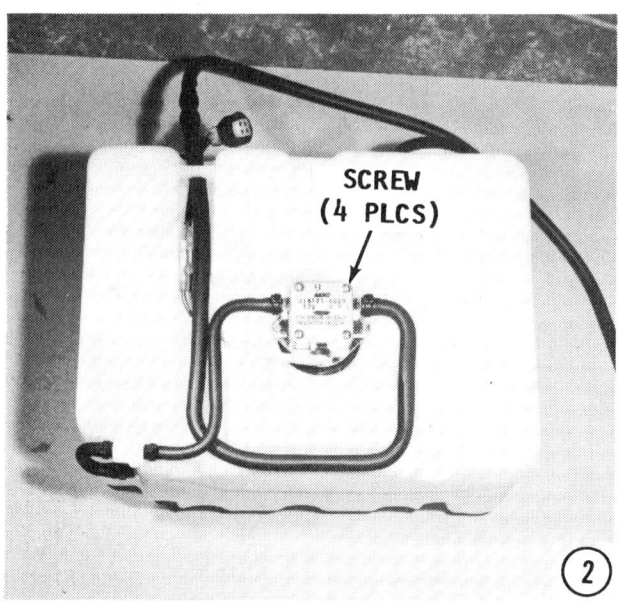

2- Secure the cover to the pump with the four Phillips head screws. Tighten the screws securely. If questionable test result were obtained in Step 2 and the pump was subsequently overhauled, go back and repeat the test outlined in Step 2. If the flow of oil is still not within specifications and the lines have been checked for leaks, obstructions, and kinks, then the feed pump must be removed and replaced.

INSTALLATION

3- Install the tank into the boat. Slide a new hose clamp onto the remote tank hose, and then connect the hose to the hose from the main tank. Tighten the hose clamp at the joint. Connect the two halves of the wire harness connector together. They will only fit one way.

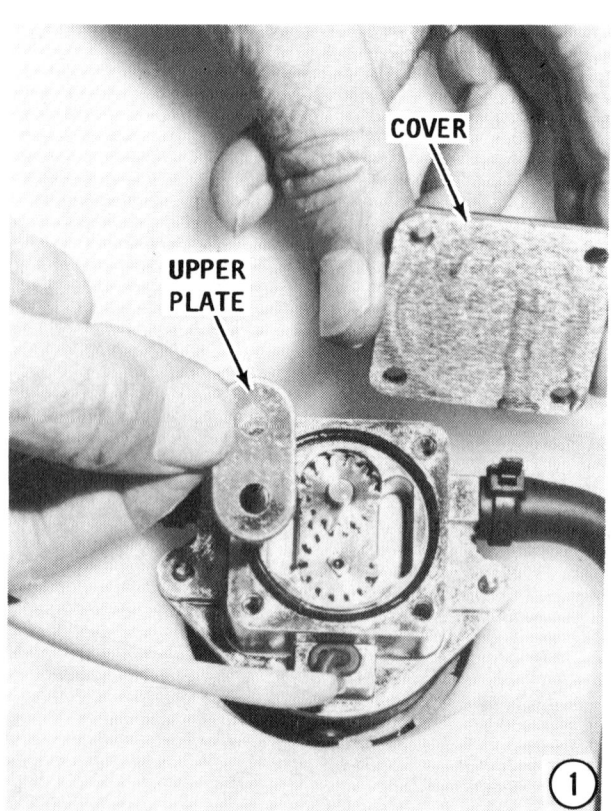

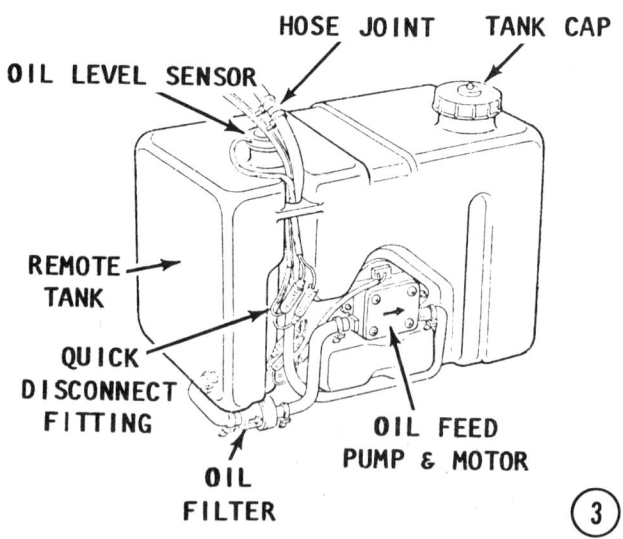

PURGING AIR   5-23

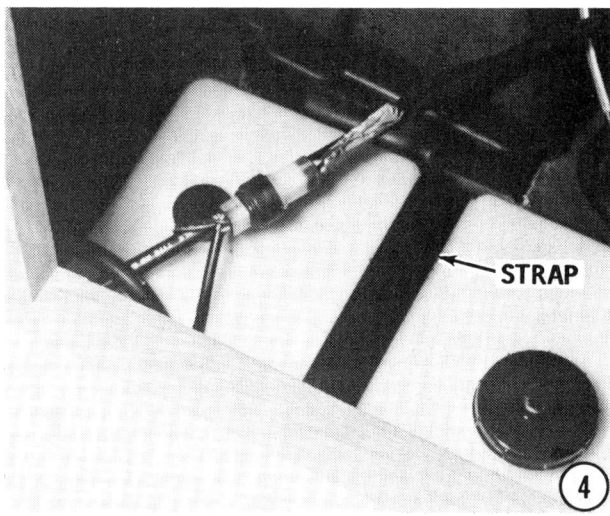

4- Bring the strap across the tank and secure the strap to the holder with the long bolt and nut. Purging air from the remote oil tank feed pump is covered in Section 5-4.

5-4 PURGING AIR FROM OIL INJECTION SYSTEM

The following procedures are to be performed any time the oil injection system has been opened (other than to add oil to the tank) and air has entered the system.

Oil Injection Pump

1- Prepare about a ten minute supply of of 50:1 premix using Yamalube, or other quality two-stroke oil. "Jury-rig" a setup to provide this mixture to the fuel pump during this purging procedure. Check the level of oil in the oil tank and replenish as necessary.

Connect a flush device to the lower unit.

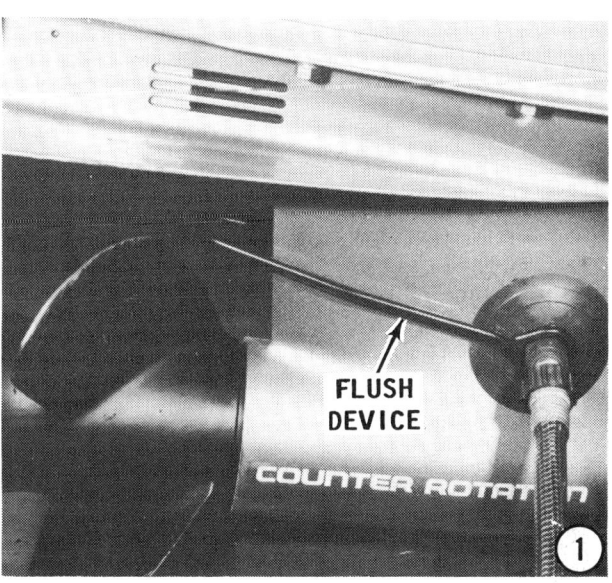

CAUTION
Never operate the powerhead over 1000 rpm with a flush device attached, because the engine may **RUNAWAY** due to no load on the propeller. A runaway engine could be severely damaged.

Start the engine.

CAUTION
Water must circulate through the lower unit to the powerhead anytime the powerhead is operating to prevent damage to the water pump in the lower unit. Just five seconds without water will damage the water pump impeller.

2- Place a suitable cloth under the air bleed screw. Remove the air bleed screw and allow oil to flow from the opening until a bubble free flow of oil is obtained. On models equipped with a check valve ball on the main tank breather, a long thin object such as a nail or a toothpick can be used to depress the check valve ball to quicken the flow of oil through the bleed screw.

SPECIAL WORDS
Because the fuel and oil premix is being fed through the carburetor in addition to oil from the oil injection system being fed directly into the cylinders, the powerhead will be operating on a heavy mixture of oil.

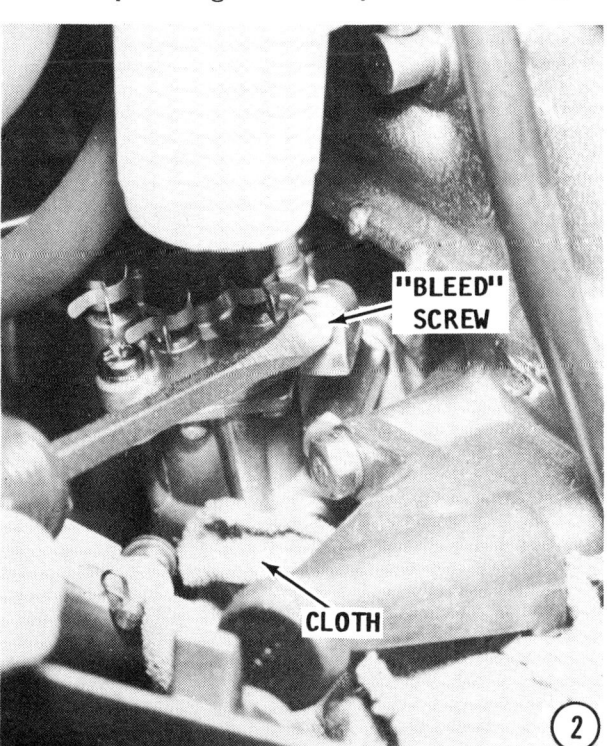

5-24 OIL INJECTION

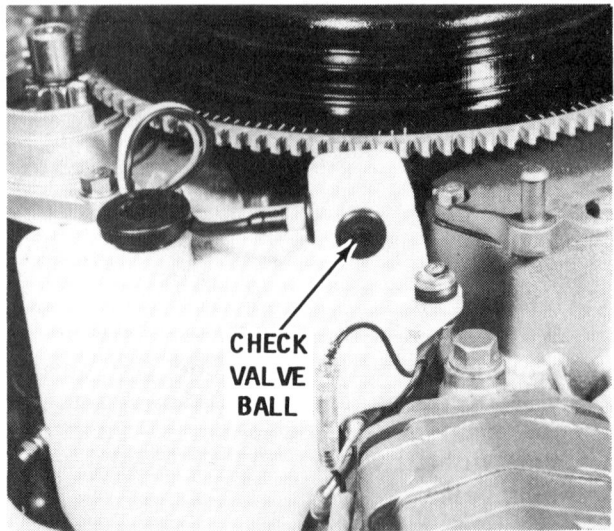

The check valve ball on the main tank breather may be depressed to quicken the flow of oil, while "bleeding" (purging) the system of air.

3- Install and tighten the bleed screw.

4- Pull out the tiny cotter pin from the oil pump shaft. Slide the plastic spacer from the shaft. Pry the oil pump link rod from the ball joint on the pump shaft. Turn the pump lever all the way **CLOCKWISE** to set the pump stroke to maximum.

5- Obtain a small container and have a shop towel handy to catch any oil drips. Using a pair of needlenose pliers, squeeze the clamp on cylinder No. 1 delivery hose. Push the clamp up the hose. Gently pull the hose from the fitting on the intake manifold and immediately hold it over the container.

Observe the flow of oil for five full minutes to verify the pump is functioning correctly.

SPECIAL WORDS

The oil/fuel mixture prepared in Step 1 will provide adequate lubrication while the system is being purged.

A steady slow pulsing flow with no air bubbles may be expected. The manufacturer

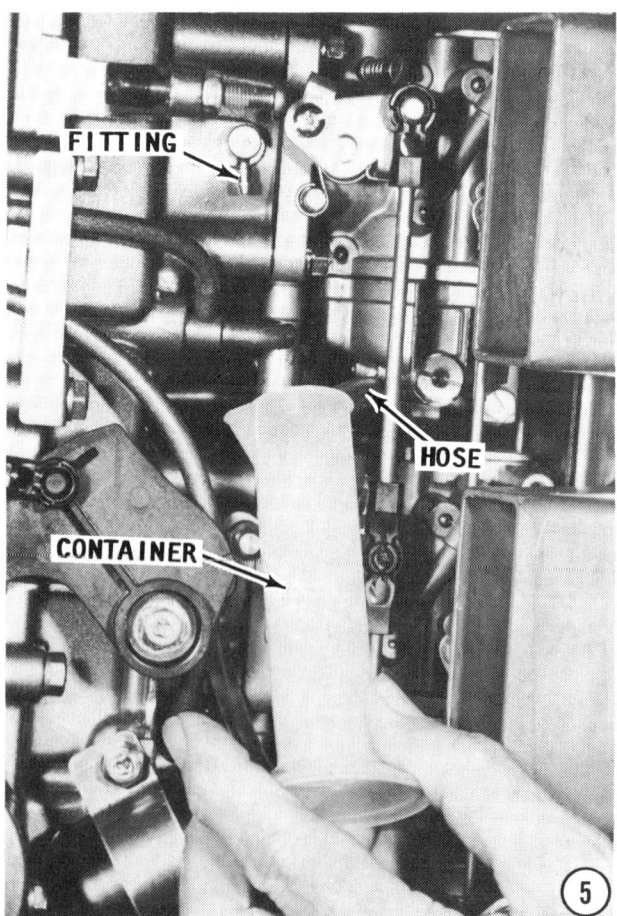

gives no specified amount per minute in this case. Reconnect the line. Shut down the powerhead and remove the flushing device.

6- Snap the oil injection link rod back onto the ball joint on the pump shaft. Slide the plastic spacer onto the shaft and install the tiny cotter pin. Check the adjustment, with the throttle shut. The oil injection pump lever should be at the minimum stroke position. If the length of the rod was accidentally altered, refer to the procedures in Section 5-5 to adjust the rod to the specified length.

Oil Feed Pump

7- Remove the oil filler cap from the remote oil tank. Place some shop cloths around the oil supply line hose joint to catch any oil which may drain during removal. Snip the clamp with a pair of dykes. Squeeze the oil line to restrict the flow of oil while pulling it free of the fitting. Insert the line quickly into the remote tank fill neck.

With the main switch in the **ON** position, but **NOT** the start position, hold up the emergency switch on the control box beneath the oil tank. Keep holding the switch up for about a minute or two to clear the pump of air. Slide a new hose clamp onto the remote tank hose, and then connect the hose to the hose from the main tank. Tighten the hose clamp at the joint.

8- Replenish the remote oil tank with Yamalube, or other quality two-stroke oil.

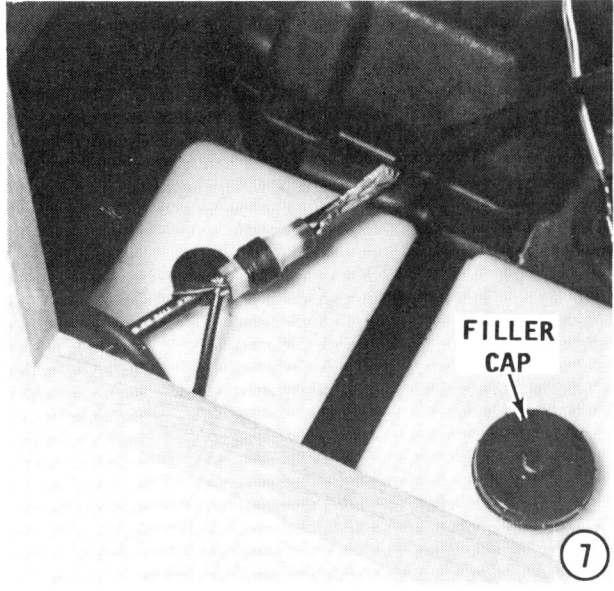

5-5 ADJUSTING OIL INJECTION CONTROL LINK ROD

Correct adjustment of the link rod from the carburetor to the oil injection pump lever is critical. If this rod is out of adjustment, an oil starvation condition could develop leading to premature wear and overheating. Incorrect adjustment of the link

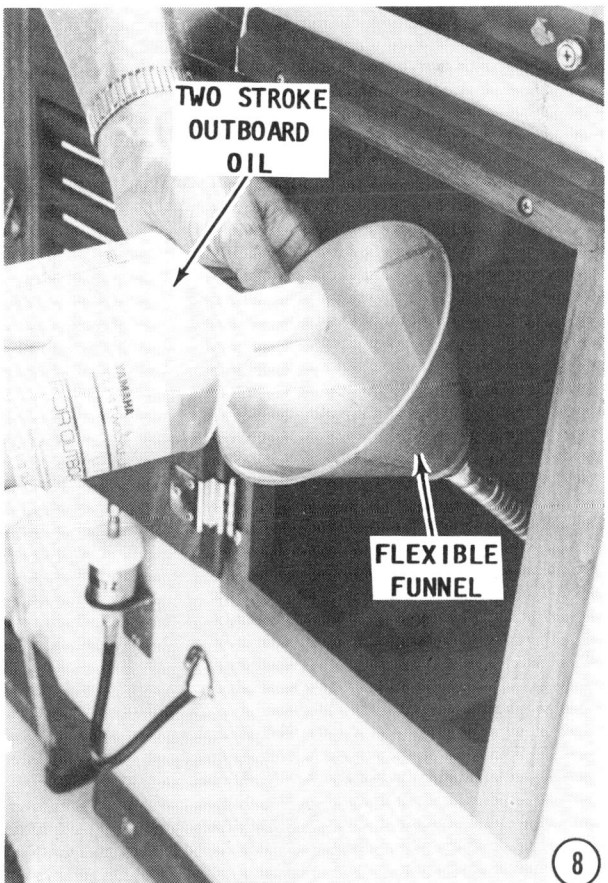

5-26 OIL INJECTION

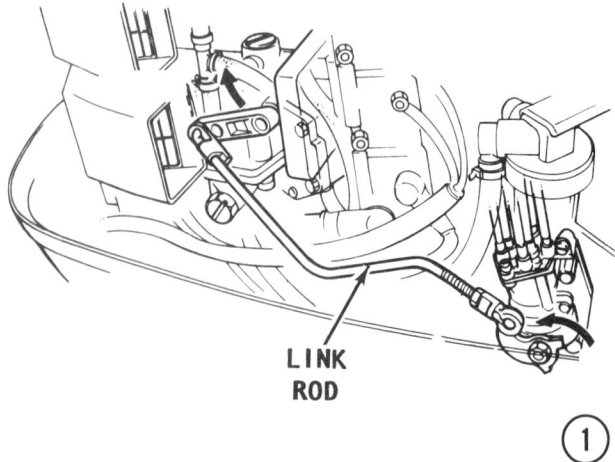

rod could also lead to excessive oil delivery resulting in a smoking powerhead, fouled spark plugs, or erratic performance.

If the oil injection system has been removed to facilitate powerhead removal or for any other reason, it is quite possible the link rod may require adjustment.

1- Close the throttle valves by pushing up on the throttle lever on the starboard side of the lower carburetor.

2- Pull out the tiny cotter pin from the oil pump shaft. Slide the plastic spacer from the shaft. Pry the oil pump link rod from the ball joint on the pump shaft. Turn the pump lever all the way **COUNTER-CLOCKWISE** to set the pump stroke to minimum.

3- Loosen the locknut on the link rod. Adjust the length of the link rod until the

rod can be snapped back onto the ball joint on the oil pump **WITHOUT** any movement at the other end of the rod. Snap the link rod in place and tighten the locknut to hold the adjustment. Install the plastic spacer and the cotter pin.

5-6 OPERATIONAL CHECK OIL INJECTION WARNING SYSTEM

The following procedures check the operation of the buzzer and the Green, Yellow, and Red lights to ensure they come on at the proper time. Correct operation of the buzzer and lights is essential to warn the operator of a low oil level in the tank. This simple procedure should be performed at the start of each season to verify the warning system is functioning correctly as a protection against expensive powerhead service work. Inadequate lubrication can almost destroy a powerhead.

After these tests are completed, the injection system should be purged of air ("bled") -- per the instructions outlined in Section 5-4, this chapter.

CRITICAL WORDS

The following tests are performed with the powerhead **NOT** running, as both oil tanks are to be drained. Running the powerhead with inadequate lubrication will cause damage to internal moving parts and may cause seizure. The battery must be connected for the tests, but the spark plugs should be removed and the high tension leads grounded.

1- Obtain a suitable container with a large enough capacity to hold all the oil in the main tank. Place some shop cloths in the bottom of the lower cowling to catch any oil which may drain during the removal of the oil supply line. Remove the oil supply line leading from the main oil tank to the oil pump beneath the tank by compressing the wire clamp with a pair of needlenose pliers and pushing the clamp up along the oil line. Squeeze the oil line to restrict the flow of oil while pulling it free of the fitting. Allow the tank to completely drain into the container. Install the line back onto the center fitting on the oil pump.

2- Perform Steps 1 and 2 outlined on Page 5-21 to remove the remote tank from the boat. Once the tank is removed from the boat, completely drain the oil from the tank. Perform Steps 3 and 4 outlined on Page 5-22 to install the empty tank into the boat.

Test A

Turn the main switch to the **ON** position. The Red light should come on and the buzzer should sound. The oil feed pump should not be operating.

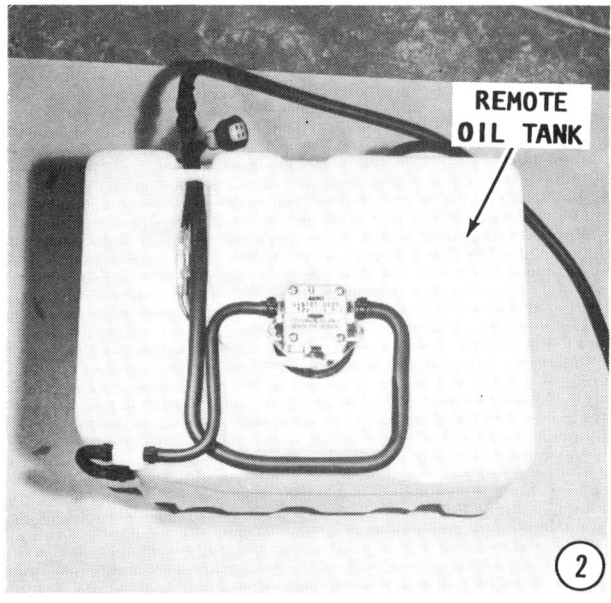

3- Begin to add oil to the remote tank. Fill the tank to a little over 1.6 U.S. quarts (1.5 litres).

Test B

Turn the main switch to the **ON** position. The Green and the Red light should come on. The buzzer should sound and the oil feed pump should operate.

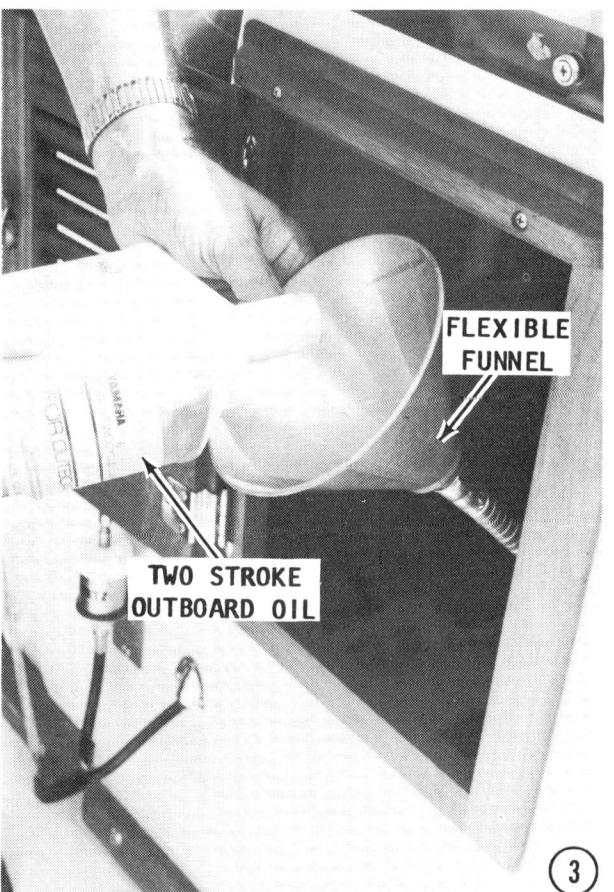

5-28 OIL INJECTION

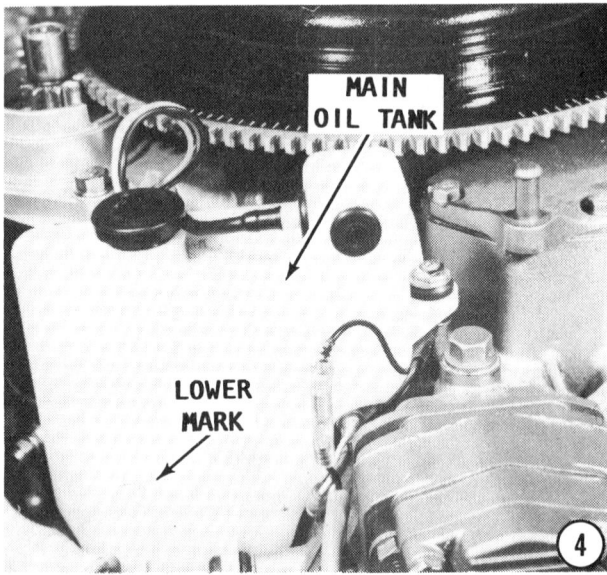

Continue to add oil to the remote tank until a total of just over 5.3 U.S. quarts (5.0 litres) have been added. Repeat Test B and check to see if the oil is being transferred from the remote tank to the main tank.

4- Check the oil level in the main tank. Continue to hold the main switch in the **ON** position until the oil level reaches the lower mark embossed on the exterior of the tank. Release the main switch.

Test C

With the main switch in the **ON** position, the Red light should go out. The Green light should be the only light lit. The buzzer should be silent and the feed pump should not operate. Release the main switch.

5- Remove the remote tank oil level sensor.

Test D

Turn the main switch to the **ON** position. Only the Yellow light should come on. The buzzer should not sound and the feed pump should not operate.

Hold the emergency switch in the **UP** position. The oil feed pump should operate. Release the emergency switch and the main switch.

Install the remote oil tank sensor.

6- Turn the main switch to the **ON** position. Tilt the outboard to 35° from the vertical position.

Test E

With the main switch in the **ON** position, hold the emergency switch **UP**. The oil feed pump should not operate, if the tilt switch in the control box is functioning correctly. No lights should come on and the buzzer should not sound.

Bring the outboard down to less than 20° from the vertical position.

Test F

With the main switch in the **ON** position, hold the emergency switch **UP**. The oil feed

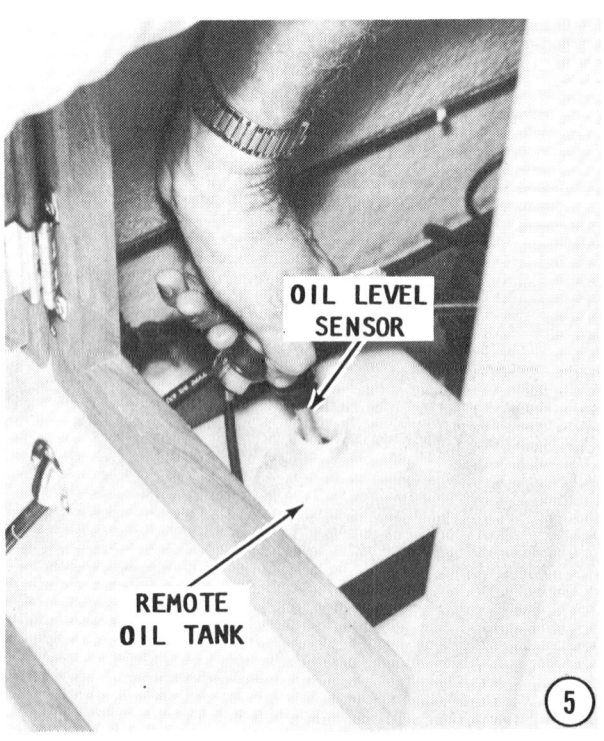

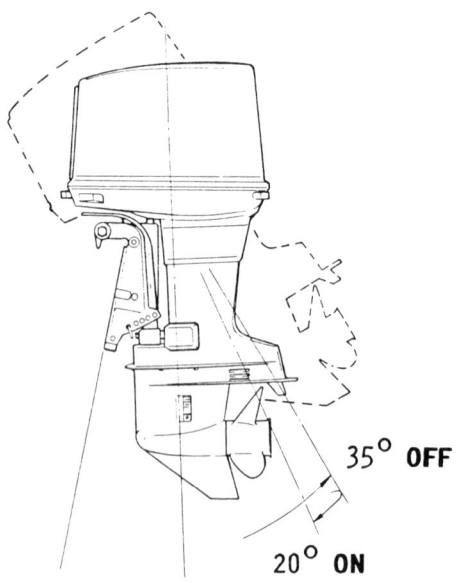

TESTING WARNING SYSTEM 5-29

pump should now operate, if the tilt switch in the control box is functioning correctly. No lights should come on and the buzzer should not sound.

7- Return the outboard to the vertical position, and continue to keep the main switch in the **ON** position while the feed pump transfers oil to the main tank. Continue holding the switch until the oil level reaches the upper mark embossed on the exterior of the tank.

Test G

When the oil level in the main tank reaches the upper mark, the feed pump should stop. No lights should come on and the buzzer should not sound.

8- Lift out the main tank oil level sensor half way.

CRITICAL WORDS

Perform the next test quickly, as the feed pump will be pumping oil into a full tank and may overflow.

Test H

With the main switch in the **ON** position, the feed pump should pump oil up to the

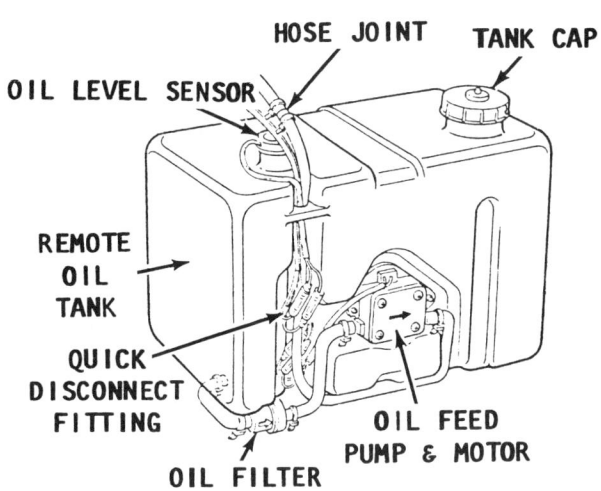
Cutaway drawing of the remote oil tank to depict associated parts in and on the tank.

main tank. No lights should come on and the buzzer should not sound.

Install the main tank oil level sensor

SPECIAL WORDS

After the foregoing tests have been completed, the operation of the warning lights, both oil level sensors, and the oil feed pump have been checked. If the system failed to give the expected results during any of the tests, refer to Section 5-2 Troubleshooting for further testing of each electrical component.

CAUTION

During these tests, the remote oil tank was completely drained allowing air to enter the feed pump. Therefore, the feed pump **MUST** be purged ("bled") of any trapped air. Failure to "bleed" the system could lead to powerhead seizure due to lack of adequate lubrication. Therefore, "bleed" the system per the instructions outlined in Section 5-4.

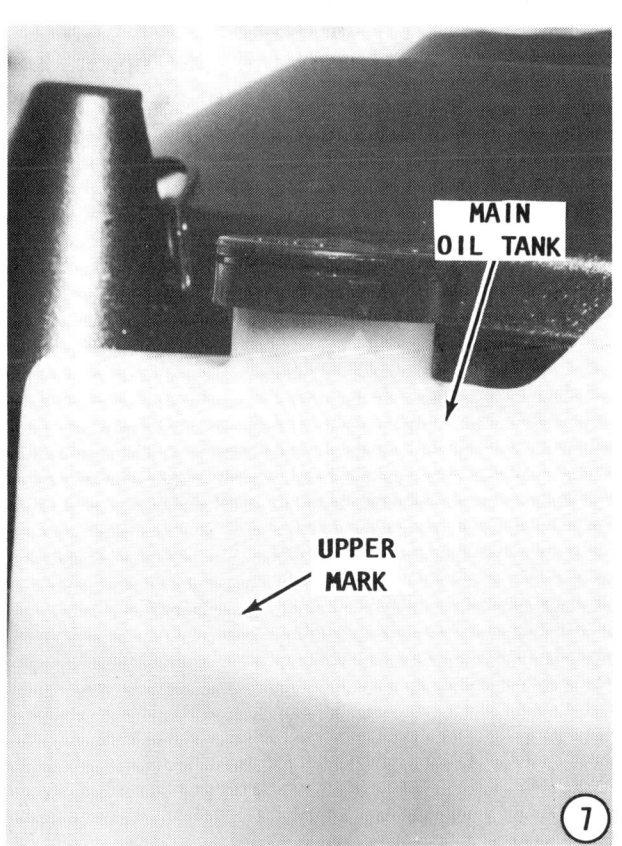

5-30 OIL INJECTION

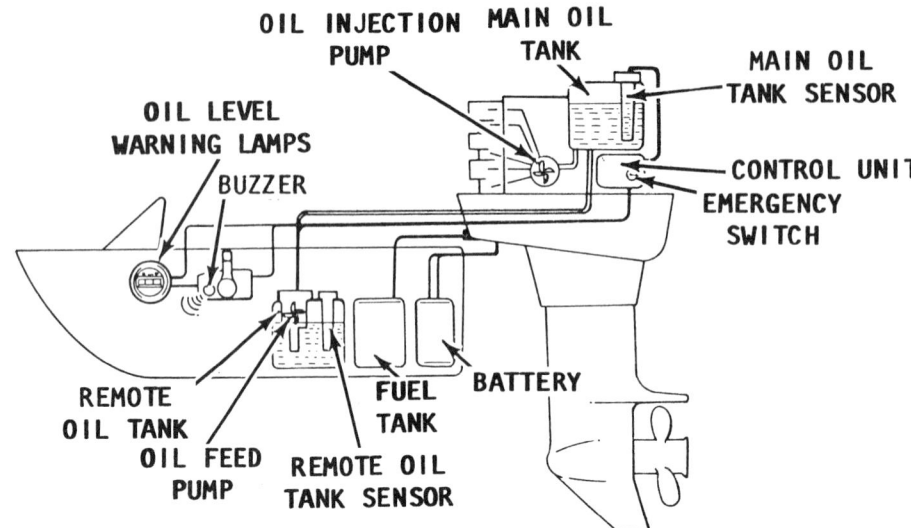

Simplified functional diagram to depict the relative location of components in a typical oil injection system.

RUNAWAY POWERHEAD

Never operate the powerhead above a fast idle with a flush attachment connected to the lower unit. Operating the powerhead at a high rpm with no load on the propeller shaft could cause the powerhead to **RUNAWAY** causing extensive damage to the unit.

REMEMBER, the powerhead will **NOT** start without the emergency tether in place behind the "kill" switch knob.

Start the powerhead and allow it to warm to operating temperature.

CAUTION

Water must circulate through the lower unit to the powerhead anytime the powerhead is operating to prevent damage to the water pump in the lower unit. Just five seconds without water will damage the water pump impeller.

Check the completed work. If the system is functioning properly, only the Green light should be on and the buzzer should not sound.

Check for oil leaks around all hose connections which were disturbed. Replenish both tanks with Yamalube two stroke outboard oil (or an equivalent oil having a certified BIA rating of TCW. This oil is suitable for use through a temperature range of $14°F$ to $140°F$ ($-10°C$ to $60°C$).

Fill the main tank to the upper line embossed on the tank exterior. Fill the remote tank to a point where the motion of the boat will not cause the tank to over flow. Check the water/dust trap on the bottom of the main tank. This check applies only to models 1988 and on.

Check the oil filter at the remote tank periodically and replace, as necessary.

6
IGNITION

6-1 INTRODUCTION AND CHAPTER COVERAGE

The less an outboard engine is operated, the more care it needs. Allowing an outboard engine to remain idle will do more harm than if it is used regularly. To maintain the engine in top shape and ready for efficient operation at any time, the engine should be operated every three to four weeks throughout the year.

The carburetion and ignition principles of two-cycle engine operation **MUST** be understood in order to perform a proper tune-up on an outboard motor.

If you have any doubts concerning your understanding of two-cycle engine operation, it would be best to study the operation theory section in the first portion of Chapter 8, before tackling any work on the ignition system.

One ignition system, a capacitor discharge ignition (CDI) is used on Yamaha V4 and V6 powerheads. Basically, this system contains an ignition coil and spark plug for each cylinder, plus a charge coil, plusar coil, and CDI control box.

In addition to these components, the Models 220hp and the 225hp units also have a microcomputer connected into the CDI unit. This additional component is termed, Yamaha Microcomputer Ignition System, and is commonly abbreviated: YMIS The YMIS microcomputer advises the CDI unit of ignition settings and provides the powerhead with optimum ignition timing under all operational conditions and loads.

The YMIS is covered in detail in Section 6-12 of this chapter.

6-2 SPARK PLUG EVALUATION

Removal

Remove the spark plug wires by pulling and twisting on only the molded cap. NEVER pull on the wire or the connection inside the cap may become separated or the boot damaged. Remove the spark plugs and keep them in order. **TAKE CARE** not to tilt the socket as you remove the plug or the insulator may be cracked.

When servicing the spark plug cap on the Model 220hp and 225hp units equipped with YMIS, the cap **MUST** be rotated counterclockwise without pulling to remove the cap from the high tension lead and rotated clockwise to install the cap to the high tension lead. The reason for this movement is because the end of the lead is threaded and the mating end of the cap has a threaded screw type connector, as shown in the accompanying illustration.

Examine

Line the plugs in order of removal and carefully examine each to determine the

The carburetion and ignition system MUST be properly adjusted and synchronized for optimum powerhead performance.

6-2 IGNITION

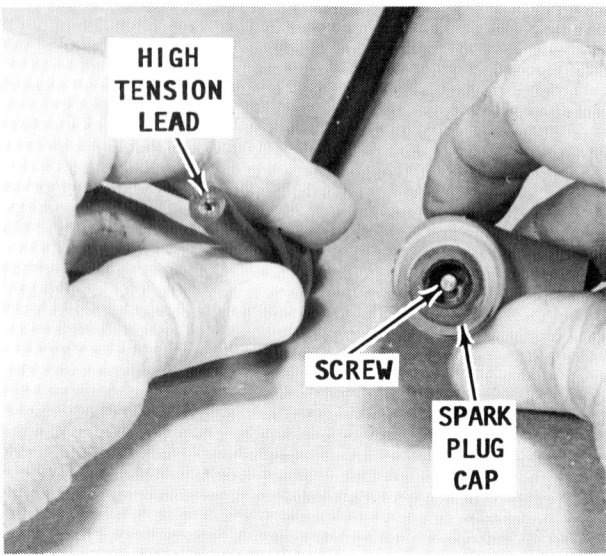

The end of the high tension lead is threaded onto the screw on the spark plug cap.

firing conditions in each cylinder. If the side electrode is bent down onto the center electrode, the piston is traveling too far upward in the cylinder and striking the spark plug. Such damage indicates the piston pin or the rod bearing is worn excessively. In most cases, an engine overhaul is required to correct the condition.

To verify the cause of the problem, rotate the flywheel by hand, using a wrench on the flywheel nut. As the piston moves to the full up position, push on the piston crown with a screwdriver inserted through the spark plug opening and at the same time rock the flywheel back and forth. If any "play" in the piston is detected, the powerhead must be rebuilt.

Correct Color

A proper firing plug should be dry and powdery. Hard deposits inside the shell indicate too much oil is being mixed with the fuel. The most important evidence is the light gray to tan color of the porcelain, which is an indication this plug has been running at the correct temperature. This means the plug is one with the correct heat range and also that the air-fuel mixture is correct.

Rich Mixture

A black, sooty condition on both the spark plug shell and the porcelain is caused by an excessively rich air-fuel mixture, both at low and high speeds. The rich mixture lowers the combustion temperature so the spark plug does not run hot enough to burn off the deposits.

Deposits formed only on the shell is an indication the low-speed air-fuel mixture is too rich. At high speeds with the correct mixture, the temperature in the combustion chamber is high enough to burn off the deposits on the insulator.

Too Cool

A dark insulator, with very few deposits, indicates the plug is running too cool. This condition can be caused by low compression or by using a spark plug of an incorrect heat range. If this condition shows on only one plug it is most usually caused by low compression in that cylinder. If all of the plugs have this appearance, then it is probably due to the plugs having a too low heat range.

Always pull and twist on the high tension cap, not on the lead. Pulling on the lead may separate the connection inside the cap.

This spark plug is foul from operating with an overrich air/fuel mixture, possibly caused by an improper carburetor adjustment.

Fouled

A fouled spark plug may be caused by the wet oily deposits on the insulator shorting the high tension current to ground inside the shell. The condition may also be caused by ignition problems which prevent a high-tension pulse from being delivered to the spark plug.

Carbon Deposits

Heavy carbon-like deposits are an indication of excessive oil in the fuel. This condition may be the result of worn piston rings or excessive ring end gap.

Overheating

A dead white or gray insulator, which is generally blistered, is an indication of overheating and pre-ignition. The electrode gap wear rate will be more than normal and in the case of pre-ignition, will actually cause the electrodes to melt. Overheating and pre-ignition are usually caused by overadvanced timing, detonation from using too low an octane rating fuel, an excessively lean air-fuel mixture, or problems in the cooling system.

Electrode Wear

Electrode wear results in a wide gap and if the electrode becomes carbonized it will form a high resistance path for the spark to jump across. Such a condition will cause the engine to misfire during acceleration. If all of the plugs are in this condition, it can cause an increase in fuel consumption and very poor performance at high speed operation. The solution is to replace the spark plugs with a rating in the proper heat range and gapped to specification.

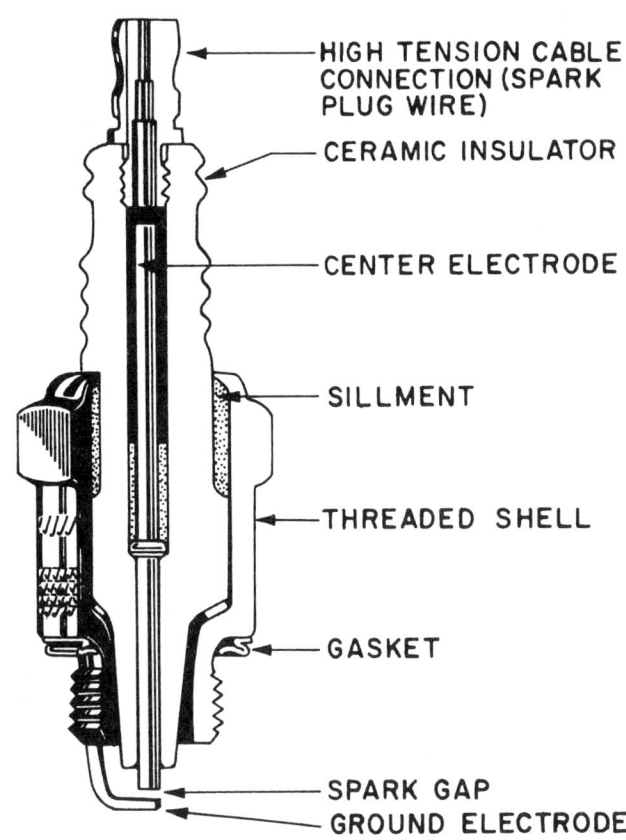

Cutaway drawing of a typical spark plug with principle parts identified.

Red rust-colored deposits on the entire firing end of a spark plug can be caused by water in the cylinder combustion chamber. This can be the first evidence of water entering the cylinders through the exhaust manifold because of an accumulation of scale. This condition **MUST** be corrected at the first opportunity. Refer to Chapter 8, Powerhead.

This spark plug has been operating too-cold, because it is rated with a too-low heat range for the powerhead.

*Damaged spark plugs. Notice the broken electrode on the left plug. The missing part **MUST** be found and removed before returning the powerhead to service, to prevent serious damage to expensive internal parts.*

6-4 IGNITION

A fairly accurate determination of the firing condition of each cylinder can be made by careful examination of the spark plugs.

SPECIAL WORDS
SPARK PLUG GAP

The spark plug gap is a specification given by the manufacturer. The spark plug gap is the same, 0.039" (1.000mm), for all powerheads covered in this manual.

Under normal operating conditions, this gap will increase by 0.004" (0.1mm), due to electrical errosion and chemical corrosion with each ten hours of powerhead operation. A new spark plug with an electrode gap of 0.039" needs approximately 15,000 volts to produce adequate spark. Each additional 0.004" (0.1mm) increase in gap (air space) requires an additional 2,000 volts.

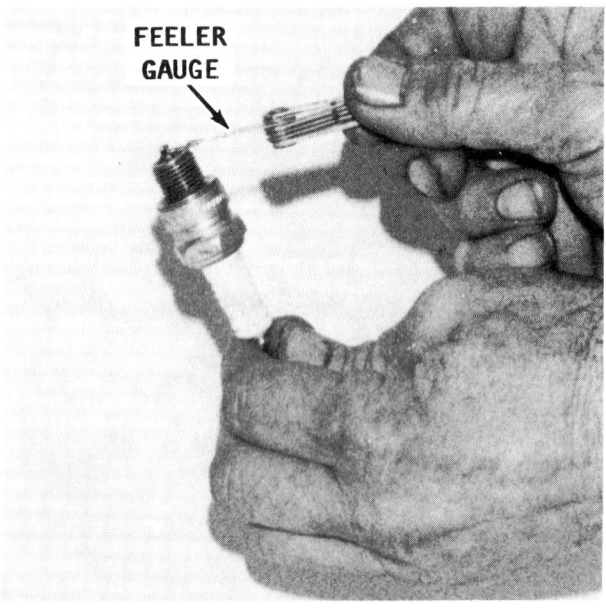

The spark plug gap should always be checked with a wire-type feeler gauge before installing new or used plugs.

An excessive gap can soon place an unnecessary strain on the ignition coil and thus shorten its useful life.

Because each combustion chamber of a two-stroke powerhead is fired at double the frequency of a four-stroke unit, the life expectancy of an outboard spark plug is approximately one-half the life of an automotive plug. Actually, spark plug life is shortened even further because a two-stroke plug operates at a much higher temperature than an automotive plug. The automotive plug has a microsecond chance to cool because of the intake and exhaust valve arrangement. This cooling chance is not present in a two-stroke unit.

6-3 POLARITY CHECK

Coil polarity is extremely important for proper battery ignition system operation. If a coil is connected with reverse polarity, the spark plugs may demand from 30 to 40 percent more voltage to fire, or on most CDI (Capacitor Discharge Ignition) systems, there will be **NO** spark. Under such demand conditions, in a very short time the coil would be unable to supply enough voltage to fire the plugs. Any one of the following three methods may be used to quickly determine coil polarity.

1- The polarity of the coil can be checked using an ordinary D.C. voltmeter set on the maximum scale. Connect the positive lead to a good ground. With the engine running, momentarily touch the negative lead to a spark plug terminal. The needle should swing upscale. If the needle swings downscale, the polarity is reversed.

2- If a voltmeter is not available, a pencil may be used in the following manner: Disconnect a spark plug wire and hold the metal connector at the end of the cable about 1/4" (6.35mm) from the spark plug terminal. Now, insert an ordinary pencil tip

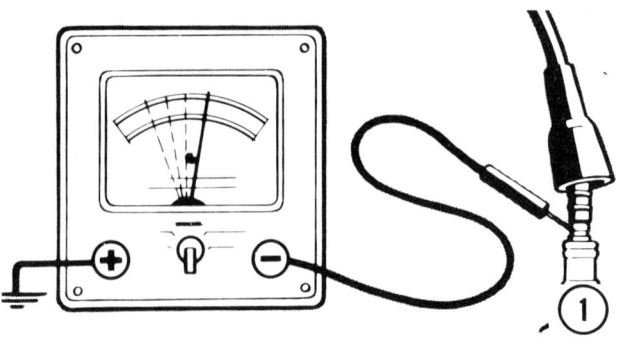

CAPACITOR DISCHARGE 6-5

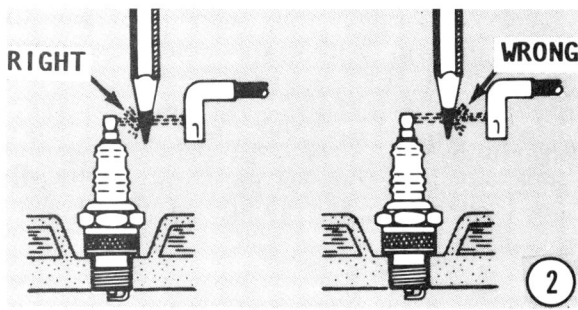

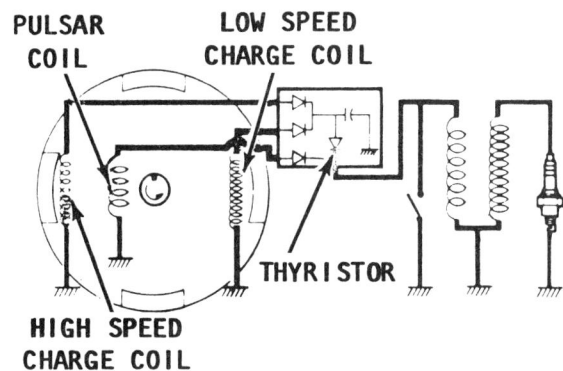

A simplified circuit diagram of the CDI (Capacitor Discharge Ignition) system.

between the terminal and the connector. Crank the engine with the ignition switch ON. If the spark feathers on the plug side and has a slight orange tinge, the polarity is correct. If the spark feathers on the cable connector side, the polarity is reversed.

3- The firing end of a used spark plug can give a clue to coil polarity. If the ground electrode is "dished", it may mean polarity is reversed.

6-4 CAPACITOR DISCHARGE IGNITION

The CDI (capacitor discharge ignition) is used on all V4 and V6 powerheads covered in this manual. The system utilizes a mechanical advance on all but the Model 220hp and 225hp. These two units are electronically advanced by the microcomputer (YMIS). The microcomputer receives signals from sensors positioned at strategic locations on the powerhead and then instantly makes a correct timing determination based on operational conditions and powerhead loads.

DESCRIPTION AND OPERATION

The V4 model has four magnets housed under the flywheel and the V6 model has six such magnets.

The following components are mounted on the stator plate: Two pulsar coils for the V4 models, or three pulsar coils for the V6 models; two charge coils; and one lighting coil. As the flywheel magnet passes a coil, a voltage is induced in the coil. As the flywheel continues to rotate and the coil is no longer influenced by the magnet, the magnetic field collapses. Therefore, an alternating voltage is produced at the coil. This AC voltage is changed to DC by a diode inside the CDI unit.

A diode is a solid state unit which permits voltage to flow in one direction but

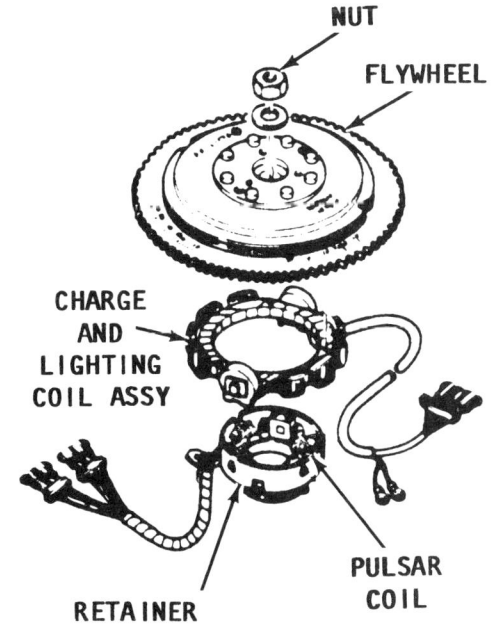

Exploded drawing of the components housed under the flywheel of a CDI system.

prevents flow in the opposite direction. A diode may also be known as a rectifier.

After voltage passes through the diode, the current then passes to a capacitor, also located inside the CDI unit, where it is stored. The unit stores the DC current, then releases it and automatically controls the timing.

Pulsar Coil Circuit

On the V4 models, the two pulsar coils are mounted to produce a pulse every $90°$ revolution of the crankshaft. Therefore four pulses are produced each crankshaft revolution.

On the V6 models the three pulsar coils are evenly spaced at $120°$ around the stator plate. The combination of these three pulsars will produce a pulse every $60°$ revolution of the crankshaft. Therefore six pulses are produced each crankshaft revolution.

Charge Coil Circuit

Both the V4 and V6 models have two charge coils, one for high speed operation and one for low speed operation. A voltage is induced in the windings of each coil. The magnitude of the voltage depends upon the number of windings (turns) in each coil and the engine rpm.

The voltages generated by these two charge coils are limited by a voltage regulator and stored separately in two condensers. The condensers are charged in stages, each time a magnet passes a charge coil.

Thyristor Circuits

The V4 models have four thyristors and the V6 models have six thyristors in the CDI circuit. A thyristor is a solid state electronic switching device which permits voltage to flow only after it is triggered by another voltage source. When energized by a voltage from one of the pulsar coils, the thyristor allows the discharge of voltage from the condensers. This voltage, which was originally generated by the charge coils, is now passed on to one of the ignition coils, where it is stepped up to operating voltage.

Secondary Circuits

Ignition coils, one per cylinder, boost the DC voltage instantly to approximately 20,000 volts to the spark. This completes the primary side of the ignition circuit.

Once the voltage is discharged from the ignition coil the secondary circuit begins and only stretches from the ignition coil to the spark plugs via extremely large high tension leads. At the spark plug end, the voltage arcs in the form of a spark, across from the center electrode to the outer electrode, and then to ground via the spark plug threads. This completes the ignition circuit.

Timing Basics

At the point in time when the ignition timing marks align, an alternating current is induced in the pulsar coil, in the same manner as previously described for the charge coil. This current is then passed to a second diode located in the CDI unit where it becomes DC current and flows on to the thyristor. This voltage triggers the thyristor to permit the voltage stored in the capacitor to be discharged. The capacitor voltage passes through the thyristor and on to the primary windings of the ignition coil.

In this manner, a spark at the plug may be accurately timed by the timing marks on the flywheel relative to the magnets in the flywheel and to provide as many as 100 sparks per second for a powerhead operating at 6000 rpm.

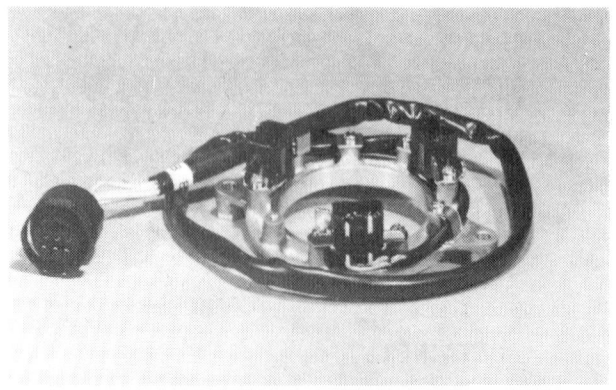

The pulsar coil removed from a V6 powerhead for bench testing.

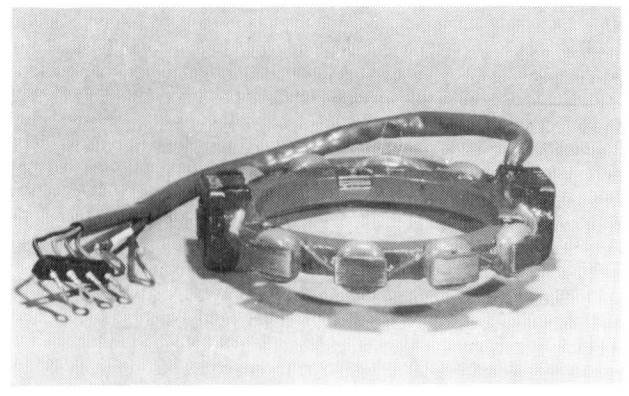

The charge and lighting coil removed from a V6 powerhead for bench testing.

Timing Advance

As stated earlier in this section, all units covered in this manual except those with YMIS are equipped with a mechanical advance type CDI system and use a series of link rods between the carburetor and the ignition base plate assembly. When the throttle is opened, the ignition base plate assembly is rotated by means of the link rod, thus advancing the timing.

On models equipped with YMIS, the microcomputer "decides" when to advance or retard the timing, based on input from various sensors. Therefore there is no link rod between the magneto control lever and the stator assembly.

Charging System

The charging system consists of the flywheel magnets, a lighting coil (alternator), a rectifier/voltage regulator, and the battery.

When the lighting coil, so named because it may be used to power the lights on the boat, is used in conjunction with a voltage regulator, the coil allows the powerhead to generate additional electrical current to charge the battery.

As the flywheel magnets rotate, voltage is induced in the lighting coil located next to the pulsar coils. This alternating current passes through a series of diodes and emerges as DC current. Therefore, it may be stored in the battery. A lighting coil may be identified by its clean laminated copper windings. All other coils are wrapped in tape and their windings are not visible.

However, lighting coil and the charging coils are an integral part of the stator assembly. Therefore, if any one coil is determined to be defective, the entire stator assembly **MUST** be replaced.

The rectifier/regulator converts the alternating current into DC, which may then be stored in the battery. The rectifier/regulator is a sealed unit and contains four diodes. If one of the diodes is defective, the entire unit **MUST** be replaced.

The voltage regulator part of the rectifier/regulator stabilizes the power output of the coils and extends the life of electrical appliances by preventing power surges.

6-5 TROUBLESHOOTING CDI SYSTEM AND CHARGING SYSTEM

Always attempt to proceed with the troubleshooting in an orderly manner. The "shot in the dark" approach will only result in wasted time, incorrect diagnosis, unnecessary replacement of parts, and frustration.

Begin the ignition system troubleshooting with the spark plug and continue through the system until the source of trouble is located.

Spark Plugs

1- Check the plug wires to be sure they are properly connected. Check the entire length of the wires from the plugs to the magneto under the stator plate. If the wires are to be removed from the spark plug, **ALWAYS** use a pulling and twisting motion as a precaution against damaging the connection.

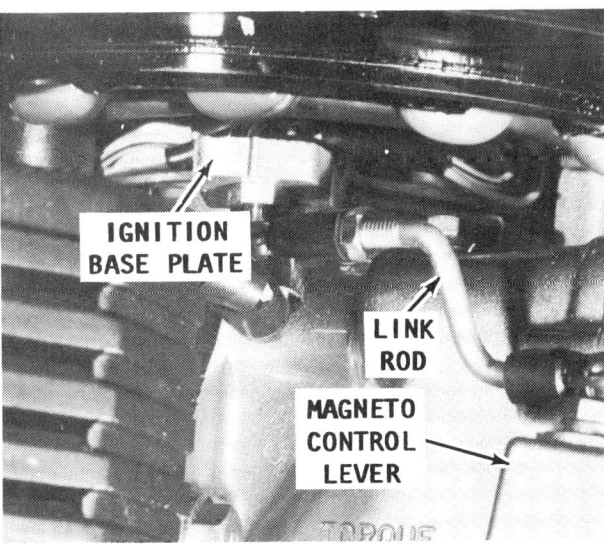

On all models NOT equipped with YMIS (Yamaha Microcomputer Ignition System), a link rod between the magneto control lever and the ignition base plate advances and retards the timing.

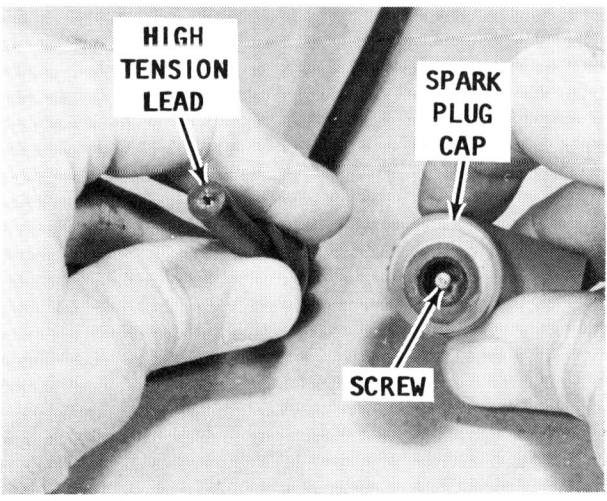

The end of the high tension lead is threaded onto the screw on the spark plug cap.

6-8 IGNTION

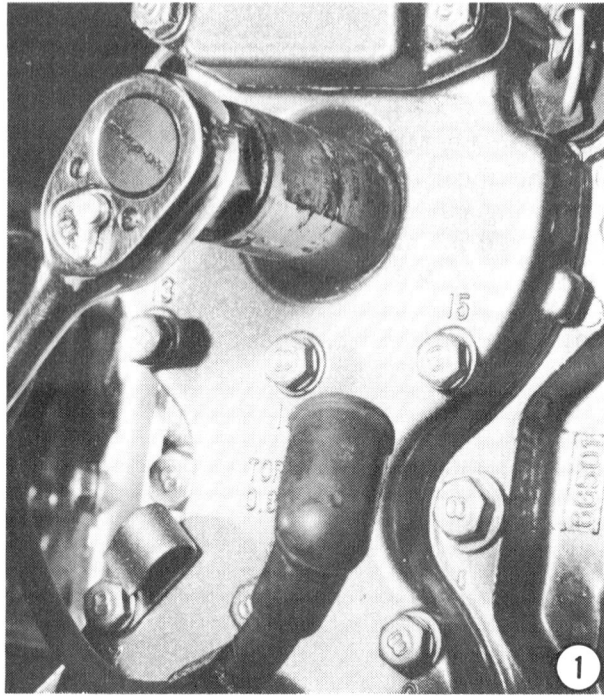

When servicing the spark plug cap on the Model 220hp and 225hp units equipped with YMIS, the cap **MUST** be rotated counterclockwise without pulling to remove the cap from the high tension lead and rotated clockwise to install the cap to the high tension lead. The reason for this movement is because the end of the lead is threaded and the mating end of the cap has a threaded screw type connector, as shown in the accompanying illustration.

Attempt to remove the spark plug by hand. This is a rough test to determine if the plug is tightened properly. The attempt to loosen the plug by hand should fail. The plug should be tight and require the proper socket size tool. Remove the spark plug and evaluate its condition as described in Section 6-2.

2- Use a spark tester and check for spark. If a spark tester is not available, hold the plug wire about 1/4" (6.4mm) from the powerhead. Rotate the flywheel with the electric cranking motor and check for spark. A strong spark over a wide gap must be observed when testing in this manner, because under compression, a strong spark is necessary in order to ignite the air-fuel mixture in the cylinder. This means it is possible to think a strong spark is present, when in reality the spark will be too weak when the plug is installed. If there is no spark, or if the spark is weak, the trouble is most likely under the flywheel in the magneto.

Compression

3- Before spending too much time and money attempting to trace a problem to the ignition system, a compression check of each cylinder should be made. If a cylinder does not have adequate compression, troub-

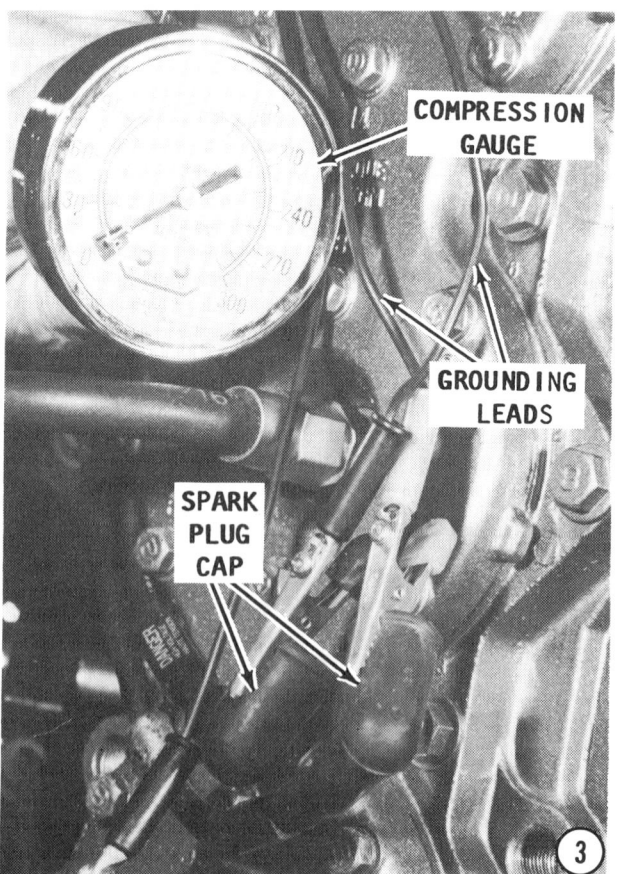

leshooting and attempted service of the ignition or fuel system will fail to give the desired results of satisfactory engine performance.

Remove the spark plug wire by pulling and twisting **ONLY** on the molded cap. **NEVER** pull on the wire because the connection inside the cap may be separated or the boot may be damaged. Remove the spark plug. Insert a compression gauge into the cylinder spark plug opening. Ground the spark plug leads to prevent damage to the ignition coil. If a lead is not grounded, the coil will attempt to match the demand created by the spark trying to jump from the electrode to the nearest ground.

Crank the powerhead through several crankshaft revolutions with the electric cranking motor. Note the compression reading.

An acceptable pressure reading for a powerhead covered in this manual is 110 psi (760 kPa) or more.

SPECIAL WORDS
V6 MODELS

In recent years the manufacturer has modified the cylinder head design on some models in an attempt to keep the temperature of each head as equal as possible. This action has changed the shape of the combustion chamber, and therefore the volume and compression pressure of each cylinder.

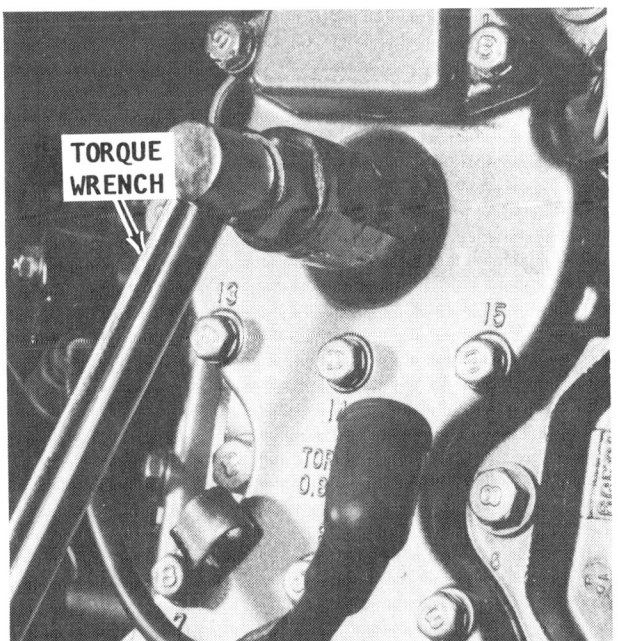

A torque wrench should always be used to tighten the spark plugs to the required torque value of 18 ft lbs (25 Nm).

As a general rule, the pressure between pairs of cylinders which share the same crankshaft throw, should be approximately the same. Cylinder No. 1 should be the same as cylinder No. 2; cylinder No. 3 should be the same as cylinder No. 4 and so on.

Normally, on a V6 powerhead, cylinder No. 1 and No. 2 will have the highest compression pressure and cylinders No. 5 and No. 6 will have the lowest compression pressure. The design modification has brought about one exception to this rule as follows for the 175hp and 200hp in 1986 only: cylinders No. 3 and No. 4 will have an even higher compression pressure than cylinders No. 1 and No. 2; cylinders No. 5 and No. 6 will still have the lowest compression pressure.

TESTING CDI COMPONENTS
GENERAL INFORMATION

Due to the complex nature of the circuitry and high energy output pulses of microsecond duration, conventional testing devices such as a Volt/Ohm/Ammeter will not measure electrical output with the degree of accuracy required. A Yamaha Ignition Tester, as used by the local Yamaha dealer is an electrical device capable of measuring the peak energy output of capacitor discharge ignition units, magneto charge and the pulsar coils.

This instrument was designed to troubleshoot the ignition system. Unfortunately this tester may not be easily accessible. Therefore, this chapter includes only those tests which may be performed with instruments normally and easily available.

If a component must be replaced, the flywheel must first be "pulled", see Section 6-7, this chapter.

WORDS FROM EXPERIENCE

During the tests, if a reading is slightly different from the specifications, but the powerhead still operates, then there is no real need to replace the affected component, until it actually fails. Bear in mind, in **MOST** cases electrical components are not returnable, once the item leaves the store. Therefore, make every attempt to avoid the **BUY** and **TRY** method of troubleshooting. Such a practice will lead only to wasted time, unnecessary cash outlay, and frustration.

6-10 IGNITION

6-6 TESTING CDI COMPONENTS ON All MODELS INCLUDING YMIS

PARTS REMAIN INSTALLED ON POWERHEAD

The manufacturer suggests the following tests be performed as close to 68°F (20°C) as possible, unless otherwise stated. If the ambient temperature is significantly higher, then slightly higher resistance values may be acceptable. If the temperature is significantly lower, then slightly lower resistance values may be acceptable. Both leads are disconnected at the battery terminals while the following tests are conducted.

Pulsar Coil Test

Models not equipped with YMIS: Identify the wire harness lead from the stator to the CDI unit.

Models equipped with the YMIS: Identify the wire harness lead from the stator to the microcomputer.

In all cases, with or without YMIS, this harness will consist of the following leads:

V4 Models:
White/Red
White/Yellow
White/Black
White/Green

V6 Models:
White/Red
White/Yellow
White/Black
White/Green
White/Blue
White/Brown

1- Disconnect the two halves of the connector. Obtain an ohmmeter and select the Rx1000 ohm scale. Measure the resistance between the female leads inside the connector half from the pulsar coil. Compare the test results with those given below for the model being serviced.

V4 Models:
280 to 430 ohms between the following terminals:
White/Red and White/Yellow
White/Black and White/Green

V6 Models:
280 and 430 ohms between the following terminals:
White/Red and White/Green
White/Black and White/Blue
White Yellow and White/Brown

If the resistance is not within the limits given and the powerhead does not operate, the pulsar coils must be replaced.

To replace the pulsar coils, the flywheel must be "pulled", see Section 6-7.

If the tests are successful, connect the two halves of the connector together. The two halves will only fit one way.

Charge Coil Test
All Models

2- Remove the bolts and then the cover over the CDI unit and rectifier/regulator.

3- Disconnect the following leads from their terminals on the CDI unit: Brown, Red, Blue, and Black/Red. These leads can be moved away from the edge of the unit but will still be retained in a small retaining block.

4- Select the Rx1000 scale on the ohmmeter. The resistance between the Brown and Red leads should be between 840 and 1260 ohms.

The resistance between the Blue and Black/Red leads should be between 100 and 150 ohms.

The charge and lighting coils are manufactured as an assembly. Therefore, if either coil is defective, the entire assembly must be replaced.

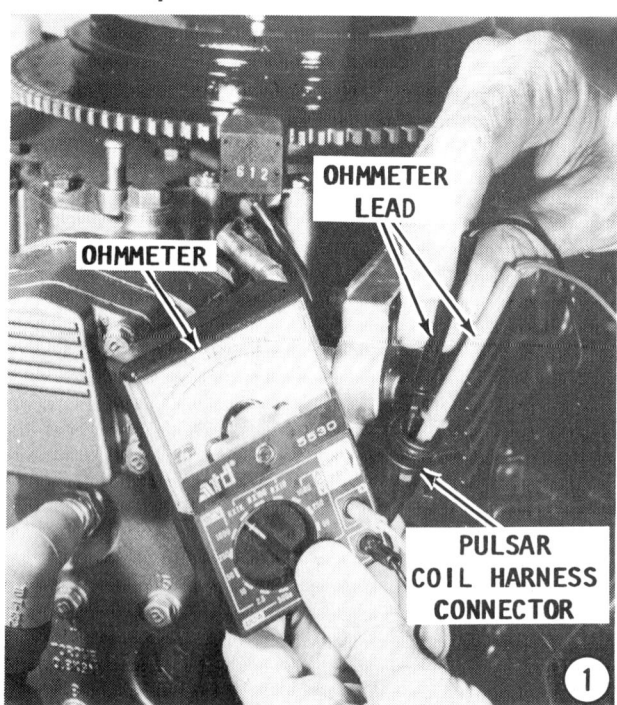

TESTING CDI INCL. YMIS 6-11

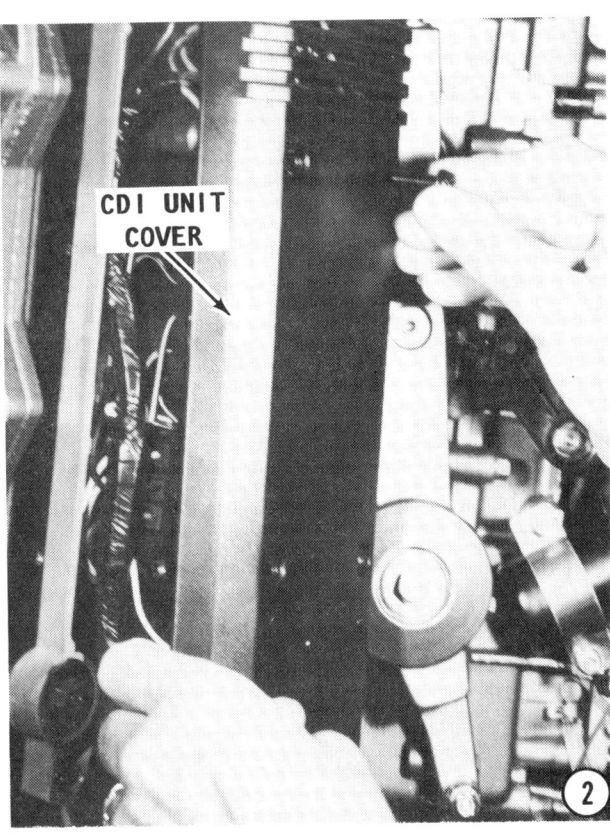

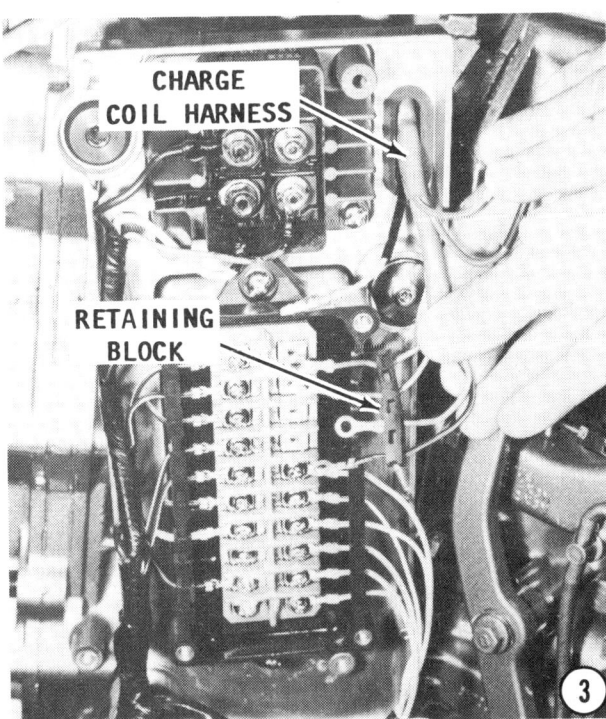

If the resistance is not within the limits listed and the powerhead does not operate, the charge coils must be replaced.

To replace the charge coils, the flywheel must be "pulled", see Section 6-7, this chapter.

If the tests are successful, connect the leads back to their original terminals on the

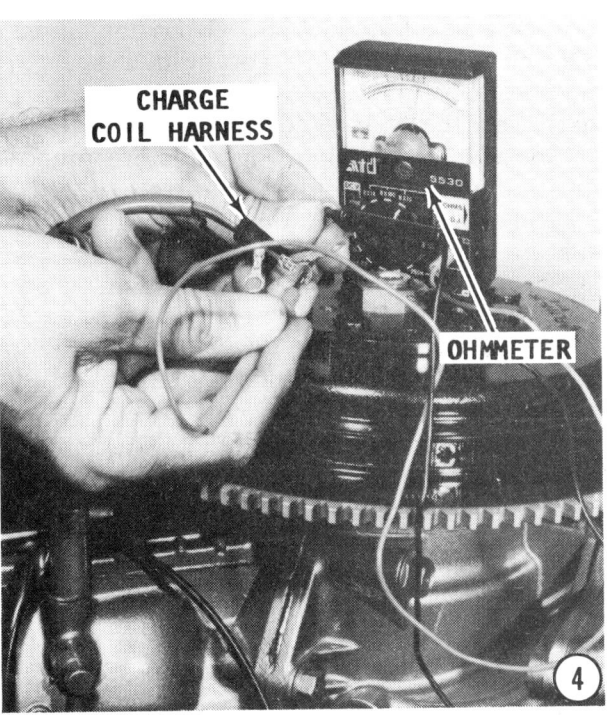

CDI unit using the appropriate chart on Page 6-14, 6-15, or 6-16, as a guide.

Lighting Coil Test
All Models

5- Select the Rx1 scale on the ohmmeter. Disconnect the two Green leads between the stator and the rectifier at the rectifier. Measure the resistance between the two Green leads. The meter should register between 0.50 and 0.75 ohms.

Reconnect the two leads to the rectifier (either lead to either terminal).

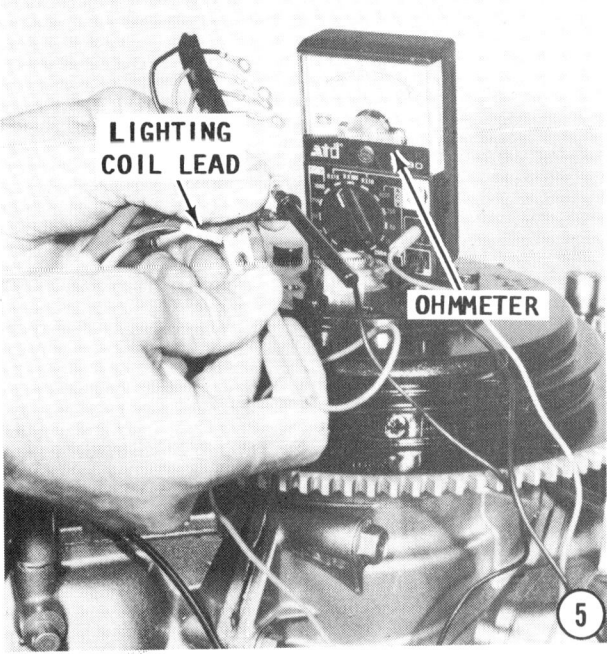

6-12 IGNITION

If the resistance is not within the limits listed; the battery will not hold a charge; and the boat accessories depend on this coil for power; the charge coil must be replaced.

The charge and lighting coils are manufactured as an assembly. Therefore if either coil is defective, the entire assembly must be replaced.

CRITICAL WORDS

Never attempt to verify the charging circuit by operating the powerhead with the battery disconnected. Such action would force current (normally directed to charge the battery), back through the rectifier and damage the diodes in the rectifier.

To replace the charge coil, the flywheel must be "pulled", see Section 6-7, this chapter.

Rectifier/Regulator Testing

6- Obtain an ohmmeter. Select the Rx100 scale. Identify one Green lead as No. 1 Green and the other as No. 2 Green. Make contact with the Black meter lead to No. 1 Green lead. Keep this lead in place for the next three resistance tests.

Make contact with the Red meter lead to the No. 2 Green lead. The meter should indicate no continuity. Next, make contact with the Red meter lead to the Black rectifier lead. Again, the meter should indicate no continuity. Now, move the Red meter lead to make contact with the Red rectifier lead. The meter should indicate continuity.

Move the Black meter lead to No. 2 Green lead and keep this lead in place for the next three resistance tests. Make contact with the Red meter lead to No. 1 Green lead. The meter should indicate no continuity. Next, make contact with the Red meter lead to the Black rectifier lead.

Again, the meter should indicate no continuity. Now, move the Red meter lead to make contact with the Red rectifier lead. The meter should indicate continuity.

Move the Black meter lead to the Red rectifier lead and keep it in place for the next three resistance tests. Make contact with the Red meter lead to No. 1 Green lead. The meter should indicate no continuity. Next, move the Red meter lead to make contact with the No. 2 Green lead. Again, the meter should indicate no continuity. Now, move the Red meter lead to the Black rectifier lead. The meter should indicate no continuity.

Move the Black meter lead to the Black rectifier lead and keep it in place for the next three resistance tests. Make contact with the Red meter lead to No. 1 Green lead. The meter should indicate continuity. Next, move the Red meter lead to make contact with the No. 2 Green lead. Again, the meter should indicate continuity.

Finally, move the Red meter lead to make contact with the Red rectifier lead. The meter should indicate continuity.

If the rectifier should fail any of the above resistance tests, it should be removed and replaced with a new unit. Service or adjustment is not possible. If a new unit is to be installed, observe the same color-to-color connections at the quick connect fittings.

Primary Winding Ignition Coil Test
All Models

7- Select the Rx1 scale on the ohmmeter. Disconnect the White/Black lead

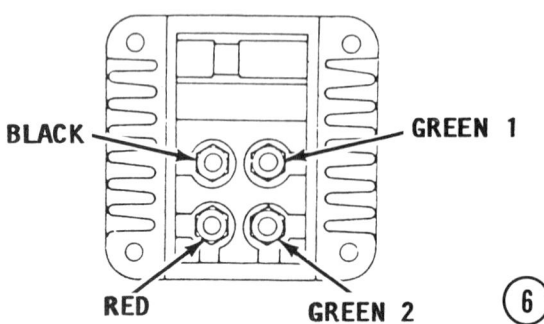

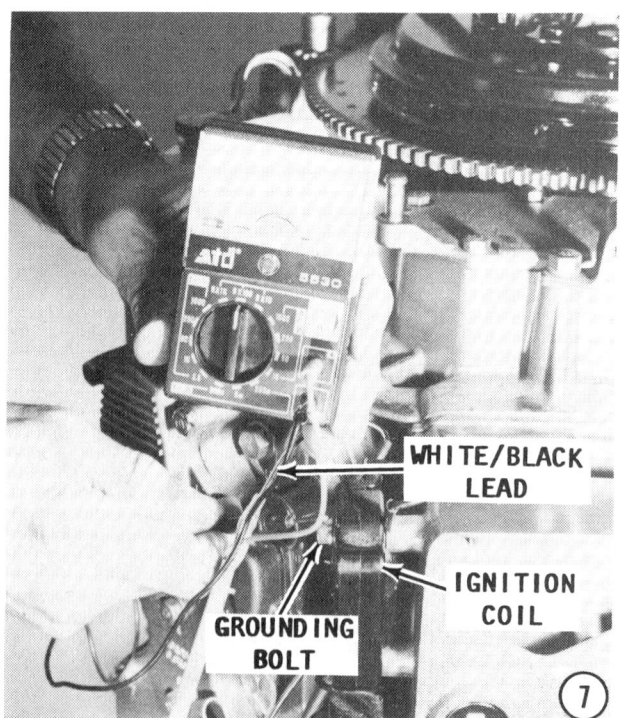

from one of the ignition coils. Make contact with the Red meter lead to the White/Black lead and the Black meter lead to the grounding bolt of the small Black lead of the ignition coil. The meter should register 0.25 ohms. Reconnect the White/Black lead.

If the reading is questionable but the powerhead operates, install a spark plug tester and check the strength of the spark.

If the spark is acceptable, it is not necessary to replace the coil. However, keep this area in mind for possible future trouble.

If the spark is weak it will get weaker, therefore, replace the coil.

Repeat this test for all the ignition coils.

Secondary Winding Ignition Coil Test

All Models

8- Select the Rx1000 scale on the ohmmeter. Disconnect the high tension leads from all the spark plugs.

Make contact with the Red meter lead to the high tension lead terminal inside the spark plug boot.

Make contact with the Black meter lead to the grounding bolt of the small Black lead. The meter should register between 2450 and 2550 ohms.

Reconnect each high tension lead to the proper spark plug.

If the reading is questionable but the powerhead operates, install a spark plug tester and check the strength of the spark.

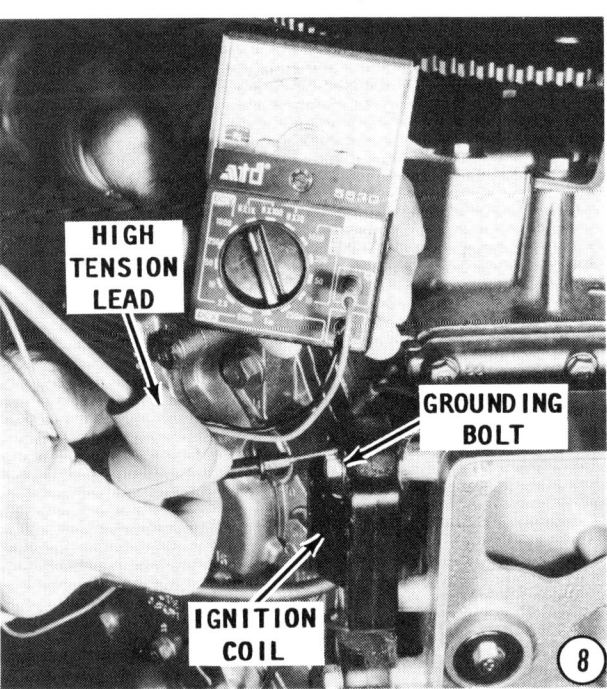

If the spark is acceptable, it is not necessary to replace the coil. However, keep this area in mind for possible future trouble.

If the spark is weak it will get weaker, therefore, replace the coil.

CDI Unit Resistance Tests

The following diagrams and charts outline testing procedures for the CDI unit. The unit may remain installed on the powerhead as long as the leads are disconnected from the screw terminals of the unit, or it may be removed for testing. In either case, the testing procedures are identical.

9- Select the Rx1000 ohm scale on the ohmmeter. Make contact with the Red meter lead to the leads called out in the horizontal heading. Make contact with the Black meter lead to the leads called out in the vertical list of leads.

Proceed slowly and **CAREFULLY** in the order given. The asterisk (*) denotes the meter needle should swing toward continuity (zero ohms), and then return to stay at the specified value.

Install both leads at the battery terminals, the Red lead to the positive terminal and the Black lead to the negative terminal.

Control Unit Installed Only on V6 Models

The safety devices which protect the powerhead from overrevving and overheating are an integral part of the CDI unit on V4 Models. On V6 Models, this circuitry is housed in a separate module, a "Black Box" mounted under the CDI unit. This component has Pink, White, Yellow, Black and Brown leads. These leads are all connected to the lower left-hand side of the CDI unit. Disconnect these leads for the following tests.

The unit may remain installed on the powerhead, provided the leads are disconnected from the screw terminals of the CDI unit. The unit may be removed for testing. In either case, the testing procedures are identical.

10- Select the Rx1000 ohm scale on the ohmmeter. Make contact with the Red meter lead to the leads called out in the horizontal heading. Make contact with the Black meter lead to the leads called out in the vertical list of leads.

V4 MODEL CDI UNIT RESISTANCE TESTS

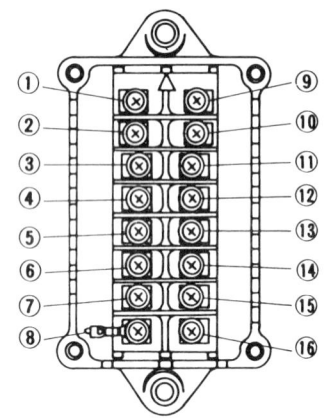

① Brown
② Blue
③ Red
④ Black/Red
⑤ White/Red
⑥ White/Black
⑦ White/Yellow
⑧ White/Green
⑨ Black/White
⑩ Black/White
⑪ Black/White
⑫ Black/White
⑬ White
⑭ Pink
⑮ Black
⑯ Over. rev.

Measuring resistance

Unit : kΩ

Tester ⊖ \ Tester ⊕		Charge				Pulser				Ignition				Stop	Ground		Over-rev.	
		Br	Bl	R	B/R	W/R	W/B	W/Y	W/G	Coil 1 (B/W)	Coil 2 (B/W)	Coil 3 (B/W)	Coil 4 (B/W)	W	P	B	OR	OR LE104
①	Br		120~180	160~240	120~180	120~180	120~180	120~180	120~180	120~180	120~180	120~180	120~180	160~240	200~300	80~120	160~240	∞
②	Bl	3.6~5.4		*120~180	*72~108	*72~108	*72~108	*72~108	*72~108	*72~108	*72~108	*72~108	*72~108	120~180	72~108	*40~60	48~72	∞
③	R	160~240	120~180		120~180	120~180	120~180	120~180	120~180	120~180	120~180	120~180	120~180	160~240	160~240	80~120	160~240	∞
④	B/R	*120~180	*72~108	3.6~5.4		*72~108	*72~108	*72~108	*72~108	*72~108	*72~108	*72~108	*72~108	120~180	72~108	40~60	48~72	∞
⑤	W/R	96~134	44~66	96~134	44~66		44~66	44~66	44~66	28~42	44~66	44~66	44~66	12~18	*160~240	24~36	*129~180	∞
⑥	W/B	120~180	72~108	120~180	72~108	72~108		72~108	72~108	72~108	32~48	72~108	72~108	12~18	*160~240	40~60	*160~240	∞
⑦	W/Y	130~180	72~108	120~180	72~108	72~108	72~108		72~108	72~108	72~108	32~48	72~108	12~18	*160~240	40~60	*160~240	∞
⑧	W/G	120~180	72~108	120~180	72~108	72~108	72~108	72~108		72~108	72~108	72~108	32~48	12~18	*160~240	40~60	*160~240	∞
⑨	Coil 1 (B/W)	∞	∞	∞	∞	∞	∞	∞	∞		∞	∞	∞	∞	∞	∞	∞	∞
⑩	Coil 2 (B/W)	∞	∞	∞	∞	∞	∞	∞	∞	∞		∞	∞	∞	∞	∞	∞	∞
⑪	Coil 3 (B/W)	∞	∞	∞	∞	∞	∞	∞	∞	∞	∞		∞	∞	∞	∞	∞	∞
⑫	Coil 4 (B/W)	∞	∞	∞	∞	∞	∞	∞	∞	∞	∞	∞		∞	∞	∞	∞	∞
⑬	W	∞	∞	∞	∞	∞	∞	∞	∞	∞	∞	∞	∞		∞	∞	∞	∞
⑭	P	∞	∞	∞	∞	∞	∞	∞	∞	∞	∞	∞	∞	∞		∞	∞	∞
⑮	B	12~18	3.6~5.4	12~18	3.6~5.4	3.6~5.4	3.6~5.4	3.6~5.4	3.6~5.4	3.6~5.4	3.6~5.4	3.6~5.4	3.6~5.4	24~36	*56~84		*20~30	∞
⑯	OR	∞	∞	∞	∞	∞	∞	∞	∞	∞	∞	∞	∞	∞	∞	∞		∞
	OR LE104	∞	∞	∞	∞	∞	∞	∞	∞	∞	∞	∞	∞	∞	∞	∞	∞	

V6 MODEL CDI UNIT RESISTANCE TESTS
UNITS WITHOUT YMIS

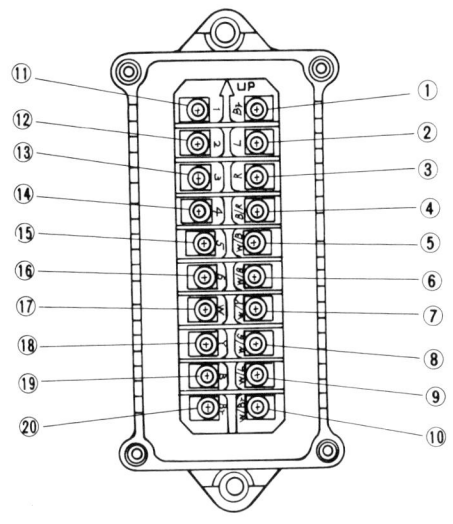

① Brown
② Blue
③ Red
④ Black/Red
⑤ White/Red
⑥ White/Black
⑦ White/Yellow
⑧ White/Green
⑨ Black/Blue
⑩ Black/Brown
⑪ 1.
⑫ 2.
⑬ 3.
⑭ 4.
⑮ 5.
⑯ 6.
⑰ White
⑱ Yellow
⑲ Black
⑳ Brown

Measuring resistance

Unit: kΩ

(+) \ (−)		Charge				Pulser						Ignition						Stop	Ground & Control		
		Br	L	R	B/R	W/R	W/B	W/Y	W/G	W/L	W/Br	1	2	3	4	5	6	W	Y	B	Br
1	Br		0.6	100~∞	100~∞	100~∞	100~∞	100~∞	100~∞	100~∞	100~∞	∞	∞	∞	∞	∞	∞	100~∞	26	2.5	∞
2	L	100~∞		100~∞	100~∞	100~∞	100~∞	100~∞	100~∞	100~∞	100~∞	∞	∞	∞	∞	∞	∞	100~∞	6.8	0.6	∞
3	R	100~∞	100~∞		0.6	100~∞	100~∞	100~∞	100~∞	100~∞	100~∞	∞	∞	∞	∞	∞	∞	100~∞	26	2.5	∞
4	B/R	100~∞	100~∞	100~∞		100~∞	100~∞	100~∞	100~∞	100~∞	100~∞	∞	∞	∞	∞	∞	∞	100~∞	6.8	0.6	∞
5	W/R	100~∞	100~∞	100~∞	100~∞		100~∞	100~∞	100~∞	100~∞	100~∞	∞	∞	∞	∞	∞	∞	100~∞	6.8	0.6	∞
6	W/B	100~∞	100~∞	100~∞	100~∞	100~∞		100~∞	100~∞	100~∞	100~∞	∞	∞	∞	∞	∞	∞	100~∞	6.8	0.6	∞
7	W/Y	100~∞	100~∞	100~∞	100~∞	100~∞	100~∞		100~∞	100~∞	100~∞	∞	∞	∞	∞	∞	∞	100~∞	6.8	0.6	∞
8	W/G	100~∞	100~∞	100~∞	100~∞	100~∞	100~∞	100~∞		100~∞	100~∞	∞	∞	∞	∞	∞	∞	100~∞	6.8	0.6	∞
9	W/L	100~∞	100~∞	100~∞	100~∞	100~∞	100~∞	100~∞	100~∞		100~∞	∞	∞	∞	∞	∞	∞	100~∞	6.8	0.6	∞
10	W/Br	100~∞	100~∞	100~∞	100~∞	100~∞	100~∞	100~∞	100~∞	100~∞		∞	∞	∞	∞	∞	∞	100~∞	6.8	0.6	∞
11	1	100~∞	100~∞	100~∞	100~∞	45	100~∞	100~∞	100~∞	100~∞	100~∞		∞	∞	∞	∞	∞	100~∞	6	0.5	∞
12	2	100~∞	100~∞	100~∞	100~∞	100~∞	45	100~∞	100~∞	100~∞	100~∞	∞		∞	∞	∞	∞	100~∞	6	0.5	∞
13	3	100~∞	100~∞	100~∞	100~∞	100~∞	100~∞	45	100~∞	100~∞	100~∞	∞	∞		∞	∞	∞	100~∞	6	0.5	∞
14	4	100~∞	100~∞	100~∞	100~∞	100~∞	100~∞	100~∞	45	100~∞	100~∞	∞	∞	∞		∞	∞	100~∞	6	0.5	∞
15	5	100~∞	100~∞	100~∞	100~∞	100~∞	100~∞	100~∞	100~∞	45	100~∞	∞	∞	∞	∞		∞	100~∞	6	0.5	∞
16	6	100~∞	100~∞	100~∞	100~∞	100~∞	100~∞	100~∞	100~∞	100~∞	45	∞	∞	∞	∞	∞		100~∞	6	0.5	∞
17	W	100~∞	100~∞	100~∞	100~∞	0.6	0.6	0.6	0.6	0.6	0.6	∞	∞	∞	∞	∞	∞		17	1.6	∞
18	Y	100~∞	100~∞	100~∞	100~∞	100~∞	100~∞	100~∞	100~∞	100~∞	100~∞	∞	∞	∞	∞	∞	∞	100~∞		10	∞
19	B	100~∞	100~∞	100~∞	100~∞	100~∞	100~∞	100~∞	100~∞	100~∞	100~∞	∞	∞	∞	∞	∞	∞	100~∞	2.5		∞
20	Br	0.6	2.5	100~∞	100~∞	100~∞	100~∞	100~∞	100~∞	100~∞	100~∞	∞	∞	∞	∞	∞	∞	100~∞	75	14	

V6 MODEL CDI UNIT RESISTANCE TESTS
UNITS WITH YMIS

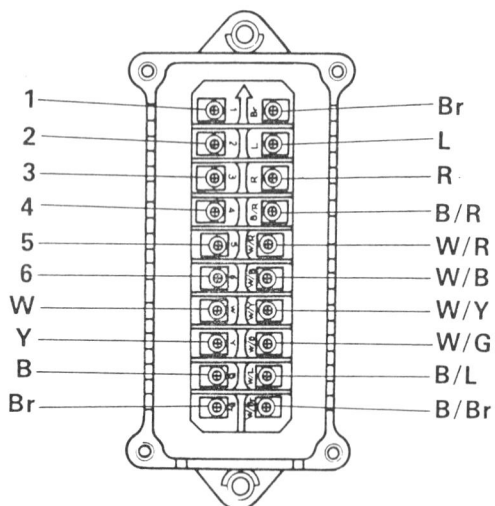

Br: Brown
L: Blue
R: Red
B/R: Black/Red
W/R: White/Red
W/B: White/Black
W/Y: White/Yellow
W/G: White/Green
B/L: Black/Blue
B/Br: Black/Brown

1: 1. Ignition
2: 2. Ignition
3: 3. Ignition
4: 4. Ignition
5: 5. Ignition
6: 6. Ignition
W: White
Y: Yellow
B: Black
Br: Brown

① Measuring resistance
② Tester ⊖
③ Tester ⊕
④ Charge
⑤ Pulser
⑥ Ignition
⑦ Stop
⑧ Ground & control
⑨ Unit: kΩ

		④ Charge				⑤ Pulser					⑥ Ignition						⑦ Stop	⑧ Ground & control		
③	②	Br	L	R	B/R	W/R	W/B	W/Y	W/G	W/L	W/Br	1	2	3	4	5	6	W	Y	B	Br
1	Br		0.56	100~∞	100~∞	100~∞	100~∞	100~∞	100~∞	100~∞	100~∞	∞	∞	∞	∞	∞	∞	100~∞	27	2.1	∞
2	L	100~∞		100~∞	100~∞	100~∞	100~∞	100~∞	100~∞	100~∞	100~∞	∞	∞	∞	∞	∞	∞	100~∞	7	0.56	∞
3	R	100~∞	100~∞		0.56	100~∞	100~∞	100~∞	100~∞	100~∞	100~∞	∞	∞	∞	∞	∞	∞	100~∞	27	2.1	∞
4	B/R	100~∞	100~∞	100~∞		100~∞	100~∞	100~∞	100~∞	100~∞	100~∞	∞	∞	∞	∞	∞	∞	100~∞	7	0.56	∞
5	W/R	100~∞	100~∞	100~∞	100~∞		100~∞	100~∞	100~∞	100~∞	100~∞	∞	∞	∞	∞	∞	∞	100~∞	7	0.56	∞
6	W/B	100~∞	100~∞	100~∞	100~∞	100~∞		100~∞	100~∞	100~∞	100~∞	∞	∞	∞	∞	∞	∞	100~∞	7	0.56	∞
7	W/Y	100~∞	100~∞	100~∞	100~∞	100~∞	100~∞		100~∞	100~∞	100~∞	∞	∞	∞	∞	∞	∞	100~∞	7	0.56	∞
8	W/G	100~∞	100~∞	100~∞	100~∞	100~∞	100~∞	100~∞		100~∞	100~∞	∞	∞	∞	∞	∞	∞	100~∞	7	0.56	∞
9	W/L	100~∞	100~∞	100~∞	100~∞	100~∞	100~∞	100~∞	100~∞		100~∞	∞	∞	∞	∞	∞	∞	100~∞	7	0.56	∞
10	W/Br	100~∞	100~∞	100~∞	100~∞	100~∞	100~∞	100~∞	100~∞	100~∞		∞	∞	∞	∞	∞	∞	100~∞	7	0.56	∞
11	1	100~∞	100~∞	100~∞	100~∞	40	100~∞	100~∞	100~∞	100~∞	100~∞		∞	∞	∞	∞	∞	100~∞	7	0.56	∞
12	2	100~∞	100~∞	100~∞	100~∞	100~∞	40	100~∞	100~∞	100~∞	100~∞	∞		∞	∞	∞	∞	100~∞	7	0.56	∞
13	3	100~∞	100~∞	100~∞	100~∞	100~∞	100~∞	40	100~∞	100~∞	100~∞	∞	∞		∞	∞	∞	100~∞	7	0.56	∞
14	4	100~∞	100~∞	100~∞	100~∞	100~∞	100~∞	100~∞	40	100~∞	100~∞	∞	∞	∞		∞	∞	100~∞	7	0.56	∞
15	5	100~∞	100~∞	100~∞	100~∞	100~∞	100~∞	100~∞	100~∞	40	100~∞	∞	∞	∞	∞		∞	100~∞	7	0.56	∞
16	6	100~∞	100~∞	100~∞	100~∞	100~∞	100~∞	100~∞	100~∞	100~∞	40	∞	∞	∞	∞	∞		100~∞	7	0.56	∞
17	W	100~∞	100~∞	100~∞	100~∞	0.56	0.56	0.56	0.56	0.56	0.56	∞	∞	∞	∞	∞	∞		20	1.6	∞
18	Y	100~∞	100~∞	100~∞	100~∞	100~∞	100~∞	100~∞	100~∞	100~∞	100~∞	∞	∞	∞	∞	∞	∞	100~∞		12	∞
19	B	100~∞	100~∞	100~∞	100~∞	100~∞	100~∞	100~∞	100~∞	100~∞	100~∞	∞	∞	∞	∞	∞	∞	100~∞	2.8		∞
20	Br	100~∞	100~∞	0.56	22	100~∞	100~∞	100~∞	100~∞	100~∞	100~∞	∞	∞	∞	∞	∞	∞	100~∞	75	13	

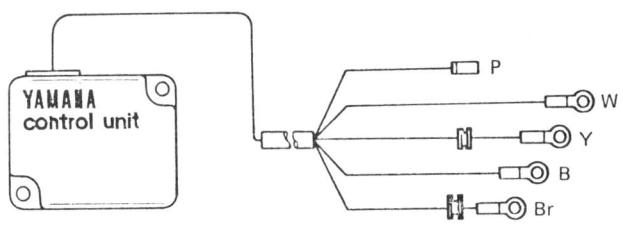

Unit: kΩ

(+) \ (−)	Brown	Black	White	Yellow	Pink
Brown		∞	∞	∞	∞
Black	20		15	22	∞
White	50	15		60	∞
Yellow	∞	∞	∞		∞
Pink	50	28	75	50	

⑩

Proceed slowly and **CAREFULLY** in the order given. If the control unit should fail any of the above resistance tests, it must be removed and replaced with a new unit. Service or adjustment is not possible. If a new unit is to be installed, observe the same connections at the CDI unit using the wiring diagram in the Appendix as a guide.

6-7 FLYWHEEL AND STATOR PLATE SERVICE

The following short section lists the procedures required to "pull" the flywheel and remove the stator plate in order to service the ignition system. Cleaning and Inspecting procedures in addition to proper assembling and installation steps are also included on the succeeding pages.

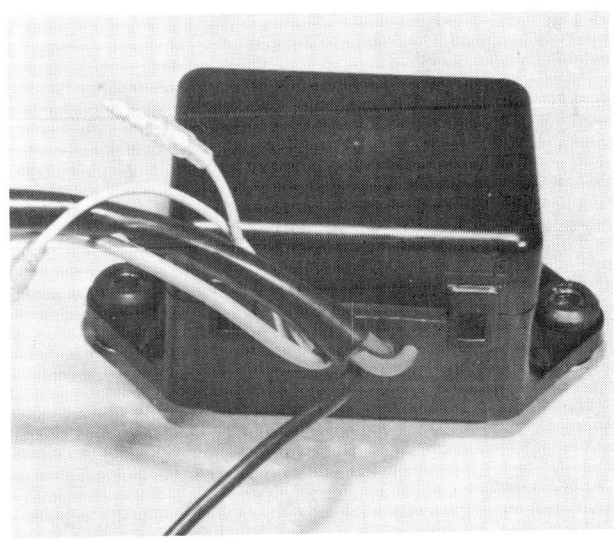

A control unit, installed only on V6 powerheads, ready for bench testing.

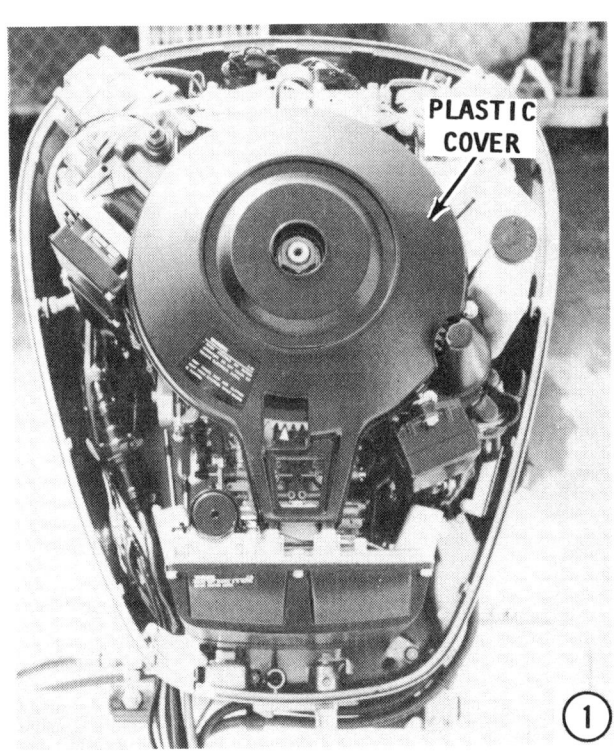

"PULLING" THE FLYWHEEL

1- Disconnect both leads at the battery terminals. Remove the plastic cover over the flywheel. Twist the spark plug leads free from the spark plugs to prevent an accident should the powerhead start inadvertently.

2- Insert a flywheel holding tool, Yamaha P/N YB6139 into the two holes provided in the flywheel and hold it steady while the flywheel nut is loosened with the correct size socket. Remove the flywheel nut and washer.

6-18 IGNITION

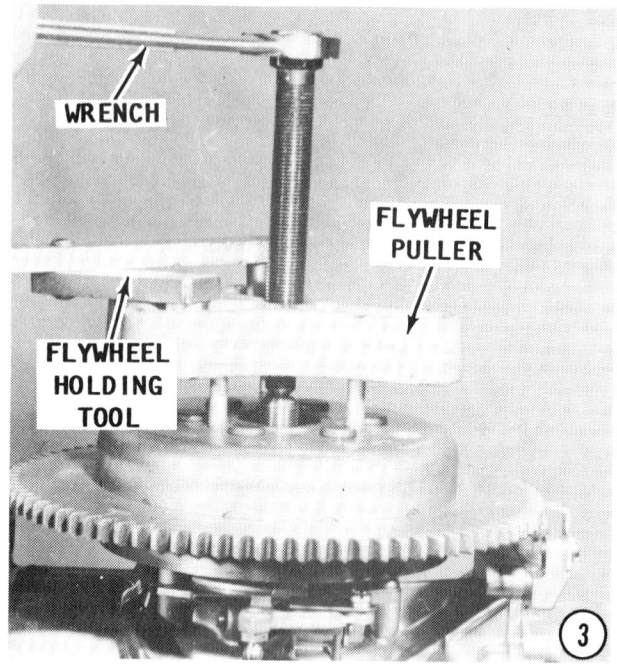

3- Obtain flywheel puller tool, Yamaha P/N YB6117. If this particular puller is not available a similar puller may be used **PROVIDED** it will pull from the bolt holes in the flywheel and **NOT** from the perimeter of the flywheel.

NEVER attempt to use a puller which pulls on the outside edge of the flywheel.

Install the puller onto the flywheel, and then using puller tool P/N YB6117 take a strain on the puller with the proper size wrench. Now, continue to tighten on the special tool and at the same time, **SHOCK** the crankshaft with a gentle to moderate tap with a hammer on the end of the special tool. This shock will assist in "breaking" the flywheel loose from the crankshaft.

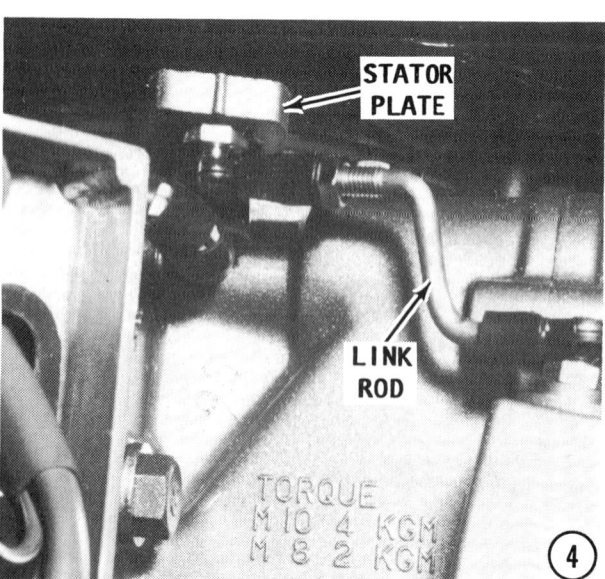

CAUTION

A "violent" strike to the center bolt may cause damage to the crankshaft oil seals. Therefore, use only a gentle to moderate tap with the hammer.

Lift the flywheel up and free of the crankshaft. The flywheel may seem heavier than it actually is due to the magnetic attraction between the flywheel magnets and the laminated cores of the coils. Remove the Woodruff key from the crankshaft.

ALL Models except those Equipped with YMIS

4- Pry the link rod from the ball joint under the stator plate. Take care not to alter the length of the link rod.

All Models

5- Remove the bolts and then the cover over the CDI unit and rectifier/regulator.

6- Disconnect the following leads from their terminals on the CDI unit: Brown, Red, Blue and Black/Red. These leads can be lifted away from the edge of the unit and still be secured in a small retainer.

Disconnect the two Green leads between the stator and the rectifier at the rectifier.

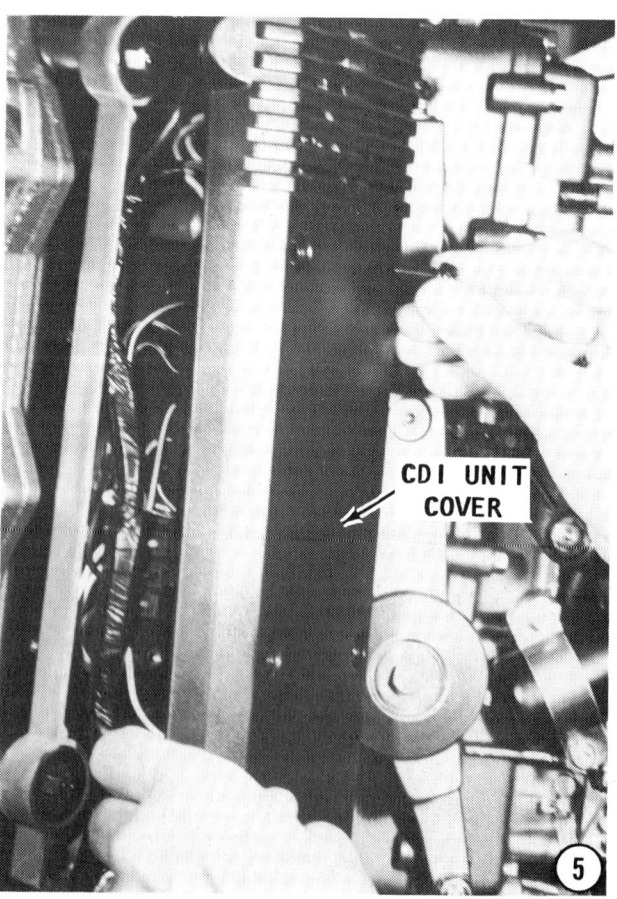

Models **NOT** equipped with YMIS: Identify the wire harness lead from the stator to the CDI unit. Models equipped with YMIS: Identify the wire harness lead from the stator to the microcomputer This harness will consist of the following leads:

V4 Models
White/Red
White/Yellow
White/Black
White/Green

V6 Models
White/Red,
White/Yellow
White/Black
White/Green
White/Blue
White/Brown.

Disconnect the two halves of the connector.

7- Mark the relative positions of the stator and pulsar assemblies to the powerhead before removal. Marking their positions will enable each to be installed in same location around the crankshaft from which they were removed. Remove the three bolts securing the stator assembly, and then lift it off the powerhead. The charge and lighting coils are considered the

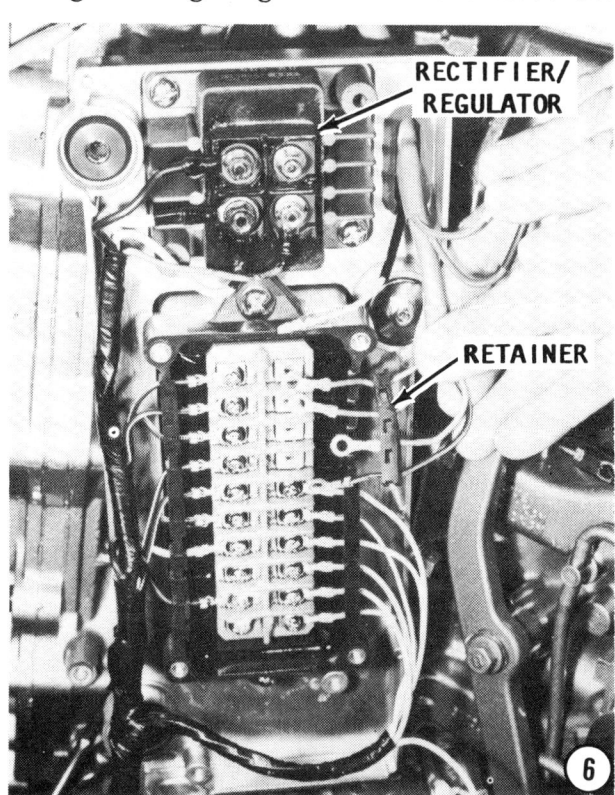

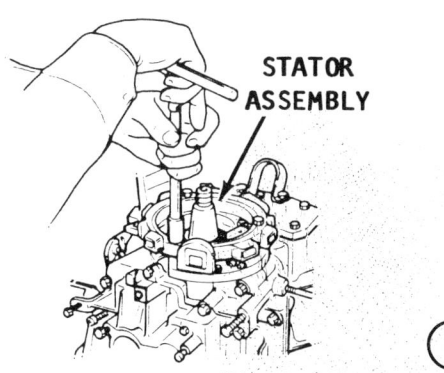

"stator assembly" and are replaced as a set. Next remove the four magneto base retainers and the pulsar assembly attaching hardware. Lift off the pulsar assembly, and remove the retaining collar from the pulsar assembly.

6-8 CLEANING AND INSPECTING

Inspect the flywheel for cracks or other damage, especially around the inside of the center hub. Check to be sure metal parts have not become attached to the magnets. Verify each magnet has good magnetism by using a screwdriver or other suitable tool.

Thoroughly clean the inside taper of the flywheel and the taper on the crankshaft to prevent the flywheel from "walking" on the crankshaft during operation.

Check the top seal around the crankshaft to be sure oil has not been leaking onto the stator plate. If there is **ANY** evidence the seal has been leaking, it **MUST** be replaced. See Chapter 8.

Test the stator assembly to verify it is not loose. Attempt to lift each side of the plate. There should be little or no evidence of movement.

Inspect the stator plate oil seal and the O-ring on the underside of the plate.

Always check the flywheel carefully to be sure particles of metal have not become stuck to the magnets.

6-9 FLYWHEEL AND STATOR PLATE INSTALLATION

The following procedures pickup the work after the flywheel and stator assembly have been cleaned, inspected, serviced, and assembled.

1- Install the retaining collar over the pulsar assembly. Place the pulsar assembly over the crankshaft, with the arm on the retaining collar on the starboard side of the powerhead. Secure the assembly to the upper bearing housing with four retainers.

Place the stator plate down over the crankshaft. The three bolt holes are unevenly spaced, therefore rotate the plate until all three holes align. Install and tighten the three securing bolts.

2- Thread the stator harness wire through the grommet next to the rectifier/regulator. Attach the two Green lighting coil leads to the two forward terminals on the rectifier/regulator. It makes no difference which lead is attached to which terminal. Connect the other four leads to the four forward terminals on the CDI terminal block, starting from the top, Brown, Blue, Red, Black/Red in order.

3- Install the CDI unit cover with the attaching hardware. Tighten the screws securely.

All Models
Except those equipped with YMIS

4- Snap the link rod from the magneto control lever back onto the ball joint on the magneto plate. If the length of this rod was accidentally altered, see Chapter 7, Section 7-2 to adjust the rod back to the specified length.

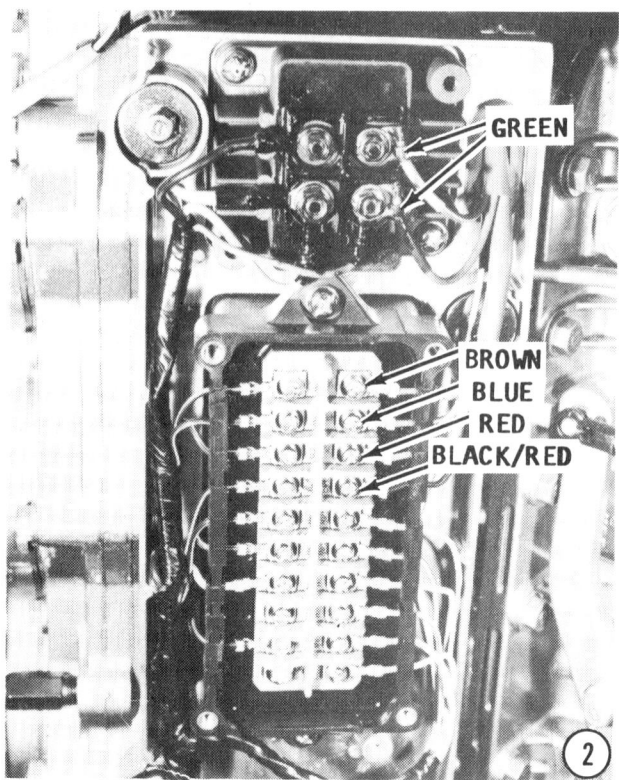

All Models
Step 4 Continues

Install the Woodruff key into the crankshaft. A tiny dab of grease will help hold the key in place while the flywheel is installed.

Lower the flywheel over the crankshaft with the Woodruff key indexing into the slot in the flywheel.

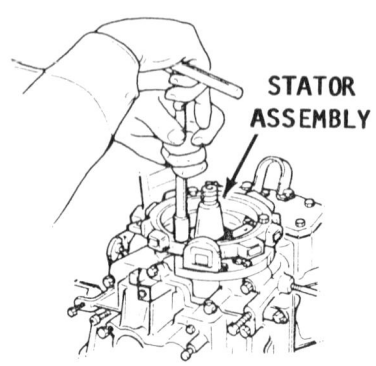

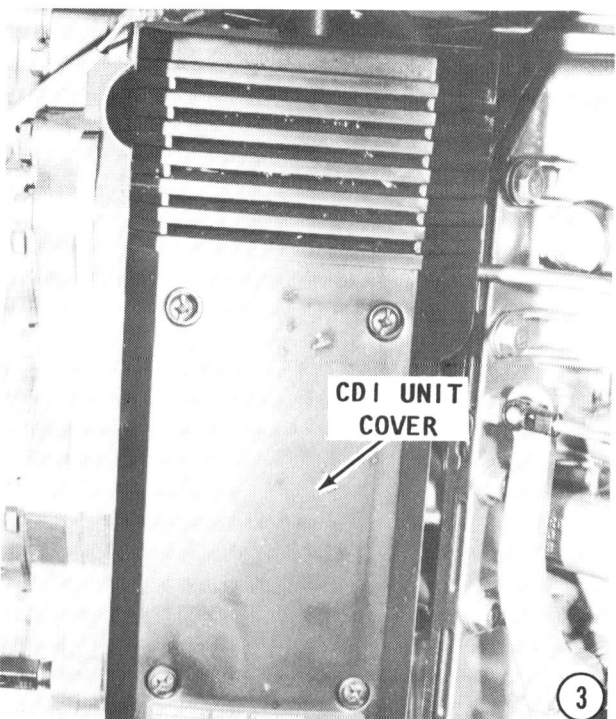

TACHOMETERS 6-21

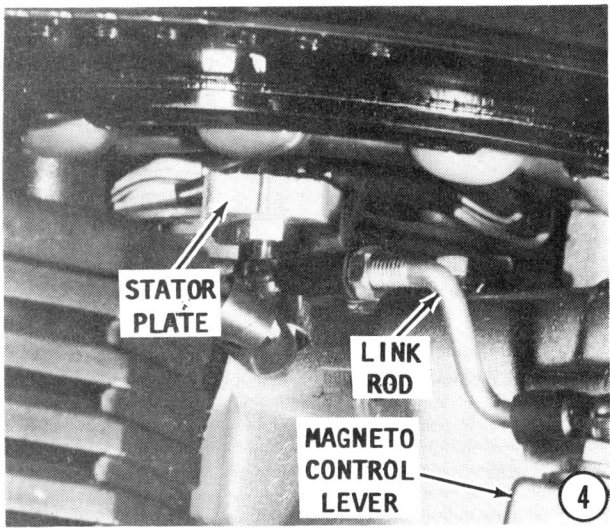

Check the action of the magneto plate. For those models equipped with a rod connecting the magneto plate with the magneto control lever, the plate should move freely within the limits of travel of the magneto control lever. If any binding is felt, remove the flywheel and check installation of the stator plate.

5- Install the washer and flywheel nut. Hold the flywheel from rotating using the holder tool, Yamaha P/N YB6139. Tighten the flywheel nut to a torque value of 115 ft lbs (160Nm).

Once again, check the action of the stator plate, as described in the previous step.

6- Connect the spark plug leads to the spark plugs.

Install the plastic cover over the flywheel.

Install the cowling to the powerhead.

Install both leads at the battery terminals, the Red lead to the positive terminal and the Black lead to the negative terminal.

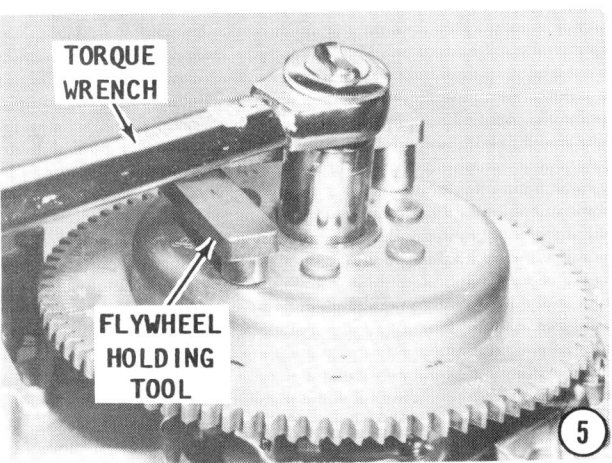

CLOSING TASKS

Mount the outboard unit in a test tank, on the boat in a body of water, or connect a flush attachment and hose to the lower unit. Connect a tachometer to the powerhead.

SPECIAL WORDS ON TACHOMETERS AND CONNECTIONS

A tachometer is installed as standard equipment on all powerheads covered in this manual.

Due to local conditions, it may be necessary to adjust the carburetor while the outboard unit is running in a test tank or with the boat in a body of water. For maximum performance, the idle rpm should be adjusted under actual operating conditions.

Under such conditions it might be necessary to attach a tachometer closer to the powerhead than the one installed on the control panel.

Open the remote control box. Disconnect the Black and Green leads. Connect the Black lead to the ground terminal of the auxiliary tachometer and the Green lead to the "input" or "hot" terminal of the auxiliary tachometer.

NEVER, AGAIN, NEVER operate the engine at high speed with a flush device at-

6-22 IGNITION

tached. The engine, operating at high speed with such a device attached, would **RUNAWAY** from lack of a load on the propeller, causing extensive damage.

Start the engine and check the completed work.

CAUTION

Water must circulate through the lower unit to the powerhead anytime the powerhead is operating to prevent damage to the water pump in the lower unit. Just five seconds without water will damage the water pump impeller.

7- Allow the powerhead to warm to normal operating temperature. Adjust the throttle stop screw until the powerhead idles at the idle speed specified in the Appendix for the model being serviced. Rotating the throttle stop screw **CLOCKWISE** increases powerhead speed, and rotating the screw **COUNTERCLOCKWISE** decreases powerhead speed.

To time and synchronize the ignition system with the fuel system, see Chapter 7.

6-10 CDI UNIT, REGULATOR/RECTIFIER AND IGNITION COIL

REMOVAL FOR TESTING, REPLACEMENT, OR FOR POWERHEAD REMOVAL

1- Disconnect both leads at the battery terminals and remove the CDI unit cover securing bolts. Remove the cover.

CDI Unit Removal
For Testing or Replacement

2- Disconnect all the leads on the CDI unit screw terminals, each set of three or four leads are grouped together in a small retaining block for ease of installation. Pull the retaining block, with leads attached from the edge of the mounting bracket. Remove the two screws, one at the top and one at the bottom of the CDI unit and lift the unit out of its mounting bracket.

Rectifier/Regulator Removal
For Testing or Replacement

Step 2 Continues

Disconnect all the leads from the rectifier/regulator and remove the two attaching screws. Lift the unit out of its mounting bracket.

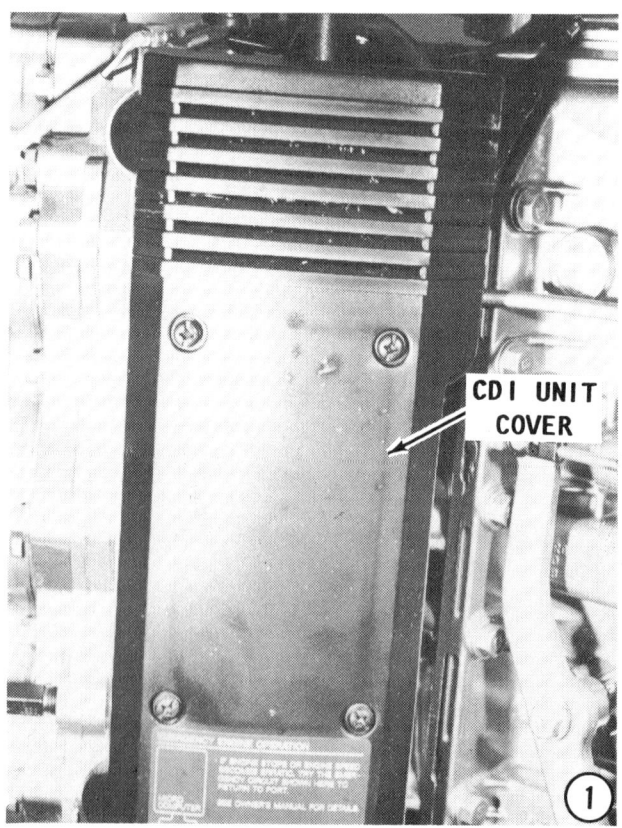

REGULATOR/RECTIFIER   6-23

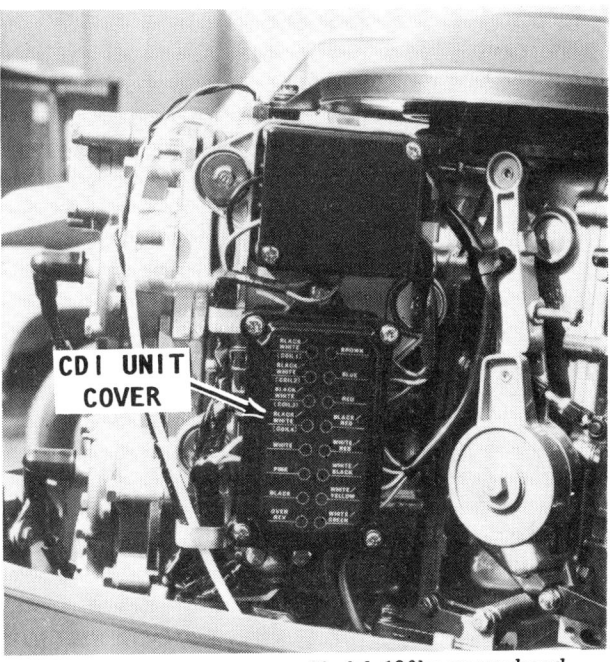

The CDI unit cover on a Model 130hp powerhead.

CDI Unit and Rectifier/Regulator Removal For Powerhead Disassembling

3- Disconnect all the leads from the CDI unit screw terminals. Each set of three or four leads are grouped together in a small retaining block for ease of installation. Pull the retaining block, with leads attached from the edge of the mounting bracket. Disconnect all the leads from the rectifier/regulator. It is not necessary to remove the two components from their mounting brack-

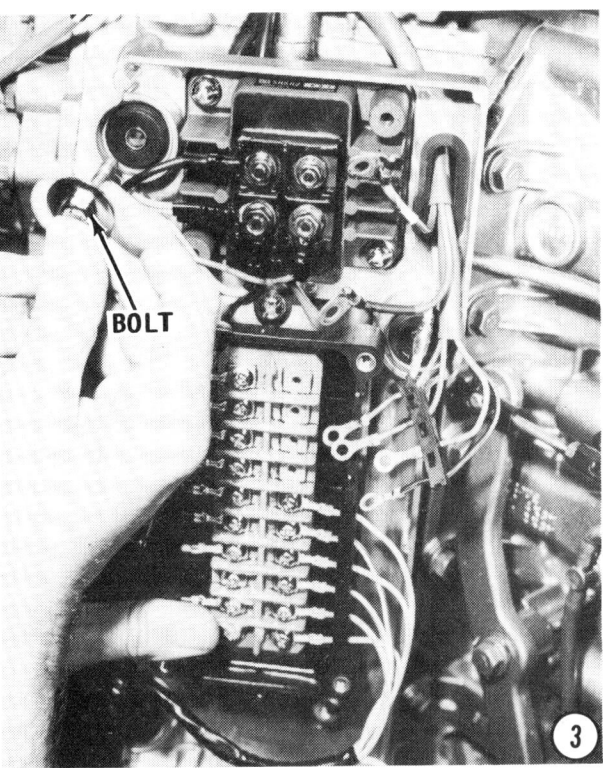

et. Remove the large attaching bolts and washers from the CDI unit mounting bracket and remove the bracket with the CDI unit and rectifier/regulator still attached.

Ignition Coil Removal

4- Remove the spark plug leads with a twisting motion, for models equipped with YMIS rotate the caps counterclockwise,

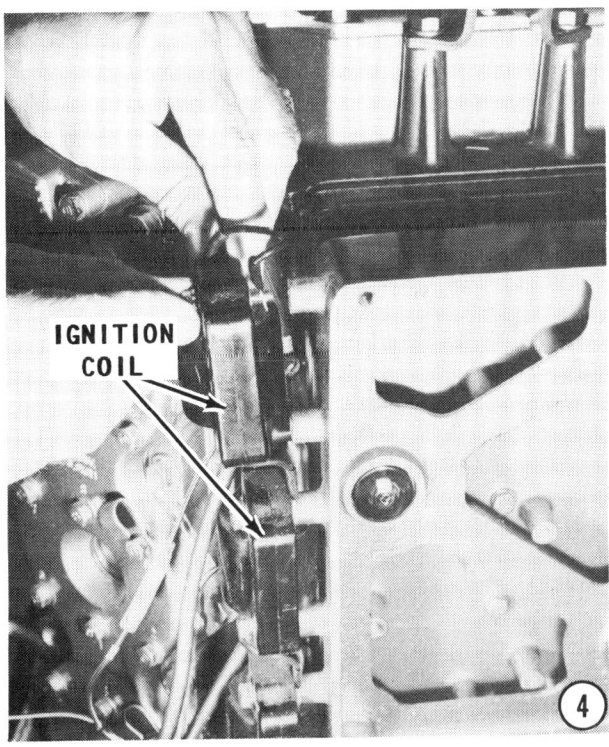

6-24 IGNITION

without pulling on them. Note the White plastic sleeves on the wires leading from the CDI unit to each ignition coil. These sleeves are numbered 1,2,3 and so on. Disconnect each wire at the quick disconnect fitting. If these sleeves are missing, then mark each wire to identify which coil the lead must be reconnected to. If these wires are not connected properly, the powerhead cylinders will not fire in the correct sequence. The powerhead may operate, but very **POORLY**.

GOOD WORDS

The **ONLY** way to determine which lead is for which cylinder other than the sleeves is by tracing the leads back to the CDI unit and comparing them to the wiring diagram. These leads are encased in a network of plastic casings which will have to be destroyed in order to trace the leads. If the leads are disconnected with no sleeves, and if the leads are not identified by markings or a piece of tape, then a trial and error method must be used to find the correct firing order for the powerhead.

Remove the two securing bolts from each coil, there will be a small Black grounding lead at each coil attached to one of the securing bolts. All ignition coils on the models covered in this manual are created equal. They may be installed to energize any spark plug.

6-11 CDI UNIT, RECTIFIER/REGULATOR AND IGNITION COIL INSTALLATION

Ignition Coil Installation

1- Thread one of the mounting bolts through the coil and through the eye of the small Black lead connector. Install this bolt and the other bolt to secure the coil to the powerhead. Connect the White/Black leads with the numbered sleeves to the leads from each coil.

CRITICAL WORDS

If the sleeves were lost and the leads not numbered as instructed during disconnecting, the correct firing order may be lost. There is **NO** other way to determine which lead is for which cylinder other than the sleeves of tracing the leads back to the CDI unit and comparing it to the wiring diagram. These leads are encased in a network of plastic casings which will have to be destroyed in order to trace back the leads. If the leads are disconnected with no sleeves,

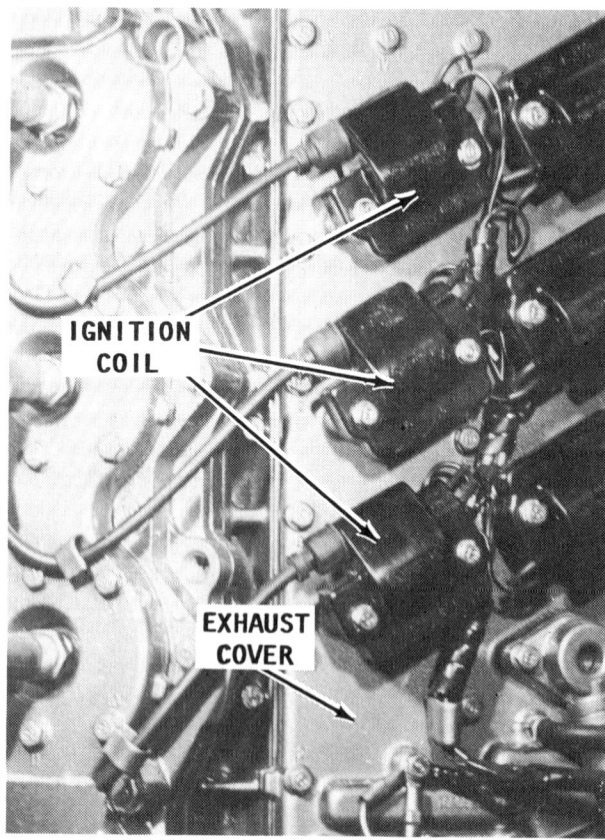

*Powerheads **NOT** equipped with YMIS (Yamaha Microcomputer Ignition System), have the ignition coils mounted on the exhaust cover. Powerheads **WITH** YMIS have the coils mounted on the side of the powerhead, as shown in the adjacent illustration in the next column.*

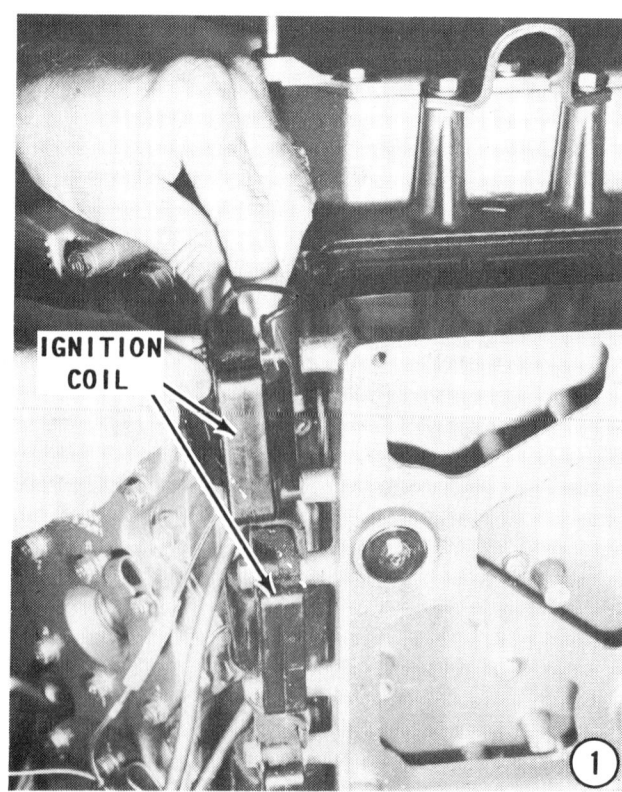

RECTIFIER/REGULATOR 6-25

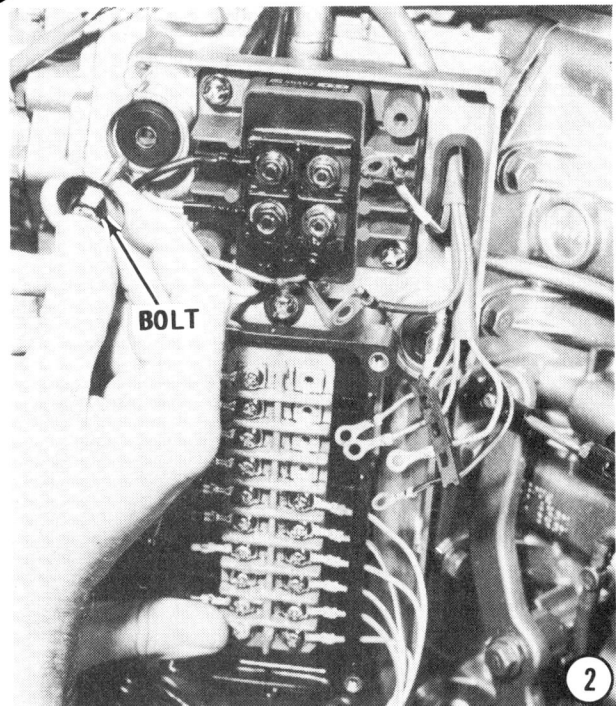

and if the leads are not identified by markings or a piece of tape, then a trial and error method must be used to find the correct firing order for the powerhead. This might even work on a powerhead equipped with YMIS, because the knocking sensor will pick up strange vibrations and send a signal to the microcomputer. At most, the timing will be retarded, but the powerhead will not be shut down.

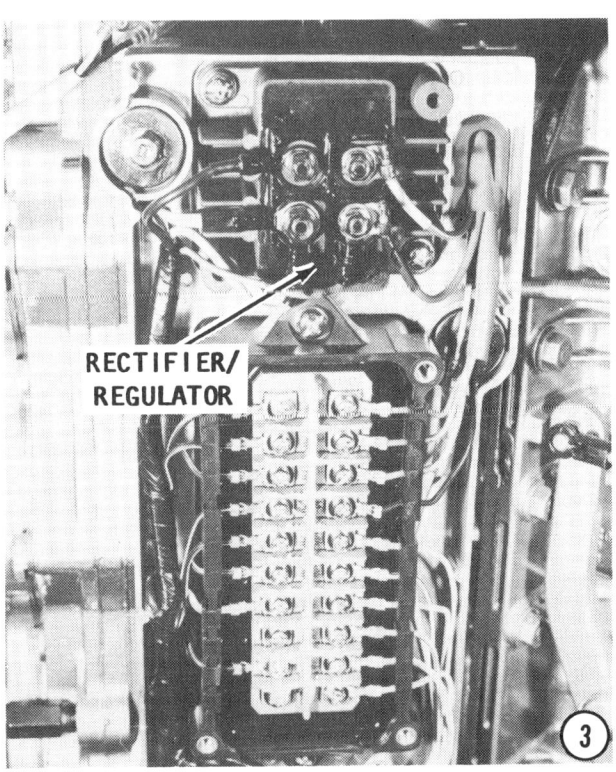

CDI Unit and Rectifier/Regulator
Installation For Powerhead Assembling

2- Install the mounting bracket, with CDI unit and rectifier/regulator still installed onto the powerhead. Refer to the appropriate illustration on Page 6-12, 6-14, 6-15, or 6-16 for correct lead connection to the screw terminals of the CDI unit and the terminals of the rectifier/regulator.

Rectifier/Regulator Installation
After Testing or Replacement

3- Position the rectifier/regulator in the mounting bracket. Install and tighten the two securing screws. Refer to the same illustrations as for Step 2 for correct lead connection to the terminals of the rectifier/regulator.

CDI Unit Installation
After Testing or Replacement
Step 3 Continues

Position the CDI unit in the mounting bracket. Install and tighten the two securing screws. Refer to the same illustrations as for Steps 2 & 3 for correct lead connection to the screw terminals of the CDI unit.

4- Install the CDI unit cover with the attaching hardware. Tighten the screws securely.

5- Connect the spark plug leads to the spark plugs.

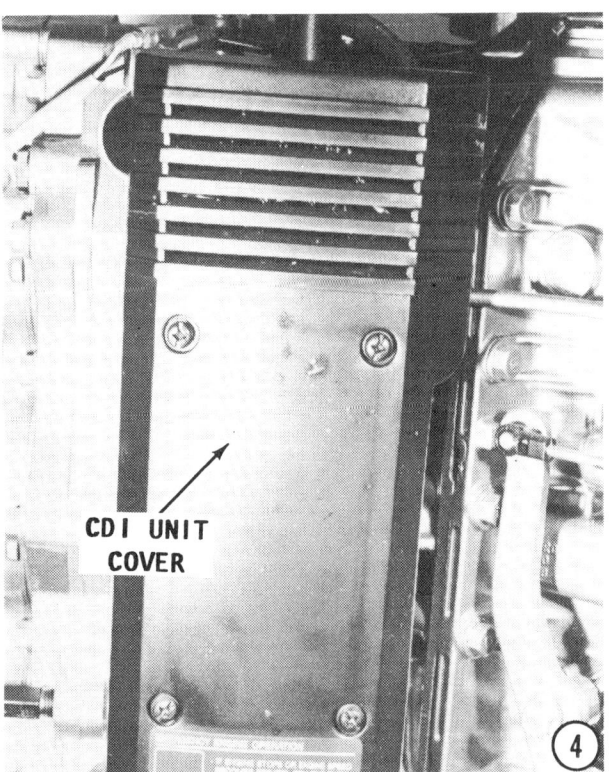

6-26 IGNITION

Install the plastic cover over the flywheel.

Install the cowling to the powerhead.

Install both leads at the battery terminals, the Red lead to the positive terminal and the Black lead to the negative terminal.

6-12 YMIS
YAMAHA MICROCOMPUTER IGNITION SYSTEM

DESCRIPTION AND OPERATION

The 220hp V6 Special and 225hp V6 Excel are equipped with an onboard computer which monitors the powerhead operating condition and regulates the ignition timing accordingly.

The onboard computor for the YMIS is mounted between the two cylinder banks.

This ignition system is identified as "Yamaha Microcomputer Ignition System", abbreviated "YMIS". This abbreviation is used throughout the chapters of this manual.

The manufacturer recommends these outboards be equipped with a battery of at least 100 to 105 amp/hour capacity.

The microcomputer receives and analyzes signals from sensors and automatically provides optimum ignition timing under any load.

The Model 220 and 225hp units have standard CDI type ignition with charge coils, pulsar coils, a CDI control box, six ignition coils and six spark plugs, as any other V6 powerhead covered in this manual. In addition to these components, these models also have a microcomputer hooked into the CDI unit to advise the unit of ignition settings.

Information gathered by four main sensors is fed into the microcomputer. These four sensors are:

Throttle position sensor
Knock sensor
Crank position sensor
Thermo sensor

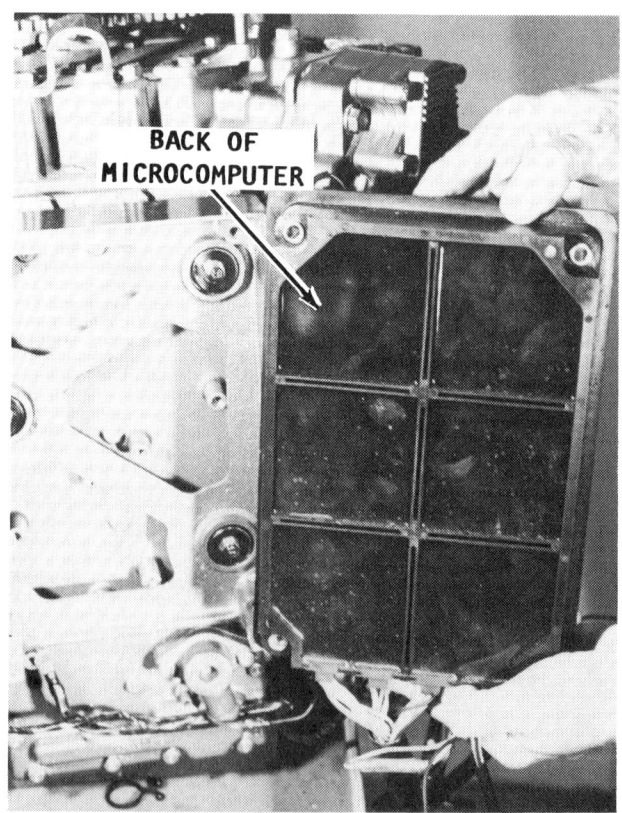

When removed, the microcomputer is found to be a sealed unit. Service is not possible, only testing.

SPECIAL WORDS

The thermo sensor must not be confused with the two thermo switches, one on each bank of cylinders. The difference between these two items will be explained in a later paragraph.

Microcomputer Control Unit

The control unit is mounted on the aft end of the powerhead. It is a "Black Box" type unit, completely sealed to prevent moisture and tampering, as shown in the accompanying illustration of a unit removed from the powerhead. A series of eight harness wires connect the following units:

Pulser coil
Both thermo switches
Main starter switch
Crank position sensor
Thermo sensor
Throttle position sensor
Knocking sensor
Battery

The microcomputer "Black Box" contains a series of eight circuits. Each of the wires from the components listed above provide one or more of the circuits with input signals by varying the voltage/resistance in the harness wires.

In addition to the eight circuits, the microcomputer contains a Central Processing Unit (CPU) an Input/Output (I/O) circuit and a memory map. The microcomputer is preprogrammed to send a specific set of instructions to the CDI unit after receiving certain signals from its sensors. The program or memory map is "engraved" in the microcomputer and differs for the Model 220hp and the Model 225hp. Therefore, the microcomputer is not interchangeable between these two models.

The pulser coil sends signals to the idling stabilizer circuit and the anti-overrev. circuit. Both the starter switch and the two thermo switches do not have special circuits attributed to them, but nevertheless their inputs are channeled directly into the inner circuitry of the microcomputer. The crank position sensor sends a signal to the pulse signal check circuit.

Both the thermo sensor and the throttle sensor send signals to the voltage signal check circuit. The knocking sensor sends a signal to the knock signal check circuit.

The power source circuit senses the condition of the battery. The "sampling interval generate circuit" (so named by the manufacturer), is designed to sense the operating voltage supplied to the microcomputer. If the voltage falls below a specified level, the microcomputer cannot function normally. Therefore, the interval generate circuit signals the CDI unit to set and keep the timing at a safe level until proper operating voltage can be restored.

Each of the seven listed circuits is integrated with the main control circuit to provide the most efficient ignition timing in response to the signals from the sensors and switches.

An "output" wire harness from the microcomputer to the CDI unit provides the unit with the recommended advance and retard settings of the spark -- eliminating some of the procedures needed to time and synchronize the fuel with the timing.

Throttle Position Sensor

The throttle position sensor is mounted portside on the top carburetor. It has Black, Red, and White leads attached to it. The sensor is a small potentiometer. The body of the sensor is stationary with a small shaft emerging from the center of the sensor. This shaft is connected to the throttle shaft via a piece of plastic coupling. As the throttle shaft is rotated, the movement is transferred to the sensor and the resistance changes. Therefore, the variable voltage signal sent to the microcomputer is directly proportional to the throttle valve angle.

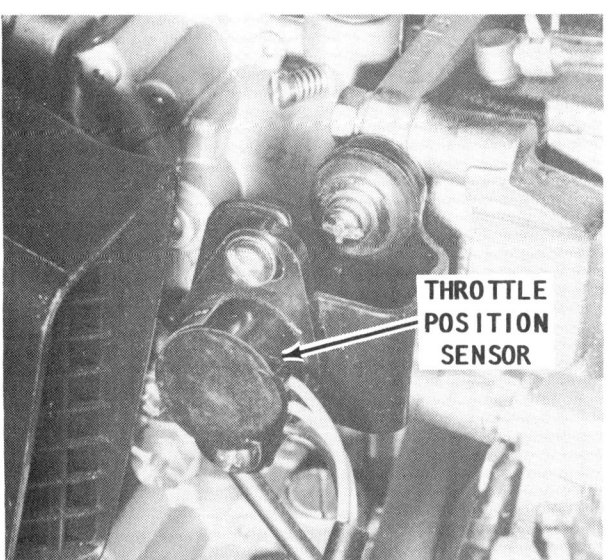

Location of the throttle position sensor on the portside of the top carburetor.

Crank Position Sensor

The crank position sensor is mounted on the cylinder block and is positioned a critical (specified) distance from the outer rim of the flywheel. It has two Blue leads attached to it. The inside face of the sensor has an electronic pickup device which is sensitive to changes in magnetic flux. The sensor "counts" the number of flywheel teeth passing the pickup in a specified time and is therefore able to calculate engine rpm. The sensor is also able to detect "crank phase angle" by "counting" the number of teeth which pass the pickup from a specified point on the flywheel.

Knocking Sensor

The knocking sensor is threaded into the portside cylinder head of the No. 6 cylinder on early models or the starboard cylinder head of the No. 5 cylinder on later models 1987 and on. The sensor has a single Black wire connected to the center terminal. The sensor is able to detect the frequency of vibrations associated with pre-ignition and detonation. If either of these conditions occur due to fuel of an insufficient octane rating (less than 90), or a sudden change in loading of the powerhead, the sensor would be activated. A signal would be sent to the microcomputer, which in turn would retard the ignition timing. If the knocking continues after the reduction in timing has taken place, the sensor will signal the microcomputer again and a further reduction will take place.

If no more knocking is detected, the ignition timing will gradually revert back to the original setting before the knocking occurred as determined by the microcomputer under the given operating conditions. The microcomputer will only take action on the input signal from the knocking sensor if there is no accompanying signal from the throttle position sensor or the crank position sensor.

Thermo Sensor

The thermo-sensor is located in the starboard cylinder head bank. It has two Black wires connected to it. This sensor consists of a **Thermistor**, an electronic device which acts in the exact opposite manner to a resistor. A resistor increases resistance (decreases voltage) with an increase in temperature. A thermistor varies resistance (increases or decreases voltage) with a change in temperature. Actually, the resistance is highest in mid range. Between $0°$ and $60°F$ the resistance is 25 ohms, between $60°$ and $100°F$ the resistance increases to 85 ohms and between $100°$ and $150°F$ the resistance decreases to 30 ohms. All values given are approximate.

If the powerhead temperature exceeds a specified level at a wide open throttle set-

Location of the crank position sensor on the cylinder block. The sensor "counts" the number of flywheel teeth passing an exact point in a specific time interval. In this manner, the sensor indicates powerhead rpm.

Location of the knocking sensor threaded into the portside cylinder head of the No. 6 cylinder on early models and the starboard cylinder head of the No. 5 cylinder on Models 1987 and on.

THERMO SENSORS & SWITCHES 6-29

ting, the thermo-sensor sends a signal to the microcomputer. In turn, the microcomputer sends two signals: one to the buzzer in the remote control box to "sound the alarm" and another signal to the CDI unit to misfire the cylinders and reduce engine rpm to a safer level. The buzzer continues to sound until the temperature drops to an acceptable level, but the signal to misfire the engine stays in effect until the throttle is backed off.

SPECIAL WORDS
THERMO SENSORS
AND THERMO SWITCHES

As promised, the difference between the thermo-sensor and the thermo switch is now explained. In the previous paragraph, the thermo-sensor was identified as a thermistor -- an electronic device capable of detecting changes in temperature and transmitting these changes as voltage signals to the microcomputer. A thermo switch is a bimetal type switch. It consists of two different type metals attached together in a small strip. Because each metal has a different coefficient of expansion -- each will expand a different amount within a given temperature range, and the strip will bend rather than extend. If a contact arrangement were to be placed at the end, when heated, the end would bend away from the contact points breaking the electrical circuit. In this manner the bimetal strip serves as a switching device. As soon as a specific temperature is reached, electrical contact is broken.

The thermo switch therefore only sends an on or off signal, not a variable voltage as in the thermo sensor signal. On powerheads covered in this manual, two thermo switches are installed -- one on each of the cylinder banks. A thermo switch is required on each cylinder bank, because it is possible for the temperature of one bank to vary considerably from the other bank. The cause for such a difference may be due to blockage of a cooling passage or blockage of an oil hose leading to one cylinder. Either one of these conditions will contribute to increased friction, raising the temperature in or around a particular cylinder. The failure of a mechanical part within a cylinder, such as a broken ring, will cause added friction and increased temperature in the area.

6-13 MICROCOMPUTER
ENGINE CONTROLS

The four sensors are able to inform the microcomputer of powerhead speed, (rpm), crank phase angle, temperature and operational loads. The microcomputer uses this information to control several functions.

Engine Start Control

When the key is switched to the **ON** position, and the engine is cranked by the cranking motor, a signal is sent from the starter relay to the microcomputer. The microcomputer sends back a signal to the CDI unit specifying the ideal setting for the engine to "catch" and start at cranking rpm.

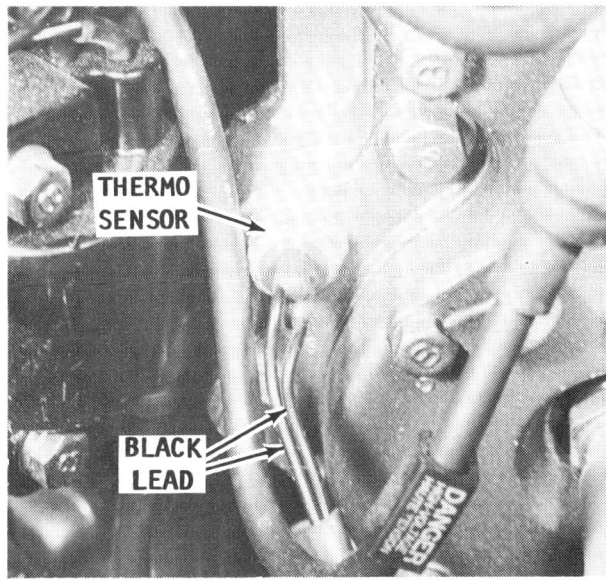

Location of the thermo sensor on the cylinder head of the starboard bank. This sensor has two Black leads.

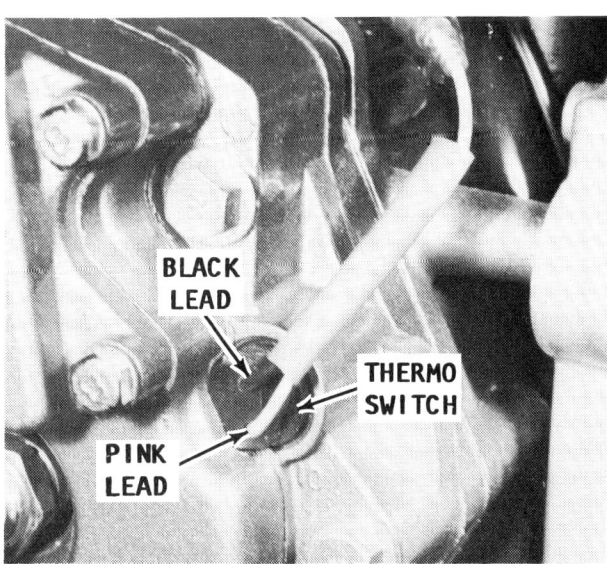

A thermo switch is installed on each cylinder bank, because the temperature may vary from port to starboard bank.

Engine Warm Up Control

From a cold start, the thermo sensor detects and reports the engine temperature to the microcomputer. The throttle position sensor detects and reports the throttle opening to the microcomputer. When the microcomputer receives these two signals simultaneously, the microcomputer overrides the throttle position sensor signal and sends a signal to the CDI unit specifying the ideal setting for ignition timing at a cool engine temperature.

The microcomputer disregards throttle opening in a cold start situation, but it does not and cannot change the throttle opening. As the engine temperature rises, the microcomputer readjusts the signal to the CDI unit to maintain the ideal ignition timing in relation to engine temperature.

During cranking of a "warm" powerhead, a different signal is sent to the microcomputer. The microcomputer will take all factors into consideration and the throttle opening will affect the timing.

Idle Stabilizing Control

Should powerhead rpm be unusually low, on the verge of stalling, this condition would be detected and reported by the crank position sensor. If the throttle opening is too small -- as detected by the throttle position sensor -- the time delay in providing the engine with an adequate supply of fuel might be too long to prevent the engine from stalling. Therefore, when the microcomputer, receives information on engine condition from these two sensors (the crank position sensor and the throttle position sensor), the computer immediately advances the ignition timing for the duration of six crankshaft revolutions. This amount of time is usually sufficient for the engine to recover from a near stall condition.

Ignition Timing Control

Under normal operating conditions, the ignition timing control is the most used. The throttle position sensor and the crank position sensor supply the microcomputer with information on throttle opening and powerhead rpm. The micocomputer sets the ignition timing which is ideal for the present conditions of operation, and alters the setting accordingly with variations in powerhead rpm. Signals from other sensors could be considered "bad news", indicating a condition not favorable to powerhead efficiency. Therefore, the microcomputer will re-evaluate and make corrections as required.

Overheating Control

If powerhead temperature exceeds a specified level at a wide open throttle setting, the thermo sensor sends a signal to the microcomputer. In turn, the microcomputer sends two signals: one to the buzzer in the remote control box to "sound the alarm" and another signal to the CDI unit to misfire the cylinders and bring down powerhead rpm to a safer level. The buzzer continues to sound until the temperature drops to an acceptable level, but the signal to misfire the engine stays in effect until the throttle is backed off.

Overrev. Control

If rough water or other conditions cause the propeller to rise out of the water, cavitation occurs. A no load situation at the propeller, or a propeller mismatch will cause the powerhead to overrev. Powerhead and lower unit life will be significantly shortened if overrevving for a prolonged period of time is allowed to occur. The crank position sensor detects and informs the microcomputer when powerhead rpm exceeds 6,000 rpm. In turn, this overrev. condition will cause a command to be sent to the CDI unit to misfire the cylinders in sequence, until powerhead rpm falls to an acceptable level.

Knocking Control

Knocking in the engine usually takes the form of pre-ignition or detonation. Both these conditions are mechanically very damaging.

The knocking sensor is able to detect the frequency of vibrations associated with pre-ignition and detonation. If either condition occurs due to fuel of an insufficient octane rating (less than 90), or a sudden change in loading of the powerhead, the sensor is activated. A signal is sent to the microcomputer which in turn retards ignition timing.

If the knocking continues after the reduction in timing has taken place, the sensor will signal the microcomputer again and a further reduction will take place. If no more knocking is detected, the ignition timing will gradually revert back to the original setting before the knocking occurred as determined by the microcomputer under given operating conditions. The microcomputer will only take action on the input signal from the knocking sensor if there is no accompanying signal from the throttle position sensor or the crank position sensor.

Reverse Direction Control

If the operator attempts a sudden shift into reverse gear while at cruising speed, the inertial momentum of the propeller would force the **crankshaft** (and naturally the driveshaft) to abruptly stop and rotate counterclockwise for an instant. The crank position sensor detects and reports such a signal to the microcomputer. The microcomputer immediately sends a signal to the CDI unit to misfire all six cylinders until the powerhead stalls.

Operating Voltage Control

A minimum operating voltage of 7 volts is necessary for the microcomputer to function normally. If the voltage falls below this value because a battery cable becomes loose or disconnected, the microcomputer would be subjected to an on/off situation and could fluctuate powerhead speed by as much as 1,500 rpm.

Therefore, the YMIS control circuit upon receipt of a low (or excessively high) voltage signal limits the timing to 7° BTDC and sends a signal to the CDI unit to misfire the cylinders in sequence, until the powerhead rpm falls to an acceptable level; a maximum of 4,800 rpm.

6-14 TROUBLESHOOTING

The following paragraphs describe some common symptoms and areas to investigate to solve problems which arise and are associated with YMIS Check all terminals for corrosion. Disconnect, clean, and reconnect the leads. Pay particular attention to the ground terminals.

If tests on components are recommended, perform both resistance and operational tests.

Hard Starting
Check battery voltage.
Check battery terminals for corrosion and tightness.
Perform tests on crank position sensor.

Poor or Erratic Powerhead Idle rpm
Perform tests on crank position sensor.
Perform tests on throttle position sensor.
Check battery connections.

High Idle
Perform tests on crank position sensor.
Check battery connections.

Poor Acceleration
Perform tests on throttle position sensor.

Engine Speed Limited to 2000 rpm
Overheat circuit activated -- investigate overheat condition, see Page 6-30.
Low oil circuit activated -- check oil levels.

Engine Loses Power and Misses at High rpm
Perform tests on crank position sensor.
Perform tests on throttle position sensor.

6-15 YMIS BYPASS PROCEDURE

If the powerhead should:
Suddenly stall through no apparent reason.
Experience power surges.
Lack acceleration.
Show any such symptoms which could be identified with a malfunction in the YMIS.

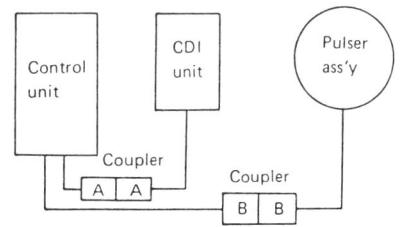

NORMAL CONNECTION

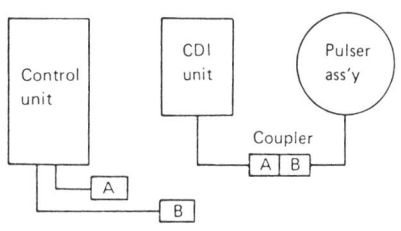

EMERGENCY CONNECTION

In an emergency condition, the microcomputer may be bypassed, as explained in the text and this diagram.

6-32 IGNITION

The microcomputer input on the CDI unit may be bypassed temporarily to resume powerhead operation while the outboard heads home.

If the YMIS is completely bypassed, the powerhead will still operate, but at a preset ignition timing irrespective of throttle opening. There is no link between the stator assembly and the magneto control lever. Therefore, advance or retard of the timing is not possible. However, if the fault lies in the microcomputer, or one of the sensors and every other system is functioning normally, the powerhead will operate at an acceptable level to permit returning the boat to its departure point.

A series of eight harness wires provide input signals to the microcomputer, but only one harness wire provides any output signal. This wire harness is connected to the CDI unit. If a signal transmitted from one of the sensors, is causing the powerhead to exhibit the symptoms listed above, a simple rearrangment of connections will isolate the microcomputer from the CDI unit.

CAUTION
THE FOLLOWING PROCEDURE MUST BE PERFORMED WITH THE POWERHEAD SHUT DOWN

First, identify the wire harness from the pulsar assembly to the microcomputer. This harness consists of six wires: White/Red, White/Black, White/Yellow, White/Blue, White/Brown and White/Green. Disconnect the two halves of the connector.

Next, identify the wire harness from the microcomputer to the CDI unit. This harness consists of seven wires, all the ones listed above, plus one Black. Disconnect the two halves of the connector.

Now, connect the six-wire harness leading from the pulsar assembly to the seven-wire harness leading to the CDI unit. The two connectors **DO** match, and can only fit together one way. After this connection is performed correctly, two wire harnesses will be left dangling and both of these should lead from the microcomputer.

The powerhead may now be started and the boat moved at reduced engine performance.

Before any checks or tests are performed on the YMIS the emergency connection just described must be disconnected and connected back as it was for testing purposes.

6-16 RESISTANCE TESTS

FIRST, THESE WORDS

Nothing can be more frustrating than an intermittent failure of a component. The following resistance tests provide a procedure to test components in a "laboratory situation" mostly at room temperature in ideal working conditions. Unfortunately, this does not provide 100% reliable test results. The sensor, switch, wire harness, and the connector, are all subjected to vibration and temperature extremes during actual operation of the powerhead. If the component could be tested in these operating conditions it might register readings which boarderline the specifications. If the heat or the vibration becomes excessive, the part may temporarily fail. Perhaps even if tested as outlined in the following section -- Operational Testing, the conditions may not be severe enough to induce the part to fail. The best advice is to attempt to duplicate the conditions under which the part failed before troubleshooting can be successful.

YMIS Power Supply Test

1- Obtain a voltmeter with probe leads, able to pierce through wire insulation in order to obtain a reading without discon-

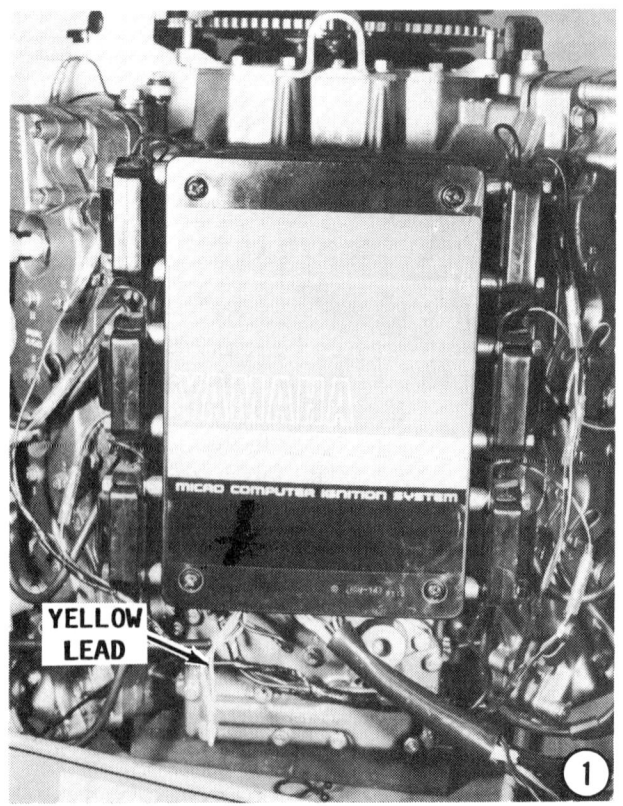

necting the leads or damaging the insulation. Select the Vx100 scale on the meter. Identify the Yellow lead from the microcomputer. Do **NOT** disconnect this lead. Make contact with the Red meter lead to the Yellow lead by probing through the insulation. Make contact with the Black meter lead to a suitable ground on the powerhead. Turn the key switch to the **ON** position, but do **NOT** start the powerhead. The meter should register between 10 and 12 volts. If the required voltage is not obtained, trace the voltage path back from the Yellow lead through the 10 prong connector, to the harness, to the key switch and on back to the battery if necessary. Keep the Black meter lead in place and use the Red meter lead to probe all connections using the wiring diagram as a guide.

Checking Crank Position Sensor Gap

2- Measure the gap between the inner face of the crank position sensor and the outer edge of the flywheel teeth. This distance should be between 1/32" and 1/16" (0.5 and 1.5mm). If the distance is not as specified, loosen the two securing screws and adjust the gap. Tighten the two screws securely to hold this new position.

Checking Crank Position Sensor Resistance

3- Obtain an ohmmeter. Select the RX1000 ohm scale. Disconnect the two Blue leads from the sensor at their harness connector. Make contact with the ohmmeter leads to the two Blue sensor leads at the connector half. The meter should register between 640 to 780 ohms. If the reading is not within these limits, the sensor must be replaced. Adjustment or service is not possible.

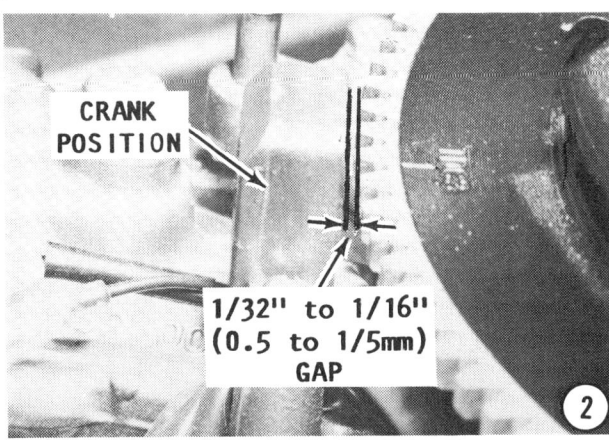

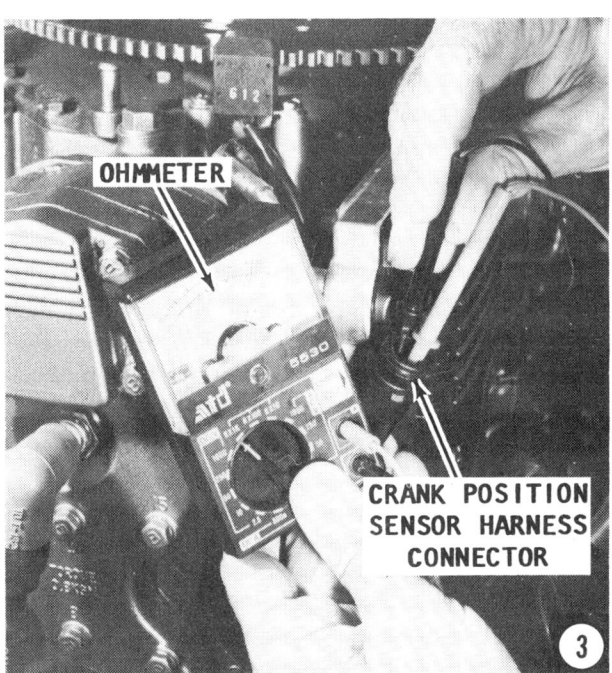

To replace the sensor, simply remove the two securing screws and remove and replace the sensor. Repeat Step 2 to properly install the sensor.

Checking Knocking Sensor Resistance

4- Remove the single Black lead from the microcomputer to the knocking sensor at the screw terminal of the sensor. Set this Black lead aside.

Obtain an ohmmeter and set the scale to Rx1000 ohms. Make contact with the Red meter lead to the screw terminal of the sensor. Make contact with the Black meter lead to a suitable ground on the powerhead. A reading of no continuity (infinite resis-

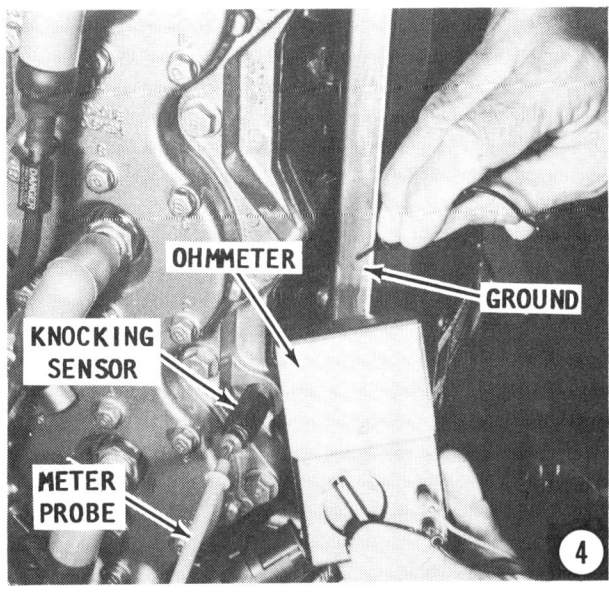

tance) should register on the meter. If the reading is less than infinity, there is a short in the sensor and the sensor **MUST** be replaced.

Checking Thermo Sensor Resistance

5- Disconnect the wire harness connector consisting of two Black leads from the thermosensor, at the quick disconnect fitting. Resistance tests cannot be performed while the sensor is installed in the powerhead. Therefore, remove the sensor from the block.

The resistance of the sensor is monitored at different temperatures while heating the "sensing" end of the sensor in a body of water. Obtain a thermometer, an ohmmeter, and a suitable pan in which to boil water.

Place the container of water, at room temperature, on a stove. Secure the thermometer in the water in such a manner to prevent the bulb from contacting the sides or bottom of the pan.

Immerse the "sensing" ends of the sensor into the water up to the shoulder. Make contact with the Red meter lead to one of the Black wire terminals and the Black meter lead to the other Black wire terminal. Make contact with the meter leads to the sensor leads. Between $0°$ and $60°F$, the meter should register 25 ohms; between $60°$ and $100°F$ the meter should register an increase in resistance to 85 ohms; between $100°$ and $150°F$ the meter should register a decrease in resistance to 30 ohms. All values given are approximate. The thermo sensor is not a resistor. It is a thermistor which characteristically varies resistance (increases or decreases voltage) with a change in temperature. In this instance, the resistance is highest in mid range.

Checking Throttle Position Sensor Resistance

Mount the engine in a test tank, or on a boat in a body of water.

Connect a tachometer to the powerhead per the instructions with the instrument. Check the "Special Words on Tachometers and Connections" on Page 7-2.

RUNAWAY POWERHEAD

Never operate the powerhead above a fast idle with a flush attachment connected to the lower unit. Operating the powerhead at a high rpm with no load on the propeller shaft could cause the powerhead to **RUNAWAY** causing extensive damage to the unit.

REMEMBER, the powerhead will **NOT** start without the emergency tether in place behind the kill switch knob.

Start the engine and allow it to warm to operating temperature and run at 750 rpm.

CAUTION
Water must circulate through the lower unit to the powerhead anytime the powerhead is operating to prevent damage to the water pump in the lower unit. Just five seconds without water will damage the water pump impeller.

Do not remove or change the throttle sensor position during the following test. Disconnect the wire harness from the sensor consisting of Black, Red, and White leads at the quick disconnect fitting. Obtain an ohmmeter and select the Rx1000 scale. Make contact with the Red meter lead to the Red sensor terminal. Make contact with the Black meter lead to the Black sensor terminal. The meter should register between 800 to 1200 ohms at idle speed. Retain the Red meter lead in place, and move the Black meter lead to the White sensor terminal. The meter should register the same reading.

Select the Rx100 scale on the meter. Make contact with the Red meter lead to the White sensor lead. Make contact with the Black meter lead to the Black sensor terminal. The meter should register between 16 to 24 ohms at idle speed. If the sensor resistance readings are not within

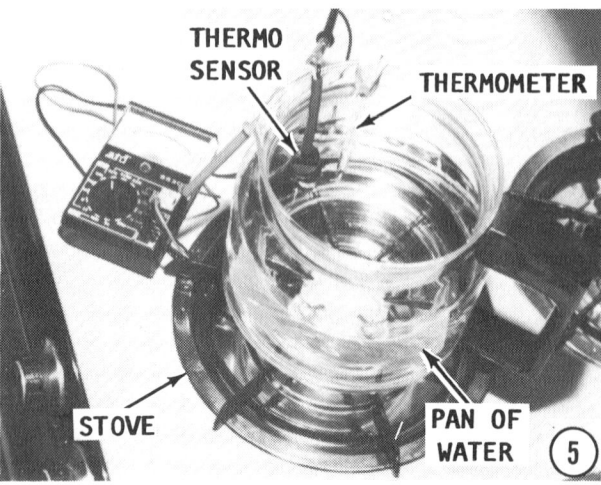

specifications, mark the original location of the throttle position sensor **BEFORE** removal. Should this sensor be misaligned during installation, a **DIGITAL** type voltmeter is needed to correctly reset the sensor on its mounting bracket, because voltages must be accurately measured to within 1/10 volt. A misaligned sensor could send misleading signals to the microcomputer and consequently affect the ignition timing.

After marking the position of the throttle sensor, remove the two securing screws and lift the sensor free of its plastic coupling. Slide the plastic coupling from the throttle shaft.

To install a new sensor, first, slide the plastic coupling over the throttle shaft extension and the throttle position sensor shaft and engage the pins of both shafts into the slots of the plastic coupling. Next, install the sensor onto the bracket on the top carburetor, with the attaching hardware, in **EXACTLY** the same location, matching the marks made during removal. If no locating marks were made during removal, **CAREFULLY** examine the bracket for traces of an outline of the sensor and attempt to mount the sensor as near to the original location as possible. A difference of 0.04" (1mm) corresponds to $2°$ throttle angle.

Throttle position sensor adjustment is covered in the following step using a **DIGITAL** type voltmeter, Yamaha P/N YU-34899-A.

Do not attempt this adjustment unless an instrument is available which can measure voltage accurately to within 1/10 volt. If such an instrument is not available, it is strongly recommended qualified technicians at the local Yamaha dealership install and calibrate this sensor using the correct equipment.

Throttle Position Sensor Adjustment

The throttle position sensor **MUST** be adjusted with the throttle plates fully closed. Therefore, ensure the plates are fully closed and adjust as required with the throttle stop screw. If it is necessary to disturb the setting of the throttle stop screw, adjust the throttle stop after the calibration of the sensor is complete. Refer to Chapter 7, Section 7-3 to readjust the setting on the throttle stop screw.

6- With the powerhead **NOT** running, loosen the throttle stop screw until the tip of the screw backs away from the throttle arm stopper. The throttle stop screw is located close to the throttle roller and is the **ONLY** vertical screw equipped with a spring on the entire carburetor assembly.

Lightly pull up on the accelerator link rod to close the throttle valves. Do **NOT** disconnect the wire harness from the microcomputer to the throttle position sensor.

SPECIAL METER

The meter used in this test must have either a special harness for measuring the output voltage or have probes capable of piercing through the insulation of the wires to make contact without disconnecting the leads or damaging the insulation of the leads.

Select the Vx1 scale on the voltmeter. Ensure the two mounting screws of the sensor are snug enough to permit the sensor to be rotated a few degrees either way and still hold its position.

Make contact with the Red voltmeter lead to the White sensor lead. Make contact with the Black voltmeter lead to the Black sensor lead.

Turn the key to the **ON** position **WITHOUT** starting the powerhead. Rotating the sensor clockwise will increase the voltage reading on the meter. When a voltage of between 0.40 and 0.42 is obtained, tighten

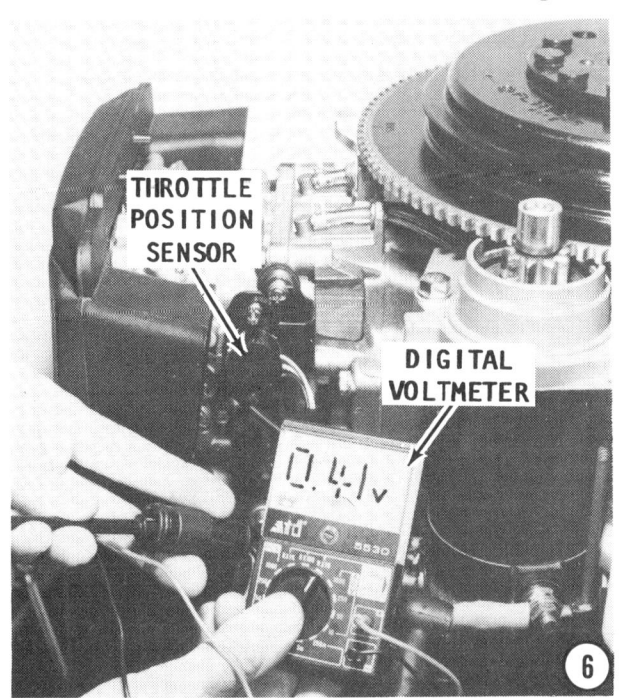

the mounting screws on the sensor to hold this position. A difference of 0.04" (1mm) corresponds to $2°$ of throttle angle.

With the voltmeter still connected, open and close the throttle valve a few times to be sure the output voltage fluctuates with different throttle angles. Also, check to see if the voltage reverts back to the specified level when the throttle valves are fully closed.

MARK the position of the new sensor in relation to the bracket as a preparation for possible future service.

Tighten the throttle stop screw until it contacts the throttle arm stopper, then continue to tighten screw 1-1/8 turns more, as a preliminary adjustment.

See Chapter 7, Section 7-3 for a finer adjustment of the throttle stop screw.

FINAL WORDS

The operation of this sensor can be compared to the volume knob on a radio. If the listener has a favorite position for the knob, in time a "flat spot" will wear on the shaft and interference may result.

If the operator of the outboard has a favorite throttle position, in time a "flat spot" will wear on the sensor shaft and not provide the microcomputer with accurate information. This condition could lead to ignition timing which varies by as much as $20°$ with no change in throttle position.

OTHER IGNITION COMPONENT TESTS

Tests for the CDI unit, pulsar coils, charge coils, lighting coil, and ignition coils are identical to the tests for the same unit on powerheads covered in this manual without YMIS.

See Section 6-6 for the appropriate tests for the component in question.

6-17 OPERATIONAL TESTS

Thermo Sensor Test

1- Remove the flywheel cover and make a mark on the flywheel at $26°$ BTDC, $7°$ BTDC, and $5°$ ATDC, which can be easily read while the powerhead is operating. The cover may now be installed.

Mount the engine in a test tank, or on a boat in a body of water.

Obtain a timing light and clip the pickup lead to the No. 1 spark plug lead. Connect the other two leads to the battery terminals. Observe any markings on the pickup clip of the timing light lead. Many have an arrow indicating the direction TOWARD the spark plug. Take time to make sure the clip is affixed properly.

Connect a tachometer to the powerhead per the instructions with the instrument. Check the "Special Words on Tachometers and Connections" on Page 7-2.

RUNAWAY POWERHEAD

Never operate the powerhead above a fast idle with a flush attachment connected to the lower unit. Operating the powerhead at a high rpm with no load on the propeller shaft could cause the powerhead to RUNAWAY causing extensive damage to the unit.

REMEMBER, the powerhead will NOT start without the emergency tether in place behind the kill switch knob.

Start the engine and allow it to warm to operating temperature and run at idle speed.

CAUTION

Water must circulate through the lower unit to the powerhead anytime the powerhead is operating to prevent damage to the water pump in the lower unit. Just five seconds without water will damage the water pump impeller.

Take care not to come in contact with any moving parts, such as the flywheel, or any high voltage components, such as the CDI unit, or the ignition coils, while making the following tests.

Aim the timing light at the timing pointer. The initial timing at idle will be $5°$ ATDC.

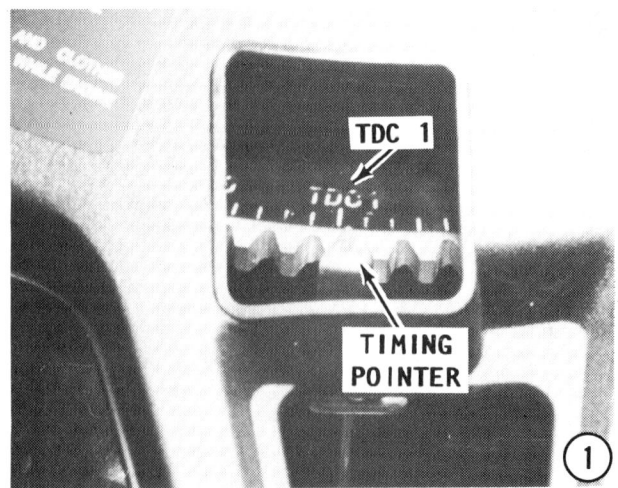

2- Disconnect the thermo sensor wire harness at the quick disconnect fitting. This harness consists of two Black leads. Use a small jumper cable and connect the two Black leads together. Aim the timing light at the timing pointer. The pointer sould align with the 7° BTDC mark at idle speed.

Obtain an 18mm open end wrench and remove the thermo sensor from the head. Reconnect the two Black leads at their quick connect fittings. Hold the "sensing" end of the sensor against an ice cube. Aim the timing light at the timing pointer, it should align with 0° TDC 1 embossed on the flywheel. Powerhead rpm should increase slightly. Wipe off any moisture from the sensor and install it back into the head. Keep the powerhead operating while performing the next test.

Crank Position Sensor Test

3- Verify the ignition timing is still 5° ATDC, with all sensors connected and the powerhead at idle rpm. Identify the harness connector from the crank position sensor. The connector consists of two Blue wires. Disconnect the two halves of the connector. The powerhead should stall. Reconnect the two halves of the connector and start the powerhead.

SPECIAL WORDS

The following four steps are performed with the powerhead running:

4- Push back the boot and disconnect the main power lead to the cranking motor from the starter relay. It is **MOST** important the cranking motor does not attempt to mesh with the flywheel teeth while the powerhead is operating.

Leave the small Brown lead on the relay in place.

Turn the key to the **START** position. Aim the timing light at the timing pointer. If the sensor is functioning correctly, the pointer should align with the 7° BTDC mark on the flywheel.

The symptoms of a defective crank position sensor are: engine rpm surges and erratic timing shift from 5° ATDC to 7° BTDC.

Knocking Sensor Test

5- Increase powerhead speed to 5,500 rpm. Aim the timing light at the timing pointer and note the number of degrees of advance.

The number of degrees should be 26° BTDC under the following conditions:
The knocking sensor is operating properly.
The fuel rating is at least 90 octane.
The internal condition of the powerhead is satisfactory.
If the timing pointer aligns with marks to the right of TDC 1.
The knocking sensor is defective and is giving a false reading.

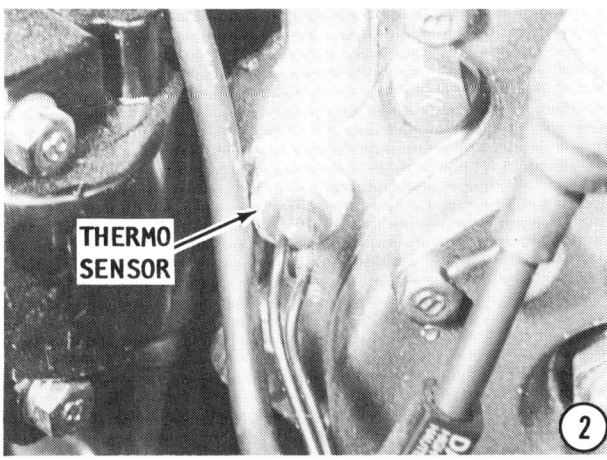

6-38 IGNITION

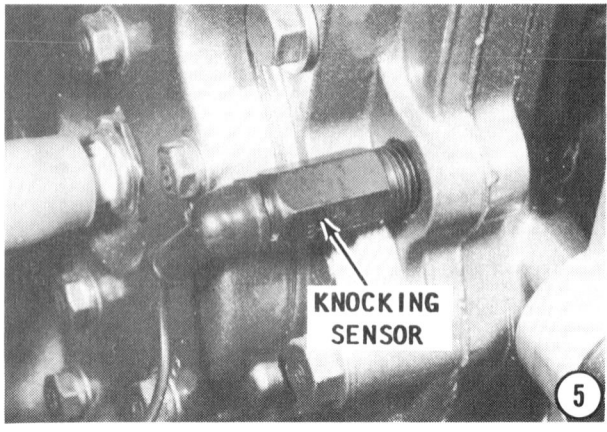

The octane rating of the fuel to too low. An internal problem exists in the powerhead.

Throttle Position Sensor Test

6- Identify the wire harness from the throttle position sensor. This harness consists of Black, Red, and White leads. Verify the ignition timing, is still 5° ATDC, with all sensors connected and the powerhead at idle rpm. Disconnect the two halves of the harness plug. There should be no change in the timing or engine rpm. Obtain a small jumper cable and connect the Red and White terminals together in the connector leading from the sensor. Powerhead rpm should remain approximately at idle, but the timing pointer should align with the 26° BTDC mark on the flywheel.

Thermo Switch Test

7- Increase engine speed to just above 2500 rpm. Identify the Pink and Black leads from one of the two thermo switches. Disconnect the two leads at the quick disconnect fittings. Touch the two leads from the switch together. The warning buzzer should sound continuously and powerhead rpm should drop to 2000 rpm. Reconnect the leads, back the throttle down to idle rpm and the buzzer should be silent. Repeat this test for the second thermo switch installed in the other cylinder bank.

Microcomputer Test

No test exists, either in resistance values or operational performance for the microcomputer. If all the sensor signals are processed correctly, all engine control circuits function, and the timing is correct then the microcomputer is alive and well.

Depressing Words

If the following tests and conditions have been performed or met:
All tests performed on the sensors.
All test performed on the switches.
Operational checks on the system have been conducted and are satisfactory.
All connectors have been checked and rechecked.
The battery is ample in size and power.
Then it might be best to take the complete outboard unit to the local Yamaha dealer for a second opinion before rushing out and buying a new, expensive, **NON-RETURNABLE** microcomputer. A second opinion might be less expensive than a new microcomputer unit. If the sad verdict indicates the outboard does need a new microcomputer, the installation labor should be minimal, the work will most likely be guaranteed, the problem definitely solved, and the owner will have "peace of mind".

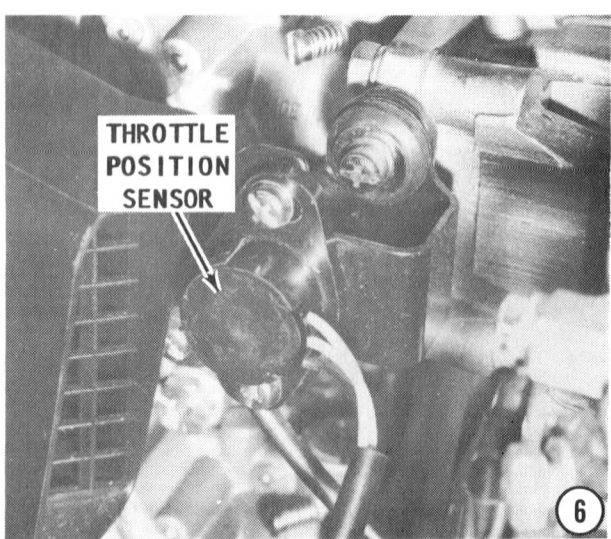

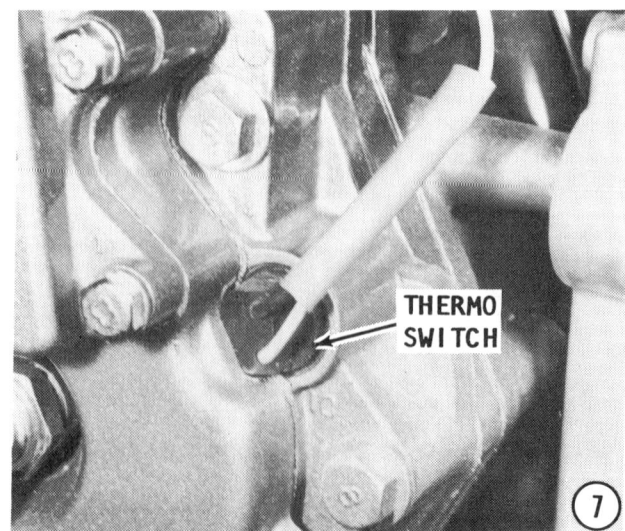

7
TIMING AND SYNCHRONIZING

7-1 INTRODUCTION AND PREPARATION

Timing and synchronization on an outboard engine is extremely important to obtain maximum efficiency. The powerhead cannot perform properly and produce its designed horsepower output if the fuel/carburetion and ignition systems have not been precisely adjusted.

Synchronization

In simple terms, synchronization is timing the carburetion to the ignition. This means, as the throttle is advanced to increase powerhead rpm, the carburetor and the ignition systems are both advanced equally and at the same rate.

Therefore, any time the fuel system or the ignition system on a powerhead is serviced to replace a faulty part, or any adjustments are made for any reason, powerhead timing and synchronization **MUST** be carefully checked and verified.

For this reason the timing and synchronizing procedures have been separated from all others and presented alone in this chapter.

Before making adjustments with the timing or synchronizing, the ignition system should be thoroughly checked according to the procedures outlined in Chapter 6, and the fuel system verified as in good working order per Chapter 4.

Timing

All outboard powerheads have some type of synchronization between the carburetion and ignition systems.

All units covered in this manual except those equipped with YMIS (Yamaha Microcomputer Ignition System), are equipped with a mechanical advance type CDI (Capacitor Discharge Ignition), system and use a series of link rods between the carburetor and the ignition base plate assembly. At the time the throttle is opened, the ignition base plate assembly is rotated by means of the link rod, thus advancing the timing.

On models equipped with YMIS, the microcomputer "decides" when to advance or retard the timing, based on input from various sensors. Therefore, there is no link rod between the magneto control lever and the stator assembly.

Many models have timing marks on the flywheel and CDI magneto base. A timing light is normally used to check the timing **DYNAMICALLY** -- with the powerhead operating.

An alternate method is to check the timing **STATICALLY** -- with the powerhead not operating. This second method requires the use of a dial indicator gauge.

Various models have unique methods of checking ignition timing. These differences are explained in detail and supported with illustrations whenever possible.

PREPARATION

Timing and synchronizing the ignition and fuel systems on an outboard motor are critical adjustments. Therefore, the following equipment is essential and is called out repeatedly in this section. This equipment must be used as described, unless otherwise instructed. Naturally, they are removed following completion of the adjustments.

Dial Indicator

Top dead center (TDC) of the No. 1 (top) piston must be precisely known before the timing adjustment can be made. TDC can only be determined through installation of a dial indicator into the No. 1 spark plug opening.

7-2 TIMING & SYNCHRONIZING

Timing Light

During many procedures in this section, the timing mark on the flywheel must be aligned with a stationary timing mark on the engine while the powerhead is being cranked, or is running. Only through use of a timing light connected to the No. 1 spark plug, can the timing mark on the flywheel be observed while the engine is operating.

Tachometer

A tachometer connected to the powerhead must be used to accurately determine engine speed during idle and high-speed adjustment.

The meter readings range from 0 to 6,000 rpm in increments of 100 rpm. Tachometers have solid state electronic circuits which eliminates the need for relays or batteries and contributes to their accuracy.

Tachometer Connections

A tachometer is installed as standard equipment on all powerheads covered in this manual.

Due to local conditions, it may be necessary to adjust the carburetor while the outboard unit is running in a test tank or with the boat in a body of water. For maximum performance, the idle rpm should be adjusted under actual operating conditions.

Under such conditions it might be necessary to attach a tachometer closer to the powerhead than the one installed on the control panel.

To connect an auxiliary tachometer, first open the remote control box. Next, disconnect the Black and Green leads. Connect the Black lead to the ground terminal of the auxiliary tachometer and the Green lead to the "input" or "hot" terminal of the auxiliary tachometer.

Flywheel Rotation

During the procedures listed in this chapter, the instructions may call for rotating the flywheel until certain marks are aligned with the timing pointer. When the flywheel must be rotated, **ALWAYS** move the flywheel in a **CLOCKWISE** direction.

If the flywheel should be rotated in the opposite direction, the water pump impeller tangs would be twisted backwards.

Should the powerhead be started with the pump tangs bent back in the wrong direction, the tangs may not have time to bend in the correct direction before they are damaged. The least amount of damage to the water pump will affect cooling of the powerhead.

Test Tank

The engine must be operated at various times during the procedures. Therefore, a test tank, or moving the boat into a body of water, is necessary.

CAUTION

Never operate the powerhead above a fast idle with a flush attachment connected to the lower unit. Operating the powerhead at a high rpm with no load on the propeller shaft could cause the powerhead to **RUNAWAY** causing extensive damage to the unit.

REMEMBER, the powerhead will **NOT** start without the emergency tether in place behind the "kill" switch knob.

CAUTION

Water must circulate through the lower unit to the powerhead anytime the powerhead is operating to prevent damage to the water pump in the lower unit. Just five seconds without water will damage the water pump impeller.

7-2 ALL V4 AND V6 MODELS NOT EQUIPPED WITH YMIS

TIMING PLATE ALIGNMENT AND PRELIMINARY ADJUSTMENTS

The following procedures are considered "static" adjustments, because they are performed with the powerhead at rest -- not operating.

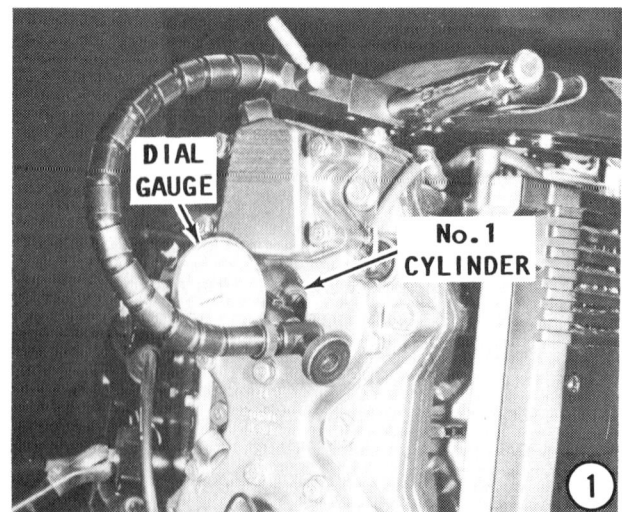

ADJUSTMENTS 7-3

1- Remove all the spark plugs from the powerhead. Install a dial indicator into the No. 1 cylinder opening.

2- Rotate the flywheel **CLOCKWISE** until the dial indicator indicates the piston is at TDC (top dead center). Check the timing pointer to be sure it aligns with the TDC mark embossed on the flywheel. If the mark is misaligned, loosen the set screw on the timing plate and align the pointer with the flywheel mark. Tighten the screw to hold the adjustment.

PRELIMINARY LINK ROD ADJUSTMENTS

3- Measure the horizontal distance from ball joint center to ball joint center of the magneto control link, between the stator plate and the magneto control lever. On V4 models, this link is perfectly straight. On V6 models this link is curved as shown in the accompanying illustration. Adjust the horizontal distance according to the model being serviced. Pry off the threaded end of the rod from the magneto plate ball joint and rotate the end until the specified distance is achieved. The rod may then be snapped in place. This is a preliminary measurement and may be changed later.

115hp	2-1/2"	(62mm)
130hp	2-3/8"	(60mm)
150hp	2-1/2"	(62mm)
175hp	2-1/2"	(62mm)
200hp	2-1/2"	(62mm)

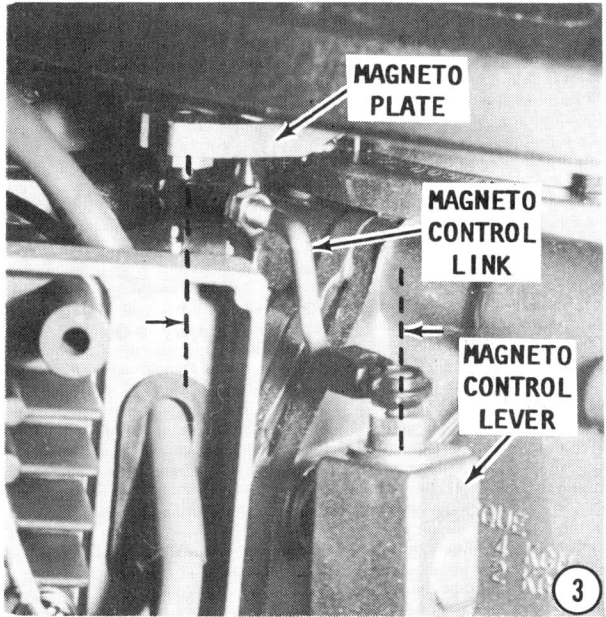

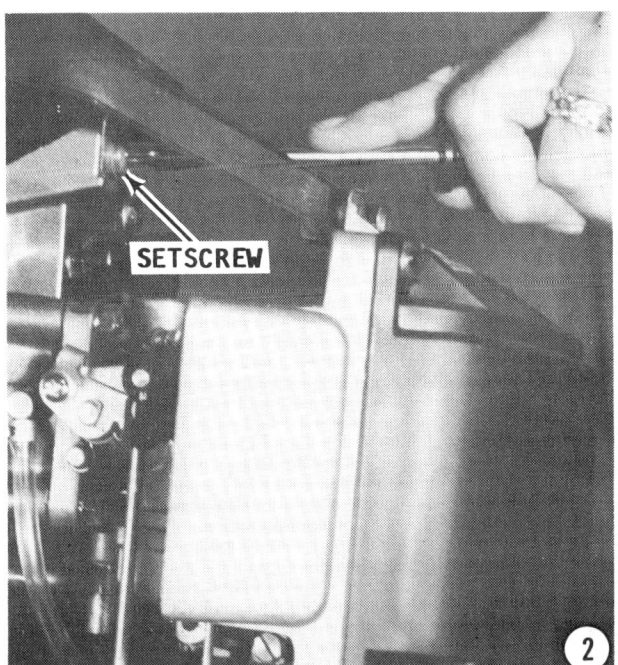

4- Measure the distance between the outer edge of the stopper, where it contacts the magneto control lever (when the lever is in the fully advanced position), and the aft edge of the threaded boss supporting the upper adjustment screw.

115hp	15/16"	(24mm)
130hp	15/16"	(24mm)
150h	23/32"	(18mm)
175hp	23/32"	(18mm)
200hp	23/32"	(18mm)

If the distance is not as specified, loosen the locknut and adjust the distance to specification. Tighten the locknut to hold this new adjustment. This is a preliminary measurement and may be changed later.

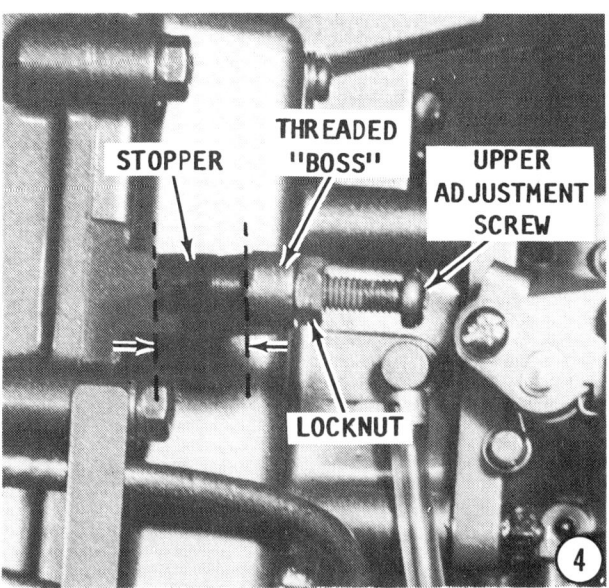

7-4 TIMING & SYNCHRONIZING

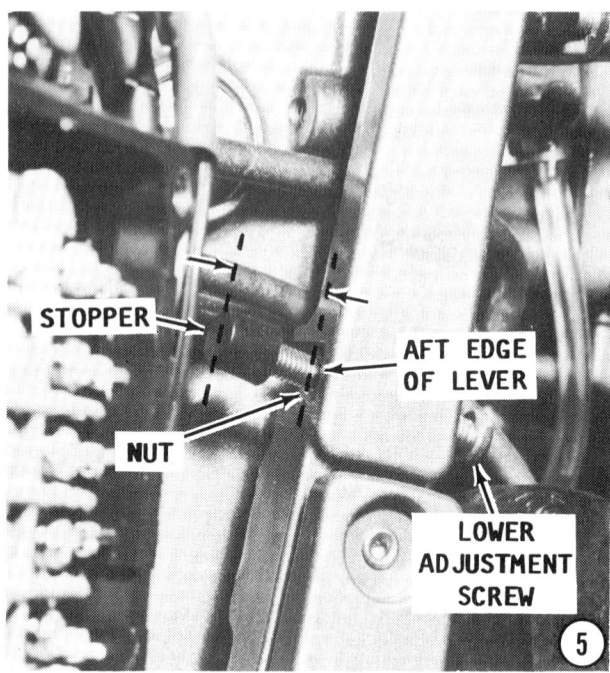

5- Two measurements and possibly adjustments are made to the lower adjustment screw. The distance between the recessed nut and the aft edge of the magneto control lever should be 5/64" (2mm) for all V4 models and 7/64" (2.5mm) for all V6 models. The recessed nut is threaded into the magneto control lever and has a spring behind it to prevent rotation. It is unlikely this dimension will ever change.

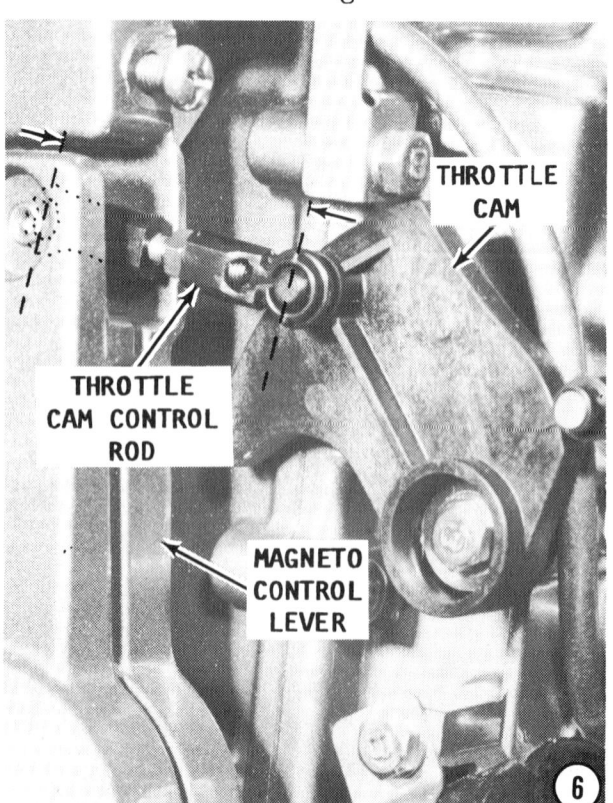

The second measurement on the lower adjustment screw is the distance between the outer edge of the stopper, which contacts a raised portion of the block (when the magneto control lever is in the fully retarded position), and the aft edge of the magneto contol lever.

115hp	27/32"	(21mm)
130hp	15/16"	(24mm)
150hp	5/8"	(16mm)
175hp	5/8"	(16mm)
200hp	5/8"	(16mm)

Adjust the position of the screw until the specified distance is achieved.

6- Measure the length of the throttle cam control link rod from ball joint center to ball joint center, between the magneto control lever and the acceleration cam.

115hp	2-7/64"	(54mm)
130hp	2-3/32"	(53mm)
150hp	1-21/32"	(42mm)
175hp	1-21/32"	(42mm)
200hp	1-21/32"	(42mm)

Adjust the distance by prying off the link from the ball joint on the accelerator cam and rotating the threaded end of the link rod until the specified distance is obtained. Snap the link rod back onto the ball joint.

STATIC TIMING
FULLY RETARDED

7- Check the Appendix for the specified number of degrees ATDC for the model being serviced. Now, rotate the flywheel **CLOCKWISE** until the timing pointer is aligned with the specified number of de-

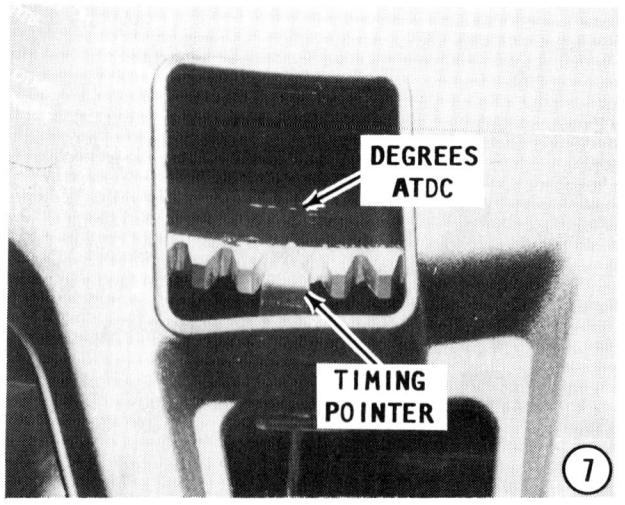

STATIC TIMING 7-5

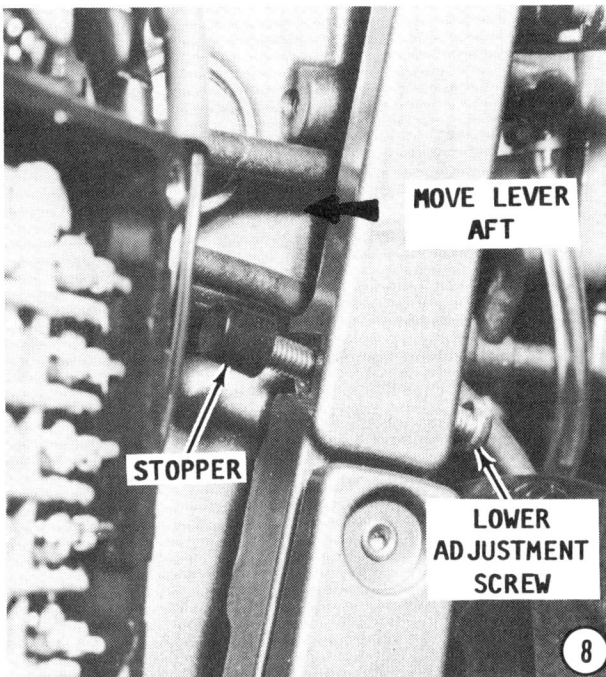

grees ATDC embossed on the flywheel This mark should come to the right of TDC 1. There are other TDC marks on the flywheel. Final position for the pointer can be between a couple of marks. Example: for $5°$, the pointer should be between the $4°$ and $6°$ marks.

8- Move the magneto control lever aft until the stopper on the lower adjustment screw contacts the stop on the block.

9- The pulsar assembly arm is located on the portside of the powerhead. This assembly has an alignment mark on its outboard surface. An embossed mark (a dot) is located on the underside of the flywheel. The mark on the pulsar assembly arm should align with the embossed dot on the flywheel. If the mark and the dot are not aligned, rotate the lower adjustment screw on the magneto control lever until the pulsar mark and the flywheel dot are aligned.

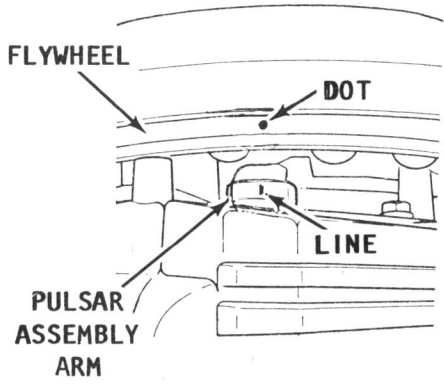

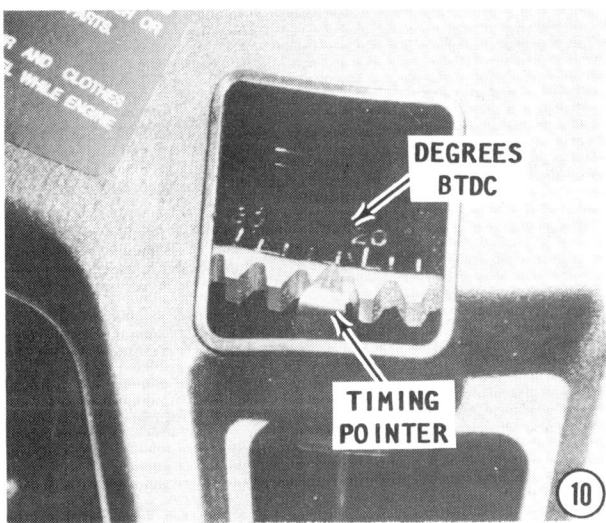

STATIC TIMING FULLY ADVANCED

10- Rotate the flywheel clockwise until the timing pointer aligns with the specified number of degrees BTDC (as listed in the Appendix for the model being seviced). This mark should come to the left of TDC 1. There are other TDC marks on the flywheel. Final position for the pointer can be between a couple of marks. Example: for $5°$, the pointer should be between the $4°$ and $6°$ marks.

11- Move the magneto control lever forward until the lever contacts the stopper on the end of the upper adjustment screw.

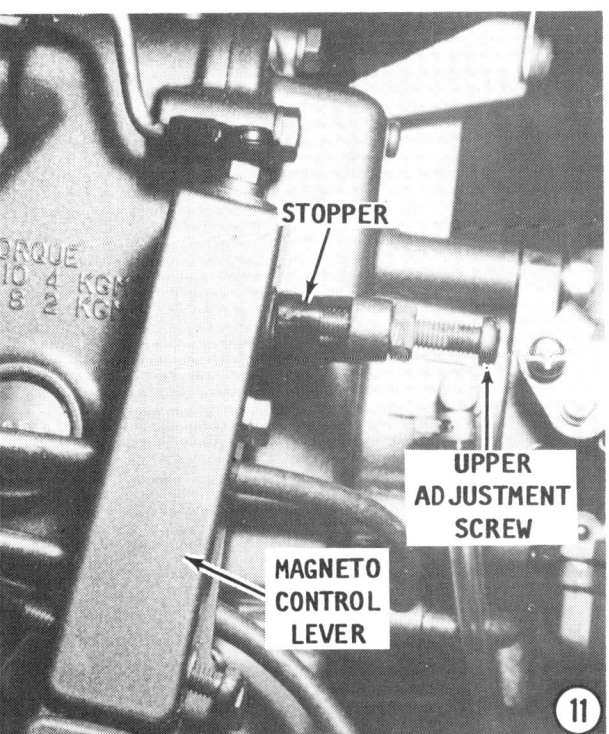

7-6 TIMING & SYNCHRONIZING

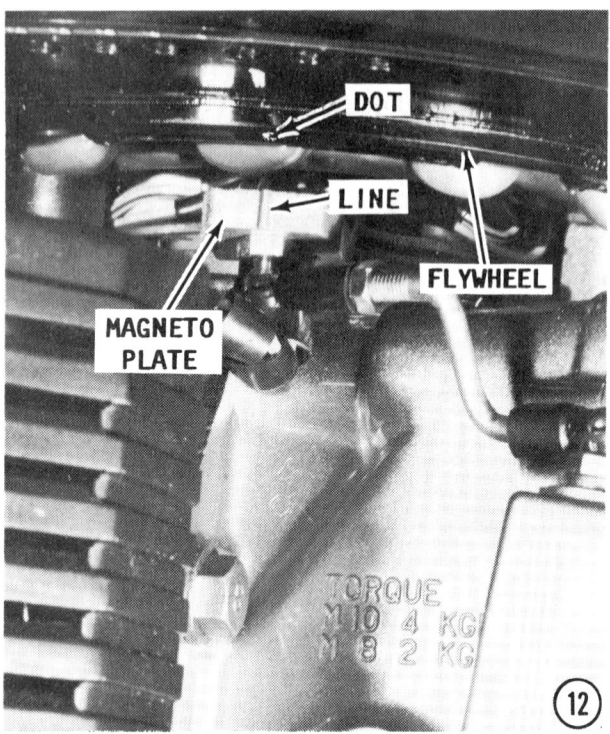

12- The magneto plate extension is located on the starboard side of the powerhead. This assembly has an alignment mark on its outboard surface. An embossed mark (a dot) is located on the underside of the flywheel. The mark on the magneto plate extension must align with the embossed dot on the flywheel. If the marks do not align, loosen the locknut on the upper adjustment screw and rotate the screw allowing the magneto control lever to move until the marks are aligned. Tighten the locknut to hold the new adjustment.

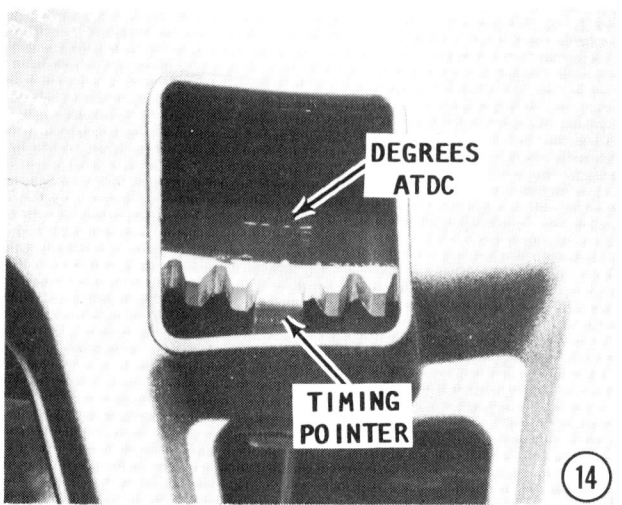

STATIC TIMING
PICKUP ADJUSTMENT

13- Pry off the throttle control link rod from the accelerator cam ball joint.

14- Continue to rotate the flywheel clockwise until the timing pointer aligns with the specified number of degrees ATDC as listed in the Appendix for the model being serviced. This mark should come to the right of TDC 1. There are other TDC marks on the flywheel. Final position for the pointer can be between a couple of marks. Example: for 5°, the pointer should be between the 4° and 6° marks.

15- Move the magneto control lever until the marks on the portside of the flywheel rotor align with the pulsar assembly arm.

16- Bring the throttle lever roller up to the accelerator cam. The line embossed on the cam is where the roller should just start to make contact -- the pickup point. If the roller contacts the cam at some other point or not at all, loosen the throttle lever adjustment screw by rotating the screw **CLOCKWISE**. This screw has **LEFT HAND**

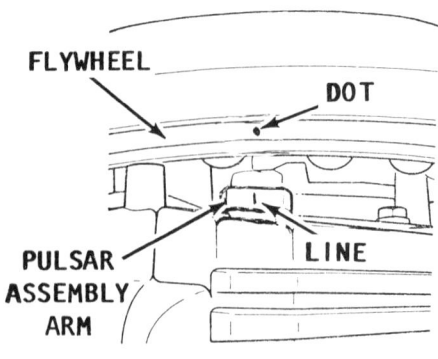

DYNAMIC TIMING 7-7

threads. Move the roller up to the line. Tighten the screw **COUNTERCLOCKWISE**.

17- Adjust the length of the throttle control link rod until the rod may be snapped back onto the ball joint on the accelerator cam, without movement of the magneto control lever or the accelerator cam.

DYNAMIC TIMING

The following procedures are considered dynamic adjustment because they are made with the powerhead operating.

Mount the engine in a test tank, or on a boat in a body of water.

CRITICAL WORDS

Never attempt the following adjustments with a flush attachment connected to the lower unit. The powerhead will be operating at fairly high speeds and with no load on the propeller, the engine could **RUNAWAY** causing extensive damage to the unit.

Obtain a timing light and clip the pickup lead to the No. 1 spark plug lead. Of course, the other two leads are connected to the battery terminals. Observe any markings on the pickup clip of the timing light lead. Many have an arrow indicating the direction **TOWARD** the spark plug. Take time to make sure the clip is affixed properly.

If working outside in bright sunlight, it would be most helpful to have an assistant create some shade to enable observing the timing light more easily.

Connect a tachometer to the powerhead per the instructions with the instrument. Check the "Special Words on Tachometers and Connections" on Page 7-2.

RUNAWAY POWERHEAD

Never operate the powerhead above a fast idle with a flush attachment connected to the lower unit. Operating the powerhead at a high rpm with no load on the propeller shaft could cause the powerhead to **RUNAWAY** causing extensive damage to the unit.

REMEMBER, the powerhead will **NOT** start without the emergency tether in place behind the "kill" switch knob.

Start the engine and allow it to warm to operating temperature and run at idle speed.

CAUTION

Water must circulate through the lower unit to the powerhead anytime the powerhead is operating to prevent damage to the water pump in the lower unit. Just five seconds without water will damage the water pump impeller.

Take care not to come in contact with any moving parts, such as the flywheel, or any high voltage components, such as the CDI unit, or the ignition coils, while making the following adjustments.

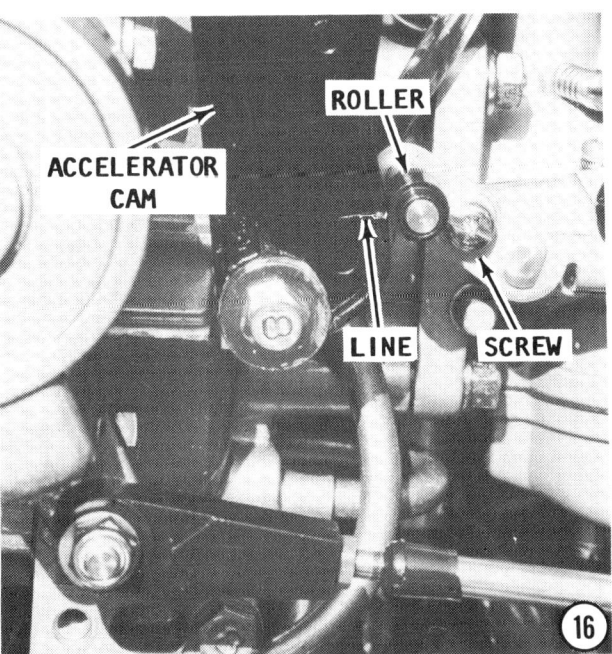

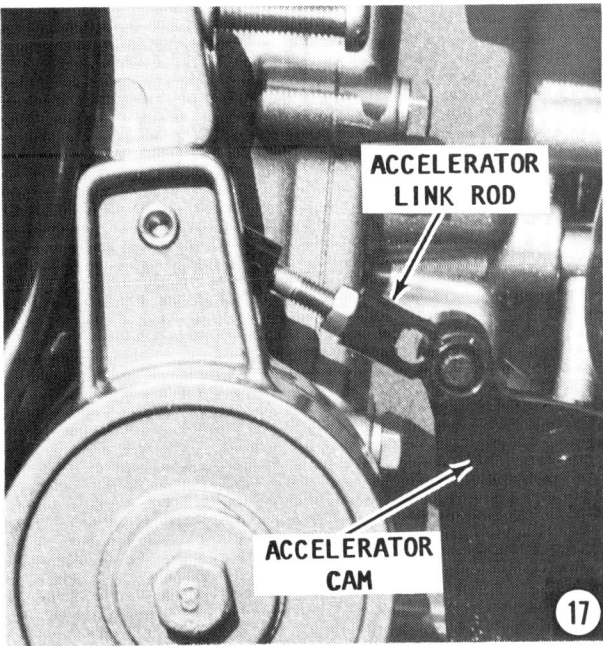

7-8 TIMING & SYNCHRONIZING

DYNAMIC TIMING
FULLY RETARDED AT IDLE SPEED

18- Move the magneto control lever aft to the fully retarded position until the stopper on the lower adjustment screw contacts the stop on the block.

19- Aim the timing light at the timing pointer and verify the number of degrees ATDC agree with the specifications listed in the Appendix for the model being serviced. This mark should come to the right of TDC 1. There are other TDC marks on the flywheel.

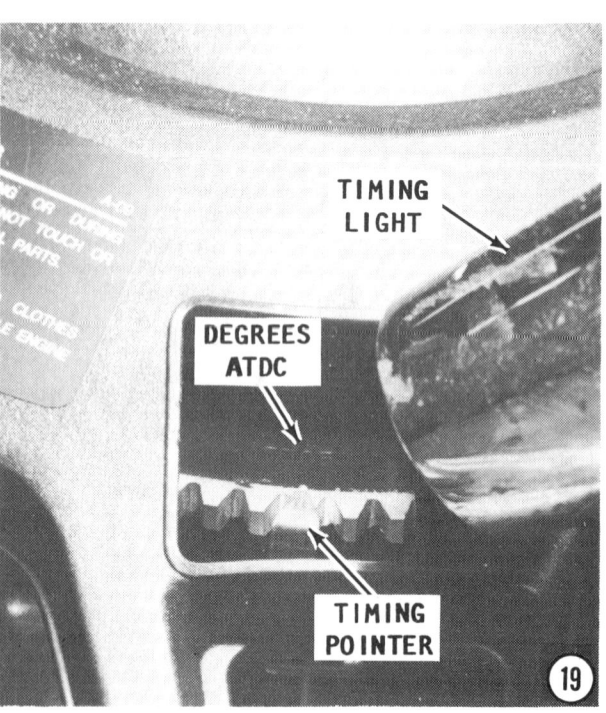

DYNAMIC TIMING
FULLY ADVANCED

20- Pry off the throttle control link rod from the ball joint on the accelerator cam. This rod will remain disconnected through Steps 21, 22, 23, 24, and 25.

21- Move the magneto control lever forward until it contacts the stopper on the upper adjustment screw.

22- Aim the timing light at the timing pointer and verify the pointer aligns with the specified number of degrees BTDC as listed in the Appendix for the model being

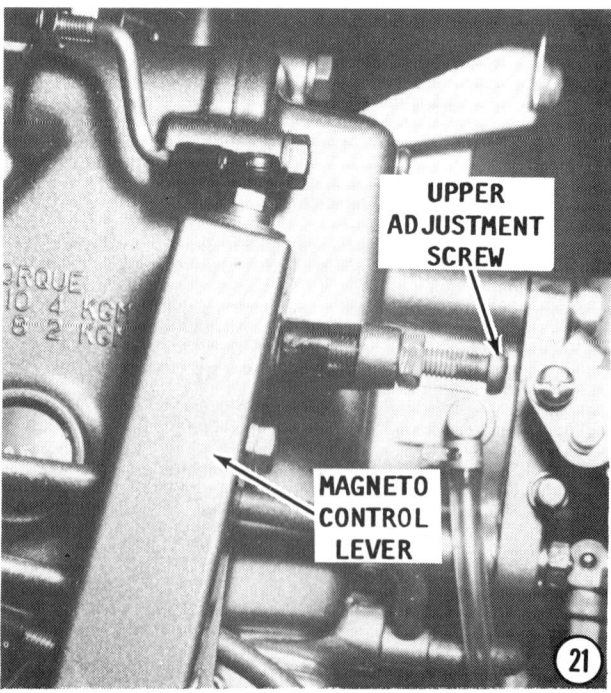

DYNAMIC TIMING 7-9

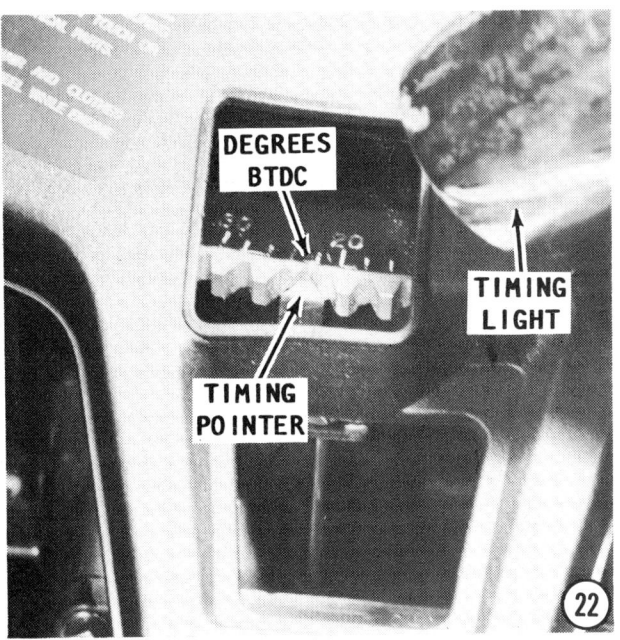

serviced. This mark should come to the left of TDC 1. There are other TDC marks on the flywheel.

If the timing pointer does not align with the desired mark, loosen the locknut on the upper adjustment screw and rotate the screw clockwise to retard the timing or counterclockwise to advance the timing. Tighten the locknut to hold this new adjustment.

WARNING
TAKE CARE NOT TO TOUCH THE ROTATING TEETH OF THE FLYWHEEL.

DYNAMIC TIMING PICKUP ADJUSTMENT

23- Move the magneto control lever slightly aft.

24- Aim the timing light at the timing pointer and continue to move the magneto control lever until the pointer aligns with the specifed number of degrees ATDC as listed in the Appendix for the model being serviced. This mark should come to the right of TDC 1. There are other TDC marks on the flywheel.

25- Bring the throttle lever roller up to the accelerator cam. The line embossed on the cam is where the roller should just start

7-10 TIMING & SYNCHRONIZING

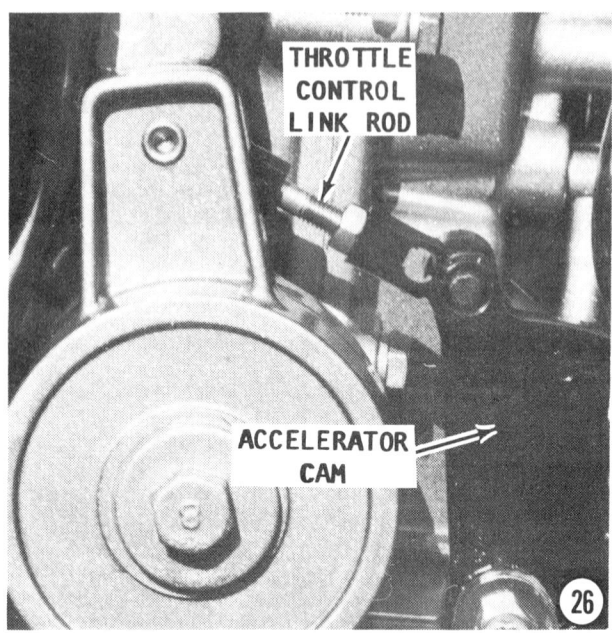

to make contact -- the pickup point. If the roller contacts the cam at some other point or not at all, loosen the throttle lever adjustment screw by rotating the screw **CLOCKWISE**. This screw has **LEFT HAND** threads. Move the roller up to the line. Tighten the screw by rotating **COUNTERCLOCKWISE**.

26- Adjust the length of the throttle control link rod until the rod may be snapped back onto the ball joint on the accelerator cam, without movement of the magneto control lever or the accelerator cam.

Shut down the powerhead.

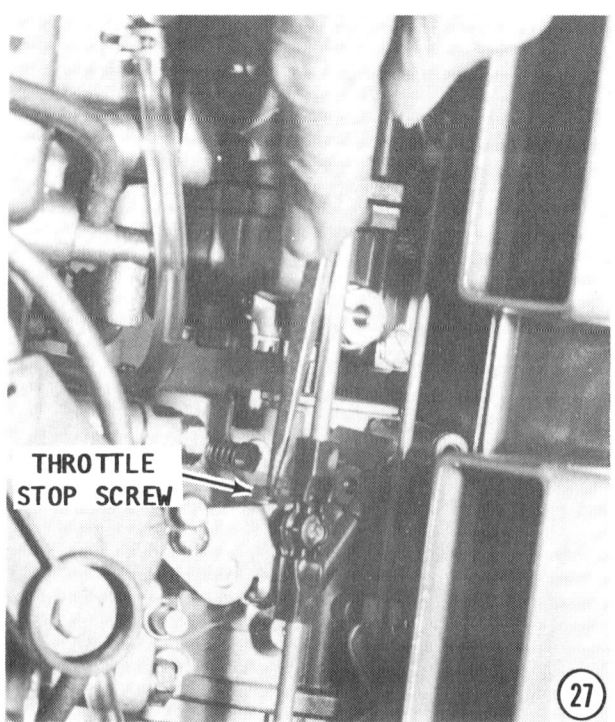

THROTTLE LINKAGE ADJUSTMENT

27- With the powerhead **NOT** running, loosen the throttle stop screw until the tip of the screw backs away from the throttle arm stopper. The throttle stop screw can be found close to the throttle roller and is the only vertical screw equipped with a spring on the entire carburetor assembly.

28- Loosen the throttle arm adjustment screws on the upper and lower carburetors, by rotating the screw clockwise. These screws have **LEFT HAND** threads.

29- Lightly pull up on the accelerator link rod to remove all play and close the throttle valves. At the same time tighten both throttle arm adjustment screws by rotating them **COUNTERCLOCKWISE**.

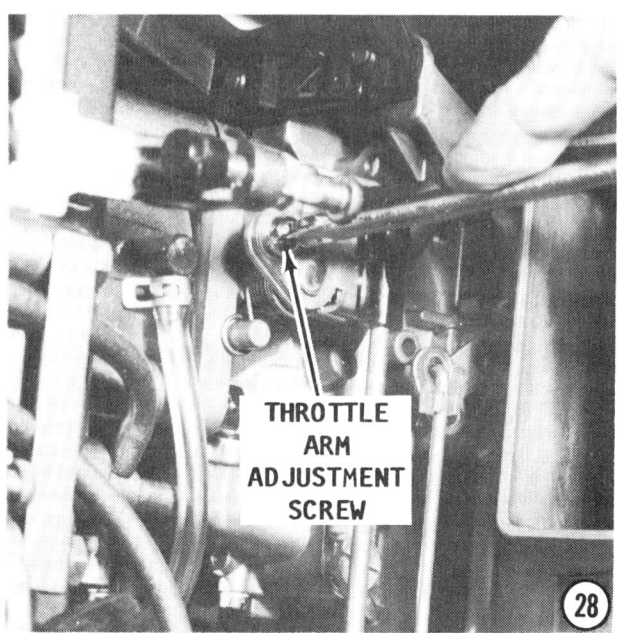

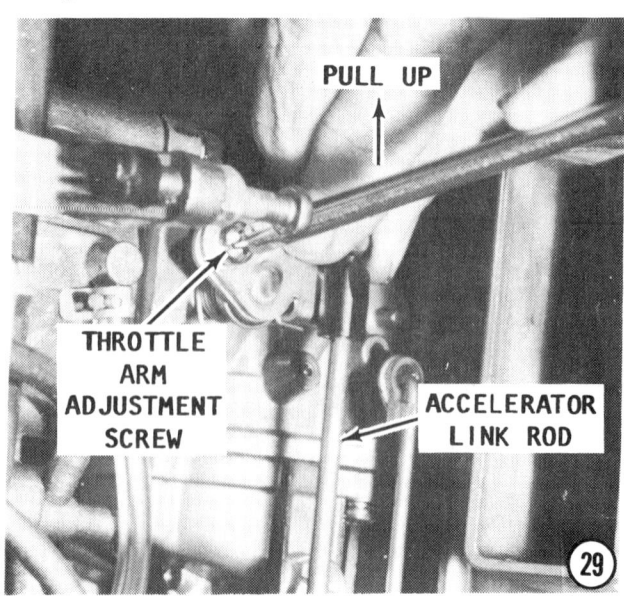

ADJUSTMENTS 7-11

30- Tighten the throttle stop screw until it contacts the throttle arm stopper, then continue to tighten one and a half turns more, as a preliminary adjustment.

IDLE ADJUSTMENT

Mount the engine in a test tank, or on a boat in a body of water.

Obtain a timing light and clip the pickup lead to the No. 1 spark plug lead. Of course, the other two leads are connected to the battery terminals. Observe any markings on the pickup clip of the timing light lead. Many have an arrow indicating the direction **TOWARD** the spark plug. Take time to make sure the clip is affixed properly.

Connect a tachometer to the powerhead per the instructions with the instrument. Check the "Special Words on Tachometers and Connections" on Page 7-2.

RUNAWAY POWERHEAD

Never operate the powerhead above a fast idle with a flush attachment connected to the lower unit. Operating the powerhead at a high rpm with no load on the propeller shaft could cause the powerhead to **RUNAWAY** causing extensive damage to the unit.

REMEMBER, the powerhead will **NOT** start without the emergency tether in place behind the kill switch knob.

Start the engine and allow it to warm to operating temperature and run at idle speed.

CAUTION

Water must circulate through the lower unit to the powerhead anytime the powerhead is operating to prevent damage to the water pump in the lower unit. Just five seconds without water will damage the water pump impeller.

Take care not to come in contact with any moving parts, such as the flywheel, or any high voltage components, such as the CDI unit, or the ignition coils, while making the following adjustments.

31- Move the magneto control lever aft to the fully retarded position until the stopper on the lower adjustment screw contacts the stop on the block.

32- Aim the timing light at the timing pointer and verify the number of degrees ATDC agree with the specifications listed in

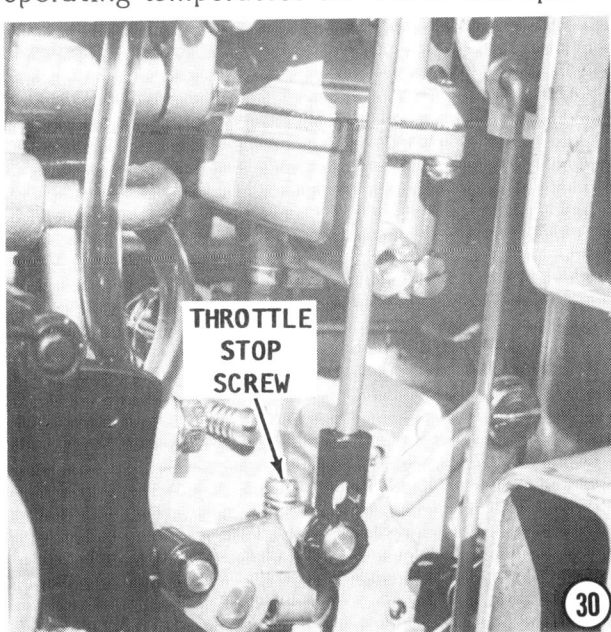

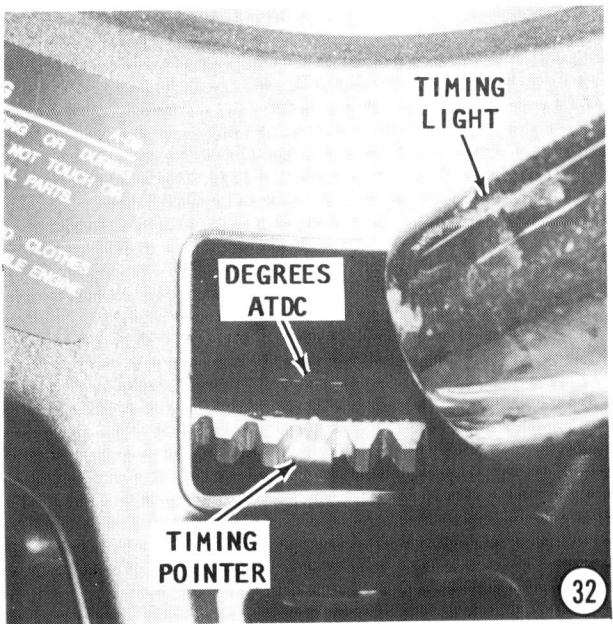

7-12 TIMING & SYNCHRONIZING

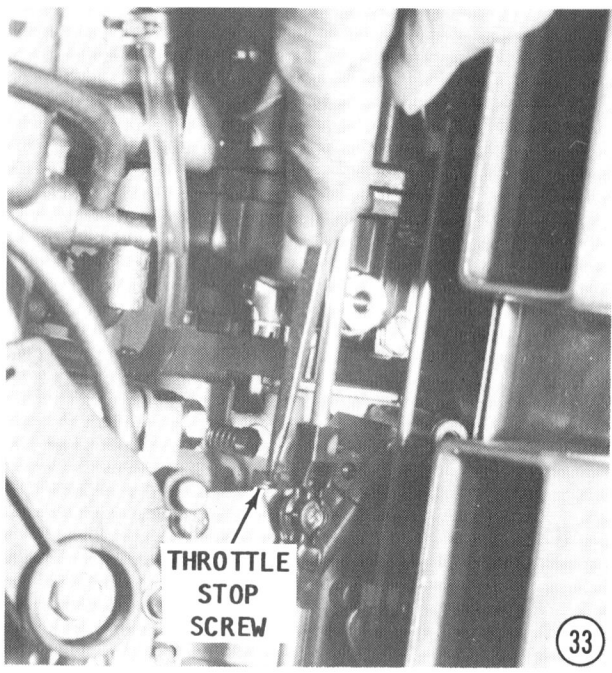

the Appendix at idle speed for the model being serviced. This mark should come to the right of TDC 1. There are other TDC marks on the flywheel.

33- Rotate the throttle stop screw until the specified idle speed, as listed in the Appendix for the model being serviced, is reached. Rotating the screw clockwise will increase engine rpm and rotating the screw counterclockwise will decrease engine rpm.

PILOT SCREW ADJUSTMENT

The pilot screw setting is not normally changed. During carburetor overhaul, the screw is set from a lightly seated position to a specified number of turns out. If the powerhead hesitates or runs roughly, check the setting of each pilot screw one by one. If the problem is not corrected, there is every reason to believe the carburetors need an overhaul.

34- Turn in the pilot screw until it is lightly seated. Now, back the screw out according to the specifications listed in the Appendix for the model being serviced. Pilot screw settings may vary from year to year and may also vary from port screw to starboard screw.

7-3 ALL MODELS EQUIPPED WITH YMIS

Ignition timing is automatically adjusted by the microcomputer, therefore unless components were disturbed during service, very little adjustment is necessary.

1- Remove all the spark plugs from the powerhead. Install a dial indicator into the No. 1 cylinder opening.

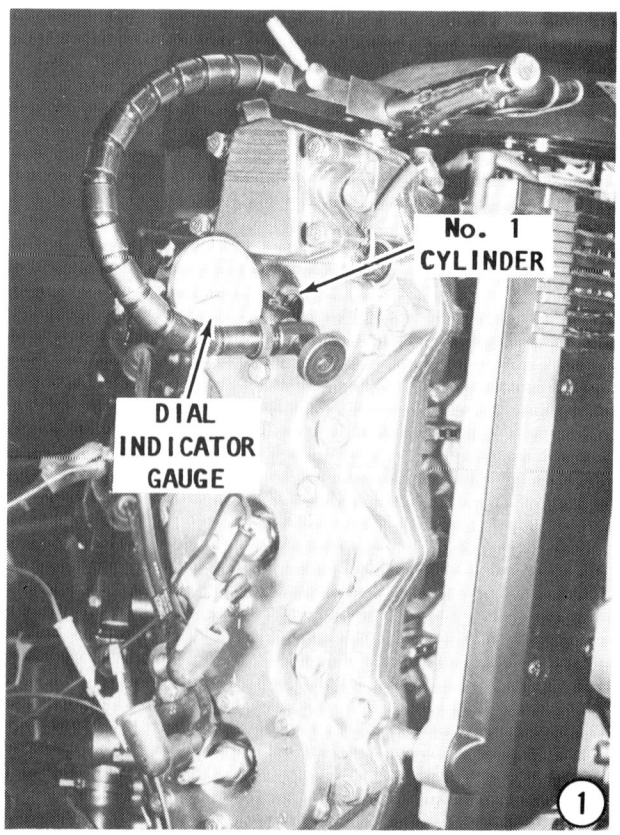

ADJUSTMENTS 7-13

2- Rotate the flywheel **CLOCKWISE** until the dial indicator indicates the piston is at TDC (top dead center). Check the timing pointer to be sure it aligns with the TDC mark embossed on the flywheel. If the mark is misaligned, loosen the set screw on the timing plate and align the pointer with the flywheel mark. Tighten the screw to hold the adjustment.

Mount the engine in a test tank, or on a boat in a body of water.

CRITICAL WORDS

Never attempt the following adjustments with a flush attachment connected to the lower unit. The powerhead will be operating at fairly high speeds and with no load on the propeller. The engine could **RUNAWAY** causing extensive damage to the unit.

Obtain a timing light and clip the pickup lead to the No. 1 spark plug lead. Of course, the other two leads are connected to the battery terminals. Observe any markings on the pickup clip of the timing light lead. Many have an arrow indicating the direction **TOWARD** the spark plug. Take time to make sure the clip is affixed properly.

Connect a tachometer to the powerhead per the instructions with the instrument. Check the "Special Words on Tachometers and Connections" on Page 7-2.

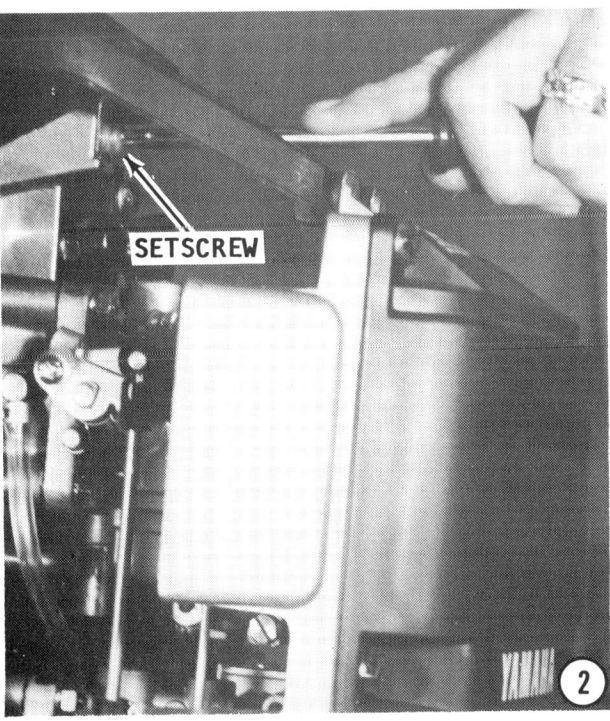

RUNAWAY POWERHEAD

Never operate the powerhead above a fast idle with a flush attachment connected to the lower unit. Operating the powerhead at a high rpm with no load on the propeller shaft could cause the powerhead to **RUNAWAY** causing extensive damage to the unit.

REMEMBER, the powerhead will **NOT** start without the emergency tether in place behind the kill switch knob.

Start the engine and allow it to warm to operating temperature and run at idle speed.

CAUTION

Water must circulate through the lower unit to the powerhead anytime the powerhead is operating to prevent damage to the water pump in the lower unit. Just five seconds without water will damage the water pump impeller.

Take care not to come in contact with any moving parts, such as the flywheel, nor any high voltage components, such as the CDI unit, or the ignition coils, while making the following adjustments.

3- Aim the timing light at the timing pointer, while allowing the powerhead to idle at 750 rpm. The timing pointer should align with the 6° ATDC mark embossed on the flywheel. This mark should come to the right of TDC 1. There are other TDC marks on the flywheel.

Increase the rpm to 5,500 or just above. The timing pointer should align with the mark 22° BTDC on the flywheel. This mark should come to the left of TDC 1.

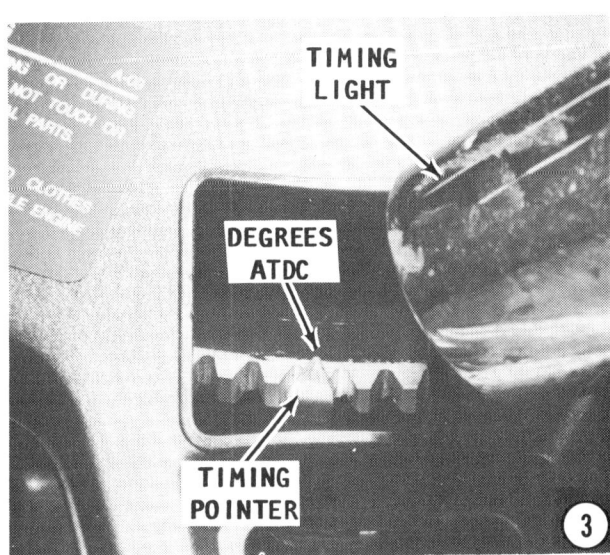

7-14 TIMING & SYNCHRONIZING

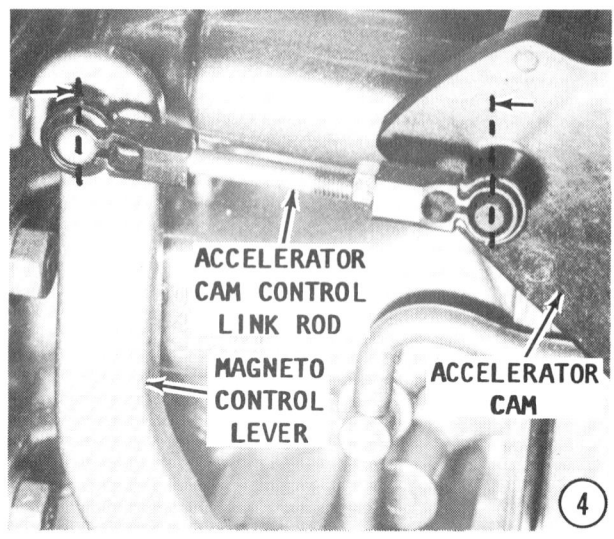

Words of Wisdom

If the timing is correct- stop right here. If it works -- don't fix it.

If the timing varies slightly, there is a good chance some minor adjustment will bring it back to specifications.

If the timing is way off, proceed with the following steps and take time to repeat Step 3 after each adjustment has been made to see if the problem has been corrected, before rushing out and buying a new, expensive, **NON-RETURNABLE** microcomputer.

If, after completing the following steps, the timing is still incorrect, then proceed to Section 6-12 in Chapter 6. This section deals with troubleshooting the YMIS and each of the sensors, until the problem can be found and corrected. If the YMIS is completely bypassed, the powerhead will still operate, but at a preset ignition timing irrespective of throttle opening. There is no link between the stator assembly and the magneto control lever. Therefore, there can be no advance or retard of the timing.

Remember to repeat Step 3 after each adjustment to see if the problem no longer exists.

GOOD WORDS

There is only one link rod to be adjusted.

THROTTLE CONTROL LINK ROD ADJUSTMENT

4- Measure the length of the throttle control link rod from ball joint center to ball joint center.

```
1984         1 21/32"  (42mm)
1985 to 1987 1 3/4"    (45mm)
1988 & ON    2 27/32"  (72mm)
```

If necessary, pry off the forward end of the rod from the ball joint on the accelerator cam, adjust the length of the rod, and then snap it back in place over the ball joint.

PICKUP TIMING ADJUSTMENT

5- Move the magneto control lever aft until the throttle lever roller just touches the accelerator cam. The line embossed on the cam is where the roller should just start to make contact -- the pickup point. If the roller contacts the cam at some other point or not at all, loosen the throttle lever adjustment screw by rotating the screw **CLOCKWISE**. This screw has **LEFT HAND** threads. Move the roller up to the line. Tighten the screw by rotating the screw **COUNTERCLOCKWISE**.

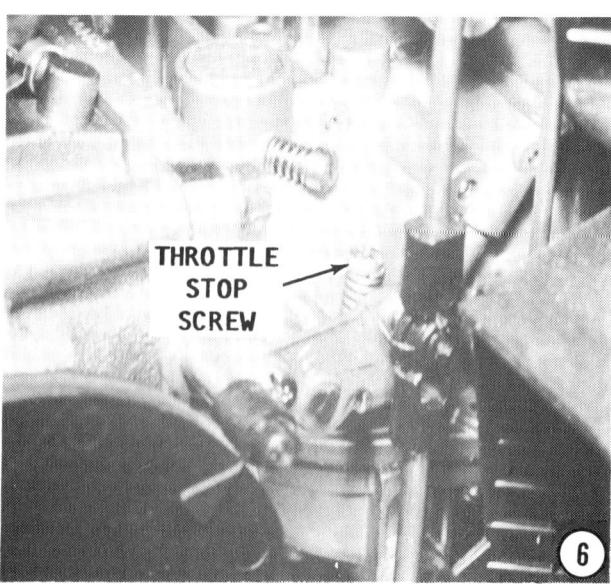

ADJUSTMENTS 7-15

THROTTLE LINKAGE ADJUSTMENT

6- With the powerhead **NOT** runnning, loosen the throttle stop screw until the tip of the screw backs away from the throttle arm stopper. The throttle stop screw can be found close to the throttle roller and is the only vertical screw equipped with a spring on the entire carburetor assembly.

7- Loosen the throttle arm adjustment screws on the upper and lower carburetors, by rotating the screws **CLOCKWISE**. These screws have **LEFT HAND** threads.

8- Lightly pull up on the accelerator link rod to remove all "play" and close the throttle valves. At the same time, tighten both throttle arm adjustment screws by rotating them **COUNTERCLOCKWISE**.

9- Tighten the throttle stop screw until it contacts the throttle arm stopper, then continue to tighten one and a half turns more, as a preliminary adjustment.

IDLE ADJUSTMENT

Mount the engine in a test tank, or on a boat in a body of water.

Obtain a timing light and clip the pickup lead to the No. 1 spark plug lead. Of course, the other two leads are connected to the battery terminals. Observe any markings on the pickup clip of the timing light lead. Many have an arrow indicating the direction **TOWARD** the spark plug. Take time to make sure the clip is affixed properly.

Connect a tachometer to the powerhead per the instructions with the instrument. Check the "Special Words on Tachometers and Connections" on Page 7-2.

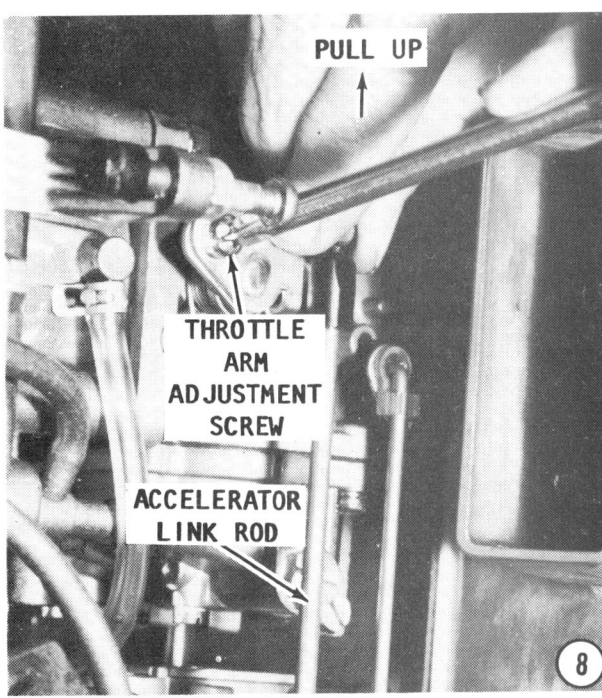

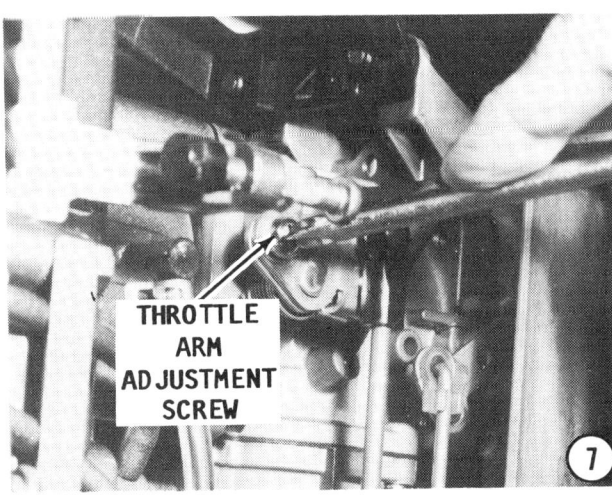

RUNAWAY POWERHEAD

Never operate the powerhead above a fast idle with a flush attachment connected to the lower unit. Operating the powerhead at a high rpm with no load on the propeller shaft could cause the powerhead to **RUNAWAY** causing extensive damage to the unit.

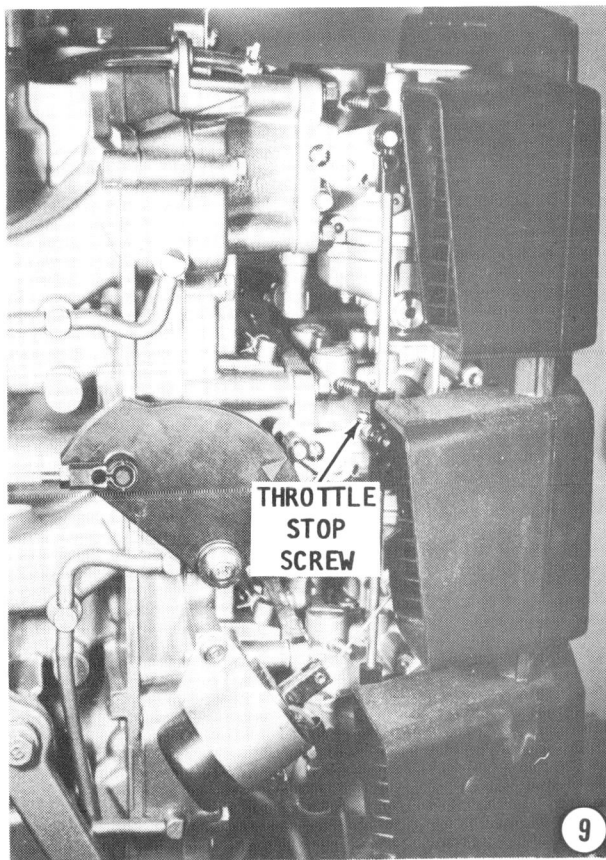

7-16 TIMING & SYNCHRONIZING

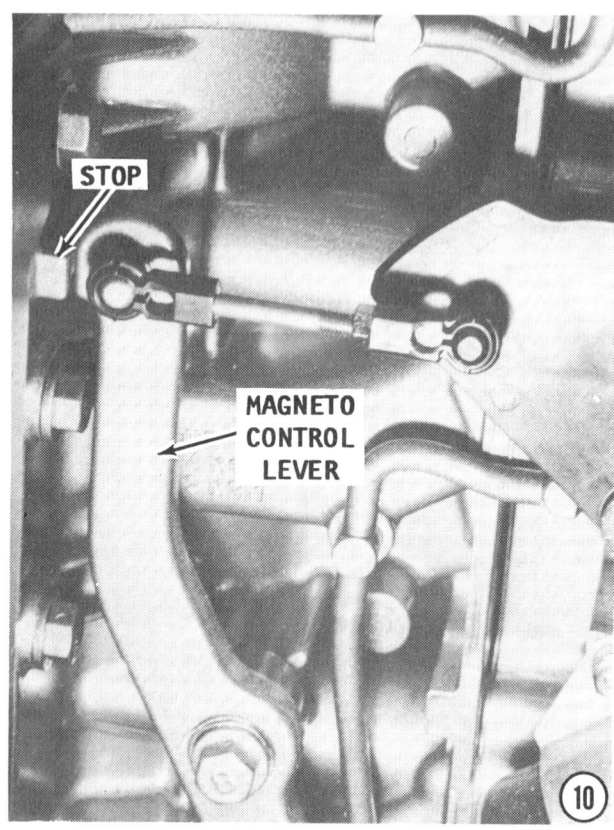

REMEMBER, the powerhead will NOT start without the emergency tether in place behind the kill switch knob.

Start the engine and allow it to warm to operating temperature and run at idle speed.

CAUTION
Water must circulate through the lower unit to the powerhead anytime the powerhead is operating to prevent damage to the water pump in the lower unit. Just five seconds without water will damage the water pump impeller.

Take care not to come in contact with any moving parts, such as the flywheel, nor any high voltage components, such as the CDI unit, or the ignition coils, while making the following adjustments.

10- Move the magneto control lever aft until the lever contacts the stop on the block.

11- Rotate the throttle stop screw until the powerhead idles at 750 rpm. Rotating the screw clockwise will increase engine rpm and rotating the screw counterclockwise will decrease engine rpm.

PILOT SCREW ADJUSTMENT

The pilot screw setting is not normally changed. During carburetor overhaul, the screw is set from a lightly seated position to a specified number of turns out. If the powerhead hesitates or runs roughly, check the setting of each pilot screw one by one. If the problem is not corrected, there is every reason to believe the carburetors need an overhaul.

12- Turn in the pilot screw until it is lightly seated. Now, back the screw out according to the specifications listed in the Appendix for the model being serviced. Pilot screw settings may vary from year to year and may also vary from port screw to starboard screw.

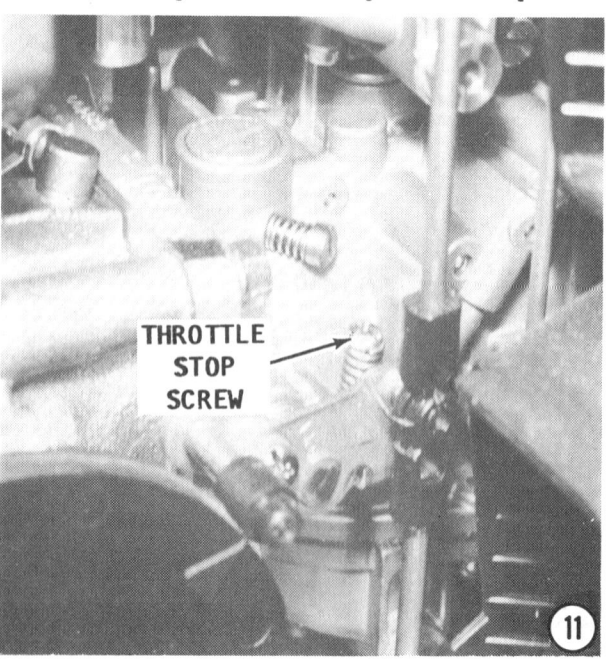

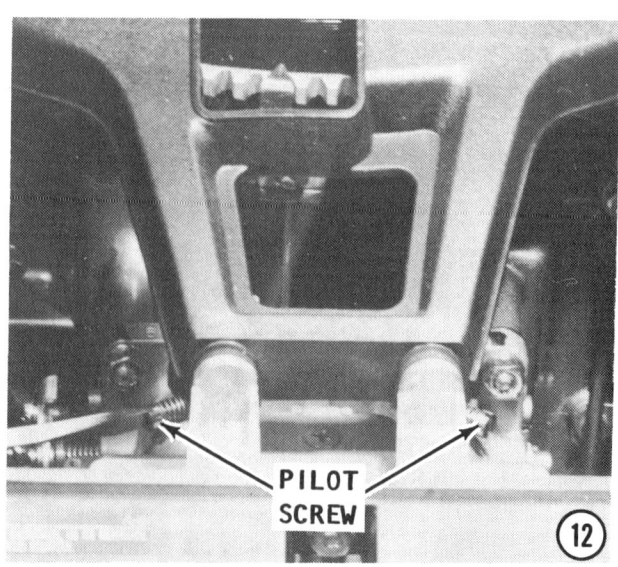

8
POWERHEAD

8-1 INTRODUCTION AND CHAPTER ORGANIZATION

This chapter is divided into five main working sections as follows:
8-1 This section, introduction and chapter organization.
8-2 Two-stroke powerhead description and operation.
8-3 Powerhead removal and disassembling.
8-4 Cleaning and Inspecting.
8-5 Powerhead assembling and installation.

The carburetion and ignition principles of two-cycle engine operation **MUST** be understood in order to perform proper service work on the outboard powerheads covered in this manual.
The two-cycle powerhead differs in several aspects from a four-stroke powerhead or engine.

Differences Include:
1- The method by which the fuel-air mixture is delivered to the combustion chamber.
2- The complete lubrication system.
3- In most cases, the ignition system.
4- The frequency of the power stroke.

Repair Procedures
Service and repair procedures will vary slightly between individual models, but the basic instructions are quite similar. Most of the differences involve the preliminary tasks to be accomplished before the powerhead is removed from the intermediate housing.
Special tools may be called out in certain instances. These tools may be purchased from the local Yamaha marine dealer. If the special tools are not available, an alternate method will be included, whenever possible.

Torque Values
All torque values must be met when they are specified. Torque values for various parts of each powerhead are given in the text.
A torque wrench is essential to correctly assemble the powerhead. **NEVER** attempt to assemble a powerhead without a torque wrench. Attaching bolts **MUST** be tightened to the required torque value in three progressive stages, following the specified tightening sequence. Tighten all bolts to 1/3 the torque value, then repeat the sequence tightening to 2/3 the torque value. Finally, on the third and last sequence, tighten to the full torque value.

The exterior and interior of the powerhead must be kept clean, well-lubricated, and properly tuned and adjusted, for the owner to receive maximum enjoyment from the unit.

8-2 POWERHEAD

Powerhead Components

Service procedures for the carburetors, fuel pumps, cranking motor and other powerhead components are given in their respective chapters of this manual. Consult the Table of Contents.

Reed Valve Service

The reeds on two-cycle powerheads covered in this manual are contained in an externally mounted reed valve block. Therefore, the powerhead need not be disassembled in order to replace a broken reed.

Cleanliness

Make a determined effort to keep parts and the work area as clean as possible. Parts **MUST** be cleaned and thoroughly inspected before they are assembled, installed, or adjusted. Use proper lubricants, or their equivalent, whenever they are recommended.

8-2 TWO-STROKE POWERHEAD DESCRIPTION AND OPERATION

Intake/Exhaust

Two-stroke cycle engines utilize an arrangement of port openings to admit fuel to the combustion chamber and to purge the exhaust gases after burning has been completed. The ports are located in a precise pattern in order for them to be opened and closed at an exact moment by the piston as it moves up and down in the cylinder. The exhaust port is located slightly higher than the fuel intake port. This arrangement opens the exhaust port first as the piston starts downward and therefore, the exhaust phase begins a fraction of a second before the intake phase.

Actually, the intake and exhaust ports are spaced so closely together that both open almost simultaneously. For this reason, the pistons of most two-stroke engines have a deflector type top. This design of the piston top serves two purposes very effectively.

First, it creates turbulence when the incoming charge of fuel enters the combustion chamber. This turbulence results in more complete burning of the fuel than if the piston top were flat.

The second effect of the deflector type piston crown is to force the exhaust gases from the cylinder more rapidly.

This system of intake and exhaust is in marked contrast to individual valve arrangement employed on four-cycle powerheads.

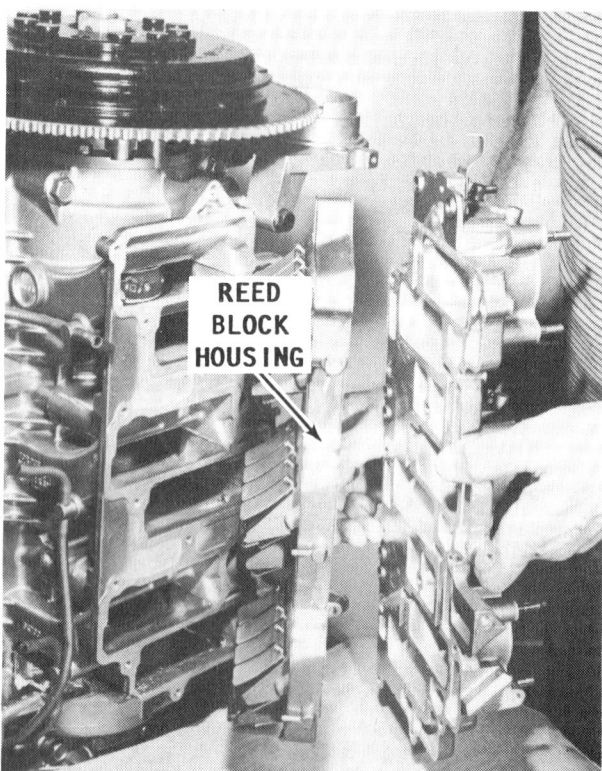

The reeds for the V4 and V6 powerheads covered in this manual are contained in an externally mounted housing. The powerhead need not be disassembled to replace a reed.

Lubrication

A two-stroke engine is lubricated by mixing oil with the fuel. Therefore, various parts are lubricated as the fuel mixture passes through the crankcase and the cylinder. Four-stroke engines have a crankcase containing oil. This oil is pumped through a circulating system and returned to the crankcase to begin the routing again.

Physical Laws

The two-stroke engine is able to function because of two very simple physical laws.

One: Gases will flow from an area of high pressure to an area of lower pressure. A tire blowout is an example of this principle. The high pressure air escapes rapidly if the tube is punctured.

Two: If a gas is compressed into a smaller area, the pressure increases, and if a gas expands into a larger area, the pressure is decreased.

If these two laws are kept in mind, the operation of the two-stroke engine will be easily understood.

ACTUAL OPERATION
TWO-STROKE POWERHEAD

Beginning with the piston approaching top dead center on the compression stroke: The intake and exhaust ports are closed by the piston; the reed valve is open; the spark plug fires; the compressed fuel-air mixture is ignited; and the power stroke begins. The reed valve was open because as the piston moved upward, the crankcase volume increased, which reduced the crankcase pressure to less than the outside atmosphere.

As the piston moves downward on the power stroke, the combustion chamber is filled with burning gases. As the exhaust port is uncovered, the gases, which are under great pressure, escape rapidly through the exhaust ports.

The piston continues its downward movement. Pressure within the crankcase increases, closing the reed valves against their seats. The crankcase then becomes a sealed chamber. The air-fuel mixture is compressed ready for delivery to the combustion chamber. As the piston continues to move downward, the intake port is uncovered. A fresh air/fuel mixture rushes through the intake port into the combustion chamber striking the top of the piston where it is deflected along the cylinder wall. The reed valve remains closed until the piston moves upward again.

When the piston begins to move upward on the compression stroke, the reed valve opens because the crankcase volume has been increased, reducing crankcase pressure to less than the outside atmosphere. The intake and exhaust ports are closed and the fresh fuel charge is compressed inside the combustion chamber.

Pressure in the crankcase decreases as the piston moves upward and a fresh charge of air flows through the carburetor picking up fuel. As the piston approaches top dead center, the spark plug ignites the air-fuel mixture, the power stroke begins and one full cycle has been completed.

TIMING -- TWO-STROKE
POWERHEAD

The exact time of spark plug firing depends on engine speed. At low speed the spark is retarded, fires later than when the piston is at or beyond top dead center (TDC).

At high speed, the spark is advanced, fires earlier than when the piston is at top dead center.

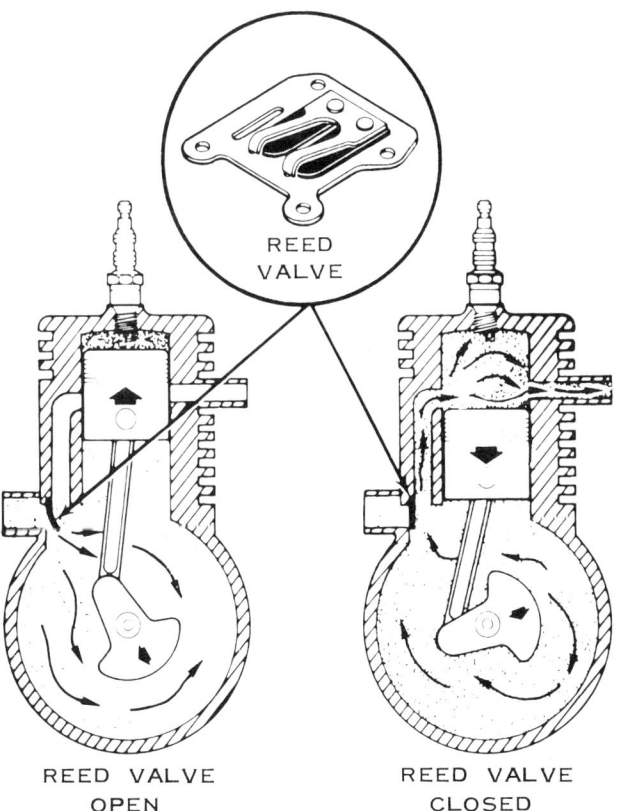

Reed valves are used to control the flow of air/fuel into the crankcase and eventually into the cylinder. As the piston moves upward in the cylinder, the resulting suction in the crankcase overcomes the spring tension of the reed. The reed is pulled free from its seat and the air/fuel mixture is drawn into the crankcase.

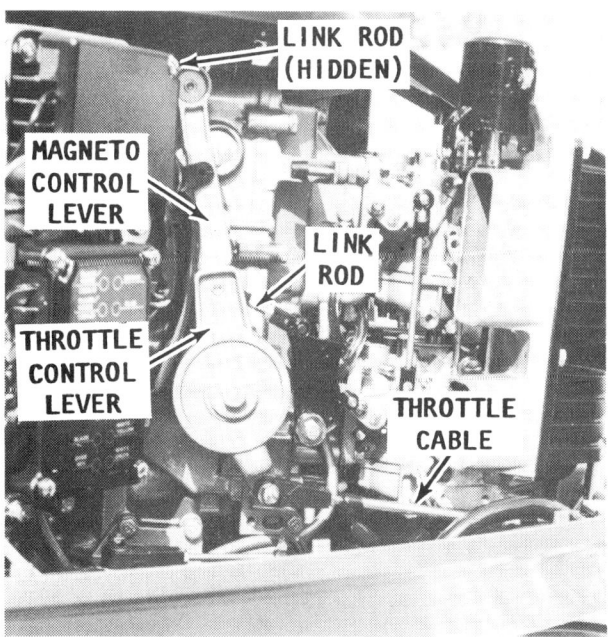

The throttle cable is linked to the stator through two link rods and two levers to provide timing advance on powerheads NOT equipped with YMIS (Yamaha Microcomputer Ignition System).

8-4 POWERHEAD

Summary

More than one phase of the cycle occurs simultaneously during operation of a two-stroke engine. On the downward stroke, power occurs above the piston while the ports are closed. When the ports open, exhaust begins and intake follows. Below the piston, fresh air-fuel mixture is compressed in the crankcase.

On the upward stroke, exhaust and intake continue as long as the ports are open. Compression begins when the ports are closed and continues until the spark plug ignites the air-fuel mixture. Below the piston, a fresh air-fuel mixture is drawn into the crankcase ready to be compressed during the next cycle.

8-3 POWERHEAD SERVICE

ADVICE

Before commencing any work on the powerhead, an understanding of two-cycle engine operation will be most helpful. Therefore, it would be well worth the time to study the principles of two-cycle engines, as outlined briefly in Section 8-2. A Polaroid, or equivalent instant type camera is an extremely useful item, providing the means of accurately recording the arrangement of parts and wire connections BEFORE the disassembly work begins. Such a record is most valuable during assembling.

Preliminary Tasks

Remove the cowling.
Remove the electric cranking motor and relay, see Chapter 9.
Remove the flame arrestor, choke, and carburetors, see Chapter 4.
Remove the oil injection system, see Chapter 5.
Remove the YMIS microcomputer unit, the various sensors and switches (if equipped). Remember to MARK the exact location of the throttle position sensor in relation to its mounting bracket.
Next, remove the CDI unit, the control unit (on V6 only) and the ignition coils, see Chapter 6.
Remove the flywheel and stator assembly, see Chapter 6.

REMOVAL AND DISASSEMBLING

The following instructions pickup the work after the preliminary tasks listed above have been accomplished.

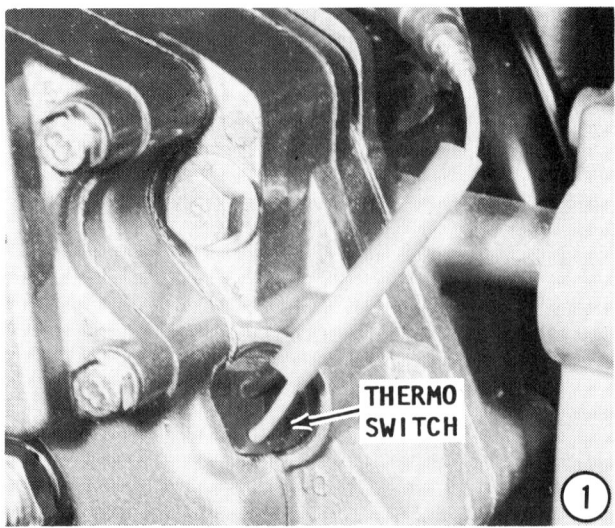

1- Gently pull the two thermo switches from the cylinder head covers and set them aside. It is not necessary to disconnect the switches from the leads.

2- Pull the water drain hose free of the fitting at the bottom of the exhaust cover and set the hose aside. On V6 models, snip the plastic clamp securing the water hose to the bypass cover, and then pull the hose free of the fitting on the cover.

3- Remove the locknut and washer from the top of the shift rod.

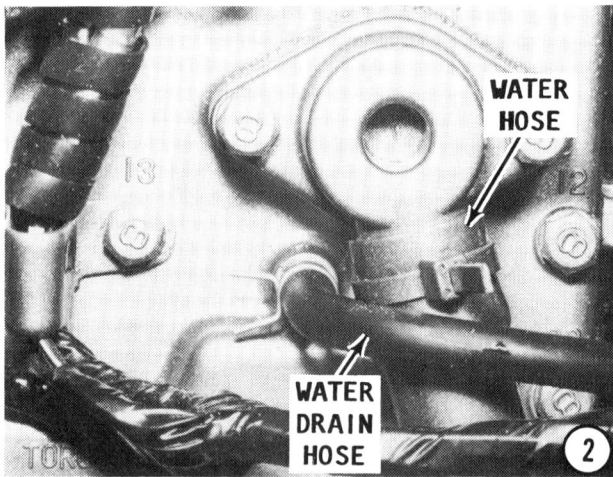

REMOVAL 8-5

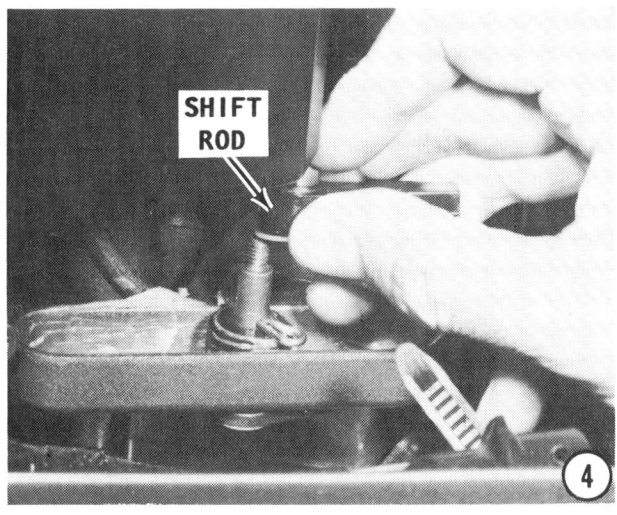

4- Lift off the shift rod and set it aside.

5- Use a pair of needle nose pliers and pull out the cotter pin. Remove the shift guide from the block.

6- Remove the locknut and washer, and then lift off the throttle cable from the throttle arm.

7- Disconnect the ground leads attached to the lower cowling. Some leads are also attached to the exhaust cover.

8- Pry the link rods from the throttle control arm (on all models except those equipped with YMIS), and the magneto control arm. Try not to disturb the length of these rods.

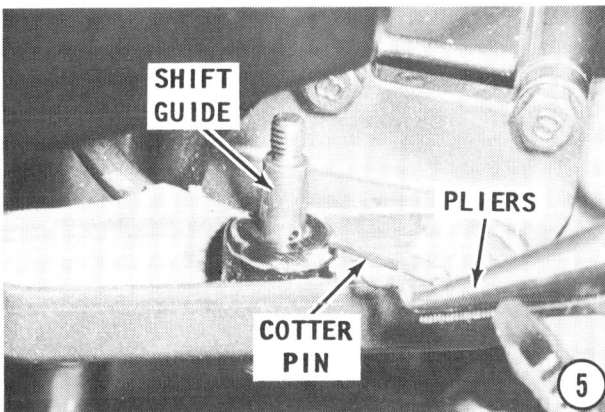

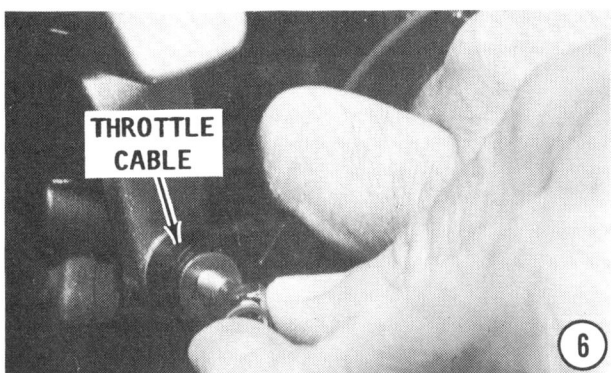

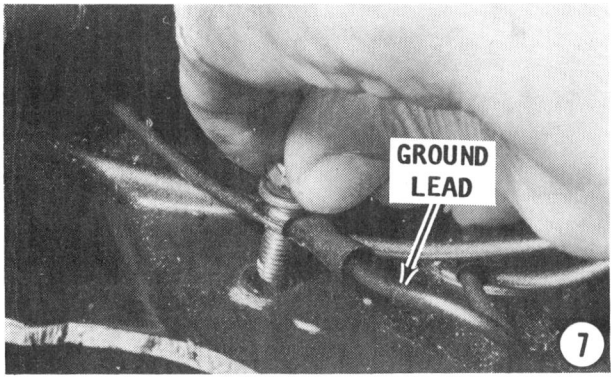

9- Remove the center bolt from the throttle control arm, lift off the washer, the insert, and the plastic sleeve. Lift off the throttle control arm and torsion spring (on all models except those equipped with YMIS), then lift off the magneto control lever and two washers. Place these items in order on the workbench as they are removed, as an assist during assembling. Do **NOT** disturb any adjusting screw settings.

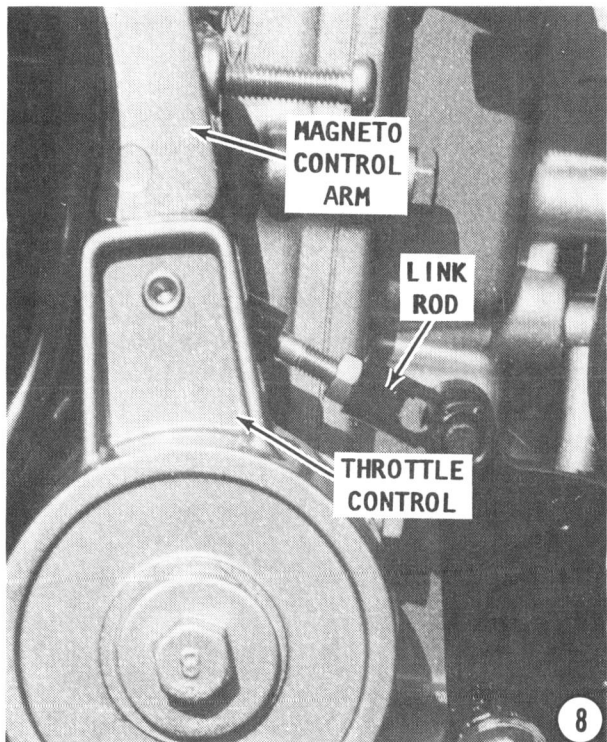

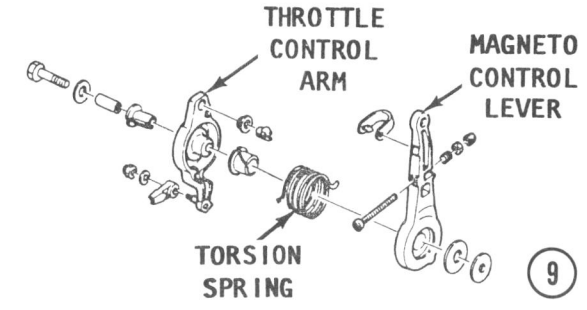

8-6 POWERHEAD

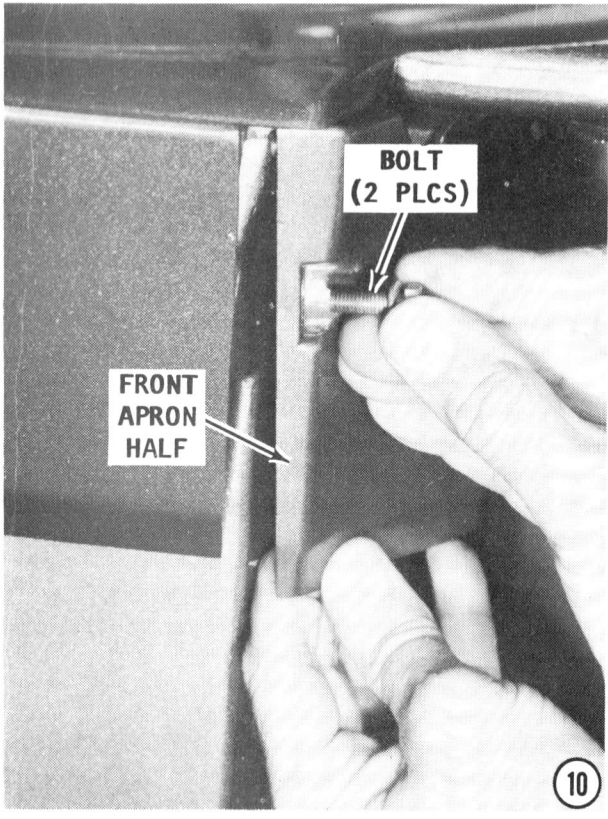

10- Remove the two bolts securing the front apron half around the lower cowling. Remove the forward apron half.

11- Remove the two bolts securing the aft apron half around the lower cowling. Remove the aft apron half.

12- Remove a total of ten bolts and two nuts securing the powerhead to the intermediate housing: six long bolts, two medium bolts and two short bolts. The nuts are

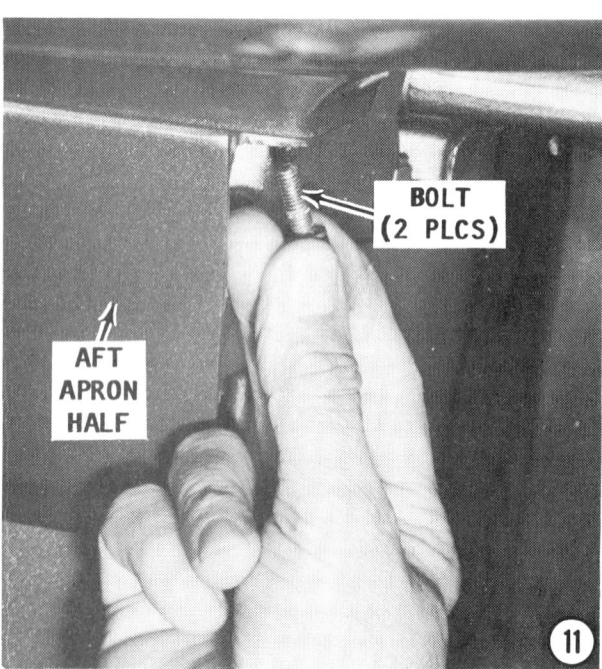

located on the forward side of the intermediate housing, just below the lower cowling pan.

GOOD WORDS

The powerhead may be difficult to dislodge from the intermediate housing because of a tight sealing gasket and joining compound. Prying up on the powerhead by using a long piece of wood and leverage on the edge of the lower cowling is an acceptable method by the manufacturer. If this method is employed, **CARE** must be exercised not to damage either the powerhead or the cowling. Once the powerhead has "broken" free of the intermediate housing, proceed to the next step.

BAD NEWS

If the unit is several years old, or if it has been operated in salt water, or has not had proper maintenance, or shelter, or any number of other factors, then separating the powerhead from the intermediate housing may not be a simple task. An air hammer may be required on the bolts to shake the corrosion loose; heat may have to be applied to the casting to expand it slightly; or other devices employed in order to remove the powerhead. One very serious condition would be the driveshaft "frozen" with the lower end of the crankshaft. In this case a circular plug type hole must be drilled and a torch used to cut the driveshaft.

Let's assume the powerhead will come free on the first attempt.

TAKE CARE not to lose the two dowel pins. The pins may come away with the powerhead or they may stay in the inter-

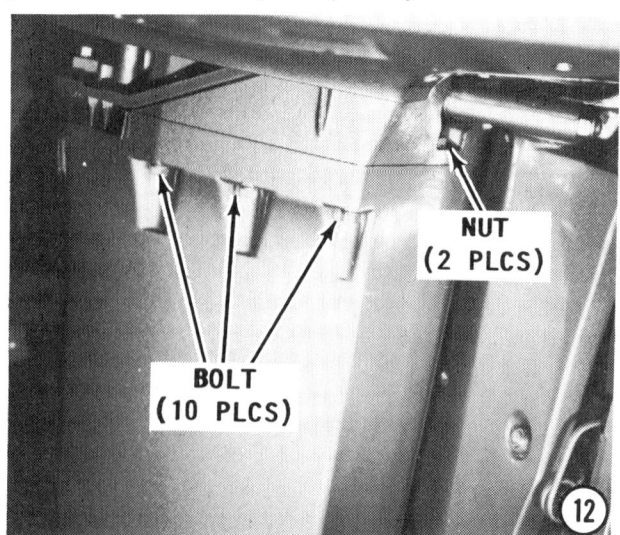

mediate housing. Be **ESPECIALLY** careful not to drop them into the lower unit.

13- Raise the powerhead free of the intermediate housing. An alternate method, and easier on your back muscles, is to use a hook through the "eye" provided and a lifting device, such as a chain hoist.

Place the powerhead on a suitable work surface. **AGAIN**, take care not to loose the alignment dowel pins. The pins may come away with the powerhead, or they may remain in the intermediate housing.

POWERHEAD DISASSEMBLING

VERY GOOD WORDS

The following procedures cover removal and installation of virtually all components of the powerhead. **HOWEVER**, if the determination can be made certain seals, bearings, etc. are fit for further service, "leave a sleeping dog lie", simply skip the steps involved for such items and proceed with the necessary work. In certain instances, the item would be destroyed or at least rendered unfit for service during the removal process.

To avoid juggling the powerhead, the first component to be removed should be the lower oil seal housing leaving a fairly flat surface on the bottom of the powerhead. A couple of 1" x 2" pieces of wood wedged

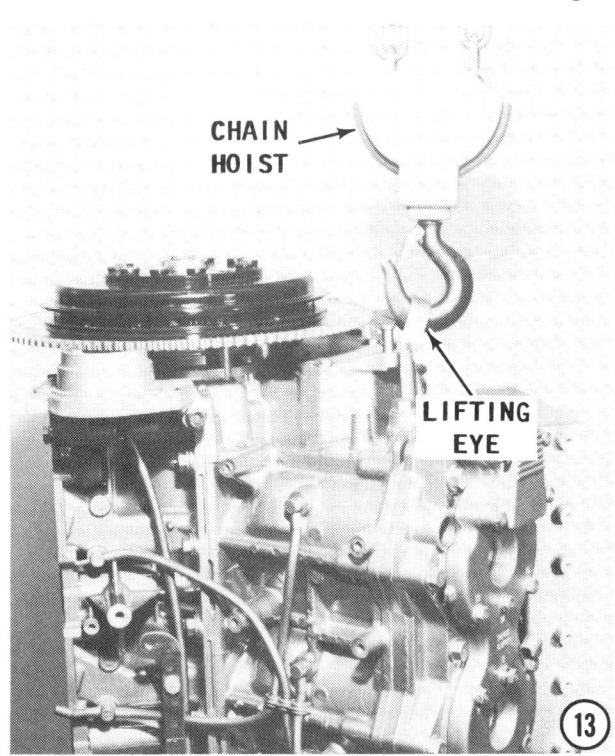

under the powerhead at strategic locations will stabilize the block.

LOWER OIL SEAL HOUSING REMOVAL

14- Remove the four bolts securing the lower oil seal housing to the block.

15- Use a couple of screwdrivers and pry the housing away from the block or tap the housing **LIGHTLY** with a soft head mallet to jar it loose. Remove and discard the outer O-ring.

CRITICAL WORDS

Perform Step 16 only if the seals have been damaged and are no longer fit for service. Removal of the seals destroys their sealing qualities. Therefore, they cannot be installed a second time. Be absolutely sure a new seal is available **BEFORE** removing the old seal in the next step.

8-8 POWERHEAD

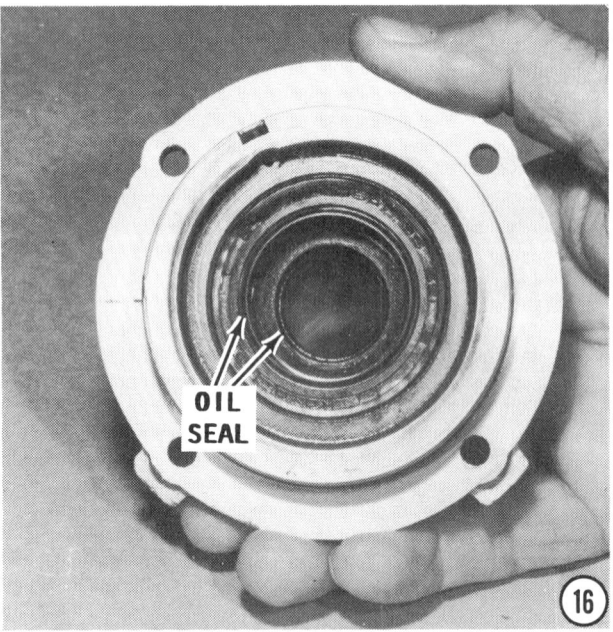

16- Inspect the condition of the two seals in the lower oil seal housing. If the seals appear to be damaged and replacement is required, proceed with removal. Obtain special tool Yamaha P/N YB6096 slide hammer and jaw attachment and remove the seals.

Set the powerhead upright to perform the following work.

INTAKE MANIFOLD REMOVAL

17- Identify the oil supply lines at each intake port, these lines are of different lengths. Remove the line retainer on the portside of the block and remove all the lines.

SPECIAL WORDS

Take care in the next step when handling reed valve assemblies. Once the assembly is removed, keep it away from sunlight, moisture, dust, and dirt. Sunlight can deteriorate valve seat rubber seals. Moisture can easily rust stoppers overnight. Dust and dirt -- especially sand or other gritty material can break reed petals if caught between stoppers and reed petals.

Make special arrangements to store the reed valves to keep them isolated from elements while further work is being performed on the powerhead.

18- Remove the bolts securing the intake manifold/reed valve assembly to the block. Obtain a long wide screwdriver to separate both pieces from the block and from each other. Insert the screwdriver between the tabs provided for this purpose, and pry the two surfaces apart. **NEVER** pry at a gasket sealing surface. Such action would very likely damage the sealing surface of an aluminum powerhead.

Remove and **DISCARD** the gasket. Set the reed valves aside. Further work on the reed valves will be performed later in the Cleaning and Inspection section of this chapter.

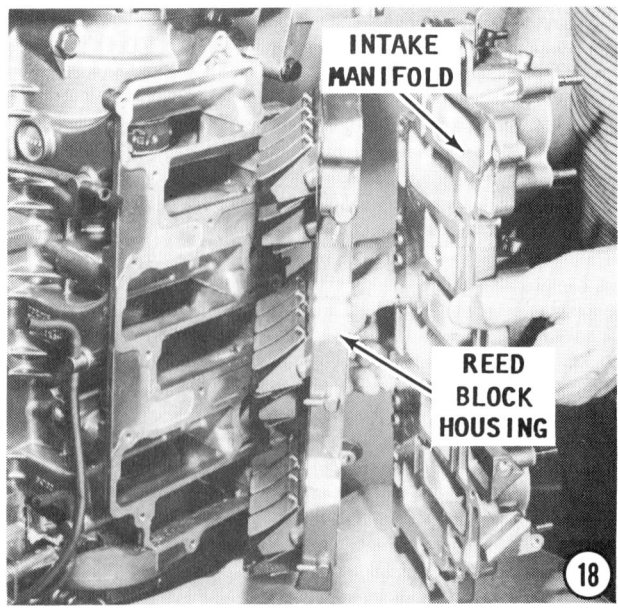

DISASSEMBLING 8-9

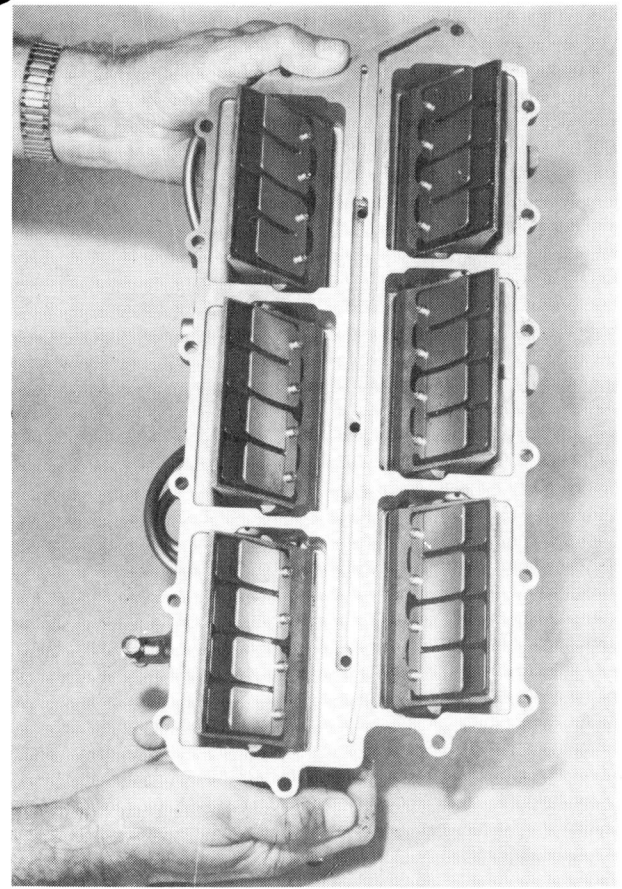

Once the reed valve housing has been removed, the reed valve assemblies require special handling and placement, as outlined in the "Special Words" on the previous page.

EXHAUST COVER REMOVAL

ADVICE

The exhaust cover should always be removed during a powerhead overhaul. Many times, water in the powerhead is caused by a leaking exhaust cover gasket or plate.

GOOD WORDS

If the inner or outer cover is stuck to the powerhead, insert a slotted screwdriver between the tabs provided for this purpose, and pry the two surfaces apart. **NEVER** pry at a gasket sealing surface. Such action would very likely damage the sealing surface of an aluminum powerhead.

19- Remove the bolts securing the outer cover.

20- Remove the cover and the gasket. Remove the inner cover and gasket. **DISCARD** the outer and inner cover gaskets.

V6 POWERHEAD

21- Remove the two bolts securing the

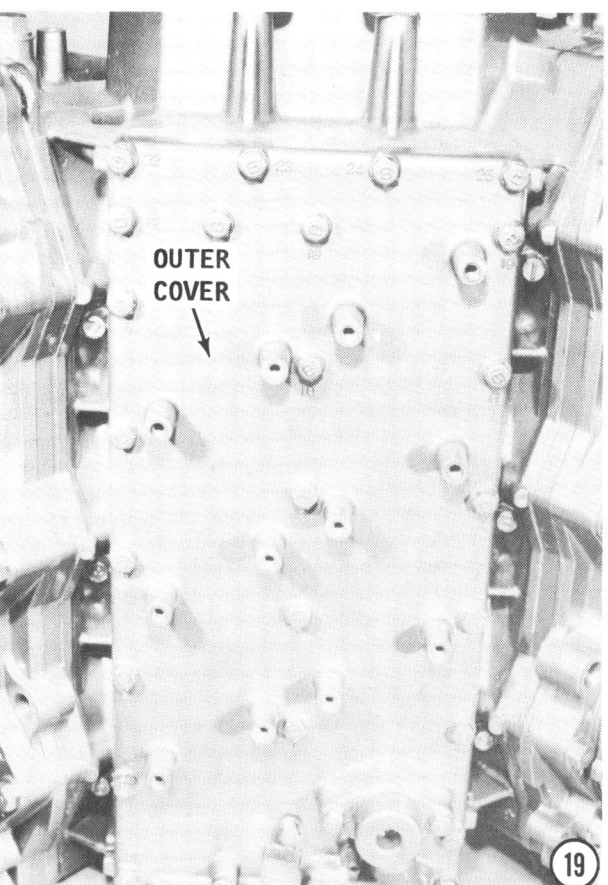

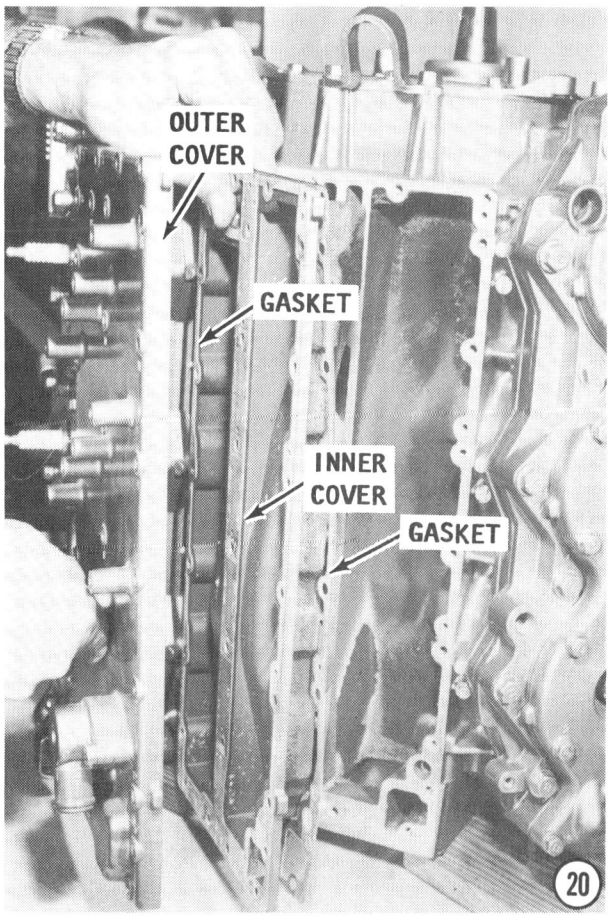

8-10   POWERHEAD

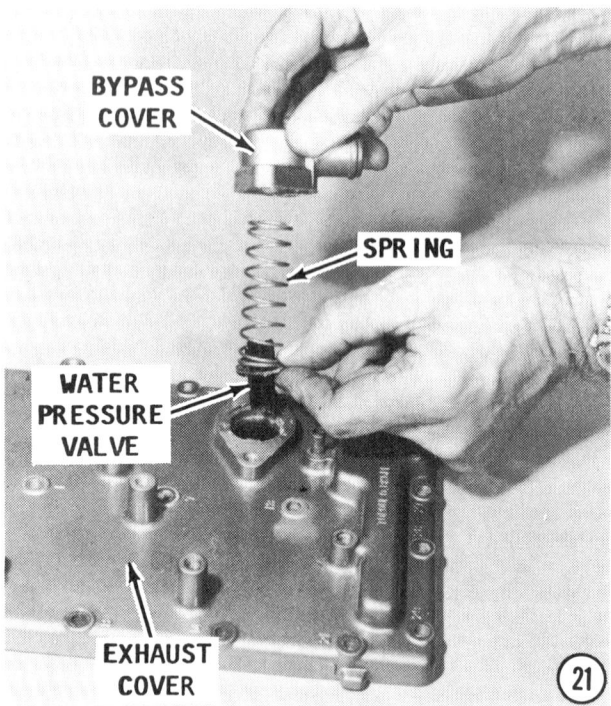

bypass cover to the exhaust cover. Lift the cover up, and then remove the spring and water pressure valve. Observe how the short end of the insert faces upward, as an assist during assembling.

ALL MODELS
SPECIAL INSTRUCTIONS

Steps 22 thru 26 apply to both banks of cylinders, therefore each step should be repeated for the other bank.

THERMOSTAT REMOVAL

22- Remove the four bolts securing the thermostat cover to the cylinder head cover.

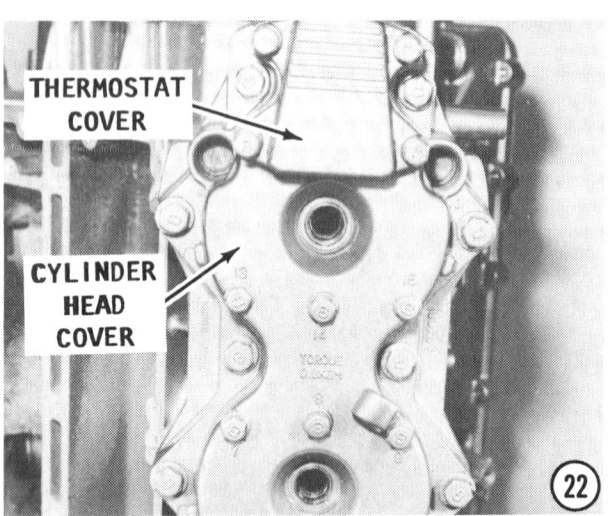

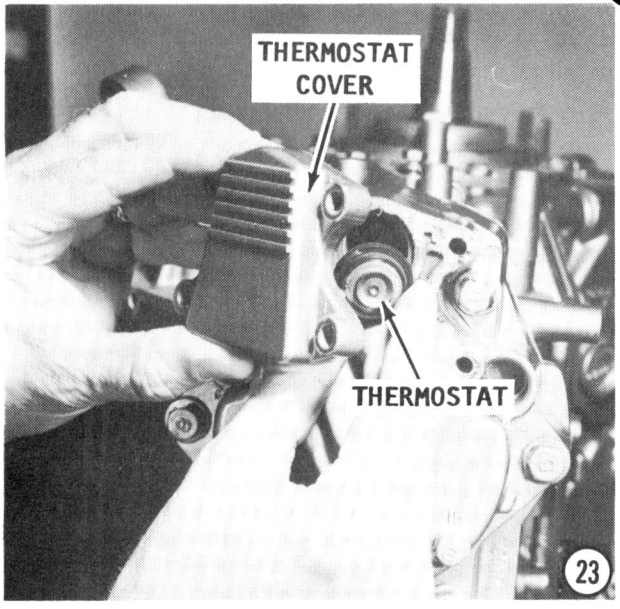

23- Remove the cover and the thermostat. If the cover is stuck to the powerhead, tap it **LIGHTLY** with a soft head mallet to jar it free. Remove and discard the gasket. As an assist during assembling, observe how the spring end of the thermostat was inserted into the cylinder head.

On V4 powerheads, the water pressure relief valve is also located under the therm-

DISASSEMBLING 8-11

ostat cover. Remove the spring and water pressure valve. As an assist during assembling, observe how the short end of the insert faces upward.

CYLINDER HEAD REMOVAL

24- Remove the cylinder head cover bolts, and then remove the cover. Remove and **DISCARD** the gasket. If the cover is stuck to the cylinder head, insert a slotted screwdriver between the tabs provided for this purpose and pry the two surfaces apart. Again, **NEVER** pry at the gasket sealing surfaces, because the sealing surface of an aluminum powerhead could be damaged.

25- Insert a long wide screwdriver between the tabs on the cylinder head and block. Pry the two surfaces apart using the tabs provided for this purpose. **NEVER** pry at a gasket sealing surface. Such action would very likely damage the sealing surface of an aluminum powerhead.

If the cylinder head is stubborn and cannot be "broken free" of the block, tap the head **LIGHTLY** with a soft head mallet to jar it free.

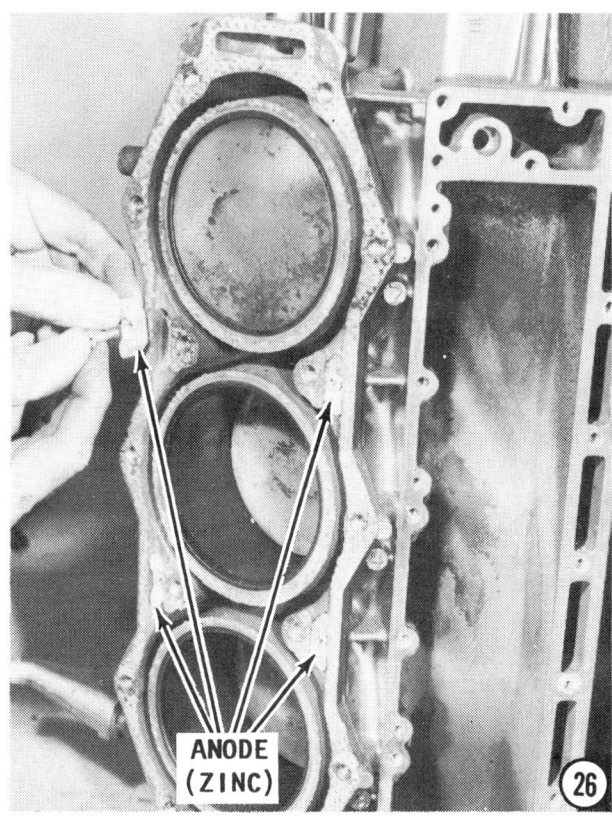

ANODE REMOVAL

26- Remove the Phillips head screws, and then remove the wedge shaped anodes located in the water jacket.

UPPER OIL SEAL HOUSING

27- **DO NOT ATTEMPT** to remove the upper oil seal housing, at this time. The housing must be removed from the crankshaft after the crankshaft has been lifted from the block. To attempt removal at this time would only damage the housing.

28- Remove the cylinder cover retaining bolts. Remove the cover, and then remove and discard the gasket.

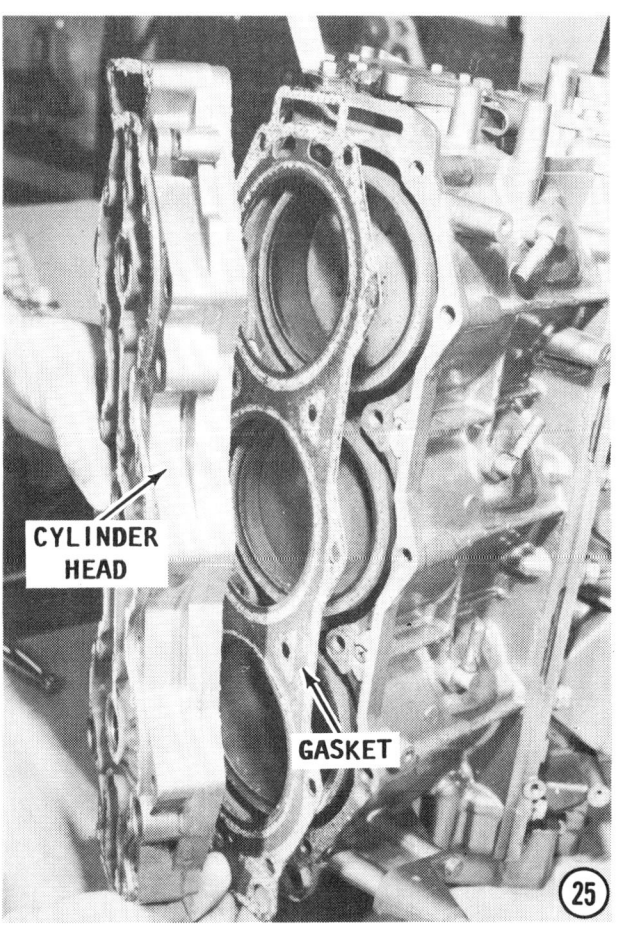

8-12 POWERHEAD

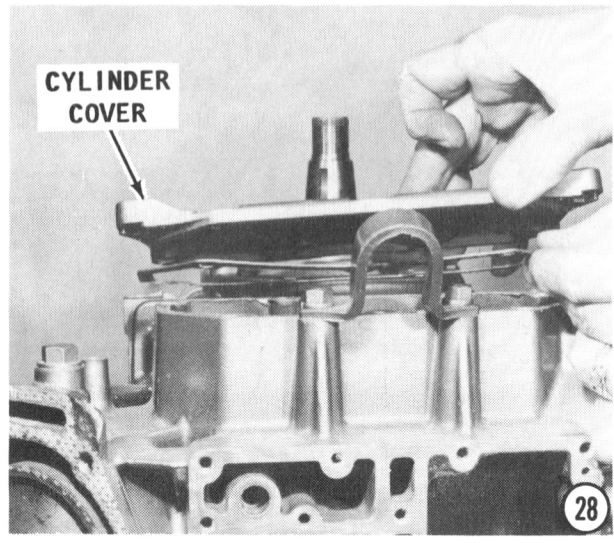

CRANKCASE SEPARATION

29- Remove the crankcase bolts. These bolts vary, depending on the unit being serviced, as follows:

V4 powerheads have 14 bolts of two different sizes.
V6 powerheads have 20 bolts of two different sizes.

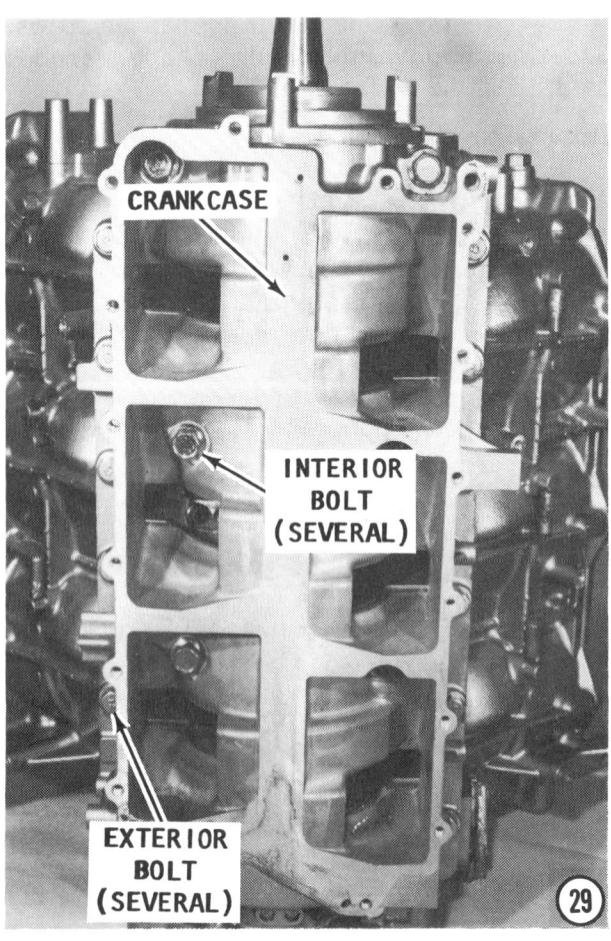

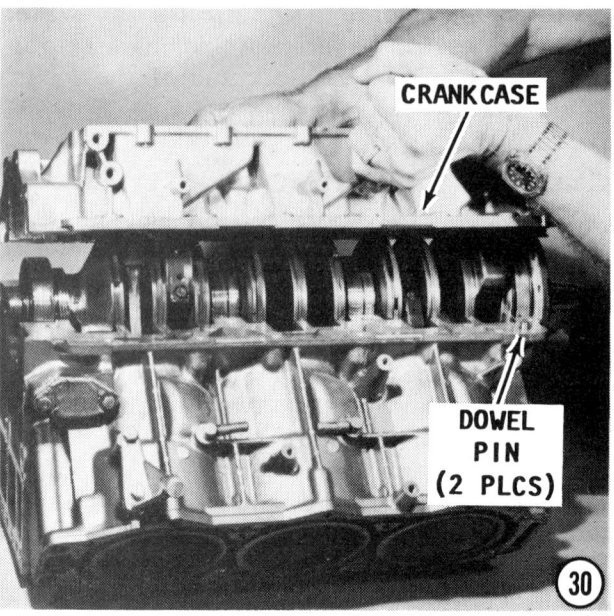

After all bolts have been removed, insert a slotted screwdriver between the tabs provided for this purpose and pry the two surfaces apart. **NEVER** pry at gasket sealing surfaces. Such action would very likely damage the sealing surface of an aluminum powerhead.

30- Separate the two halves of the crankcase. **TAKE CARE** not to lose the two dowel pins. These pins may remain in either half when the crankcase is separated.

UPPER OIL SEAL HOUSING REMOVAL

31- Tap the edge of the oil seal housing with a soft head mallet, to jar it free of the crankshaft. Remove the upper oil seal housing from the top of the crankshaft.

DISASSEMBLING 8-13

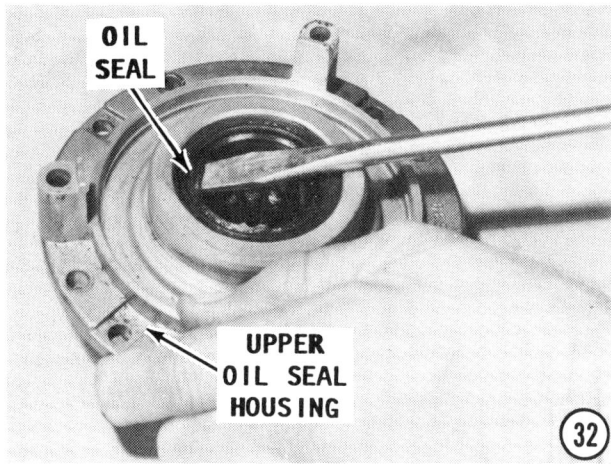

32- Remove and **DISCARD** the two O-rings around the housing.

Inspect the condition of the oil seal in the upper oil seal housing. Make a determination if the seal is fit for further service. The seal will be destroyed during removal. Therefore, remove the seal **ONLY** if it is damaged and has lost its sealing qualities.

To remove the seal, first obtain Yamaha slide hammer puller P/N YB6096 with expanding jaw attachment, and then "pull" the seal, or pry the seal free with a screwdriver.

SPECIAL WORDS

Perform Step 33 **ONLY** if the needle bearing in question is no longer fit for service.

33- The roller bearing is pressed into the upper oil seal housing. To remove the bearing, first obtain Yamaha special tool P/N YB6205 bearing installer. Using the shaft of the installer, remove the bearing from the housing using a hydraulic press. Press the bearing **DOWNWARDS** free of the housing.

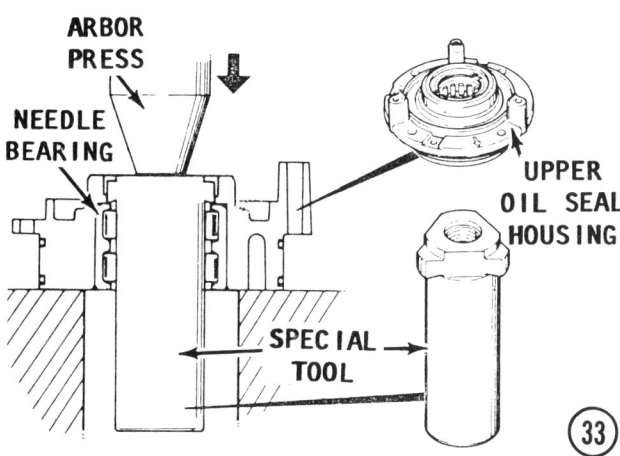

CRANKSHAFT AND CONNECTING ROD REMOVAL

34- **TAKE TIME** to mark the cylinder number on both halves of the connecting rod caps. This mark should be made with a marker, whiteout, paint, or any substance which will adhere to a metal surface. **UNDER NO CIRCUMSTANCES** should the mark be a series of notches, or gouges, or even scribed. All these can cause "stress risers", and under heavy engine load can cause parts to crack and fail.

SPECIAL WORDS

These connecting rods were cast as one piece. Holes were then drilled where they would normally separate, and then the two halves were physically broken apart. Therefore, each mating surface is unique, one of a kind. These rods can never be resized and can never be made to match another set.

35- Obtain an 8mm 12 point socket, with very narrow sides -- "a rare animal" -- not in every one's tool kit.

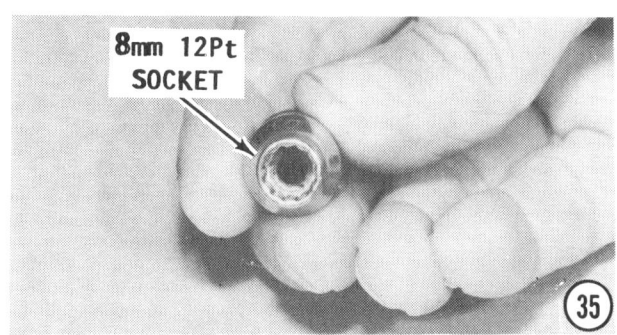

GOOD WORDS

Snap-on Tool Company markets such a socket. Without modification, this socket can be used successfully to remove the rod bolts. If this particular socket is not available, a 5/16" 12 point socket will suffice. The external diameter of most sockets, with the exception of the first one listed, will be too great to fit between the bolt and the cap.

As a last resort, an 8mm or a 5/16" 12 point socket may be ground down to fit, but not without its disadvantages: First, grinding away material from the external wall will weaken the socket and the walls may break away. Upon installation, the rod bolts are tightened to a torque value of 24 ft lb (34Nm). Secondly, if the grinding is not symmetrical, any high spots on the external socket wall may flake off when used and fall down into the cylinder bore.

36- Remove the connecting rod cap bolts. Lift off the cap and caged roller bearing beneath the cap. Slide the other half of the caged roller bearing around the journal and keep all three items together on the workbench.

WORDS FROM EXPERIENCE

CLEANLINESS is the password, when handling roller bearings. Take care to prevent any dirt, lint or other contaminents from getting onto the bearings or in the cages. If the bearings are to be used again, store them in a numbered container to ensure they will be installed with the same rod and cap from which they were removed. **NEVER** intermix roller bearings from one rod to another. **NEVER** intermix used roller bearings with new bearings. If just one

bearing is unfit for further service, the entire set **MUST** be replaced.

New bearings should be installed in the connecting rods, even though they may appear to be in serviceable condition. New bearings will ensure lasting service after the overhaul work is completed. If it is necessary to install the used bearings, keep them separate and identified to **ENSURE** they will be installed onto the same crankpin throw and with the same connecting rod from which they were removed.

Continue removing the rod caps and two bearing halves until all have been removed from the crankshaft.

37- Tap the crankshaft **LIGHTLY** with a soft head mallet to jar it free from the block. Lift the crankshaft out of the block.

38- Remove the large circlip from around each set of main bearings. Remove the two halves of the bearing shell and the two halves of the roller bearing. If the main bearings are to be reused, keep each set together, so they may be installed back into their original locations.

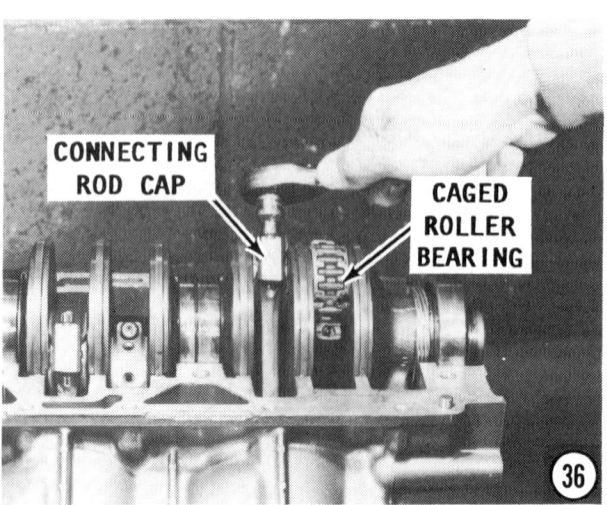

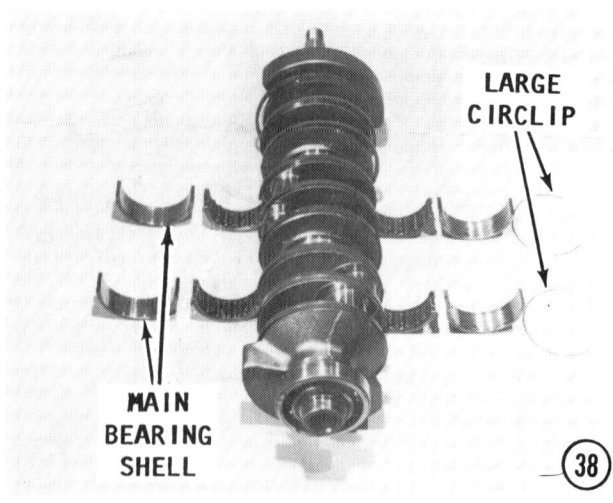

DISASSEMBLING 8-15

ADVICE

New main bearings should be installed in the connecting rods, even though they may appear to be in serviceable condition. New bearings will ensure lasting service after the overhaul work is completed. If it is necessary to install the used bearings, keep them separate and identified to **ENSURE** they will be installed in the same location from which they were removed.

PISTON REMOVAL

39- Pull each piston and connecting rod assembly from the **BOTTOM** -- not through the top of the block. A ridge might have formed on the top of the cylinder bore. This ridge may have to be removed with a ridge reamer.

IMMEDIATELY after removing the piston and rod assembly, temporarily assemble the rod cap back onto the rod to keep it as a matched set.

40- Make an identifying mark on the outside edge of each rod "I" beam and a matching mark on the inside of each piston skirt. The identification mark must match the cylinder from which the piston and rod were removed.

This mark should be made with a marker, whiteout, paint or any substance which will adhere to a metal surface. **UNDER NO CIRCUMSTANCES** should the mark be a series of notches, or gouges, or even scribed. All these can cause "stress risers", and under heavy engine load can cause parts to crack and fail.

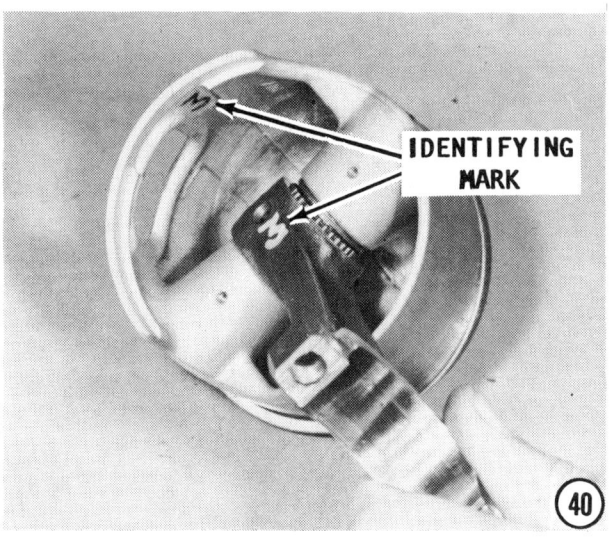

SAFETY WORDS
ALL MODELS

The piston pin C-lockrings are made of spring steel and may slip out of the pliers or pop out of the groove with considerable force. Therefore, **WEAR** eye protection glasses while removing the piston pin lockrings in the next step.

41- Remove the C-lockring from both ends of the piston pin using a pair of needle nose pliers or an awl. Discard the C-lockrings. These rings stretch during removal and cannot be used a second time.

42- Press the pin free of the piston using an arbor press or a long suitably sized socket, with the piston resting on a padded surface. Be sure to catch all the loose needle bearings and two small spacers as they fall free of the piston pin bore.

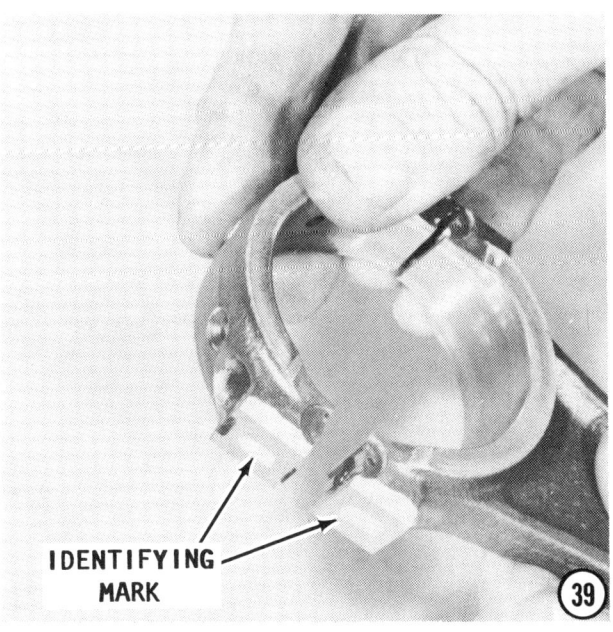

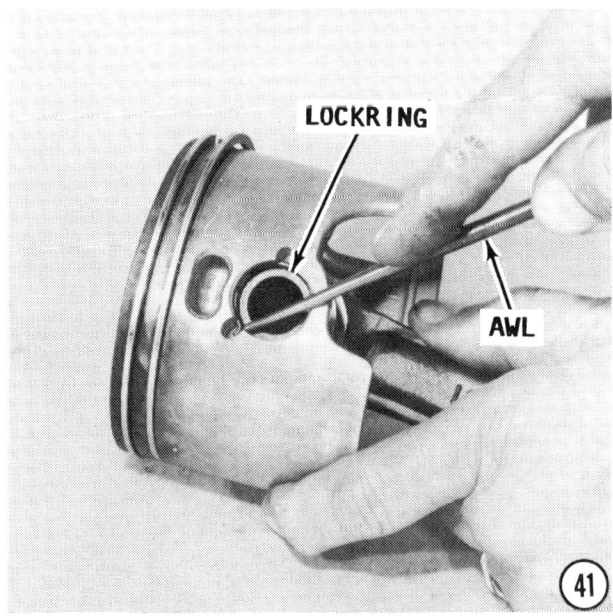

8-16 POWERHEAD

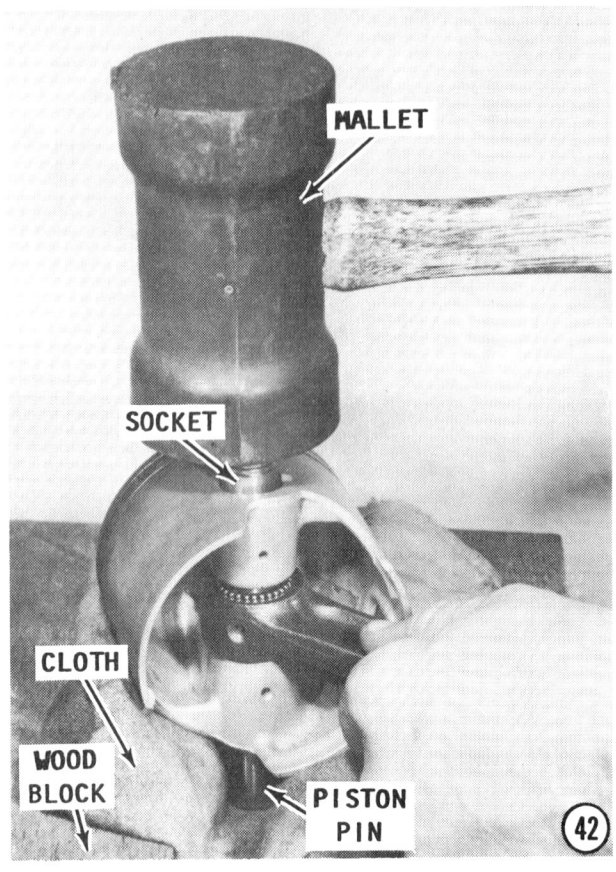

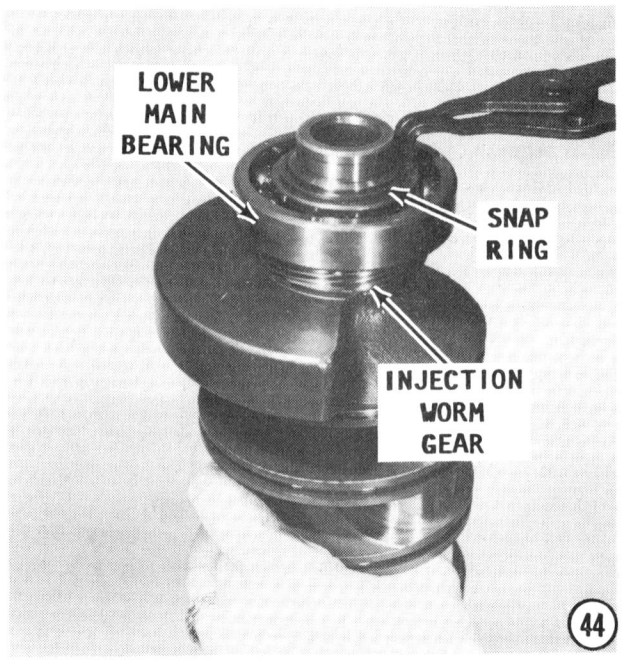

GOOD WORDS

Good shop practice dictates to replace the rings during a powerhead overhaul. However, if the rings are to be used again, expand them **ONLY** enough to clear the piston and the grooves, because used rings are brittle and break very easily.

43- Gently spread the top piston ring enough to pry it out and up over the top of the piston. No special tool is required to remove the piston rings. Remove the middle and lower rings in a similar manner. These rings are **EXTREMELY** brittle and have to be handled with care if they are intended for further service.

CRANKSHAFT DISASSEMBLY

44- Inspect the bronze oil injection worm gear at the lower end of the crankshaft. If the gear needs to be replaced, the lower main bearing must be "pulled" from

the crankshaft first. The bronze worm gear may then be "pulled".

Slowly rotate the bearing race on the crankshaft. If rough spots are felt, the bearing will have to be pressed free of the crankshaft. Obtain a pair of snap ring pliers. Remove the snap ring around the lower end of the crankshaft.

45- Obtain special bearing separator tool, Yamaha P/N YB6219 and an arbor press. Press the bearing from the crankshaft. Be sure the crankshaft is supported, because once the bearing "breaks" loose, the crankshaft is free to fall. Relocate the bearing separator and press the bronze oil injection gear from the crankshaft.

46- Gently spread each of the crankshaft sealing rings enough to ease it out and up over the top of the groove and into a journal space. Then, once again separate the two ends enough to clear the journal, and lift the ring free. Take care not to scratch the highly polished journal surface as each ring is removed.

Remove all the rings in a similar manner. These rings are **EXTREMELY** brittle and have to be handled with care if they are intended for further service.

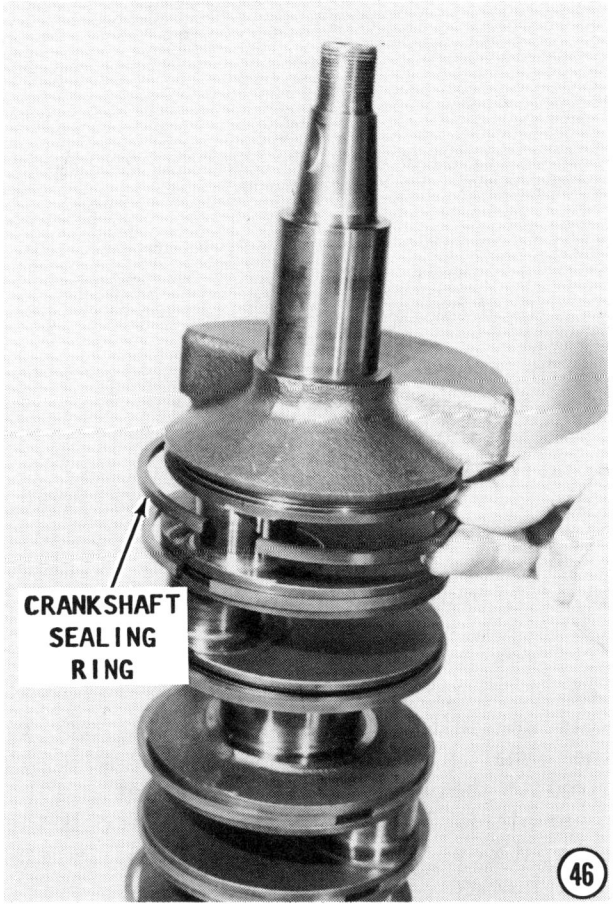

8-4 CLEANING AND INSPECTING

The success of the overhaul work is largely dependent on how well the cleaning and inspecting tasks are completed. If some parts are not thoroughly cleaned, or if an unsatisfactory unit is allowed to be returned to service through negligent inspection, the time and expense involved in the work will not be justified with peak powerhead performance and long operating life. Therefore, the procedures in the following sections should be followed closely and the work performed with patience and attention to detail.

REED BLOCK SERVICE

Disassemble the reed block housing by first removing spacers (on some models) and then the screws securing the reed stoppers and reed petals to the housing. After the screws are removed, lift the stoppers and petals from the housing.

Clean the gasket surfaces of the housing. Check the surfaces for deep grooves, cracks or any distortion which could cause leakage.

Replace the reed block housing if it is damaged. The reed petals should fit flush against their seats, and not be preloaded against their seats or bend away from their seats. The following reed bending (or reed distortion) limits are recommended by the manufacturer.

Model	Petal Clearance
V4 --	0.008" (0.2mm)
V6 --	0.035" (0.9mm)

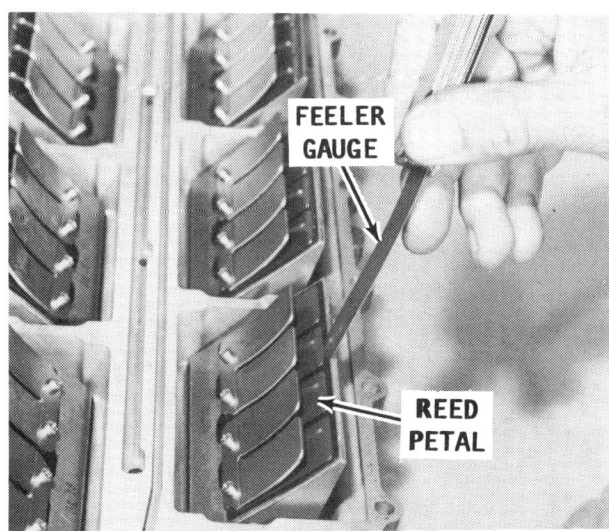

Using a feeler gauge to measure the reed petal clearance, as described in the text.

If the reed petal is distorted beyond its bending limit, rarely, if ever, can the petal be successfully straightened. **THEREFORE**, it must be replaced.

The valve stopper clearance is the distance between the bottom edge of the stopper and the top of the reed petal. This measurement **MUST** be within the specified limits of between 0.24 to 0.26" (6.2 to 6.8mm).

The valve stopper can sometimes be successfully bent to achieve the required clearance, if not, it **MUST** be replaced.

Do not remove the reed valves unless they, or the stoppers, are to be replaced.

Apply Loctite to the threads of the reed retaining screws. Tighten each screw gradually, starting in the center and working outwards across the reed block. Tighten the screws to a torque value of 0.8 ft lbs (1Nm), for all models.

EXHAUST COVER

The exhaust covers are one of the most neglected items on any outboard powerhead. Seldom are they checked and serviced. Many times a powerhead may be overhauled and returned to service without the exhaust covers ever having been removed.

One reason the exhaust covers are not removed is because the attaching bolts usually become corroded in place. This means they are very difficult to remove, but the work should be done. Heat applied to the bolt head and around the exhaust cover will

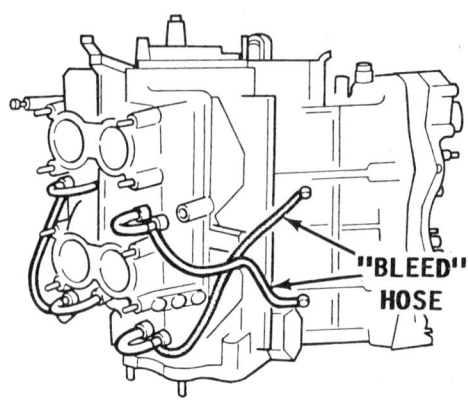

PORT SIDE

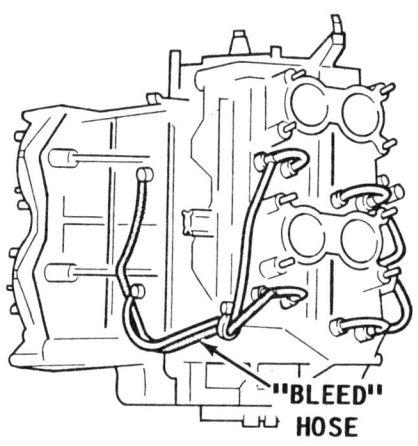

STARBOARD SIDE

Line drawing to depict the bleed hose routing, port and starboard, on a V4 powerhead.

help in removal. However, some bolts may still be broken. If the bolt is broken it must be drilled out and the hole tapped with new threads.

The exhaust covers are installed over the exhaust ports to allow the exhaust to leave the powerhead and be transferred to the exhaust housing. If the cover was the only item over the exhaust ports, they would become so hot from the exhaust gases they might cause a fire or a person could be severely burned if they came in contact with the cover.

Therefore, an inner plate is installed to help dissipate the exhaust heat. Two gaskets are installed -- one on either side of the inner plate. Water is channeled to circulate between the exhaust cover and the inner plate. This circulating water cools the exhaust cover and prevents it from becoming a hazard.

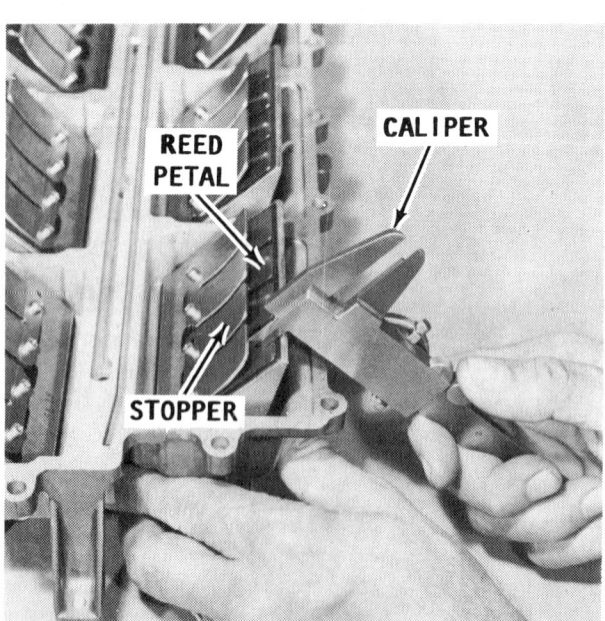

Using vernier calipers to measure the valve stopper clearance

CRANKSHAFT SERVICE 8-19

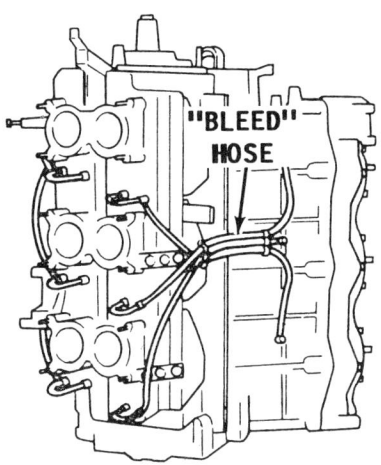

PORT SIDE

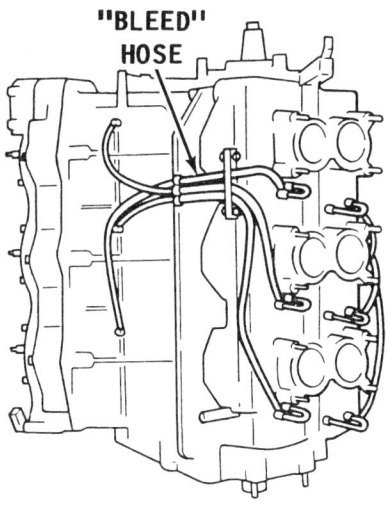

STARBOARD SIDE

Line drawing to depict the bleed hose routing, port and starboard, on a V6 powerhead.

A thorough cleaning of the inner plate behind the exhaust covers should be performed during a major powerhead overhaul. If the integrity of the exhaust cover assembly is in doubt, replace the inner plate.

BLEED SYSTEM SERVICE

The bleed system consists of one or more hoses depending on the model being serviced. These hoses transfer unburnt fuel from one cylinder to another. The fuel is pulsed by crankcase vacuum. This system prevents accumulation of unburnt fuel in the lower cylinder and transfers the fuel and oil mixture to the crankshaft upper main bearing for lubrication.

The accompanying illustrations show the routing of bleed hoses for V4 and V6 powerheads covered in this manual.

Check the condition of each rubber bleed hose. Replace the hose if it shows signs of deterioration or leakage. Check the operation of the check valves. The air/fuel mixture should be able to pass through the valve in only **ONE** direction.

Defective check valves cannot be serviced. If defective, they **MUST** be replaced.

CRANKSHAFT SERVICE

Clean the crankshaft with solvent and wipe the journals dry with a lint free cloth. Inspect the main journals and connecting rod journals for cracks, scratches, grooves, or scores. Inspect the crankshaft oil seal surface for nicks, sharp edges, or burrs, which might damage the oil seal during installation or might cause premature seal wear. **ALWAYS** handle the crankshaft carefully to avoid damaging the highly finished journal surfaces. Blow out all oil passages with compressed air. The oil passageway leads from the rod to the main bearing journal. **TAKE CARE** not to blow dirt into the main bearing journal bore.

Inspect the internal splines at one end and threads at the other end for signs of abnormal wear. Check the crankshaft for runout by supporting it on two "V" blocks at the main bearing surfaces.

Install a dial indicator gauge above the main bearing journals. Rotate the crankshaft and measure the runout (or the out-of-round) and the taper at both ends and in the two center journals.

The crankshaft runout and taper limit for all models is 0.002" (0.05mm) measured at each of the main bearing journals.

If "V" blocks or a dial indicator are not available, a micrometer may be used to

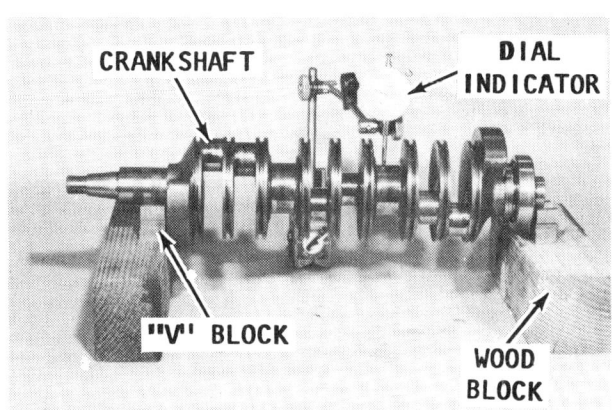

Using "V" blocks raised on blocks of wood to provide adequate clearance for crankshaft rotation while runout and taper are being measured with a dial indicator.

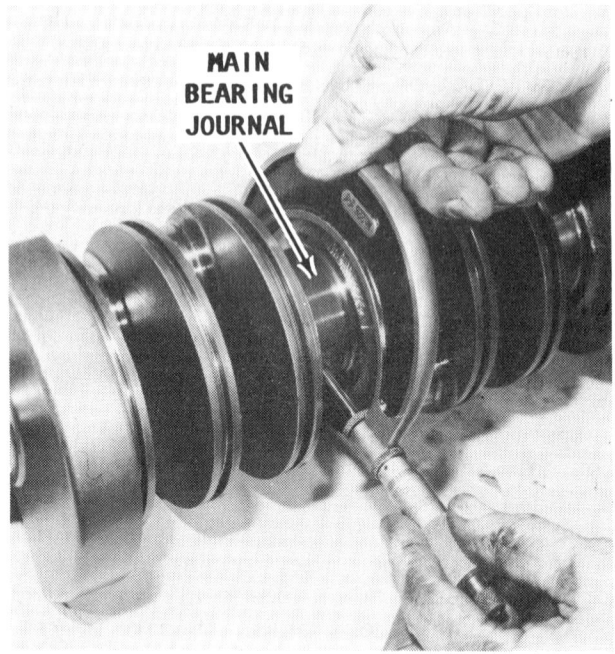

Using a micrometer to measure runout and taper on main bearing journals.

measure the diameter of the journal. Make a second measurement at right angles to the first. Check the difference between the first and second measurement for out-of-round condition. If the journals are tapered, ridged, or out-of-round by more than the specification allows, the journals should be reground, or the crankshaft replaced.

Any out-of-round or taper shortens bearing life. Good shop practice dictates new main bearings be installed with a new or reground crankshaft.

Inspect the crankshaft oil seal surfaces to be sure they are not grooved, pitted, or scratched. Replace the crankshaft if it is severely damaged or worn. Check all crankshaft bearing surfaces for rust, water marks, chatter marks, uneven wear, or overheating. Clean the crankshaft surfaces with 320-grit carborundum cloth. **NEVER** spin-dry a crankshaft ball bearing with compressed air.

Clean the crankshaft and crankshaft ball bearing with solvent. Dry the parts, but not the ball bearing, with compressed air. Check the crankshaft surfaces a second time. Replace the crankshaft if the surfaces cannot be cleaned properly for satisfactory service. If the crankshaft is to be installed for service, lubricate the surfaces with light oil. **DO NOT** lubricate the crankshaft ball bearing at this time.

After the crankshaft has been cleaned, grasp the outer race of the crankshaft ball bearing installed on the lower end of the crankshaft, and attempt to work the race back and forth. There should not be excessive "play". A very slight amount of side "play" is acceptable because there is only about 0.001" (0.025mm) clearance in the bearing.

Lubricate the ball bearing with light oil. Check the action of the bearing by rotating the outer bearing race. The bearing should have a smooth action and no rust stains. If the ball bearing sounds or feels rough or catches, the bearing should be removed and discarded.

CONNECTING ROD SERVICE

Inspect the connecting rod bearings for rust or signs of bearing failure. **NEVER** intermix new and used bearings. If even one bearing in a set needs to be replaced, all bearings at that location **MUST** be replaced.

Clean the inside diameter of the piston pin end of the connecting rod with crocus cloth.

Clean the connecting rod **ONLY** enough to remove marks. **DO NOT** continue, once the marks have disappeared.

Insert the piston pin into the upper end of the connecting rod and check for vertical "play". The piston pin should have **NO** noticeable vertical "play".

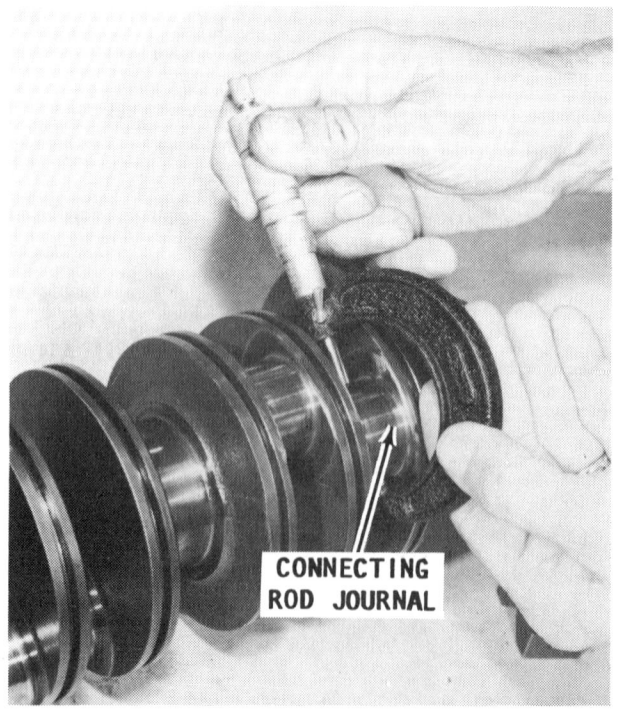

Using a micrometer to measure runout and taper on connecting rod journals.

PISTON SERVICE 8-21

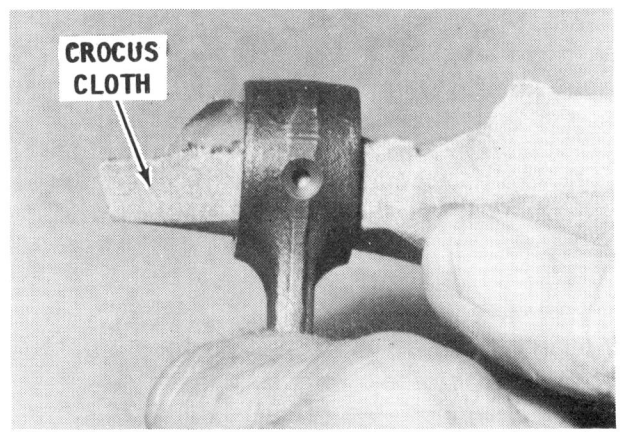

Cleaning the piston end of a connecting rod with crocus cloth.

If the pin is loose or there is vertical "play" check for and replace the worn part/s.

Inspect the piston pin and matching rod end for signs of heat discoloration. Overheating is identified as a bluish bearing surface color and is caused by inadequate lubrication or by operating the powerhead at excessive high rpm.

Check the bearing surface of the rod and rod cap for signs of chatter marks. This condition is identified by a rough bearing surface resembling a tiny washboard. The condition is caused by a combination of low-speed low-load operation in cold water. The condition is aggravated by inadequate lubrication and improper fuel.

Under these conditions, the crankshaft journal is hammered by the connecting rod. As ignition occurs in the cylinder, the piston pushes the connecting rod with tremendous force. This force is transferred to the connecting rod journal. Since there is little or no load on the crankshaft, it bounces away from the connecting rod. The crankshaft then remains immobile for a split second, until the piston travel causes the connecting rod to catch up to the waiting crankshaft journal, then hammers it. In some instances, the connecting rod crankpin bore becomes highly polished.

While the powerhead is running, a "whir" and/or "chirp" sound may be heard when the powerhead is accelerated rapidly -- say from idle speed to about 1500 rpm, then quickly returned to idle. If chatter marks are discovered, the crankshaft and the connecting rods should be replaced.

Inspect the bearing surface of the rod and rod cap for signs of uneven wear and possible overheating. Overheating is identified as a bluish bearing surface color and is caused by inadequate lubrication or by operating the powerhead at excessively high rpm.

PISTON SERVICE

Inspect each piston for evidence of scoring, cracks, metal damage, cracked piston pin boss, or worn pin boss. Be especially critical during inspection if the outboard unit has been submerged. If the piston pin is bent, the piston and pin **MUST** be replaced as a set for two reasons. First, a bent pin will damage the boss when it is removed. Secondly, a piston pin is not sold as a separate item.

Check the piston ring grooves for wear, burns distortion, or loose locating pins. During an overhaul, the rings should be replaced to ensure lasting repair and proper powerhead performance after the work is completed. Clean the piston dome, ring grooves, and the piston skirt. Clean carbon deposits from the ring grooves using the recessed end of a broken piston ring.

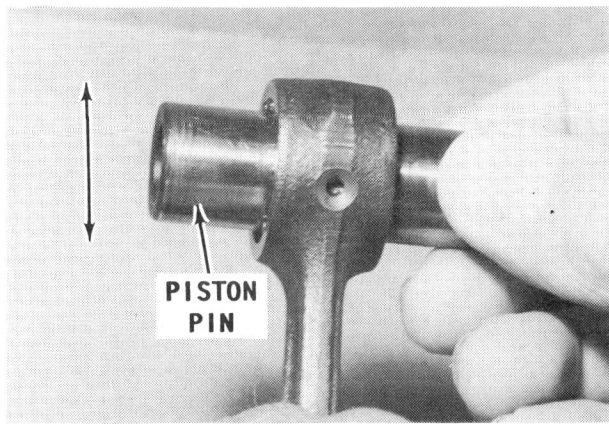

Checking the piston end of a connecting rod for vertical free "play" using a piston pin.

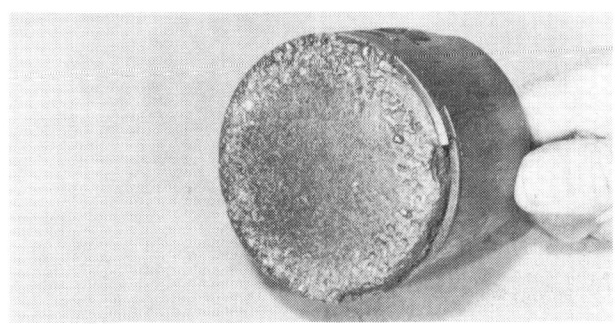

The pitted damage to this piston crown was probably caused by a broken piston ring working its way into the combustion chamber. The little "hills" then became "hot" spots on the crown, contributing to "dieseling" after the powerhead was shut down.

8-22 POWERHEAD

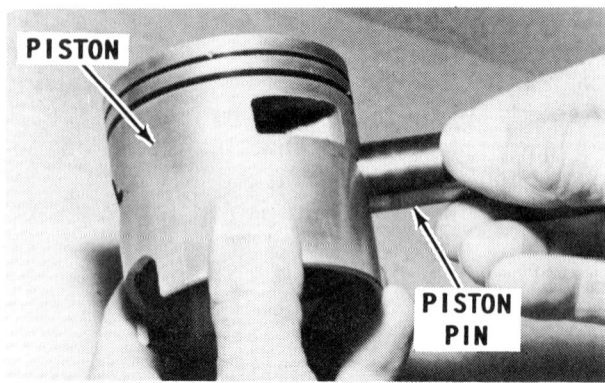

Checking free "play" between the piston pin and the piston boss. There should be NO "play".

NEVER use a rectangular ring to clean the groove for a tapered ring, or use a tapered ring to clean the groove for a rectangular ring.

NEVER use an automotive type ring groove cleaner, because such a tool may loosen the piston ring locating pins.

Clean carbon deposits from the top of the piston using a soft wire brush, carbon removal solution or by sand blasting. If a wire brush is used, **TAKE CARE** not to burr or round machined edges. Clean the piston skirt with crocus cloth.

Install the piston pin through the first boss only. Check for vertical free "play". There should be **NO** vertical free "play". The presence of "play" is an indication the piston boss is worn. The piston is manufactured from a softer material than the piston pin. Therefore, the piston boss will wear more quickly than the pin.

Excessive piston skirt wear **CANNOT** be visually detected. Therefore, good shop practice dictates, the piston skirt diameter be measured with a micrometer.

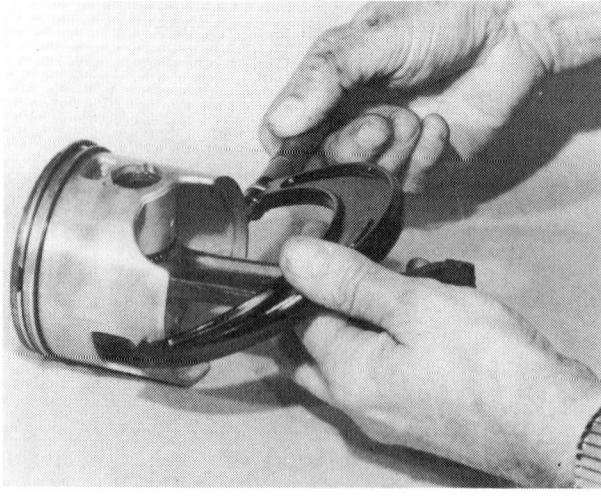

Using a micrometer to measure piston skirt diameter, as described in the text.

Piston skirt diameter should be measured at right angles to the piston pin axis at a point 3/8" (10mm), above the bottom edge of the piston.

Piston diameters are the same for V4 and V6 powerheads covered in this manual. A standard piston has a diameter of 3.54 (90mm).

OVERSIZE PISTONS AND OVERSIZE RINGS

Scored cylinder blocks can be saved for further service by reboring and installing oversize pistons and piston rings. **HOWEVER**, if the scoring is over 0.0075" (0.13mm) deep, the block cannot be effectively rebored for continued use.

Oversize pistons come in two sizes: 3.553" (90.250mm) and 3.563" (90.500mm).

If oversize pistons are not available, the local marine shop may have the facilities to "knurl" the piston, making it larger.

A honed cylinder block is just surface finished. It is not parallel and therefore if an oversized or knurled piston is installed in this bore, the piston will soon seize- at the lower narrow end of the bore.

RING END GAP CLEARANCE

FIRST, THESE IMPORTANT WORDS

Before the piston rings are installed onto the piston, the ring end gap clearance for each ring must be determined. The purpose

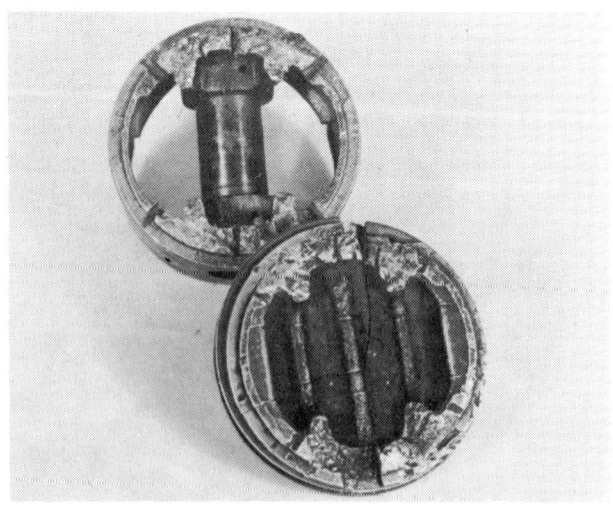

It is believed, this crown siezed with the cylinder wall when the unit was operated at high rpm and the timing was not adjusted properly. At the same instant, the rod apparently pulled the lower part of the piston downward, severing it from the crown.

of the piston rings is to prevent the blowby of gases in the combustion chamber. This cannot be achieved unless the correct oil film thickness is left on the cylinder wall.

This thin coating of oil acts as a seal between the cylinder wall and the face of the piston ring. An excessive end gap will allow blowby and the cylinder will lose compression. An inadequate end gap will scrape too much oil from the cylinder wall and limit lubrication. Lack of adequate lubrication will cause excessive heat and wear.

IDEALLY the ring end gap measurement should be taken **AFTER** the cylinder bore has been measured for wear and taper **AND** after any corrective work, such as boring or honing, has been completed.

IF the ring end gap is measured with a taper to the cylinder wall, the diameter at the lower limit of ring travel will be smaller than the diameter at the top of the cylinder.

IF the ring is fitted to the upper part of a cylinder with a taper, the ring end gap will not be great enough at the lower limit of ring travel. Such a condition could result in a broken ring and/or damage to the cylinder wall and/or damage to the piston and/or damage to the cylinder head.

IF the cylinder is to be only honed, not bored, **OR** if only cleaned, not honed, the ring end gap should be measured at the lower limit of ring travel.

The manufacturer actually gives the precise depth the ring should be inserted into the cylinder -- usually just above the ports -- and assumes the cylinder walls are **PARALLEL** with no taper.

Insert the piston ring from the top of the cylinder. Push the ring down 3/4" (20mm) from the surface using the piston crown. Once the ring is in the proper position, measure the ring end gap with a feeler gauge. The gap measured as directed should be between 0.012 to 0.020" (0.3 to 0.5mm) for all rings.

If the end gap is greater than the amount listed, replace the entire ring set.

If the end gap is less than the amount listed, carefully file the ends of the ring -- just a little at a time -- until the correct end gap is obtained.

SPECIAL WORDS

Inspect the piston ring locating pins to be sure they are tight. There is one locating pin in each ring groove. If the locating pins are loose, the piston **MUST** be replaced.

PISTON RING SIDE CLEARANCE

After the rings are installed on the piston, the clearance in the grooves needs to be checked with a feeler gauge. Check the clearance between the ring and the upper "land" and compare your measurement with the Specifications listed in the following table. Ring wear forms a step at the inner portion of the upper "land". If the piston grooves have worn to the extent to cause

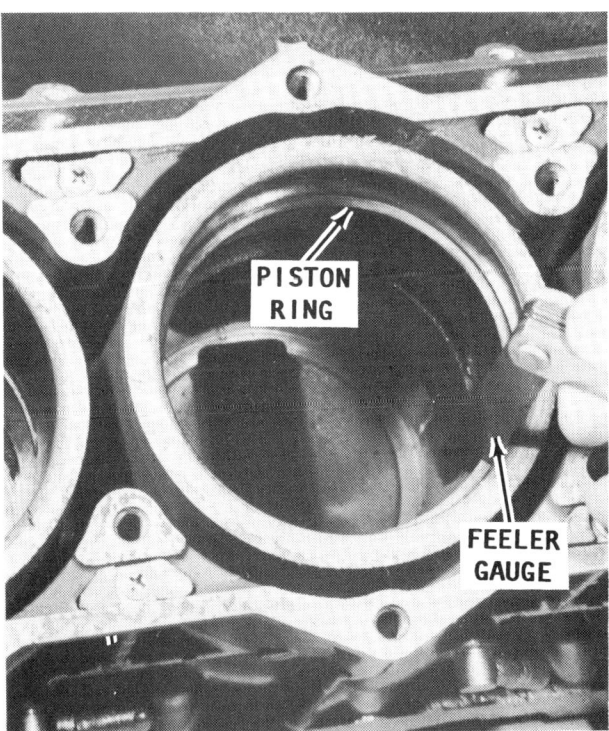

Using a feeler gauge to measure ring end gap.

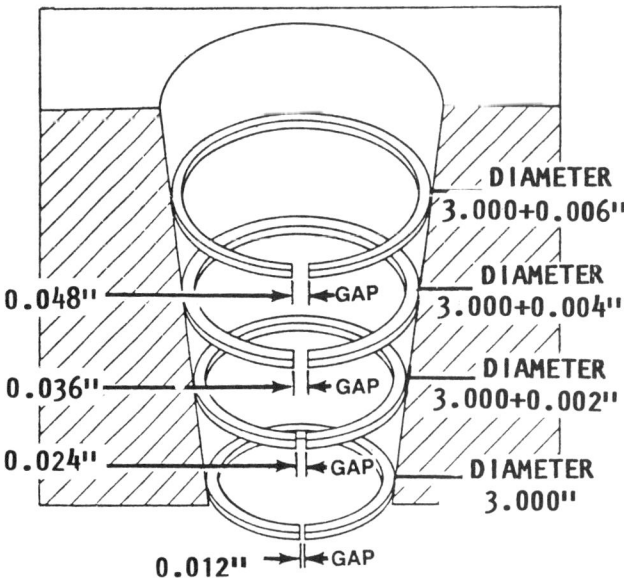

The cylinder taper drastically affects ring and end gap, as shown in this cross-section line drawing.

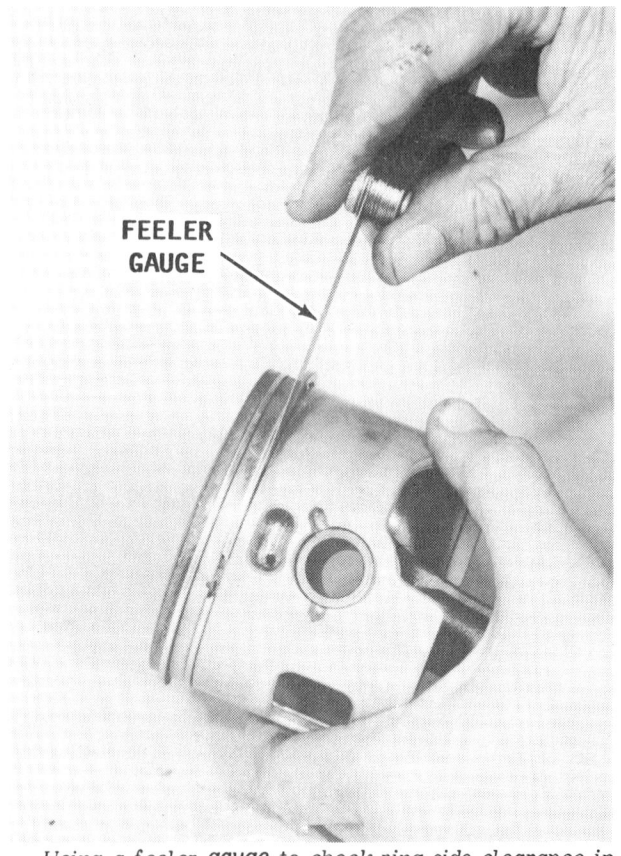

Using a feeler gauge to check ring side clearance in the piston ring groove.

high steps on the upper "land", the step will interfere with the operation of new rings and the ring clearance will be too much. Therefore, if steps exist in any of the upper "lands", the piston should be replaced.

On a Keystone ring (top ring) -- ring with a taper at the top -- the clearance **MUST** be measured below the ring. On a plain ring (2nd ring), this clearance may be measured either above or below the ring.

Model	Type Ring	Side Clearance
V4	Keystone (Top)	0.0012-0.0026"
&		(0.03-0.065mm)
V6	Plain (2nd)	0.0016-0.0030"
		(0.04-0.075mm)

CYLINDER BLOCK SERVICE

Inspect the cylinder block and cylinder bores for cracks or other damage. Remove carbon with a fine wire brush on a shaft attached to an electric drill or use a carbon remover solution.

STOP: If the cylinder block is to be submerged in a carbon removal solution, the crankcase bleed system **MUST** be removed from the block to prevent damage to hoses and check valves.

Use an inside micrometer or telescopic gauge and micrometer to check the cylinders for wear. Check the bore for out-of-round and/or oversize bore. If the bore is tapered, out-of-round or worn more than the wear limit specified by the manufacturer, the cylinders should be rebored and oversize pistons and rings installed.

GOOD WORDS

Oversize piston weight is approximately the same as a standard size piston. Therefore, it is **NOT** necessary to rebore all cylinders in a block just because one cylinder requires reboring.

Cylinder sleeves are an integral part of the die cast cylinder block and **CANNOT** be replaced. In other words, the cylinder cannot be "resleeved".

Four inside cylinder bore measurements must be taken for each cylinder to determine an out-of-round condition, the maximum taper, and the maximum bore diameter.

In the accompanying illustration, measurements D1 and D2 are diameters measured at 0.8" (20mm) from the top of the cylinder at right angles to each other. Measurements D3 and D4 are diameters measured at 2.4" (60mm) from the top of the cylinder at right angles to each other.

Out-of-Round

Measure the cylinder diameter at D1 and D2. The manufacturer requires the difference between the two measurements should

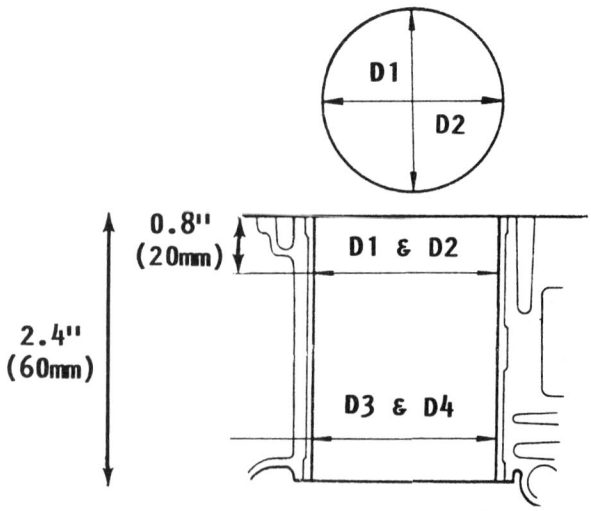

Top view diagram and cross section of a typical cylinder to indicate where measurements are to be taken for wear limit, taper, and out-of-round limits.

CYLINDER BLOCK SERVICE 8-25

be less than 0.002" (0.050mm) for **ALL** models.

Maximum Taper

Measure the cylinder diameter at D1, D2, D3, and D4. Take the largest of the D1 or D2 measurements and subtract the smallest measurement at D3 or D4. The answer to the subtraction -- the cylinder taper -- should be less than 0.003" (0.08mm) for **ALL** models.

Bore Wear Limit

The maximum cylinder diameter D1, D2, D3, and D4 must not exceed the bore wear limits specified by the manufacturer **BEFORE** the bore is rebored for the **FIRST** time. These limits are only imposed on original parts because the sealing ability of the rings would be lost, resulting in power loss, increased powerhead noise, unnecessary vibration, piston slap, and excessive oil consumption.

The limits indicated are usually 0.007" (0.1mm) above the standard bore. Therefore, if the bore is resized, it may sustain another 0.007" (0.1mm) wear before a second reboring is required.

The wear limit on a standard bore of 3.543" (90mm) is 3.547" (90.1mm).

Piston Clearance

Piston clearance is the difference between a maximum piston diameter and a minimum cylinder bore diameter. If this clearance is excessive, the powerhead will develop the same symptoms as for excessive cylinder bore wear -- loss of ring sealing ability, loss of power, increased powerhead noise, unnecessary vibration, and excessive oil consumption.

Maximum piston diameter was described earlier in this section. Minimum cylinder bore diameter is usually determined by measurement D3 or D4 also described earlier in this section.

If the piston clearance exceeds the limits specified by the manufacturer, either the piston or the cylinder block **MUST** be replaced.

Calculate the piston clearance by subtracting the maximum piston skirt diameter from the maximum cylinder bore measurement. The piston clearance for all powerheads covered in this manual is 0.0033 to 0.0035" (0.085 to 0.90mm).

HONING CYLINDER WALLS

Hone the cylinder walls lightly to seat the new piston rings, as outlined in this section. If the cylinders have been scored, but are not out-of-round or the bore is rough, clean the surface of the cylinder with a cylinder hone as described in the following procedures.

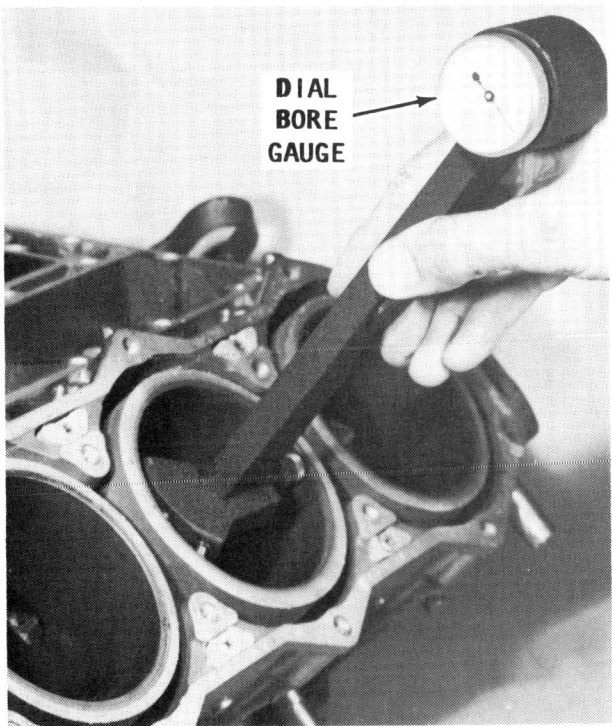

Checking the cylinder for taper using an inside micrometer. One measurement should be taken near the top and another near the bottom. The difference between the two measurements is the amount of taper.

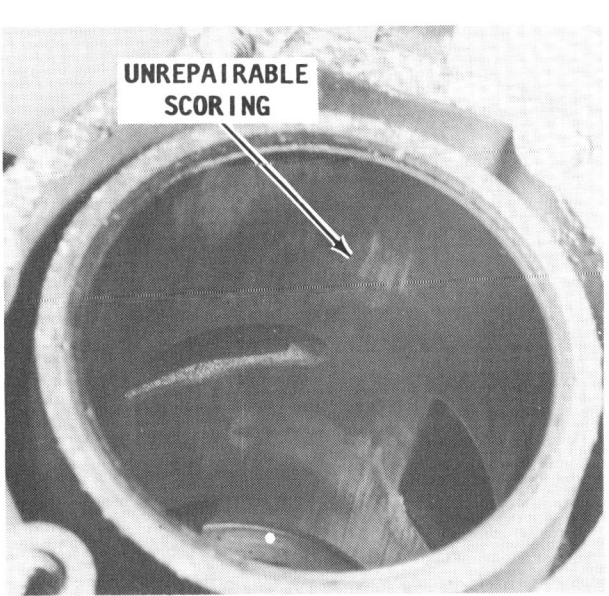

The walls of this cylinder were damaged beyond repair when a piston ring broke and worked its way into the combustion chamber.

SPECIAL WORDS

If overheating has occurred, check and resurface the spark plug end of the cylinder block, if necessary. This can be accomplished with 240-grit sandpaper and a small flat block of wood.

To ensure satisfactory powerhead performance and long life following the overhaul work, the honing work should be performed with patience, skill, and in the following sequence:

a- Follow the hone manufacturer's recommendations for use of the hone and for cleaning and lubricating during the honing operation. A "Christmas tree" hone may also be used.

b- Pump a continuous flow of honing oil into the work area. If pumping is not practical, use an oil can. Apply the oil generously and frequently on both the stones and work surface.

c- Begin the stroking at the smallest diameter. Maintain a firm stone pressure against the cylinder wall to assure fast stock removal and accurate results.

d- Expand the stones as necessary to compensate for stock removal and stone wear. The best cross hatch pattern is obtained using a stroke rate of 30 complete cycles per minute. Again, use the honing oil generously.

e- Hone the cylinder walls **ONLY** enough to deglaze the walls.

f- After the honing operation has been completed, clean the cylinder bores with hot water and detergent. Scrub the walls with a stiff bristle brush and rinse thoroughly with

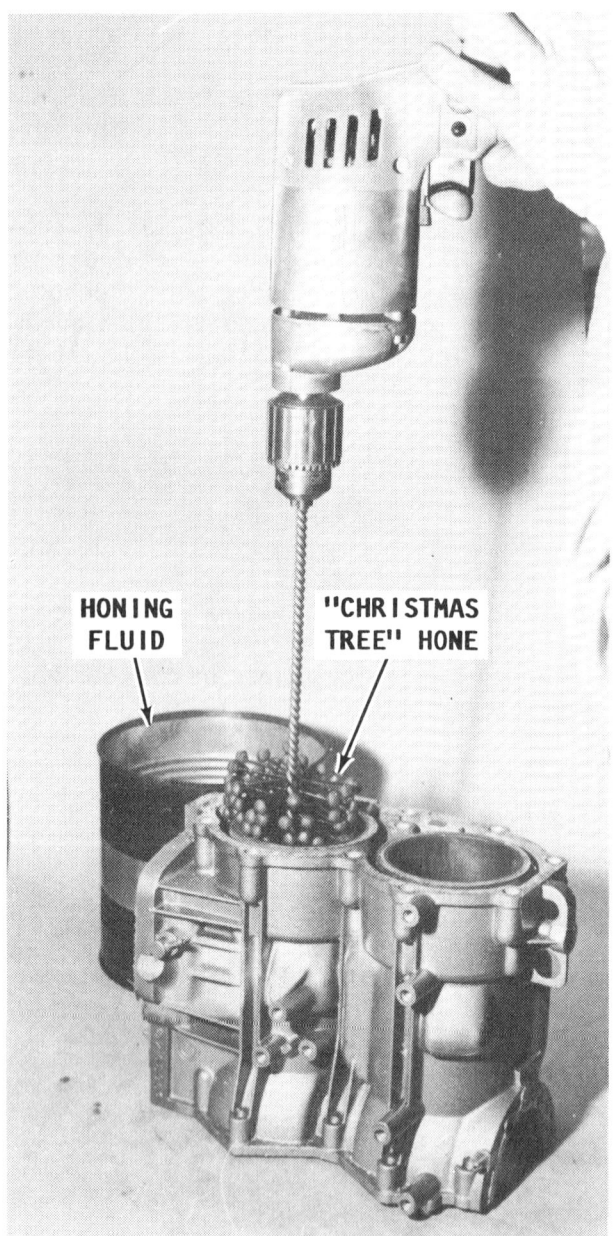

Refinishing the cylinder wall using an electric drill and "Christmas tree" hone. ALWAYS keep the tool moving in long even strokes the entire cylinder depth. Use a continuous liberal amount of honing fluid. The powerhead shown is a four stroke outboard block, but the technique is the same for the powerheads covered in this manual.

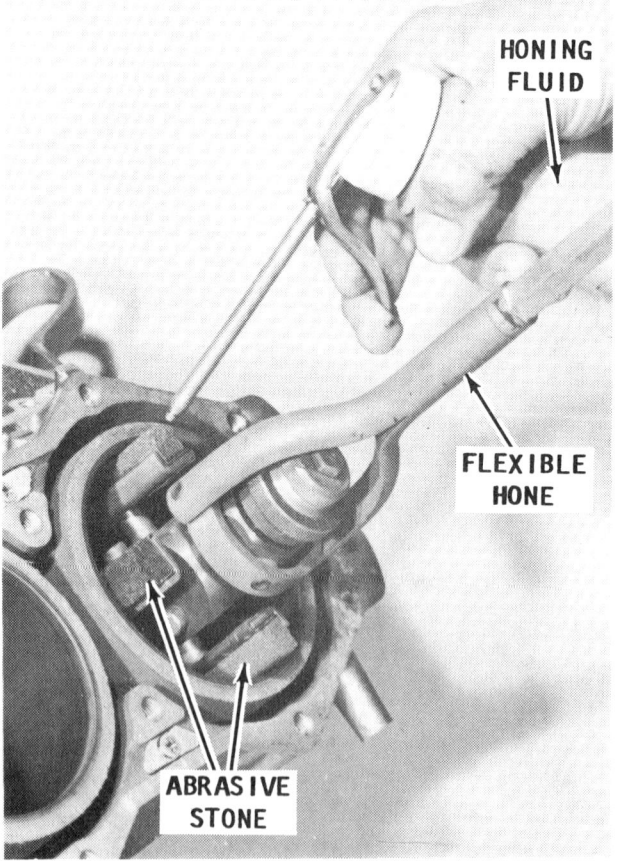

A flexible hone attached to a drill motor is the most common tool used for cylinder wall refinishing. This tool is adjustable to fit various size bore diameters.

BLOCK & HEAD SERVICE 8-27

hot water. The cylinders **MUST** be thoroughly cleaned to prevent any abrasive material from remaining in the cylinder bore. Such material will cause rapid wear of new piston rings, the cylinder bore, and the bearings.

g- After cleaning, swab the bores several times with engine oil and a clean cloth, and then wipe them dry with a clean cloth. **NEVER** use kerosene or gasoline to clean the cylinders.

h- Clean the remainder of the cylinder block to remove any excess material spread during the honing operation.

If oversize pistons are not available, the local marine shop may have the facilities to "knurl" the piston, making it larger.

A honed cylinder block is just surface finished. It is not parallel and therefore if an oversized or knurled piston is installed in this bore, the piston will soon seize at the lower narrow end of the bore.

BLOCK AND CYLINDER HEAD WARPAGE ALL MODELS

First, check to be sure all old gasket material has been removed from the contact surfaces of the block and the cylinder head. Clean both surfaces down to shiny metal, to ensure a true measurement.

Next, place a straight edge across the gasket surface. Check under the straight

Using a wire brush and drill motor to clean carbon deposits from the combustion area of a cylinder head.

edge with a 0.004" (0.1mm) feeler gauge. Move the straight edge to at least eight different locations. If the feeler gauge can pass under the straight edge -- anywhere contact with the other is made -- the surface will have to be resurfaced.

The block or the cylinder head may be resurfaced by placing the warped surface on 400-600 grit **WET** sandpaper, with the sandpaper resting on a **FLAT MACHINED** surface. If a machined surface is not available a large piece of glass or mirror may be used. **DO NOT** attempt to use a workbench or similar surface for this task. A workbench is never perfectly flat and the block or cylinder head will pickup the imperfections of the surface and the warpage will be made worse.

Sand -- work -- the warped surface on the wet sandpaper using large figure "8" motions. Rotate the block, or head, through 180° (turn it end for end) and spend an equal amount of time in each position to avoid

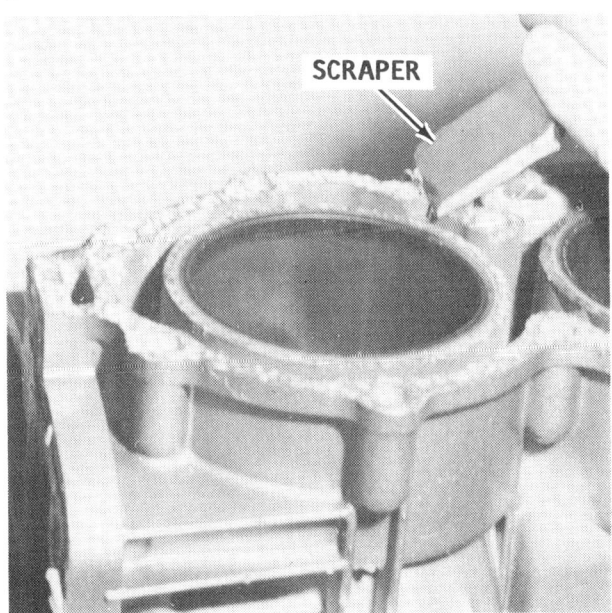
*All traces of gasket material **MUST** be removed to ensure a good seal with a new gasket. If checking for warpage, even the smallest trace of foreign material would give a false reading.*

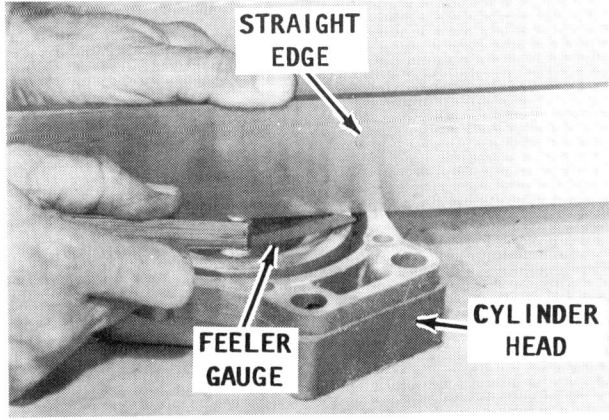
Using a straight edge and feeler gauge to check for cylinder head warpage, as described in the text. The mating surface of the cylinder block should be checked in the same manner.

removing too much material from one side.

If a suitable flat surface is not available, the next best method is to wrap 400-600 grit wet sandpaper around a **LARGE** file. Draw the file as evenly as possible in one sweep across the surface. Do not file in one place. Draw the file in many, many directions to get as even a finish as possible.

As the work moves along, check the progress with the straight edge and feeler gauge. Once the 0.004" (0.1mm) feeler gauge will no longer slide under the straight edge, consider the work completed.

If the warpage cannot be reduced using one of the described methods, the block or cylinder head should be replaced.

ONE LAST CHANCE

If the warpage cannot be reduced and it is not possible to obtain new items -- and the warped part must be assembled for further use -- there is a strong possibility of a water leak at the head gasket. In an effort to prevent a water leak, follow the instructions outlined in the following paragraphs.

SEALING SURFACE WORDS
CRITICAL READING

Because of the high temperatures and pressures developed, the sealing surfaces of the cylinder head and the block are the most prone to water leaks. No sealing agent is recommended **BECAUSE** it is almost impossible to apply an even coat of sealer. An even coat would be essential to ensure an air/water tight seal.

NEVER, NEVER, use automotive type head gasket sealer. The chemicals in the sealer will cause electrolytic action and eat the aluminum faster than you can get to the bank for money to buy a new cylinder block.

Some head gaskets are supplied with a "tacky" coating on both surfaces applied at the time of manufacture. This "tacky" substance will provide an even coating all around. Therefore, no further sealing agent is required.

HOWEVER, if a slight water leak should be noticed following completed assembly work and powerhead start up, **DO NOT** attempt to stop the leak by tightening the head bolts beyond the recommended torque value. Such action will only aggravate the problem and most likely distort the head.

FURTHERMORE, tightening the bolts, which are case hardened aluminum, may force the bolt beyond its elastic limit and cause the bolt to **FRACTURE. BAD NEWS,** very **BAD NEWS** indeed. A fractured bolt must usually be drilled out and the hole re-tapped to accommodate an oversize bolt, etc. Avoid such a situation.

Probable causes and remedies of a new head gasket leaking are:

a- Sealing surfaces not thoroughly cleaned of old gasket material. Disassemble and remove **ALL** traces of old gasket.

b- Damage to the machined surface of the head or the block. The remedy for this damage is the same as for the next case "**c**".

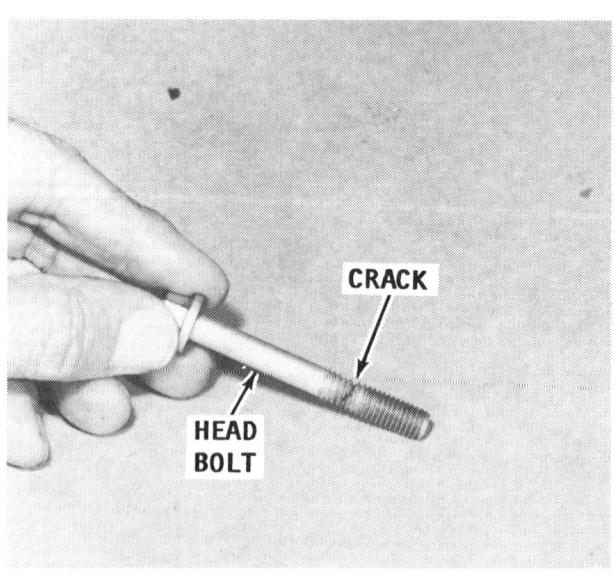

A "lucky break". This head bolt was tightened beyond the recommended torque value. The bolt cracked, but fortunately not all the way through. If the bolt had broken completely, leaving part in the block, the hole would require drilling and tapping oversize for a larger size bolt.

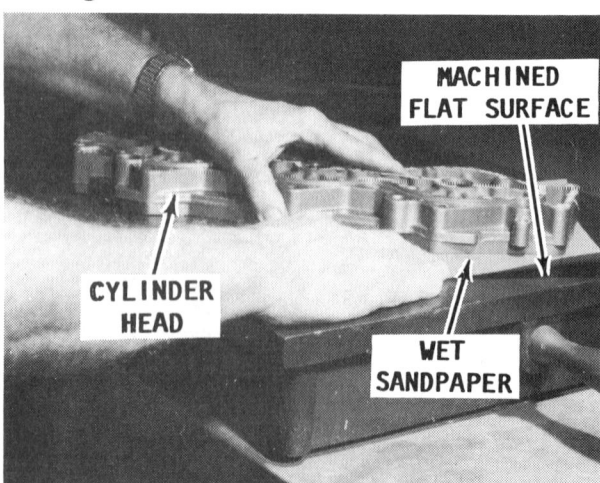

*Slight imperfections on the sealing surface of the head and block may be removed by sanding against wet sandpaper on a **PERFECTLY** smooth surface.*

ASSEMBLING 8-29

c- Permanently distorted head or block. Spray a light **EVEN** coat of any type metallic spray paint on both sides of a new head gasket. Use only metallic paint -- any color will do. Regular spray paint does not have the particle content required to provide the extra sealing properties this procedure requires.

Assemble the block and head with the gasket while the paint is till **TACKY**. Install the head bolts and tighten in the recommended sequence and to the proper torque value and **NO** more!

Allow the paint to set for at least 24 hours before starting the powerhead.

Consider this procedure as a temporary "band aid" type solution until a new head may be purchased or other permanent measures can be performed.

Under normal circumstances, if procedures have been followed to the letter, the head gasket will not leak.

End Sealing Surface Lecture

8-5 POWERHEAD ASSEMBLING AND INSTALLATION

Detailed procedures are given to assemble and install virtually all parts of the powerhead. Therefore, if certain parts were not removed or disassembled because the part was found to be fit for further service, simply skip the particular step involved and continue with the required tasks to return the powerhead to operating condition.

The following instructions pickup the work after all parts of the powerhead have been thoroughly cleaned and inspected according to the procedures outlined in Section 8-4, and new required parts have been purchased and are on hand.

CRANKSHAFT ASSEMBLY

1- Install the crankshaft sealing rings by spreading each ring enough to clear the journal adjacent to its groove. Once the ring is around the journal, spread the ends gently again and slide it up and over the upper edges and into its groove.

Perform the next step **ONLY** if the ball bearing was removed during diassembling Step 45. If the ball bearing was not disturbed, apply a coat of Yamalube to the inner bearing surfaces, and then proceed directly to Step 3.

2- Install the bronze oil injection gear over the lower end of the crankshaft. Using an arbor press. Press the gear down until the gear seats against the shoulder on the shaft.

Position the ball bearing over the shaft with the embossed marks on the bearing surface facing **UPWARD** towards the press shaft.

Now, use a suitable mandrel and press against the inner bearing race.

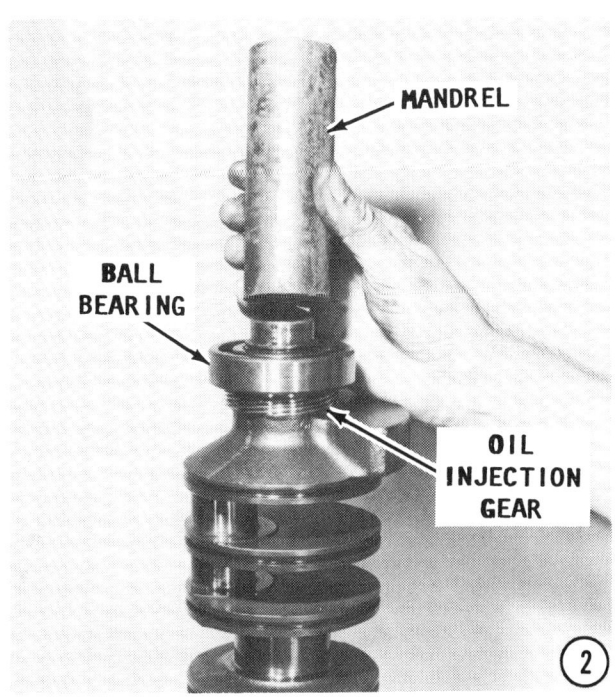

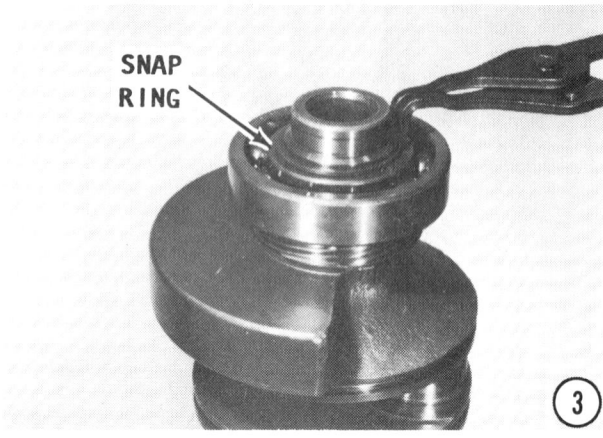

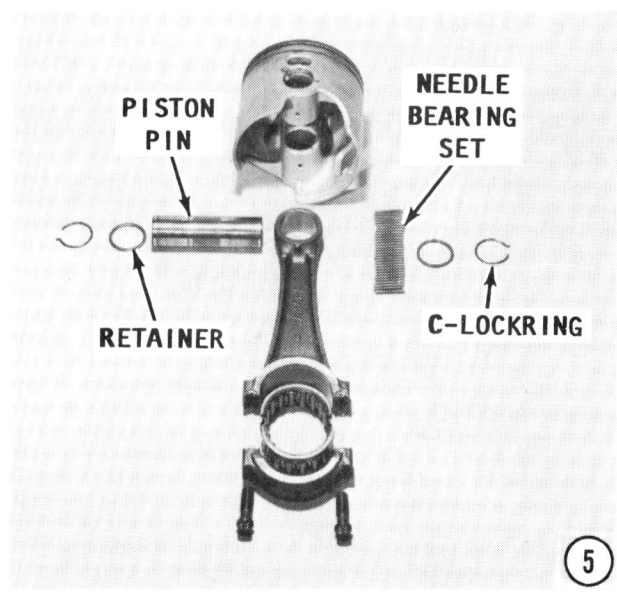

CRITICAL WORDS

Take care to ensure the mandrel is pressing on the inner race and not on the outer race or the ball bearings. Such action would destroy the bearing.

Continue to press the bearing into place until the bearing is seated against the oil injection gear

3- Using a pair of snap ring pliers, install the snap ring into the groove above the gear.

PISTON AND CONNECTING ROD ASSEMBLING

4- Install a new set of piston rings onto the piston, with the embossed marks facing **UP**. No special tool is necessary for installation. **HOWEVER**, take care to spread the ring only enough to clear the top of the piston. The rings are **EXTREMELY** brittle and will snap if spread beyond their limit. Align the ring gap over the locating pin.

SAFETY WORDS

The piston pin lockrings are made of spring steel and may slip out of the pliers or pop out of the groove with considerable force. Therefore, **WEAR** eye protection glasses while installing the piston pin lockrings in the next step.

5- Gather all the components needed to assemble the piston for installation into No. 1 cylinder bore: the piston, two C-lockrings, two retainers, the piston pin, needle bearing set (30 by count), and the connecting rod.

6- Select the set of needle bearings removed from the No. 1 piston or obtain a new set of bearings.

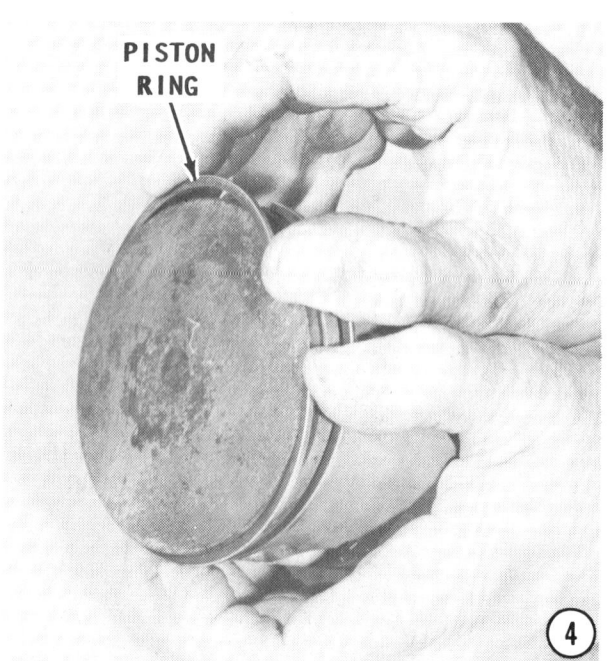

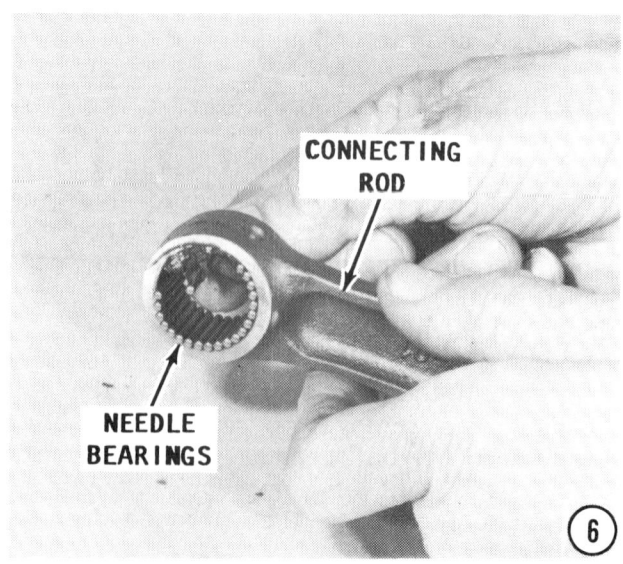

ASSEMBLING 8-31

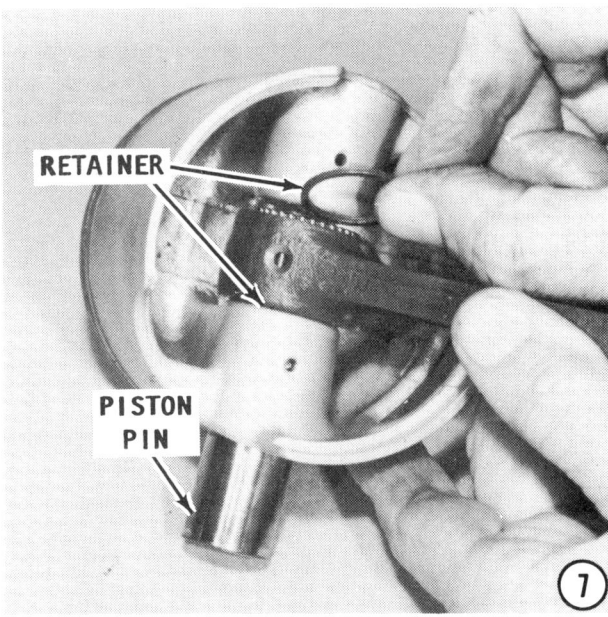

Coat the inner circumference of the small end of the No. 1 connecting rod with Yamaha "A" grease, or equivalent multipurpose water resistant lubricant. Position the needle bearings one by one around the circumference.

7- Position the end of the rod with the needle bearings and retainers in place up into the piston with the word **"YAMAHA"** facing the same direction as the word **"UP"** on the piston crown. The word **UP** on the piston crown **MUST** face **TOWARD** the tapered (upper) end of the crankshaft.

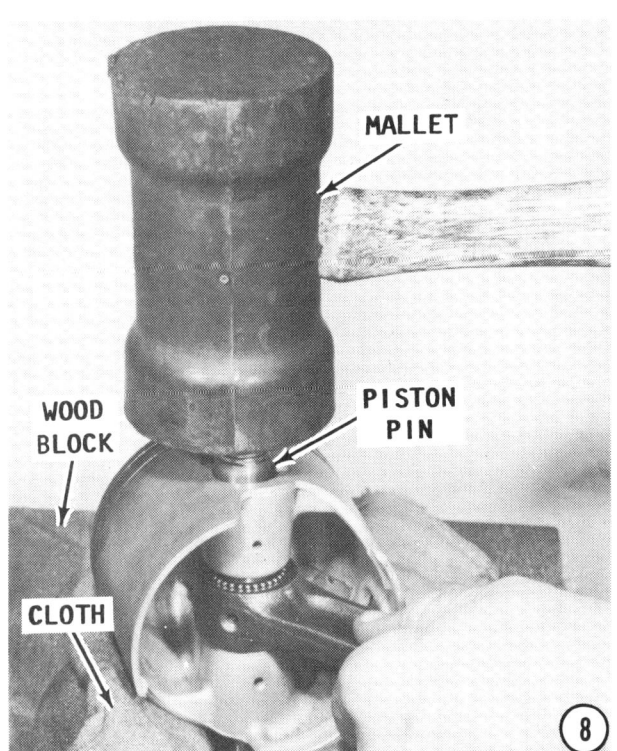

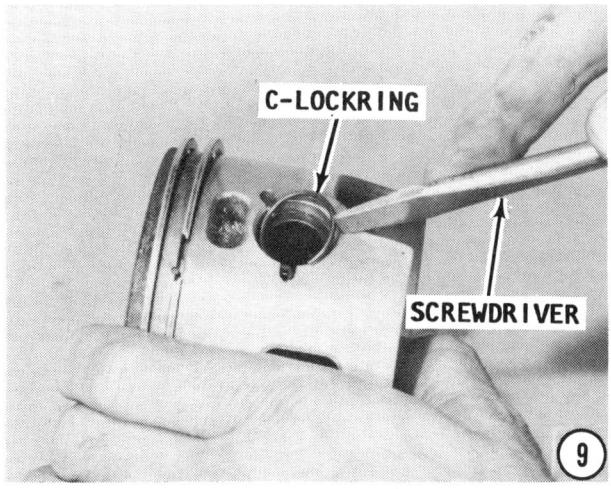

Dab some lubricant on the sides of the rod and "stick" the retainers in place.

8- Press the piston pin through the piston and connecting rod, using an arbor press or by just resting the piston on an open vise with plenty of padding under the piston. Center the pin in the piston.

9- Install the C-lockring at each end of the piston pin.

Perform the procedures in these last 5 steps to assemble the remaining connecting rods.

PISTON INSTALLATION

10- Coat the cylinder bores with a good grade of engine oil. **BEFORE** installing the the piston into the cylinder, make the following test.

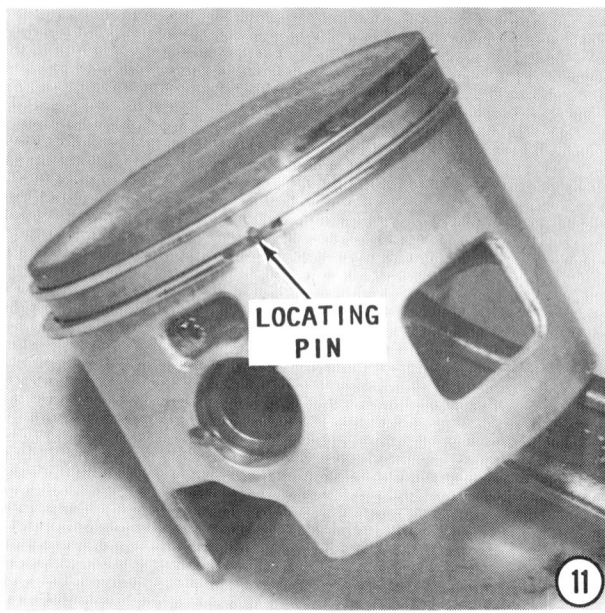

Run your finger along the top rim of the cylinder. If the surface of the block and the surface of the bore have a sharp edge, the piston may be installed from the top.

If the the slightest groove or ridge is felt on the rim, the ridge **MUST** be removed using a ridge reamer, as shown in the accompanying illustration. Attempting to install the piston from the top without removing the ridge, will not be successful because the piston ring will bottom on the ridge. If force is used the ring may very well break.

11- Align each ring gap over its locating pin.

12- Obtain special tool P/N YB34454. Compress the aligned rings and at the same time use the end of a wooden mallet handle

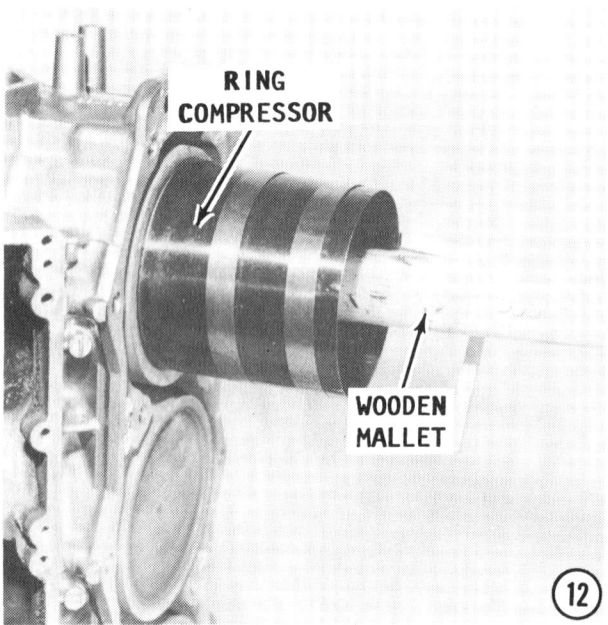

to gently tap the piston down into the cylinder bore. The word **UP** embossed on the piston crown **MUST** face toward the flywheel end of the block. The embossed letter "P" **MUST** face portside, and the embossed letter "S" **MUST** face starboard.

PISTON PIN OFFSET
A LESSON FROM THE RACERS

The piston pin is offset ever so slightly and for good reason.

When the piston is correctly installed, having the pin slightly offset will reduce the thrust, commonly termed "piston slap". This thrust is a frictional force on the cylinder walls.

As the piston approaches TDC, the piston exerts a frictional force on one side of the cylinder wall due to the rod angle. This wall is considered the "minor thrust wall". When the piston actually reaches TDC, the connecting rod pivots on the piston pin. Just a microsecond after TDC, the piston is forced downward in the cylinder under the pressure of combustion. During downward

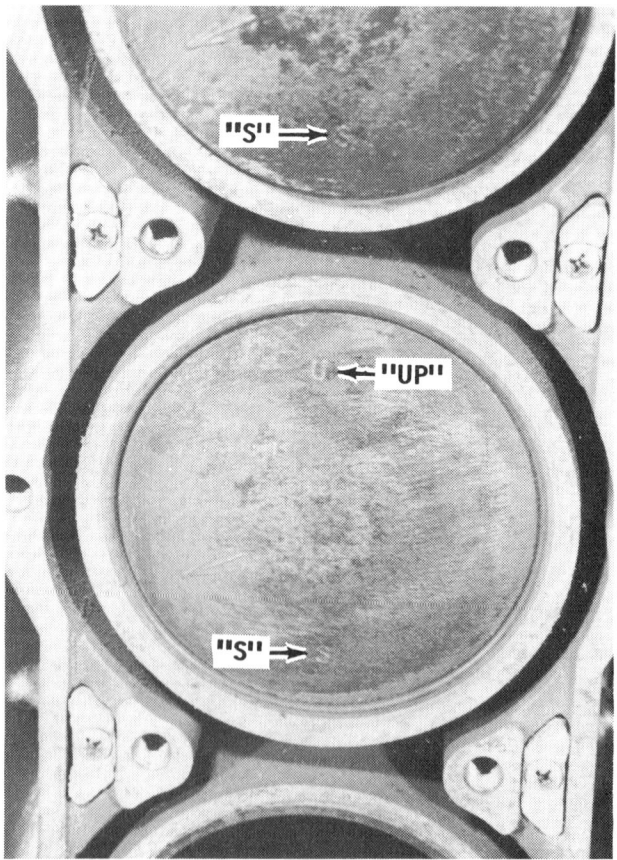

Starboard bank pistons correctly installed. The embossed word "UP" on the piston crown faces the flywheel end of the block; the letter "S" is in the starboard bank; and rings are properly seated.

movement, the piston exerts a greater frictional force on the opposite cylinder wall. This side of the cylinder is referred to as "the major thrust wall".

When the forces on the piston change, at TDC, the piston undergoes a tendency to "tilt". The offset position of the piston pin towards the "major thrust wall" tends to:

First, increase the piston to cylinder friction on the "minor thrust wall" during upward movement.

Secondly, reduce the piston to cylinder friction on the "major thrust wall" during downward movement.

If the piston is installed incorrectly into the cylinder, the powerhead will still operate, but frictional forces on the "minor thrust wall" and the contacting wall of the piston will be far greater than normal. This friction will cause accelerated wear on both the piston and the cylinder wall and "piston slap" will increase to the point where it will be audibly noticeable.

Mechanics involved in racing units will sometimes purposely install the piston with the offset on the wrong side. This position will place the piston dome further up in the combustion chamber as TDC is approached. Result: increase in compression and therefore more horsepower.

Under such conditions the powerhead will not last as long due to the extra friction created, but longevity is not the prime concern -- if they win the race.

End Piston Pin Lecture

ON WITH THE WORK

From the foregoing paragraphs, be advised, if the port and starboard pistons are inadvertently installed incorrectly, the piston assembly and the cylinder wall will be subjected to more than normal wear.

Repeat this procedure for the remaining pistons.

After all pistons have been installed, slide each piston up and down in the cylinder bore several times. Check for binding. Listen for scratching noises. Scratching, or any other "spooky" noise, may indicate a ring was broken during installation.

UPPER OIL SEAL HOUSING INSTALLATION

Perform the following two steps only if the seal and bearing were disturbed in disassembly steps 32 and 33.

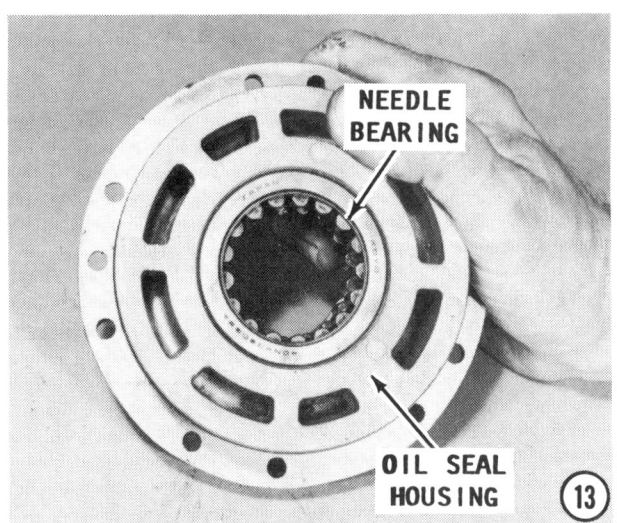

13- Obtain driver P/N YB6205. Support the upper oil seal housing in an arbor press with the O-ring end facing upward. Lower the needle bearing into the housing from the O-ring end. Install the bearing with the stamped mark on the bearing facing **DOWNWARD** — away from the driver. Seat the bearing squarely into the bore, until the shoulder of the driver seats against the shoulder of the housing. The bearing will then be correctly seated in the housing. Invert the housing and leave it in place on the arbor press for the next step.

14- Pack the lips of the seal wth Yamalube Grease "A", or equivalent water resistant lubricant.

Obtain driver P/N YB6198. Place the seal over the installed bearing with the lips of the seals facing **DOWNWARD** toward the driver.

Lower the seal **SQUARELY** into the top of the oil seal housing with the lip of the seal facing **DOWNWARD**. Place the seal installer over the seal and actuate the press until the seal is fully seated.

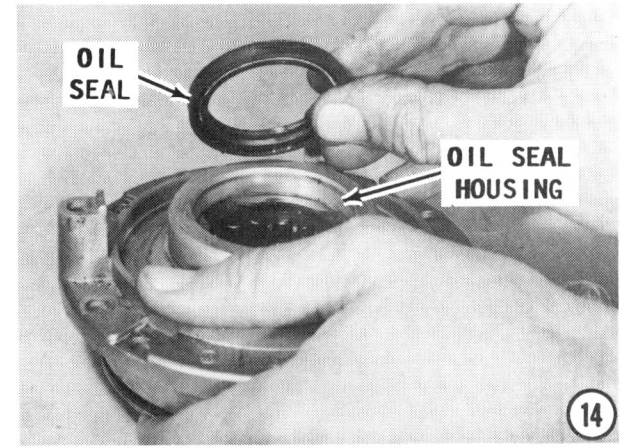

8-34 POWERHEAD

CRANKSHAFT INSTALLATION

SPECIAL BEARING WORDS

If the old roller bearings are to be installed for further service, each must be installed in the same location from which it was removed.

On a V4 crankcase there are center bearing locating holes between the No. 2 and No. 3 cylinder. On a V6 crankcase, these holes are located between the No. 2 and No. 3 cylinders, also between the No. 4 and No. 5 cylinders. These locating holes **MUST** match tangs on the main bearing shells during installation, or the crankcase will not seat properly.

15- Install two new O-rings around the housing and slide the housing over the tapered end of the crankshaft. Rotate the installed housing until the arrow embossed on the top points directly towards the small indent of the cylinder cover.

16- Place the lower half roller bearings in their original positions on the connecting rod caps.

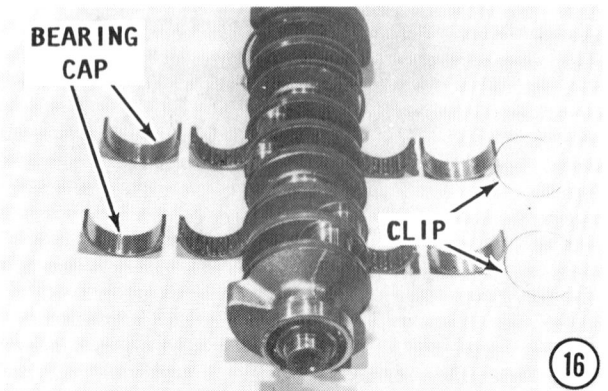

Place the roller bearing, bearing caps and clips around the main journals of the crankshaft, hold them together while lowering the crankshaft down onto the connecting rod bearings.

Place the upper caged roller bearings and connecting rod caps in their original locations.

SPECIAL ROD CAP BOLT WORDS

The manufacturer does not recommend rod cap bolts be used a second time. Also, the manufacturer recommends a specific method of tightening the rod cap bolts be followed as outlined in the next step.

Work on just one rod at a time.

17- Obtain an 8mm 12 point socket, with very narrow sides.

GOOD WORDS

Snap-on Tool Company markets such a socket. Without modification, this socket can be used successfully to install the rod bolts. If this particular socket is not available, a 5/16" 12 point socket will suffice. The external diameter of most sockets, with the exception of the first one listed, will be too great to fit between the bolt and the cap.

As a last resort, an 8mm or a 5/16" 12 point socket may be ground down to fit, but not without its disadvantages: First, grinding away material from the external wall will weaken the socket and the walls may break away. During installation, the rod bolts are tightened to a torque value of 24 ft lb (34Nm). Secondly, if the grinding is not symmetrical, any high spots on the external socket wall may flake off when used and fall down into the cylinder bore.

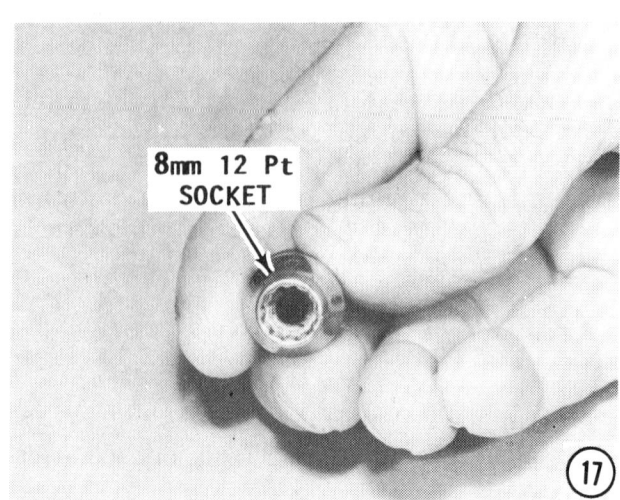

ASSEMBLING 8-35

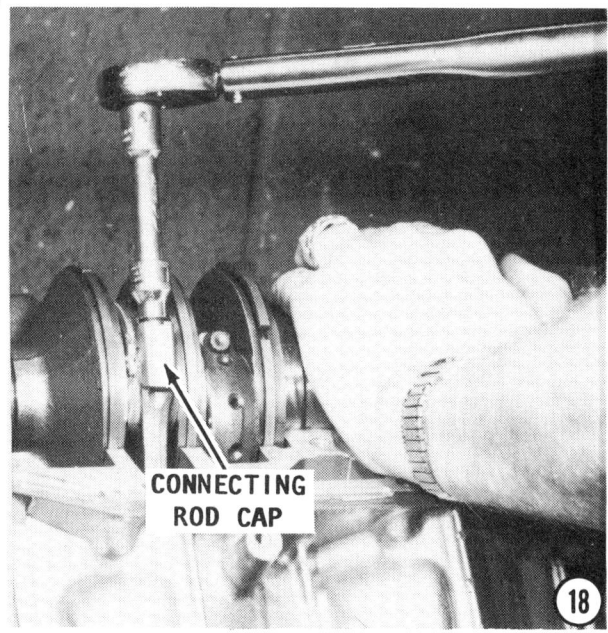

CONNECTING ROD CAP

18

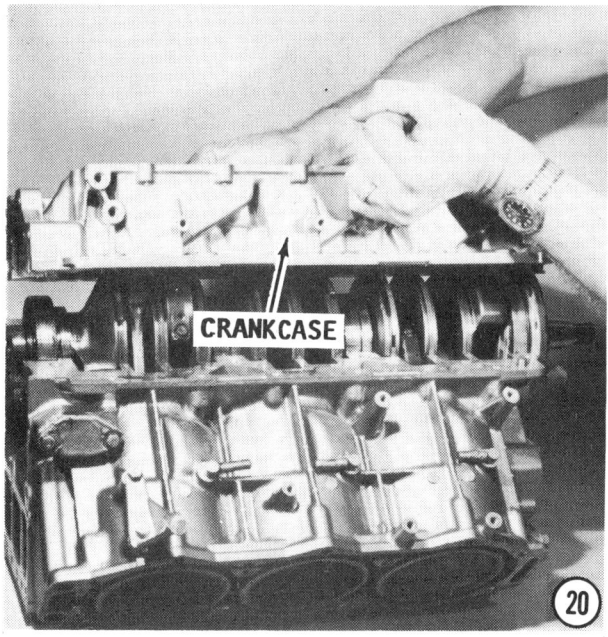

CRANKCASE

20

18- Thread in **NEW** connecting rod cap bolts a few turns each. Now, tighten the cap bolts to a torque value of 12 ft lbs (17Nm). Check to be sure the single mark on each rod cap end is centrally located between the two matching embossed marks on the rod end and on the cap.

If the marks are aligned, as described, loosen each bolt one half turn, and then tighten the bolt to a torque value of 25 ft lbs (35Nm).

If the marks are **NOT** aligned, as described, remove the cap bolts, and then the upper caged roller bearing half. Install the upper caged roller bearing half again. Check the alignment. If the second attempt fails to align the marks as described, the crankshaft **MUST** be removed and the lower caged roller bearing half checked for correct alignment.

19- Arrange the end gaps of all the crankshaft sealing rings to face **UPWARD**.

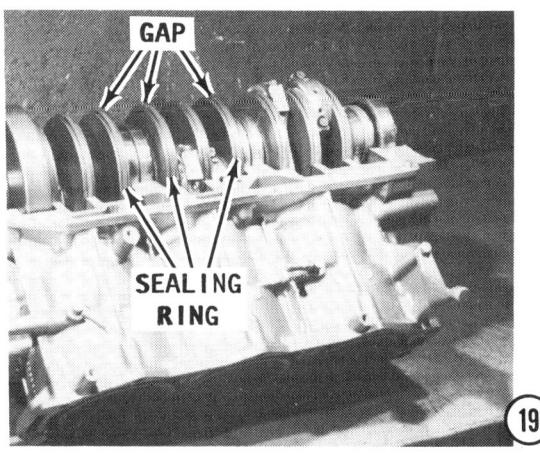

GAP

SEALING RING

19

20- Apply a thin bead of Yamabond No. 4 Permatex around both surfaces of the crankcase and block. Check to be sure the two dowel pins are in place and install the crankcase to the block.

21- Install and tighten the attaching bolts in the sequence shown in the accompanying illustrations. The numbering sequence is embossed around the mating surface of the crankcase on both sides of the block. Tighten the bolts in two rounds and to the torque value as follows:

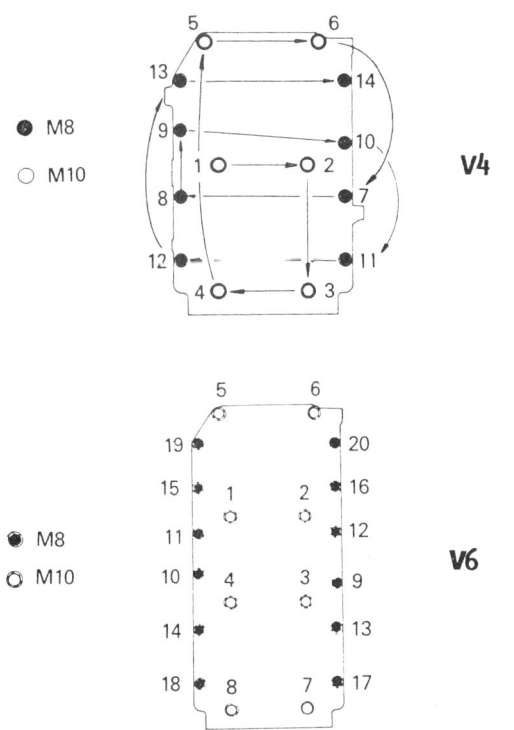

21

V4 powerheads
   First Round
       6 large bolts - 14 ft lbs (20Nm)
       8 small bolts - 7.2 ft lbs (10Nm)
   Second Round
       6 large bolts - 29 ft lbs (40Nm)
       8 small bolts - 13 ft lbs (18Nm)

V6 powerheads
   First Round
       8 large bolts - 14 ft lbs (20Nm)
       12 small bolts - 7.2 ft lbs (10Nm)
   Second Round
       8 large bolts - 29 ft lbs (40Nm)
       12 small bolts - 13 ft lbs (18Nm)

Rotate the crankshaft by hand to be sure the crankshaft does not bind.

BAD NEWS

If binding is felt, it will be necessary to remove the crankcase and reseat the crankshaft and also to check the positioning of the crankshaft sealing rings and the bearing locating pins. If binding is still a problem after the crankcase has been installed a second time, the cause might very well be a broken piston ring.

PISTON RING TENSION

22- Check to be sure each piston ring has spring tension. This is accomplished by **CAREFULLY** pressing on each ring with a screwdriver extended through the exhaust ports, as shown in the accompanying illustration. If spring tension cannot be felt (the spring fails to return to its original position), the ring was probably broken during the piston and crankshaft installation pro-

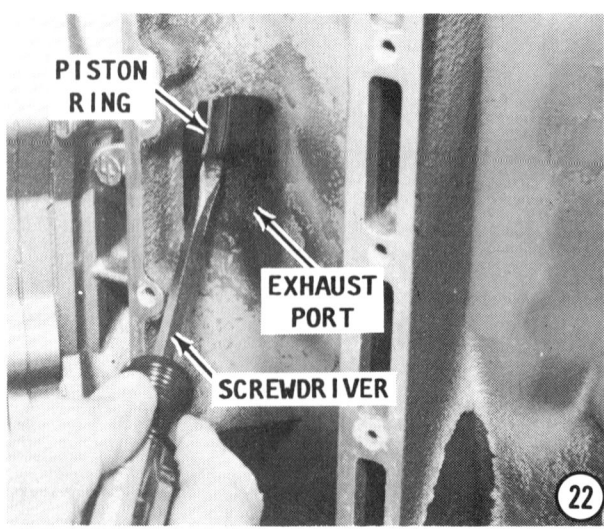

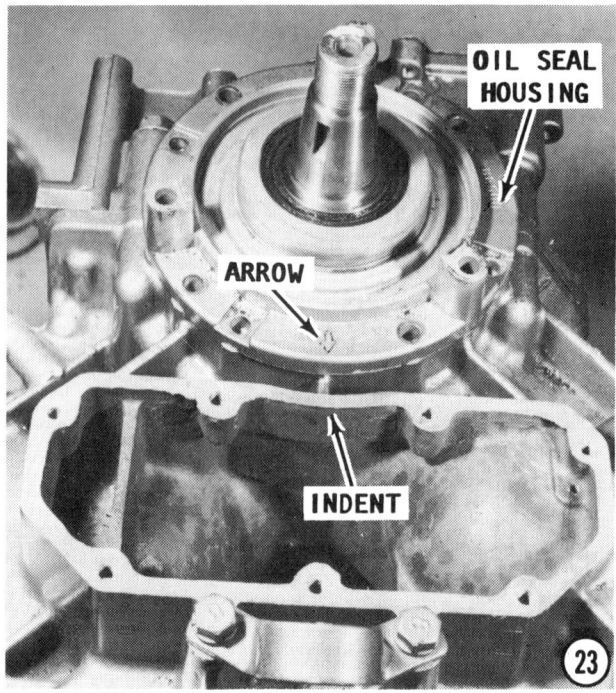

cess. **TAKE CARE** not to burr the piston rings while checking for spring tension.

23- Verify the arrow embossed on the oil seal housing points directly towards the indent of the cylinder cover opening, as shown. Install the bolts securing the upper oil seal housing. Tighten the bolts alternately (across from each other), working in a clockwise direction around the housing. Tighten the bolts to a torque value of 5.8 ft lbs (8Nm).

24- Install a new gasket and then install the cylinder cover. Tighten the attaching bolts to a torque value of 5.8 ft lbs (8Nm).

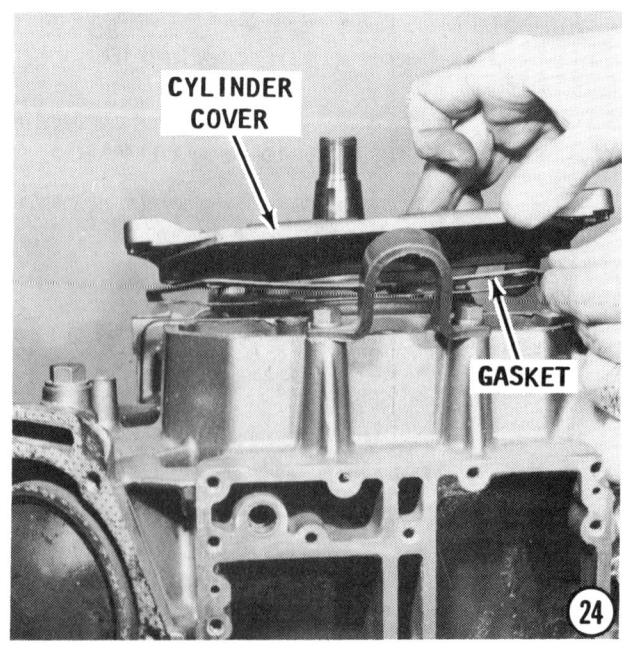

ASSEMBLING 8-37

ANODE INSTALLATION

25- Insert the wedge shaped anodes into the water jacket of the block. Secure the anodes with Phillips head screws. These anodes will only be visible and accessible the next time the powerhead is overhauled. Therefore, if the anodes show **ANY** sign of deterioration in this location, a **NEW** anode should be installed.

CYLINDER HEAD ASSEMBLY

26- Position a new head gasket in place on the powerhead.

SPECIAL WORDS

The manufacturer recommends **NO** sealing agent be used on either side of the head gasket.

Install the cylinder head onto the powerhead.

27- Install and tighten the attaching bolts in the sequence shown in the accompanying illustrations. The numbering sequence is embossed around the mating surface of the head and block. Tighten the

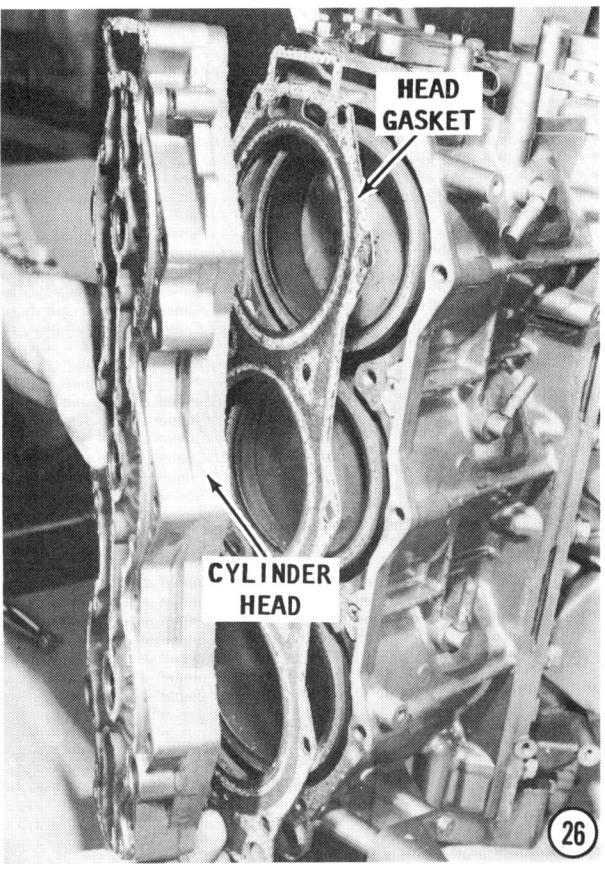

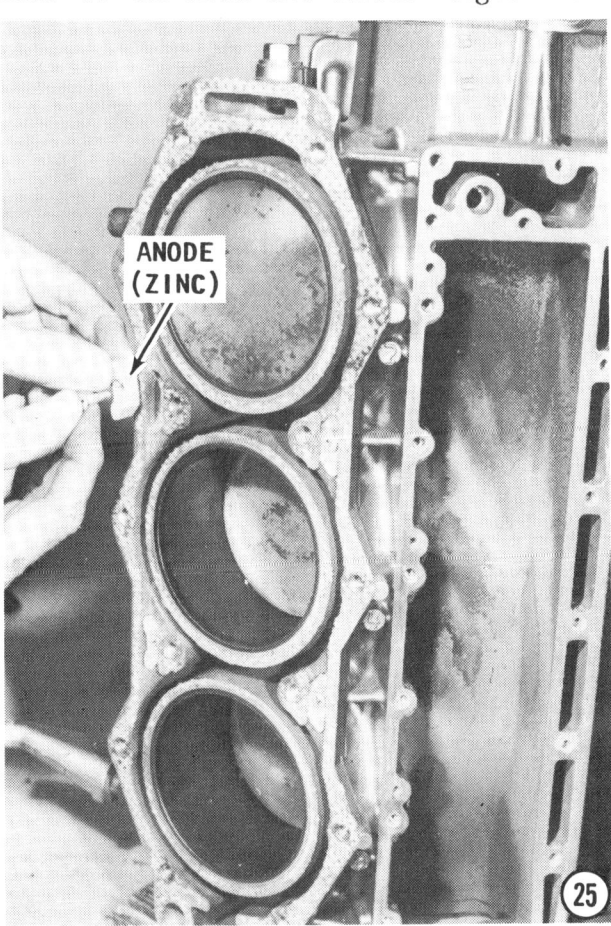

bolts in two rounds and to the torque value as follows:

First Sequence -- 11 ft lbs (15Nm)
Second Sequence -- 22 ft lbs (30Nm)

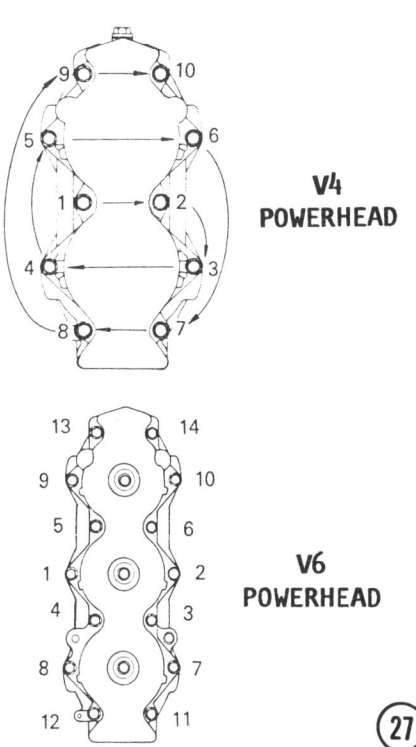

SEALING SURFACE WORDS
CRITICAL READING

Because of the high temperatures and pressures developed, the sealing surfaces of the cylinder head and the block are the most prone to water leaks. No sealing agent is recommended **BECAUSE** it is almost impossible to apply an even coat of sealer. An even coat would be essential to ensure a air/water tight seal.

Some head gaskets are supplied with a "tacky" coating on both surfaces applied at the time of manufacture. This "tacky" substance will provide an even coating all around. Therefore, no further sealing agent is required.

HOWEVER, if a slight water leak should be noticed following completed assembly work and powerhead start up, **DO NOT** attempt to stop the leak by tightening the head bolts beyond the recommended torque value. Such action will only aggravate the problem and most likely distort the head.

FURTHERMORE, tightening the bolts, which are case hardened aluminum, may force the bolt beyond its elastic limit and cause the bolt to **FRACTURE. BAD NEWS,** very **BAD NEWS** indeed. A fractured bolt must usually be drilled out and the hole retapped to accommodate an oversize bolt, etc. Avoid such a situation.

Probable causes and remedies of a new head gasket leaking are:

a- Sealing surfaces not thoroughly cleaned of old gasket material. Disassemble and remove **ALL** traces of old gasket.

b- Damage to the machined surface of the head or the block. The remedy for this damage is the same as for the next case "c".

c- Permanently distorted head or block. Spray a light **EVEN** coat of any type metallic spray paint on both sides of a new head gasket. Use only metallic paint -- any color will do. Regular spray paint does not have the particle content required to provide the extra sealing properties this procedure requires.

Assemble the block and head with the gasket while the paint is till **TACKY.** Install the head bolts and tighten in the recommended sequence and to the proper torque value and **NO** more!

Allow the paint to set for at least **24** hours before starting the powerhead.

Consider this procedure as a temporary "band aid" type solution until a new head may be purchased or other permanent measures can be performed.

Under normal circumstances, if procedures have been followed to the letter, the head gasket will not leak.

End Sealing Surface Lecture
On With The Work

28- Position a new head cover gasket in place on the cylinder head.

SPECIAL WORDS

The manufacturer recommends **NO** sealing agent be used on either side of this gasket.

Install the head cover onto the head.

29- Install and tighten the attaching bolts in the sequence shown in the accompanying illustrations. The numbering sequence is embossed around the mating surface of the head and block. Tighten the bolts in two rounds and to the torque value as follows:

First Sequence -- 2.9 ft lbs (0.4Nm)
Second Sequence -- 5.8 ft lbs (0.8Nm)

ASSEMBLING 8-39

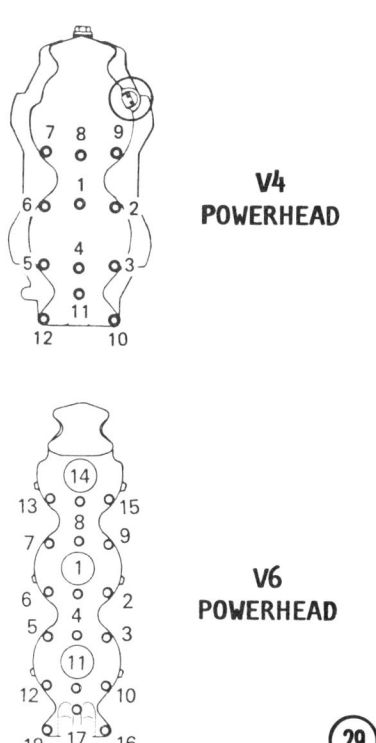

THERMOSTAT INSTALLATION

30- Insert the thermostat into the powerhead with the spring end going in **FIRST**.

On V4 powerheads, the water pressure relief valve is also located under the thermostat cover. Install the longer end of the valve into the head. Place the spring over the short end of the valve. Coat both sides of a new gasket with Permatex and install the cover with the four attaching bolts. Tighten the bolts to a torque value of 5.8 ft lbs (8Nm) starting with the lower right bolt and tightening the remaining three in a clockwise sequence.

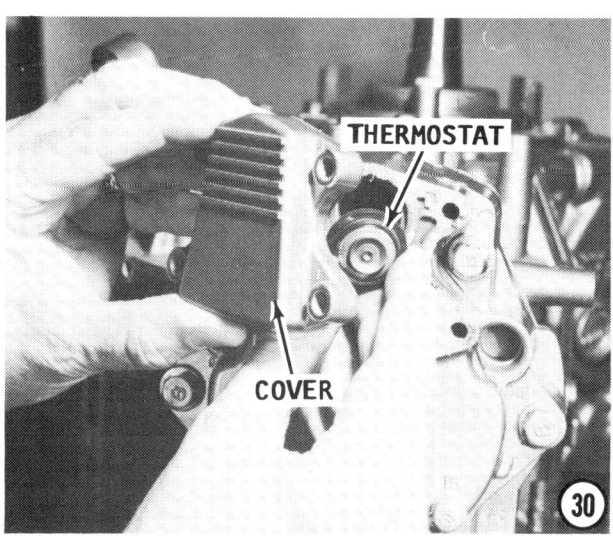

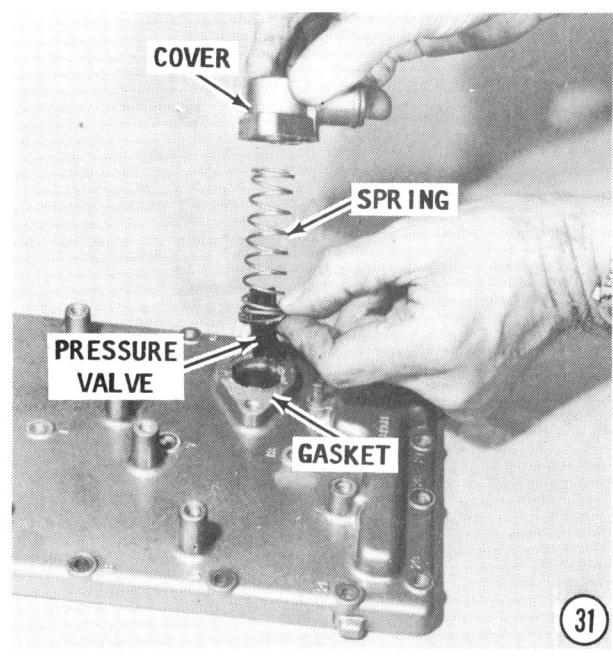

V6 POWERHEAD

31- Install the pressure valve into the exhaust cover with the longer end of the valve being inserted into the cover. Place the spring over the short end of the valve. Install a new gasket and install the cover. Tighten the bolts securely.

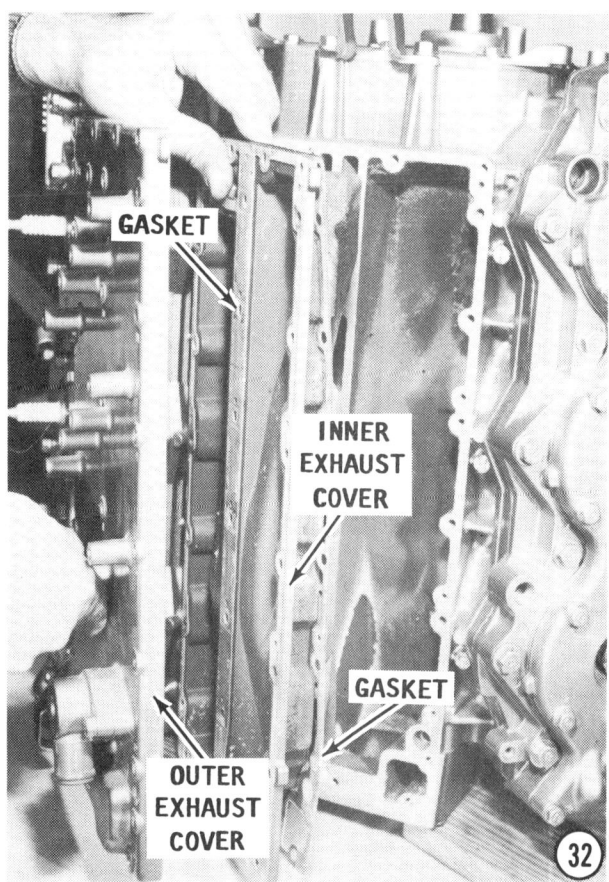

8-40 POWERHEAD

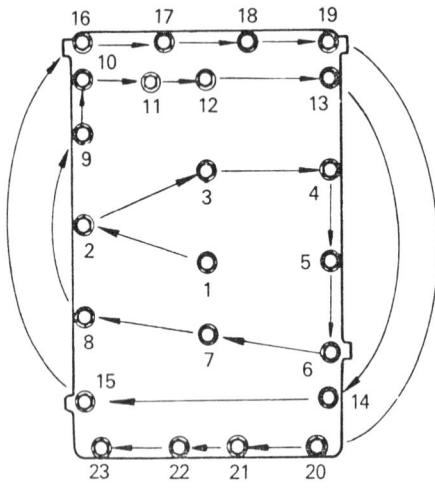

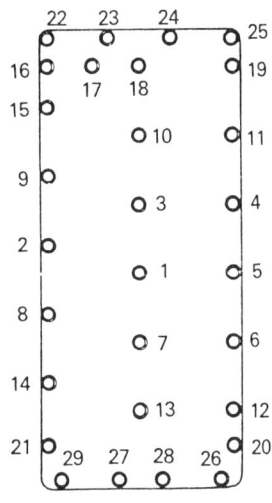

ALL MODELS
EXHAUST COVER INSTALLATION

32- Obtain VHT (very high temperature) Quick Gasket. Apply this sealant to both sides of the two exhaust manifold gaskets. Quick Gasket is a silicone rubber based adhesive sealant recommended by the manufacturer for use where rubber gaskets are used next to metal surfaces.

Install the first gasket, the inner exhaust cover, the second gasket, and then the outer exhaust cover.

33- Tighten the bolts to the following torque values and in the sequence indicated in the accompanying illustration for the unit being serviced:

First Sequence -- 2.9 ft lbs (0.4Nm)
Second Sequence -- 5.8 ft lbs (0.8Nm)

INTAKE MANIFOLD INSTALLATION

34- Install the intake manifold housing. The manufacturer recommends no sealant be applied to the gasket at this location.

35- Install and tighten the attaching bolts to the following torque values and in the sequence indicated in the accompanying illustration for the unit being serviced:

First Sequence -- 2.9 ft lbs (0.4Nm)
Second Sequence -- 5.8 ft lbs (0.8Nm)

36- Route the oil injection supply lines between the retainers and the intake manifold. Slide the lines on the fittings at each intake port. If the lines were not identified, see Chapter 5 for a more detailed installation procedure.

LOWER OIL SEAL HOUSING INSTALLATION

37- If the oil seals in the lower oil seal housing were removed during disassembling, install new oil seals in the housing using a handle and driver Yamaha P/N YB6198. Both seals are driven into the housing with the machined surface facing **UP**. Secure the lower oil seal housing in a vice equipped with soft jaws.

Both the large and the small seal are installed with lips facing **DOWNWARD**.

When the lower oil seal housing is installed onto the powerhead, all seals will face in the correct direction, as installed.

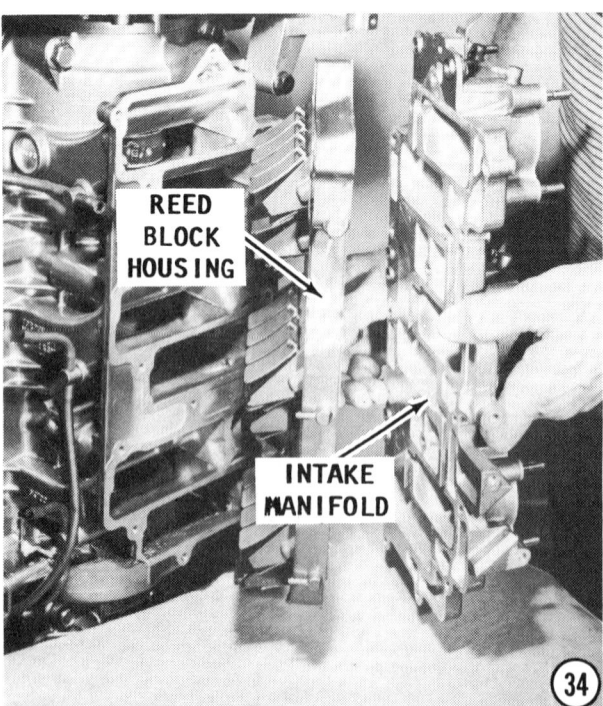

ASSEMBLING 8-41

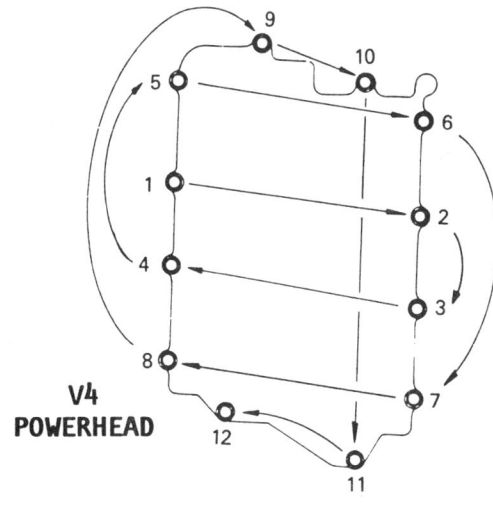

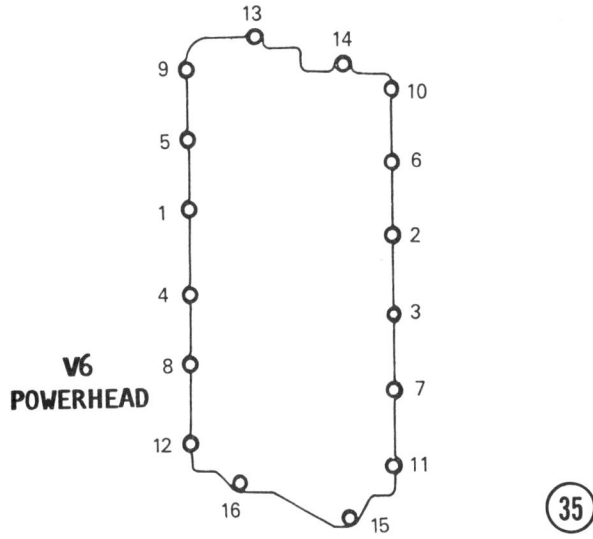

Pack each seal lip with Yamalube Grease "A", or equivalent water resistant lubricant, as soon as the seal is installed.

SPECIAL WORDS
FROM YAMAHA ENGINEERS

When two seals are installed, the lips of **BOTH** seals face in the **SAME** direction. Yamaha engineers have concluded it is better to have both seals face the same direction rather than have them "back to back" as directed by other outboard manufacturers. In this position, with both lips facing toward the water after oil seal housing installation, the engineers feel the seals will be more effective in keeping water out of the oil seal housing. Any lubricant lost will be negligible.

38- Apply engine oil around the groove in the oil housing and around the new O-ring. Install the O-ring around the housing. Install the housing over the lower end of the crankshaft. Tap LIGHTLY around the circumference to seat the housing properly into the crankcase. Align the holes, and then install and tighten the securing bolts.

8-42 POWERHEAD

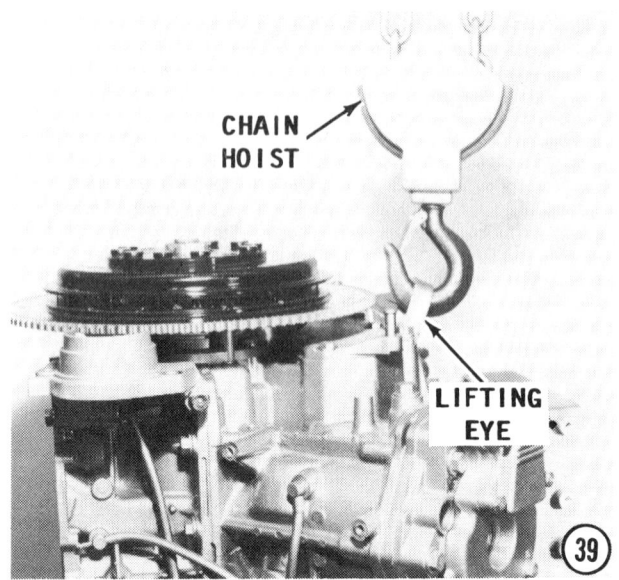

POWERHEAD INSTALLATION

39- Coat both sides of a new gasket with Permatex, or equivalent material. Place the gasket in position on the intermediate housing surface. Check to be sure the two alignment dowel pins are in place. Carefully lower the powerhead onto the intermediate housing with the dowel pins indexing into the holes in the powerhead. The splines of the crankshaft must also index into the splines of the driveshaft. If the splines do not index, shift the lower unit into forward gear; have an assistant **SLOWLY** rotate the propeller just a "whisker" **CLOCKWISE**; and the splines should index. Check to be sure the powerhead is fully down on the intermediate housing -- the mating surfaces hard against each other.

40- Apply Loctite to the bolt threads and install the ten bolts of three different lengths and two nuts securing the powerhead to the intermediate housing.

Tighten the bolts alternately and evenly to a torque value of 13 ft lbs (18Nm).

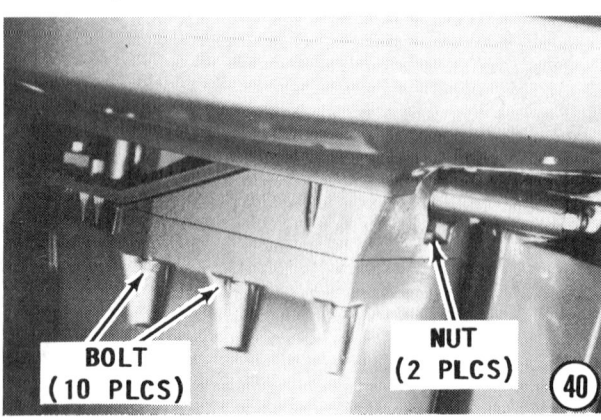

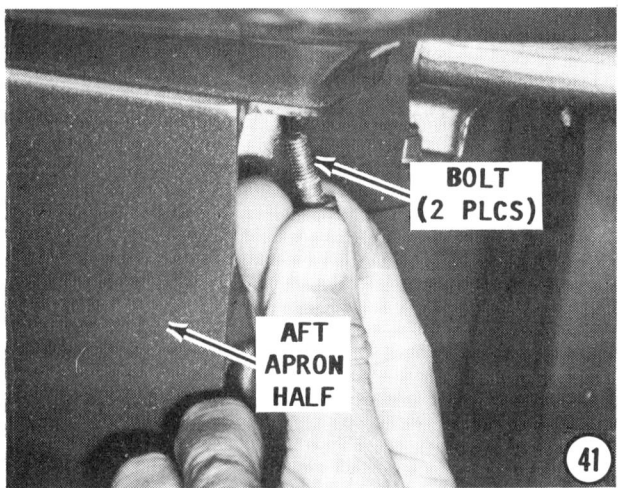

41- Wrap the aft half of the apron around the lower cowling. Secure the apron half to the cowling with the two attaching bolts at the outer edges of the apron. Tighten the bolts to a torque value of 5.8 ft lbs (8Nm).

42- Wrap the forward half of the apron around the lower cowling. Secure both halves together with the two attaching screws. Tighten the screws securely.

Models Without YMIS

43- Place the bushing on the inside of the throttle control arm. Install the torsion spring on the magneto control lever and hook the throttle control lever onto the torsion spring. Twist the throttle control lever and snap it over the magneto control lever. Insert the bolt through the throttle control arm, slide the nylon washer on first, followed by the metal washer and install the assembly to the powerhead. If the length of any control rods were altered, see Chapter 7 for procedures to bring the lengths back to specifications.

44- Install all the link rods onto both levers.

ASSEMBLING 8-43

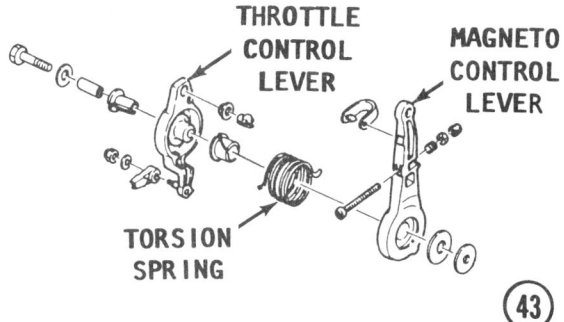

Models Equipped With YMIS

Slide the bolt through the magneto control arm, slide the nylon washer on first, followed by the metal washer and install the assembly to the powerhead.

All Models

45- Connect the ground leads attached to the lower cowling and a lower bolt on the exhaust cover.

46- Install the throttle cable onto the throttle arm. Secure the cable with a washer and a locknut.

47- Install the shift guide to the block, over the shift arm stud. Secure with attaching hardware. Using a pair of needle nose pliers, install the cotter pin through the hole in the stud.

48- Hook the shift rod over the stud.

49- Install the washer and locknut over the stud to retain the shift rod.

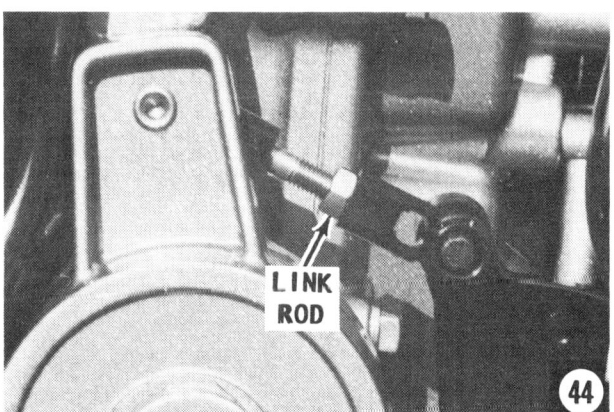

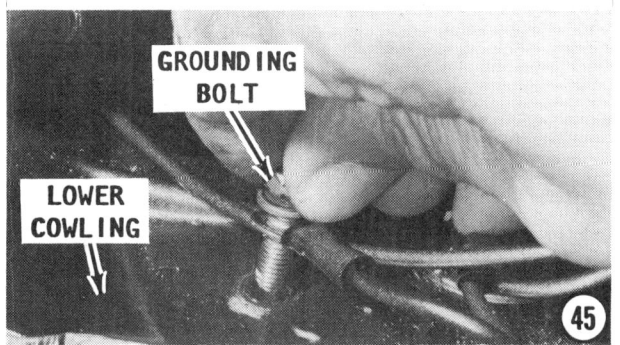

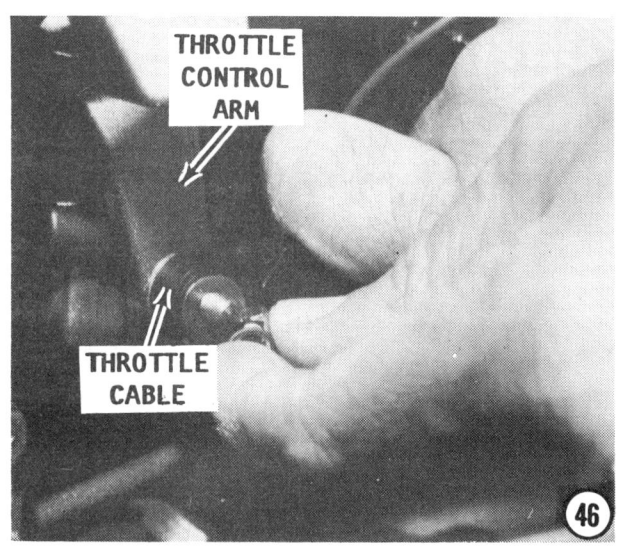

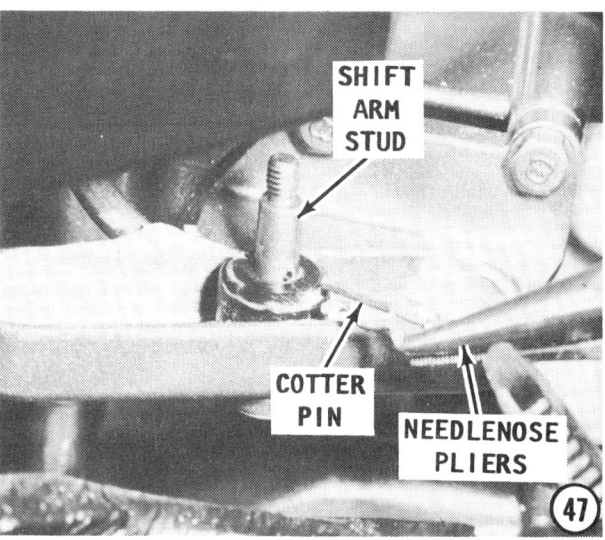

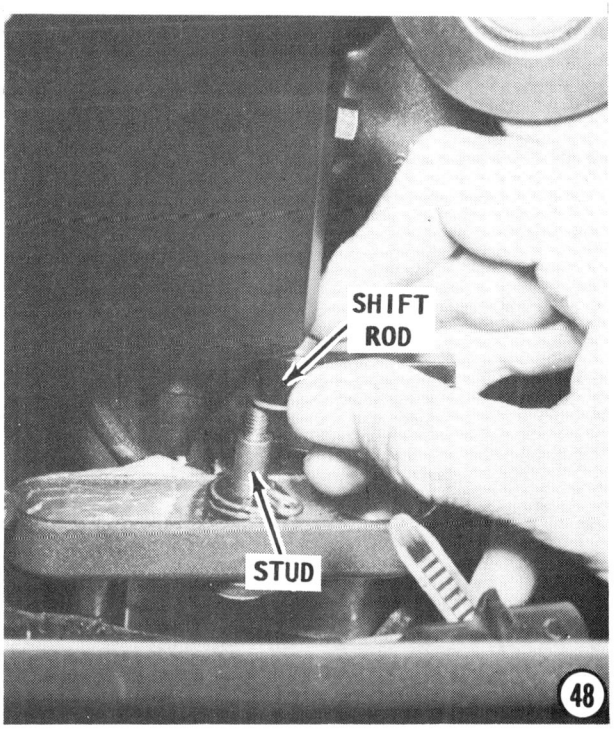

8-44 POWERHEAD

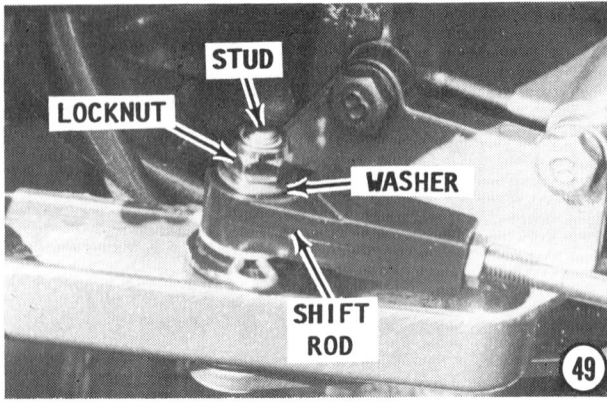

50- Push the water drain hose back onto the fitting on the base of the exhaust cover. Check to be sure the other end of the hose is firmly attached to the pilot hole fitting.

On V6 models connect the water hose to the fitting on the bypass cover.

51- Insert the thermoswitches into the cylinder head covers until they seat.

Install the flywheel and stator assembly, see Chapter 6.

Install the CDI unit and the ignition coils, see Chapter 6.

Install the oil injection system, see Chapter 5.

Install the carburetors, choke, and flame arrestor, see Chapter 4.

Install the electric cranking motor, see Chapter 9.

Move the outboard unit to a test tank or mount the unit on a boat in a body of water. Leave the cowling off to permit inspection and to make adjustments, if necessary.

CAUTION

Water must circulate through the lower unit to the powerhead anytime the powerhead is operating to prevent damage to the water pump in the lower unit. Just five seconds without water will damage the water pump impeller.

Attempt to start and run the powerhead without the cowling installed. This will provide the opportunity to check for fuel and oil leaks, and to make adjustments, if required.

After the engine is operating properly, install the cowling. Follow the break-in procedures with the unit on a boat and with a load on the propeller.

Break-in Procedures

As soon as the engine starts, **CHECK** to be sure the water pump is operating. If the water pump is operating, a fine stream will be discharged from the exhaust relief hole at the rear of the drive shaft housing.

DO NOT operate the engine at full throttle except for **VERY** short periods, until after 10 hours of operation as follows:

a- Operate at 1/2 throttle, approximately 2500 to 3500 rpm, for 2 hours.

b- Operate at any speed after 2 hours **BUT NOT** at sustained full throttle until another 8 hours of operation.

c- Mix gasoline and oil during the break-in period, total of 10 hours, at a ratio of 25:1.

d- While the engine is operating during the initial period, check the fuel, exhaust, and water systems for leaks.

e- See Chapter 2 for tuning procedures.

f- See Chapter 7 for timing and synchronizing procedures.

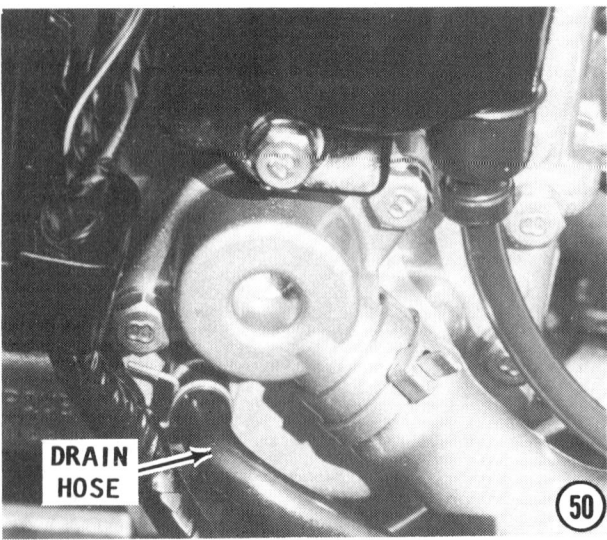

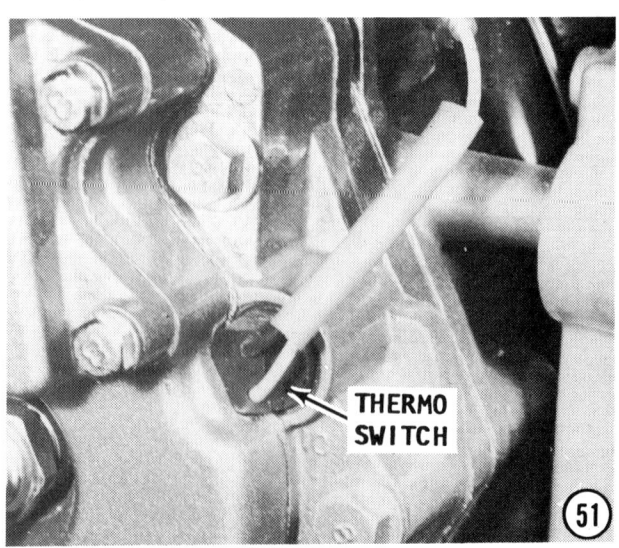

9
ELECTRICAL

9-1 INTRODUCTION

The battery, charging system, and the cranking system are considered subsystems of the electrical system. Each of these units will be covered in detail in this chapter beginning with the battery.

9-2 BATTERIES

The battery is one of the most important parts of the electrical system. In addition to providing electrical power to start the engine, it also provides power for operation of the running lights, radio, electrical accessories, and possibly the pump for a bait tank.

Because of its job and the consequences, (failure to perform in an emergency) the best advice is to purchase a well-known brand, with an extended warranty period, from a reputable dealer.

The usual warranty covers a prorated replacement policy, which means the purchaser would be entitled to a consideration for the time left on the warranty period if the battery should prove defective before its time.

Do not consider a battery of less than **70-ampere hour** or **100-minute** reserve capacity. If in doubt as to how large the boat requires, make a liberal estimate and then purchase the one with the next higher ampere rating.

The 220hp V6 Special and 225hp V6 Excel are equipped with an onboard computer, commonly referred to in the trade and through this manual as YMIS (Yamaha Microcomputer Ignition System). This computer monitors the powerhead operating condition and regulates the ignition timing accordingly.

The manufacturer recommends outboards equipped with YMIS be equipped with a battery of at least 100 to 105 amp/hour capacity.

MARINE BATTERIES

Because marine batteries are required to perform under much more rigorous conditions than automotive batteries, they are constructed much differently than those used in automobiles or trucks. Therefore, a marine battery should always be the No. 1 unit for the boat and other types of batteries used only in an emergency.

Marine batteries have a much heavier exterior case to withstand the violent pounding and shocks imposed on it as the boat moves through rough water and in extremely tight turns.

The plates in marine batteries are thicker than in automotive batteries and each plate is securely anchored within the battery case to ensure extended life.

The caps of marine batteries are "spill proof" to prevent acid from spilling into the bilges when the boat heels to one side in a

A fully charged battery, filled to the proper level with electrolyte, is the heart of the ignition and electrical systems. Engine cranking and efficient performance of electrical items depend on a full-rated battery.

tight turn, or is moving through rough water.

Because of these features, the marine battery will recover from a low charge condition and give satisfactory service over a much longer period of time than any type intended for automotive use.

NEVER use a "Maintenance Free" type battery with an outboard unit. The charging system is not regulated as with automotive installations and the battery may be quickly damaged.

BATTERY CONSTRUCTION

A battery consists of a number of positive and negative plates immersed in a solution of diluted sulfuric acid. The plates contain dissimilar active materials and are kept apart by separators. The plates are grouped into what are termed elements. Plate straps on top of each element connect all of the positive plates and all of the negative plates into groups.

The battery is divided into cells which hold a number of the elements apart from the others. The entire arrangement is contained within a hard rubber case. The top is a one piece cover and contains the filler caps for each cell. The terminal posts protrude through the top where the battery connections for the boat are made. Each of the cells is connected to its neighbor in a positive-to-negative manner with a heavy strap called the cell connector.

BATTERY RATINGS

Four different methods are used to measure and indicate battery electrical capacity:

1- Ampere-hour rating
2- Cold cranking performance
3- Reserve capacity
4- Watt hour rating

The **ampere-hour** rating of a battery refers to the battery's ability to provide a set amount of amperes for a given amount of time under test conditions at a constant temperature of $80°$ ($27°C$). Amperes x hours equals ampere-hour rating. Therefore, if the battery is capable of supplying 4 amperes of current for 20 consecutive hours, the battery is rated as an 80 ampere-hour battery.

The ampere-hour rating is useful for some service operations, such as slow charging or battery testing.

Cold cranking performance is measured by cooling a fully charged battery to $0°F$ ($-17°C$) and then testing it for 30 seconds to determine the maximum current flow. In this manner the cold cranking amperes rating is the number of amperes available to be drawn from the battery before the voltage drops below 7.2 volts.

The accompanying graphic illustration depicts the amount of power in watts available from a battery at different temperatures and the amount of power in watts

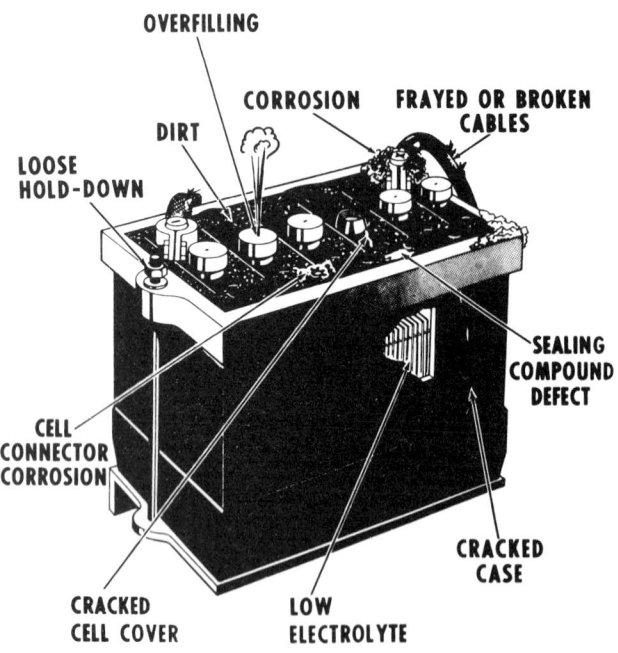

A visual inspection of the battery should be made each time the boat is used. Such a quick check may reveal a potential problem in its early stages. A dead battery in a busy waterway or far from assistance could have serious consequences.

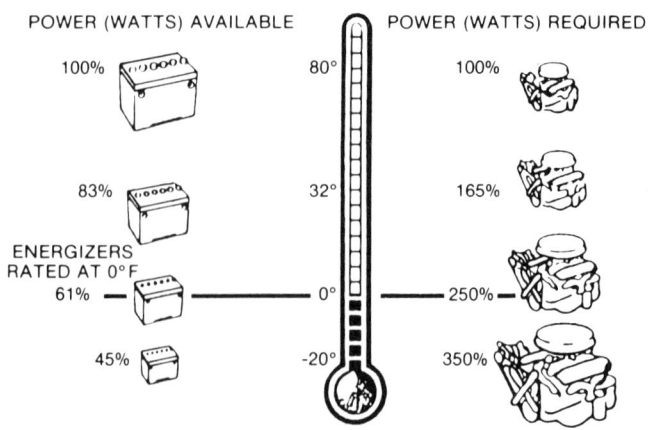

Comparison of battery efficiency and engine demands at various temperatures.

required of the engine at the same temperature. It becomes quite obvious --the colder the climate -- the more necessary for the battery to be **FULLY** charged.

Reserve capacity of a battery is considered the length of time -- in minutes -- at 80°F (27°C) a 25 ampere current can be maintained before the voltage drops below 10.5 volts. This test is intended to provide an approximation of how long the engine, including electrical accessories such as bilge pump, running lights, etc., could operate satisfactorily if the stator assembly or lighting coil did not produce sufficient current. A typical rating is 100 minutes.

Watt-hour is a very useful rating of battery power. It is determined by multiplying the number of ampere hours times the voltage. Therefore, a 12-volt battery rated at 80 ampere-hours would be rated at 960 watt-hours (80 x 12 = 960).

If possible, the new battery should have a power rating equal to or higher than the unit it is replacing.

BATTERY LOCATION

Every battery installed in a boat must be secured in a well-protected ventilated area. If the battery area lacks adequate ventilation, hydrogen gas which is given off during charging could become very explosive. This is especially true if the gas is concentrated and confined.

BATTERY SERVICE

The battery requires periodic servicing and a definite maintenance program will ensure extended life. If the battery should test satisfactorily, but still fails to perform properly, one of five problems could be the cause.

1- An accessory might have accidentally been left on overnight or for a long period during the day. Such an oversight would result in a discharged battery.

2- Slow speed engine operation for long periods of time resulting in an undercharged condition.

3- Using more electrical power than the stator assembly or lighting coil can replace would result in an undercharged condition.

4- A defect in the charging system. A faulty stator assembly or lighting coil, defective rectifier, or high resistance somewhere in the system could cause the battery to become undercharged.

5- Failure to maintain the battery in good order. This might include a low level of electrolyte in the cells; loose or dirty cable connections at the battery terminals; or possibly an excessively dirty battery top.

Electrolyte Level

The most common practice of checking the electrolyte level in a battery is to remove the cell cap and visually observe the level in the vent well. The bottom of each vent well has a split vent which will cause the surface of the electrolyte to appear distorted when it makes contact. When the distortion first appears at the bottom of the split vent, the electrolyte level is correct.

Some late model batteries have an electrolyte level indicator installed which operates in the following manner:

A transparent rod extends through the center of one of the cell caps. The lower tip of the rod is immersed in the electrolyte when the level is correct. If the level should drop below normal, the lower tip of the rod is exposed and the upper end glows as a warning to add water. Such a device is only necessary on one cell cap because if the electrolyte is low in one cell it is also low in the other cells. **BE SURE** to replace the cap with the indicator onto the second cell from the positive terminal.

During hot weather and periods of heavy use, the electrolyte level should be checked more often than during normal operation.

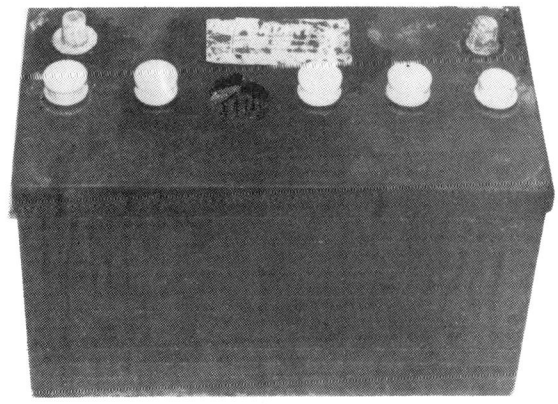

An explosive hydrogen gas is normally released from the cells under a wide range of circumstances. This battery exploded when the gas ignited from someone smoking in the area when the caps were removed. Such an explosion could also be caused by a spark from the battery terminals igniting the gas.

9-4 ELECTRICAL

Add potable (drinking) water to bring the level of electrolyte in each cell to the proper level. **TAKE CARE** not to overfill, because adding an excessive amount of water will cause loss of electrolyte and any loss will result in poor performance, short battery life, and will contribute quickly to corrosion. **NEVER** add electrolyte from another battery. Use only clean pure water.

Battery Testing

A hydrometer is a device to measure the percentage of sulfuric acid in the battery electrolyte in terms of specific gravity. When the condition of the battery drops from fully charged to discharged, the acid leaves the solution and enters the plates, causing the specific gravity of the electrolyte to drop.

It may not be common knowledge, but hydrometer floats are calibrated for use at 80°F (27°C). If the hydrometer is used at any other temperature, hotter or colder, a correction factor must be applied. (Remember, a liquid will expand if it is heated and will contract if cooled. Such expansion and contraction will cause a definite change in the specific gravity of the liquid, in this case the electrolyte.)

A quality hydrometer will have a thermometer/temperature correction table in the lower portion, as shown in the accompanying illustration. By knowing the air temperature around the battery and from the table, a correction factor may be applied to the specific gravity reading of the hydrometer float. In this manner, an accurate determination may be made as to the condition of the battery.

The following six points should be observed when using a hydrometer.

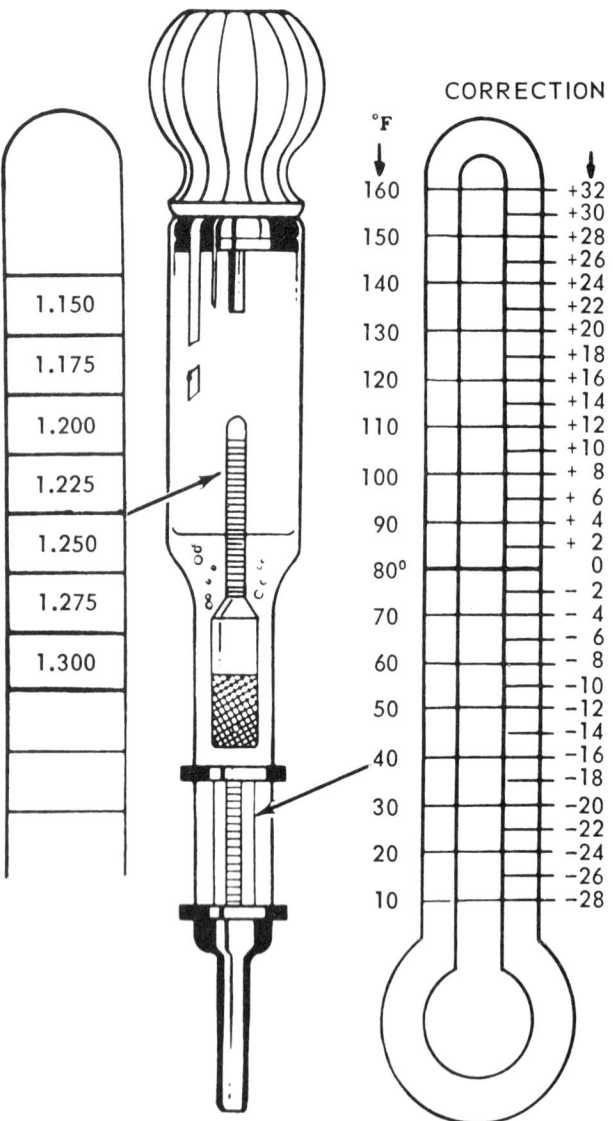

A check of the electrolyte in the battery should be on the maintenance schedule for any boat. A hydrometer reading of 1.300, or in the green band, indicates the battery is in satisfactory condition. If the reading is 1.150 or in the red band, the battery must be charged. Observe the six safety points listed in the text when using a hydrometer.

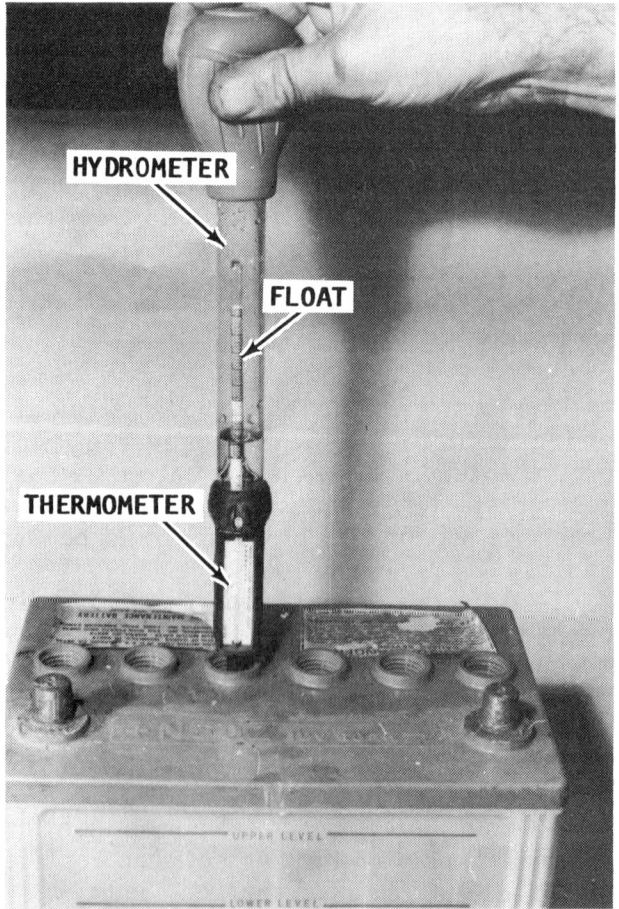

Testing the specific gravity electrolyte of a battery using a temperature corrected hydrometer.

1- **NEVER** attempt to take a reading immediately after adding water to the battery. Allow at least 1/4 hour of charging at a high rate to thoroughly mix the electrolyte with the new water. This time will also allow for the necessary gases to be created.

2- **ALWAYS** be sure the hydrometer is clean inside and out as a precaution against contaminating the electrolyte.

3- If a thermometer is an integral part of the hydrometer, draw liquid into it several times to ensure the correct temperature before taking a reading.

4- **BE SURE** to hold the hydrometer vertically and suck up liquid only until the float is free and floating.

5- **ALWAYS** hold the hydrometer at eye level and take the reading at the surface of the liquid with the float free and floating.

Disregard the slight curvature appearing where the liquid rises against the float stem. This phenomenon is due to surface tension.

6- **DO NOT** drop any of the battery fluid on the boat or on your clothing, because it is extremely caustic. Use water and baking soda to neutralize any battery liquid that does accidentally drop.

After withdrawing electrolyte from the battery cell until the float is barely free, note the level of the liquid inside the hydrometer. If the level is within the green band range for all cells, the condition of the battery is satisfactory. If the level is within the white band for all cells, the battery is in fair condition.

If the level is within the green or white band for all cells except one, which registers in the red, the cell is shorted internally. No amount of charging will bring the battery back to satisfactory condition.

If the level in all cells is about the same, even if it falls in the red band, the battery may be recharged and returned to service. If the level fails to rise above the red band after charging, the only solution is to replace the battery.

Cleaning

Dirt and corrosion should be cleaned from the battery just as soon as it is discovered. Any accumulation of acid film or dirt will permit current to flow between the terminals. Such a current flow will drain the battery over a period of time.

Clean the exterior of the battery with a solution of diluted ammonia or a soda solution to neutralize any acid which may be present. Flush the cleaning solution off with clean water. **TAKE CARE** to prevent any of the neutralizing solution from entering the cells, by keeping the caps tight.

A poor contact at the terminals will add resistance to the charging circuit. This resistance will cause the voltage regulator to register a fully charged battery, and thus cut down on the stator assembly or lighting coil output adding to the low battery charge problem.

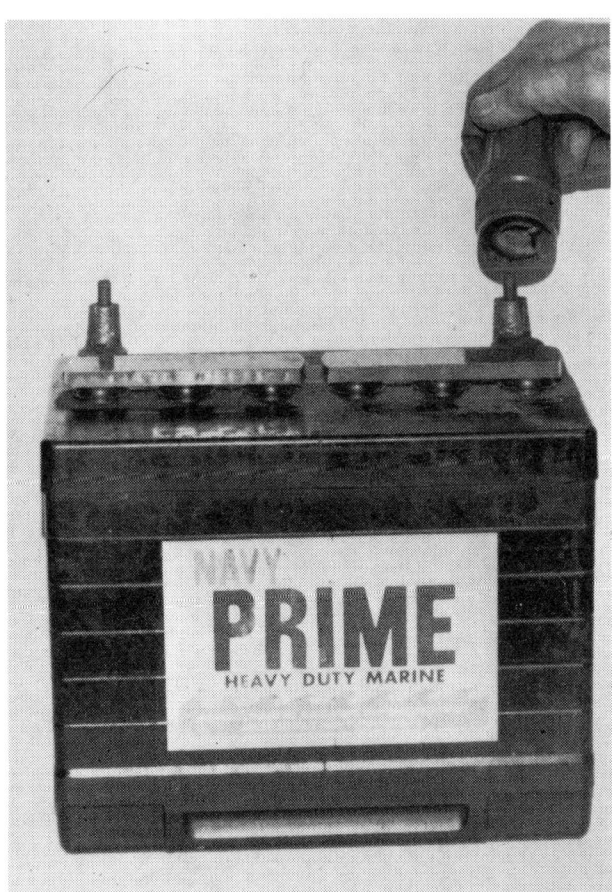

An inexpensive two-part tool will do an excellent job of cleaning the battery terminals.

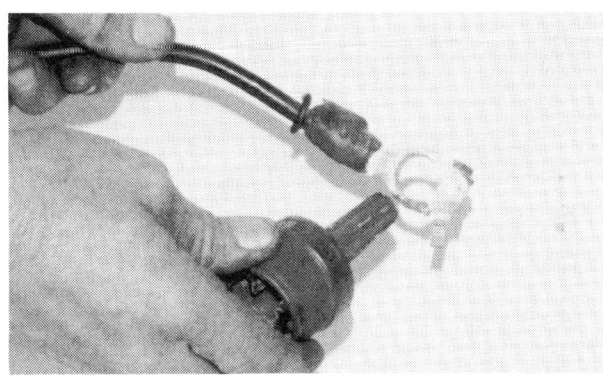

The second part of the tool shown at left is used to clean the battery lead terminals.

9-6 ELECTRICAL

Scrape the battery posts clean with a suitable tool or with a stiff wire brush. Clean the inside of the cable clamps to be sure they do not cause any resistance in the circuit.

JUMPER CABLES

If booster batteries are used for starting an engine the jumper cables must be connected correctly and in the proper sequence to prevent damage to either battery, or diodes in the circuit.

Jumper cables may be used on the 220hp and 225hp models equipped with YMIS (Yamaha Microcomputer Ignition System). No damage will occur to the microcomputer as long as the following procedure is closely followed. A sudden power surge may "fry" the electronic components and lead to a costly replacement of the "black box".

First, check to be sure the main switch is in the **OFF** position and the booster battery is fully charged.

ALWAYS connect a cable from the positive terminals of the dead battery to the positive terminal of the good battery **FIRST**.

NEXT, connect one end of the other cable to the negative terminals of the good battery and the other end to a good ground on the powerhead. By making the ground connection on the powerhead, if there is an arc when the connection is made, the arc will not be near the battery. An arc near the battery could cause an explosion, destroying the battery and causing serious personal injury.

Once all connections are secure, attempt to start the powerhead with the key switch. Keep the powerhead running at about 3,000 rpm for at least three minutes with the jumper cables **CONNECTED**. **DO NOT** attempt to disconnect the jumper cables while the powerhead is running, especially on the 220hp and 225hp models equipped with YMIS. Such action would probably lead to a power "surge" damaging the circuits in the microcomputer.

STORAGE

If the boat is to be laid up for the winter or for more than a few weeks, special attention must be given to the battery to prevent complete discharge or possible damage to the terminals and wiring. Before putting the boat in storage, disconnect and remove the batteries. Clean them thoroughly of any dirt or corrosion, and then charge them to full specific gravity reading. After they are fully charged, store them in a clean cool dry place where they will not be damaged or knocked over, preferably on a couple blocks of wood. Storing the battery up off the deck, will permit air to circulate freely around and under the battery and will help to prevent condensation.

NEVER store the battery with anything on top of it nor cover the battery in such a manner as to prevent air from circulating around the fillercaps. All batteries, both new and old, will discharge during periods of storage, more so if they are hot than if they remain cool. Therefore, the electrolyte level and the specific gravity should be checked at regular intervals. A drop in the specific gravity reading is cause to charge them back to a full reading.

In cold climates, care should be exercised in selecting the battery storage area. A fully-charged battery will freeze at about 60 degrees below zero. A discharged battery, almost dead, will have ice forming at about 19 degrees above zero.

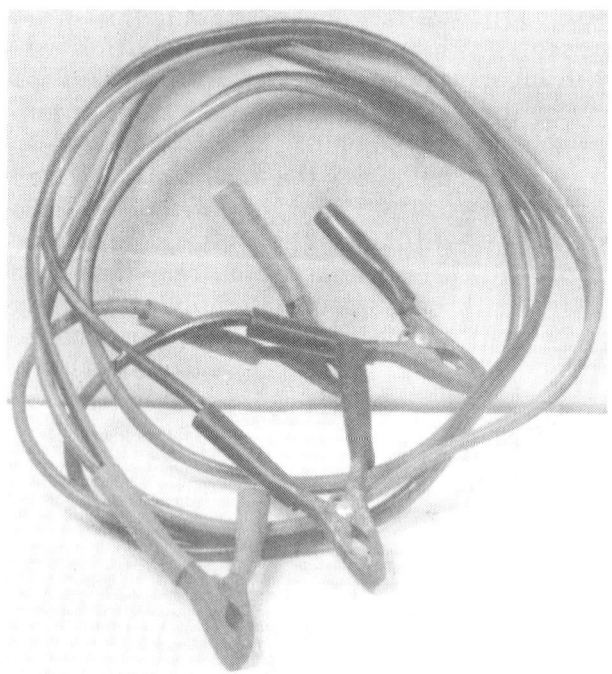

A common set of heavy-duty jumper cables. The booster battery must be connected correctly and in the proper sequence, as outlined in the text. Proper procedure will prevent damage to the battery, the diodes, or the rectifier.

9-3 THERMOMELT STICKS

Thermomelt sticks are an easy method of determining if the powerhead is running at the proper temperature. Thermomelt sticks are not expensive and are available at your local marine dealer.

Start the engine with the propeller in the water and run it for about 5 minutes at roughly 3000 rpm.

CAUTION

Water must circulate through the lower unit to the powerhead anytime the powerhead is operating to prevent damage to the water pump in the lower unit. Just five seconds without water will damage the water pump impeller.

The 140 degree stick should melt when you touch it to the lower thermostat housing or on the top cylinder. If it does not melt, the thermostat is stuck in the open position and the engine temperature is too low.

Touch the 170 degree stick to the same spot on the lower thermostat housing or on the top cylinder. The stick should not melt. If it does, the thermostat is stuck in the closed position or the water pump is not operating properly because the engine is running too hot.

See Chapter 8 to service the thermostat. See Chapter 11 to service the water pump in the lower unit.

Maximum engine performance can only be obtained through proper tuning using a tachometer.

9-4 TACHOMETER

An accurate tachometer can be installed on any engine. Such an instrument provides an indication of engine speed in revolutions per minute (rpm). This is accomplished by measuring the number of electrical pulses per minute generated in the primary circuit of the ignition system.

The meter readings range from 0 to 6,000 rpm, in increments of 100. Tachometers have solid-state electronic circuits which eliminate the need for relays or batteries and contribute to their accuracy. The electronic parts of the tachometer, susceptible to moisture, are coated to prolong their life.

SPECIAL WORDS ON TACHOMETERS AND CONNECTIONS

A tachometer is installed as standard equipment on all powerheads covered in this manual.

Due to local conditions, it may be necessary to adjust the carburetor while the outboard unit is running in a test tank or with the boat in a body of water. For maximum performance, the idle rpm should be adjusted under actual operating conditions.

Under such conditions it might be necessary to attach a tachometer closer to the powerhead than the one installed on the control panel.

Open the remote control box. Disconnect the Black and Green leads. Connect the Black lead to the ground terminal of the auxiliary tachometer and the Green lead to the "input" or "hot" terminal of the auxiliary tachometer.

9-5 ELECTRICAL SYSTEM GENERAL INFORMATION

In the early days, all outboard engines were started by simply pulling on a rope wound around the flywheel. As time passed and owners were reluctant to use muscle power, it was necessary to replace the rope starter with some form of power cranking system. Today, many small engines are still started by pulling on a rope, but others have a powered cranking motor installed.

The system utilized to replace the rope method was an electric cranking motor coupled with a mechanical gear mesh be-

9-8 ELECTRICAL

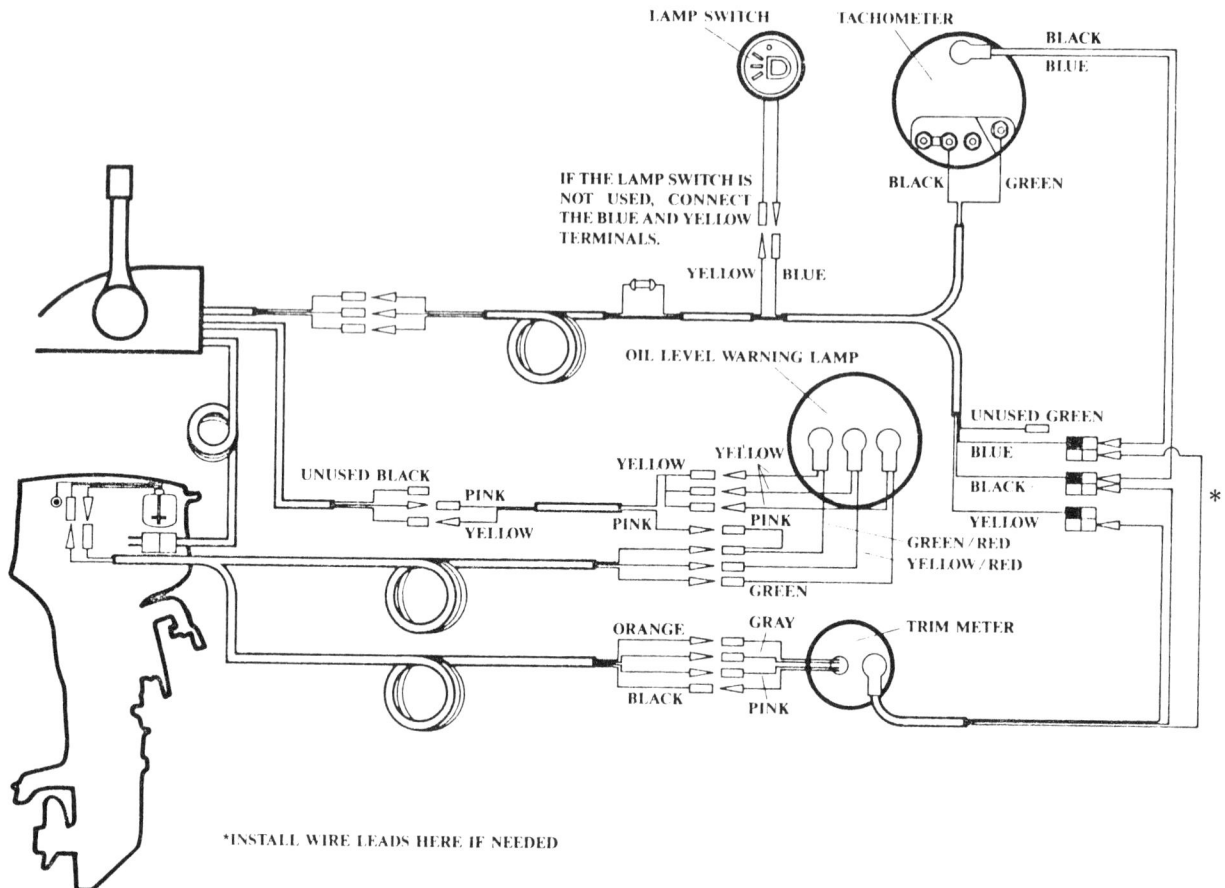

Simplified functional diagram depicting typical wire routing between the powerhead and the remote control box for the outboard units covered in this manual.

tween the cranking motor and the powerhead flywheel, similar to the method used to crank an automobile engine.

The electrical system consists of three circuits:
- a- Charging circuit
- b- Cranking motor circuit
- c- Ignition circuit

Charging Circuit

The charging circuit consists of permanent magnets and a stator located within the flywheel; a lighting coil installed on the stator plate; a rectifier located elsewhere on the powerhead; an external battery; and the necessary wiring to connect the units. The negative side of the rectifier is grounded. The positive side of the rectifier passes through the internal harness plug to the battery. The negative side of the battery is connected, through the connector, to a good ground on the engine.

The alternating current generated in the stator windings passes to the rectifier. The rectifier changes the alternating current (AC) to direct current (DC) to charge the 12-volt battery.

Cranking Motor Circuit

The cranking motor circuit consists of a cranking motor and a starter-engaging mechanism. A starter relay is used as a heavy-duty switch to carry the heavy current from the battery to the cranking motor. On most models, the starter relay is actuated by depressing the **START** button. On boats equipped with a remote control shift box, the ignition key is turned to the **START** position.

Ignition Circuit

The ignition circuit is covered extensively in Chapter 6.

9-6 CHARGING CIRCUIT SERVICE

The stator is located under, and protected by, the flywheel. Therefore, the stator, including the lighting coil, seldom causes problems in the charging circuit. Most problems in the charging circuit can be traced to the rectifier or to the battery. If either the stator or the rectifier fails the troubleshooting tests, the defective unit cannot be repaired, it **MUST** be replaced.

CRANKING MOTOR 9-9

9-7 CRANKING MOTOR CIRCUIT DESCRIPTION

As the name implies, the sole purpose of the cranking motor circuit is to control operation of the cranking motor to crank the powerhead until the engine is operating. The circuit includes a solenoid or magnetic switch to connect or disconnect the motor from the battery. The operator controls the switch with a key switch.

A neutral safety switch is installed into the circuit to permit operation of the cranking motor **ONLY** if the shift control lever is in **NEUTRAL**. This switch is a safety device to prevent accidental engine start when the engine is in gear.

The cranking motor is a series wound electric motor which draws a heavy current from the battery. It is designed to be used only for short periods of time to crank the engine for starting. To prevent overheating the motor, cranking should not be continued for more than 30-seconds without allowing the motor to cool for at least three minutes. Actually, this time can be spent in making preliminary checks to determine why the engine fails to start.

Theory of Operation

Power is transmitted from the cranking motor to the powerhead flywheel through a Bendix drive. This drive has a pinion gear mounted on screw threads. When the motor is operated, the pinion gear moves upward and meshes with the teeth on the flywheel ring gear.

When the powerhead starts, the pinion gear is driven faster than the shaft, and as a result, it screws out of mesh with the flywheel. A rubber cushion is built into the Bendix drive to absorb the shock when the pinion meshes with the flywheel ring gear. The parts of the drive **MUST** be properly assembled for efficient operation. If the drive is removed for cleaning, **TAKE CARE** to assemble the parts as shown in the accompanying illustrations in this section. If the screw shaft assembly is reversed, it will strike the splines and the rubber cushion will not absorb the shock.

The sound of the motor during cranking is a good indication of whether the cranking motor is operating properly or not. Naturally, temperature conditions will affect the speed at which the cranking motor is able to crank the engine. The speed of cranking a cold engine will be much slower than when cranking a warm engine. An experienced operator will learn to recognize the favorable sounds of the powerhead cranking under various conditions.

Faulty Symptoms

If the cranking motor spins, but fails to crank the engine, the cause is usually a corroded or gummy Bendix drive. The drive should be removed, cleaned, and given an inspection.

If the cranking motor cranks the engine too slowly, the following are possible causes and the corrective actions that may be taken:

a- Battery charge is low. Charge the battery to full capacity.

b- High resistance connections at the battery, solenoid, or motor. Clean and tighten all connections.

c- Undersize battery cables. Replace cables with sufficient size.

d- Battery cables too long. Relocate the battery to shorten the run to the solenoid.

Maintenance

The cranking motor does not require periodic maintenance or lubrication. If the motor fails to perform properly, the checks outlined in the previous paragraph should be performed.

The frequency of starts governs how often the motor should be removed and reconditioned. The manufacturer recommends removal and reconditioning every 1000 hours.

Naturally, the motor will have to be removed if the corrective actions outlined

Typical location of the cranking motor and relay for the powerheads covered in this manual.

under **Faulty Symptoms** above, does not restore the motor to satisfactory operation.

9-8 CRANKING MOTOR TROUBLESHOOTING

Before wasting too much time troubleshooting the cranking motor circuit, the following checks should be made. Many times, the problem will be corrected.

a- Battery fully charged.
b- Shift control lever in **NEUTRAL**.
c- Main 20-amp fuse is "good" (not blown).
d- All electrical connections clean and tight.
e- Wiring in good condition, insulation not worn or frayed.

Two more areas may cause the powerhead to crank slowly even though the cranking motor circuit is in excellent condition: a tight or "frozen" powerhead and water in the lower unit. The following troubleshooting procedures are presented in a logical sequence, with the most common and easily corrected areas listed first in each problem area. The connection number refers to the numbered positions in the accompanying illustrations.

Perform the following quick checks and corrective actions for following problems:

1- **Cranking Motor Rotates Slowly**
a- Battery charge is low. Charge the battery to full capacity.
b- Electrical connections corroded or loose. Clean and tighten.
c- Defective cranking motor. Perform an amp draw test. Lay an amp draw gauge on the cable leading to the cranking motor. Turn the key on and attempt to crank the engine. If the gauge indicates an excessive amperage draw, the cranking motor **MUST** be replaced or rebuilt.

2- **Cranking Motor Fails To Crank Powerhead**

Test Motor
a- Disconnect the cranking motor lead from the solenoid to prevent the powerhead from starting during the testing process.

NOTE: This lead is to remain disconnected from the solenoid during tests No. 2 thru No. 6.

b- Disconnect the black ground wire from the No. 2 terminal.
c- Connect a voltmeter between the No. 2 terminal and a common engine ground.
d- Turn the key switch to the **START** position.
e- Observe the voltmeter reading. If there is the slightest amount of reading, check the Black ground wire connection or check for an open circuit. Connect the ground wire back to the No. 2 terminal and move to Step 6. If there is no voltmeter reading, proceed with Step 3.

3- **Test Cranking Motor Solenoid**
a- Connect a voltmeter between the engine common ground and the No. 3 terminal.
b- Turn the ignition key switch to the **START** position.
c- Observe the voltmeter reading. If the meter indicates more than 0.3 volt, the solenoid is defective and must be replaced. If the meter indicates less than 0.3 volt, proceed with Step 4.

Almost 90% of cranking motor problems can be traced to high resistance connections at the battery. The battery terminals and the lead connections must be kept clean, to prevent excessive resistance.

4- Test Neutral Start Switch

a- Connect a voltmeter between the common engine ground and the No. 4. Turn the ignition key switch to the **START** position.

b- Observe the voltmeter. If there is any indication of a reading, the neutral start switch is open or the brown wire lead is open between the No. 3 and No. 4. If there is no voltmeter reading, proceed to Step 5.

5- Test for Open Wire

a- Connect a voltmeter between the common engine ground and No. 5.

b- The voltmeter should indicate 12-volts. If the meter needle flickers (fails to hold steady), check the circuit between No. 5 and common engine ground. If meter fails to indicate voltage, replace the positive battery cable.

6- Further Tests for Solenoid

a- Connect the voltmeter between the common engine ground and No. 1.

b- Turn the ignition key switch to the **START** position.

c- Observe the voltmeter. If there is no reading, the cranking motor solenoid is defective and must be replaced. If a reading is indicated and a click sound is heard, proceed to Step 7.

7- Test Large Red Cable

a- Connect the red cable to the cranking motor solenoid.

b- Connect the voltmeter between the engine common ground and No. 6.

c- Turn the ignition key switch to the **START** position, or depress the start button.

d- Observe the voltmeter. If there is no reading, check the red cable for a poor

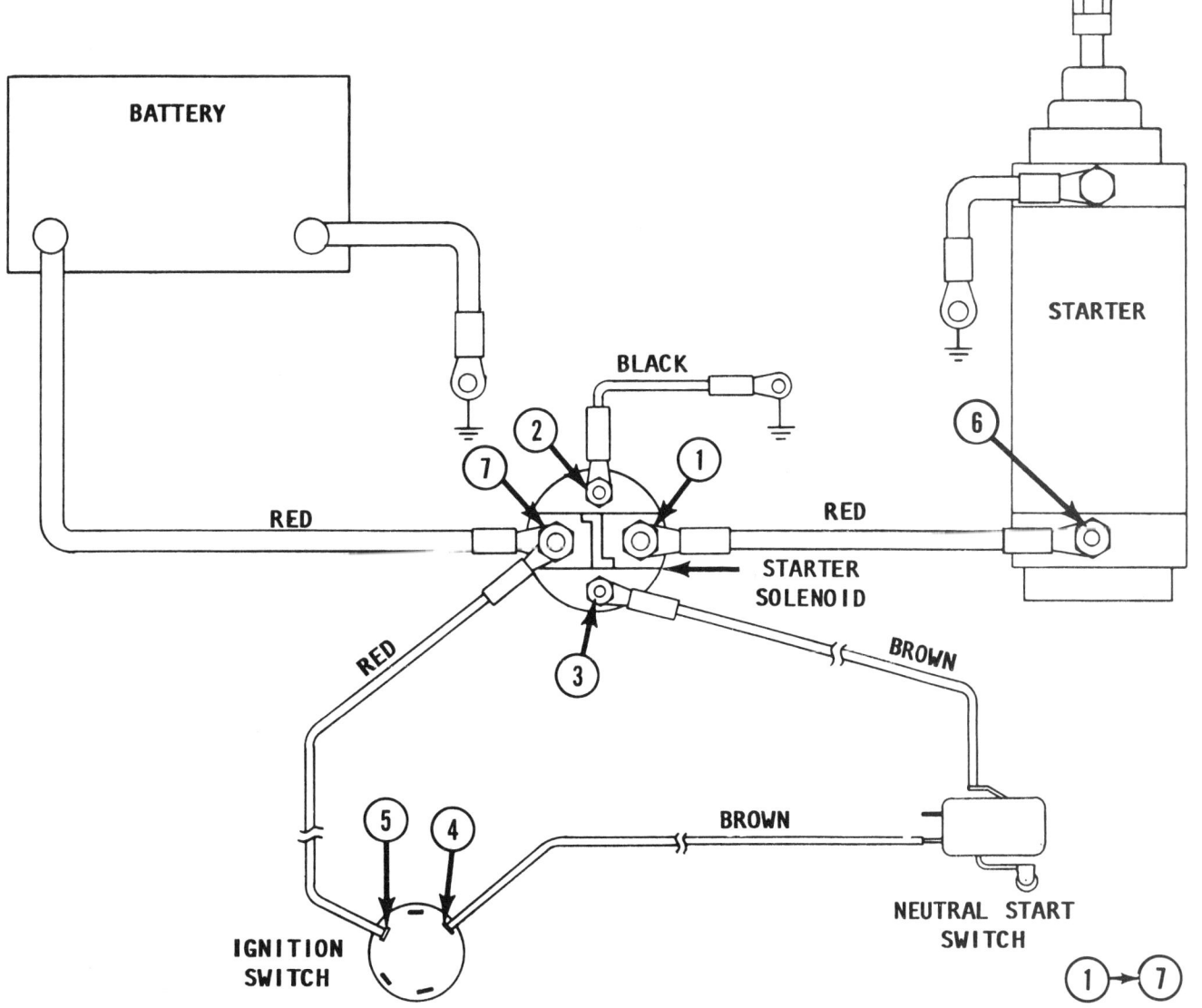

connection or an open circuit. If there is any indication of a reading, and the cranking motor does not rotate, the cranking motor must be replaced.

CRANKING MOTOR RELAY TROUBLESHOOTING

Description

The cranking motor relay is actually a switch between the battery and the powerhead. The switch cannot be serviced. Therefore, if troubleshooting indicates the switch to be faulty, it **MUST** be replaced.

Before beginning any work on the relay, disconnect the positive (+) lead from the battery terminal. Remove the cowling from the powerhead.

Check to be sure the battery cables are disconnected from the battery.

CAUTION

Disconnecting the battery leads is **MOST** important because the cranking motor relay lead will be disconnected and allowed to hang free. The other end of this lead is connected to the battery. If the leads are not disconnected from the battery and the free relay end should happen to come in contact with any metal part on the powerhead, **SPARKS WOULD FLY** and the end of the lead burned.

Removal

1- Push down the boots from the two large Red leads at the relay. Take time to

make a note of leads and to which terminal they are connected, and then remove the leads. Disconnect the small Brown lead at the quick disconnect fitting.

2- Slide the relay from its mounting bracket. Remove the grounding bolt on the bracket behind the relay to release the Black relay grounding wire.

RELAY TESTING

The following tests must be conducted with the relay **REMOVED** from the powerhead.

1- Obtain an ohmmeter. Select the Rx1000 ohm scale. Make contact with the Black meter lead to the Black relay lead and the Red meter lead to the Brown relay lead. The meter should indicate continuity. If the meter registers no continuity, the relay is defective and must be replaced. No service or adjustment is possible.

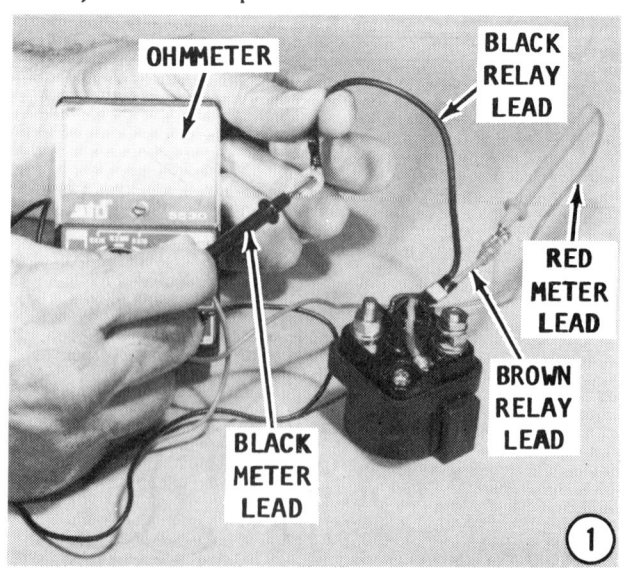

CRANKING MOTOR 9-13

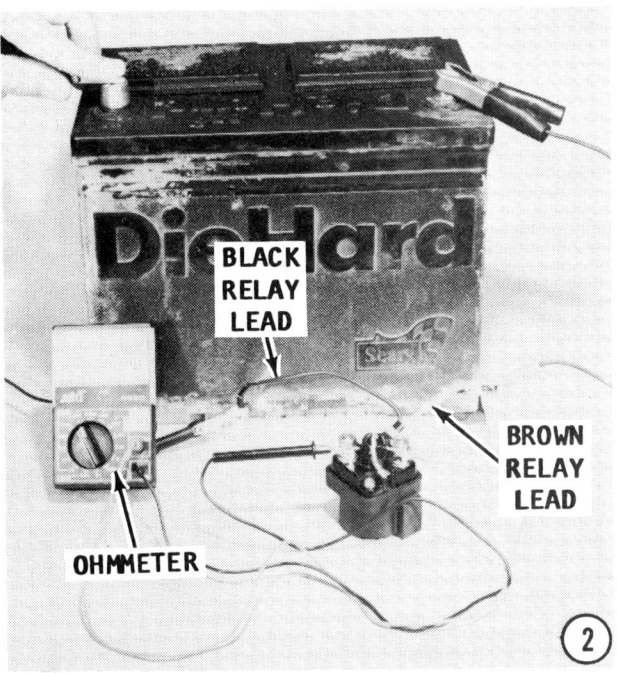

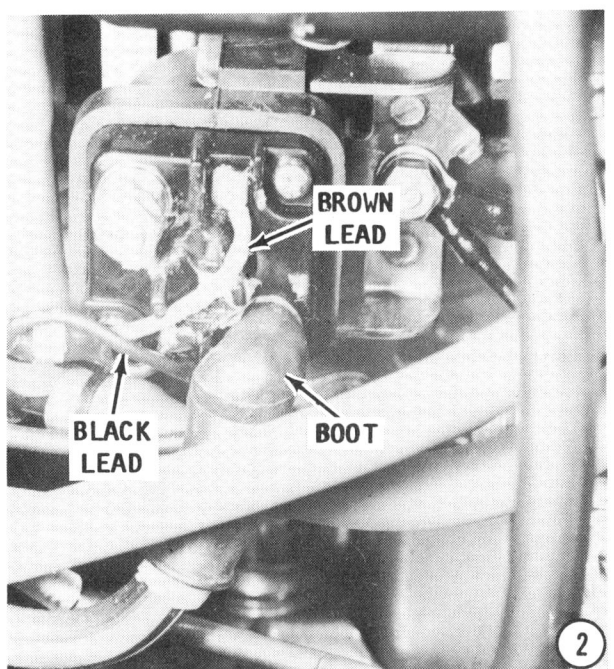

2- Connect one test lead of the ohmmeter to each of the large relay terminals. Connect the positive (+) lead from a fully charged 12-volt battery to the Brown lead. Momentarily make contact with the ground lead from the battery to the Black lead. If a loud "click" sound is heard, and the ohmmeter indicates continuity, the solenoid is in serviceable condition. If, however, a "click" sound is not heard, and/or the ohmmeter does not indicate continuity, the solenoid is defective and must be replaced **ONLY** with a **MARINE** type solenoid.

INSTALLATION

1- Slide the relay back onto its mounting bracket. Secure the Black lead behind the

relay by installing the grounding bolt on the bracket.

2- Connect the Brown lead at the quick disconnect fitting.

Connect the two large Red leads to the front of the cranking motor relay. The lead on the right is from the "hot" terminal on the cranking motor. The lead on the left is also "hot" coming from the positive battery terminal. This lead also has a small Red lead attached to the relay terminal.

Install the elbow "boots" onto both relay leads.

If the work is complete, then reconnect the battery terminals. If further work is to be carried out on the cranking system, then leave the battery cables disconnected until the work is complete.

9-9 CRANKING MOTOR SERVICE

Description

One type cranking motor is used on the powerheads covered in this manual. The exterior appearance of the cranking motor may vary. Therefore, the accompanying illustrations may differ slightly from the unit being serviced, but the procedures and maintenance instructions are valid for all cranking motors on all Yamaha powerheads covered in this manual.

Marine cranking motors are very similar in construction and operation to the units used in the automotive industry.

9-14 ELECTRICAL

All marine cranking motors use the inertia type drive assembly. This type assembly is mounted on an armature shaft with external spiral splines which mate with the internal splines of the drive assembly.

NEVER operate a cranking motor for more than 30 seconds without allowing it to cool for at least three minutes. Continuous operation without the cooling period can cause serious damage to the cranking motor.

9-10 CRANKING MOTOR REMOVAL

Before beginning any work on the cranking motor, disconnect the positive (+) lead from the battery terminal. Remove the cowling from the powerhead.

1- Check to be sure the battery cables are disconnected from the battery.

CAUTION

Disconnecting the battery leads is **MOST** important because the cranking motor relay lead will be disconnected and allowed to hang free. The other end of this lead is connected to the battery. If the leads are not disconnected from the battery and the free relay end should happen to come in contact with any metal part on the powerhead, **SPARKS WOULD FLY** and the end of the lead burned.

Disconnect both large cables from the cranking motor relay.

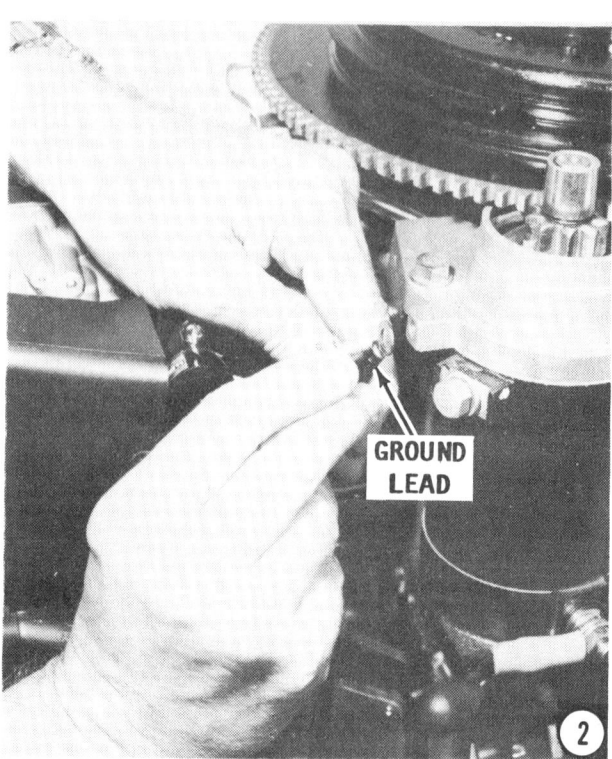

2- Disconnect the upper "ground" lead from the cranking motor.

3- Disconnect the lower "hot" lead from the cranking motor.

4- Remove the three bolts securing the cranking motor to the powerhead.

5- Lift the motor free.

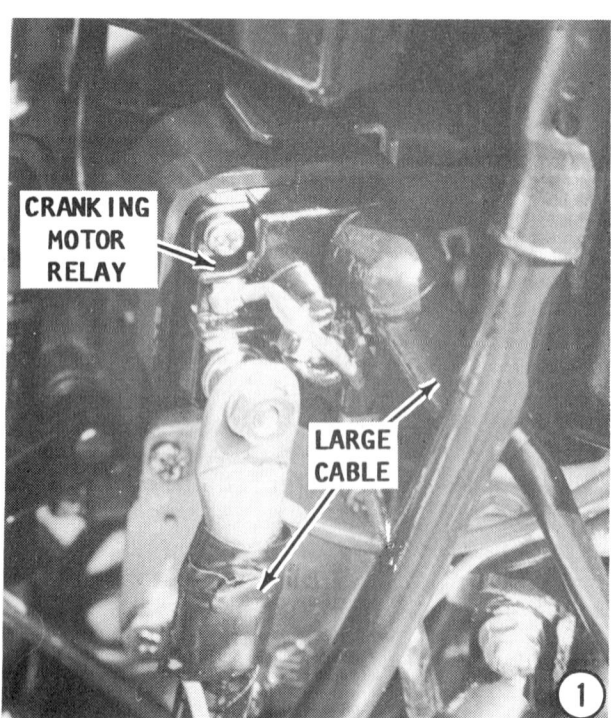

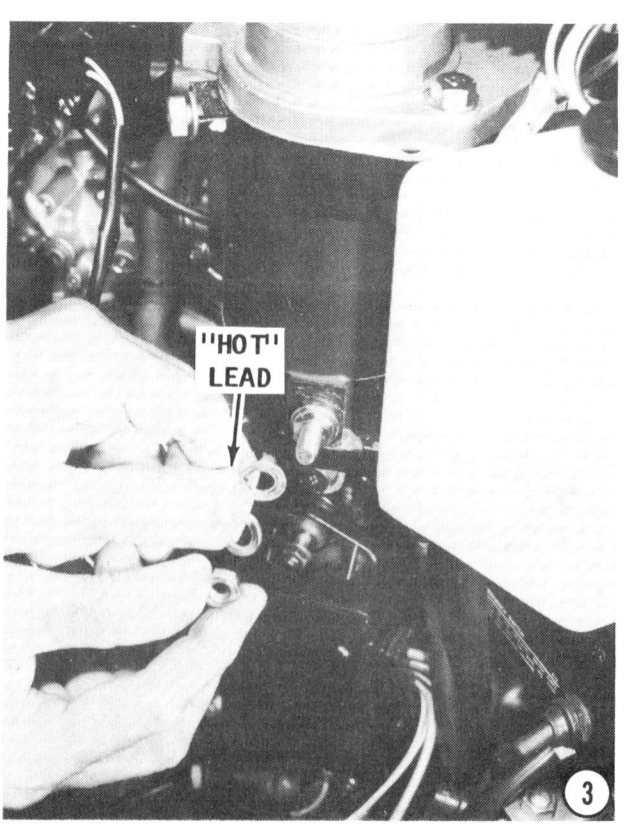

CRANKING MOTOR 9-15

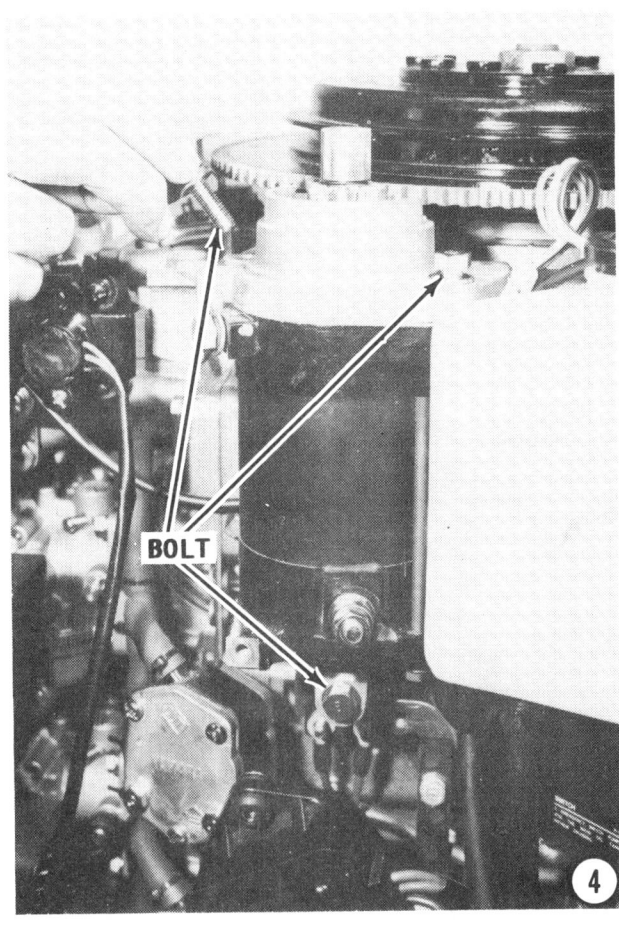

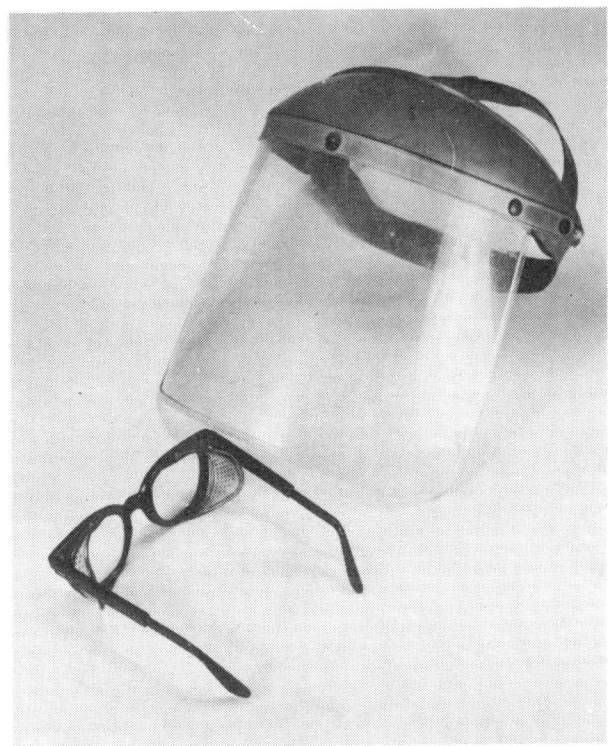

Protect your eyes with a face mask or safety glasses while working with spring steel snap rings.

9-11 CRANKING MOTOR DISASSEMBLING

SAFETY WORDS
ALL MODELS

The snap ring is made of spring steel and may slip out of the pliers or pop out of the groove with considerable force. Therefore, **WEAR** eye protection glasses while remov-

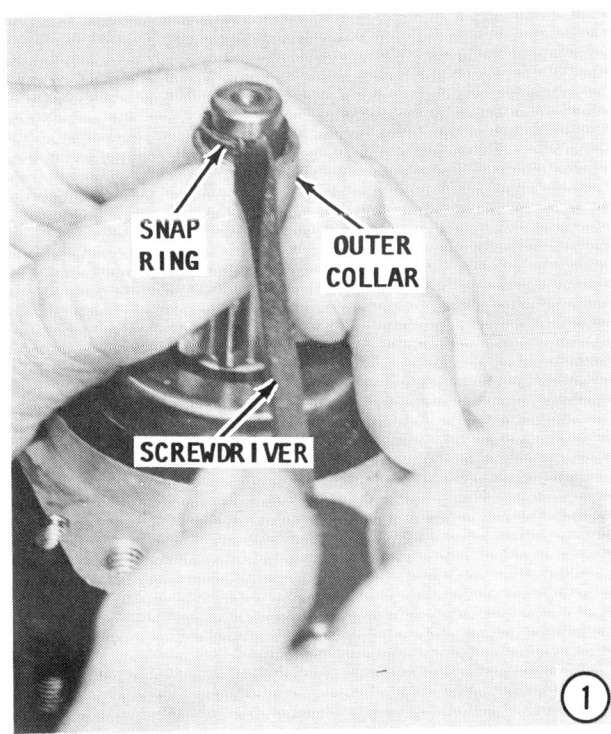

9-16 ELECTRICAL

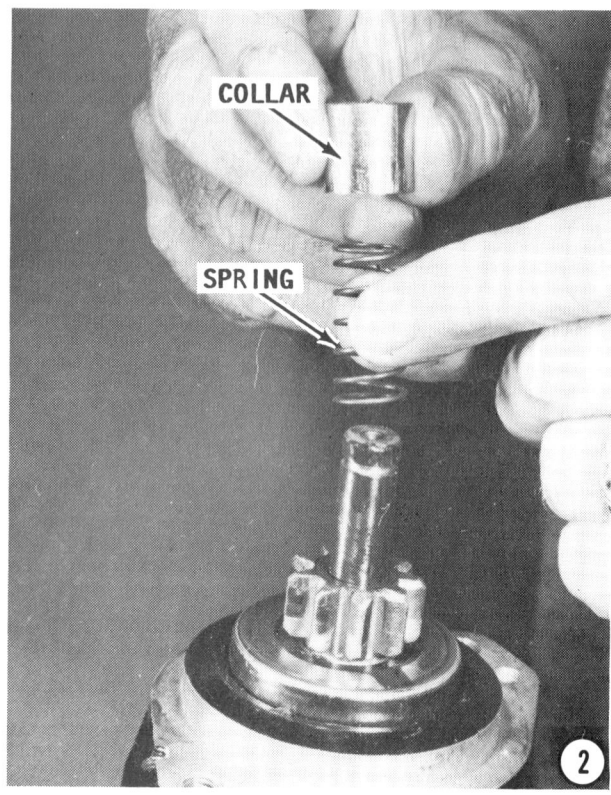

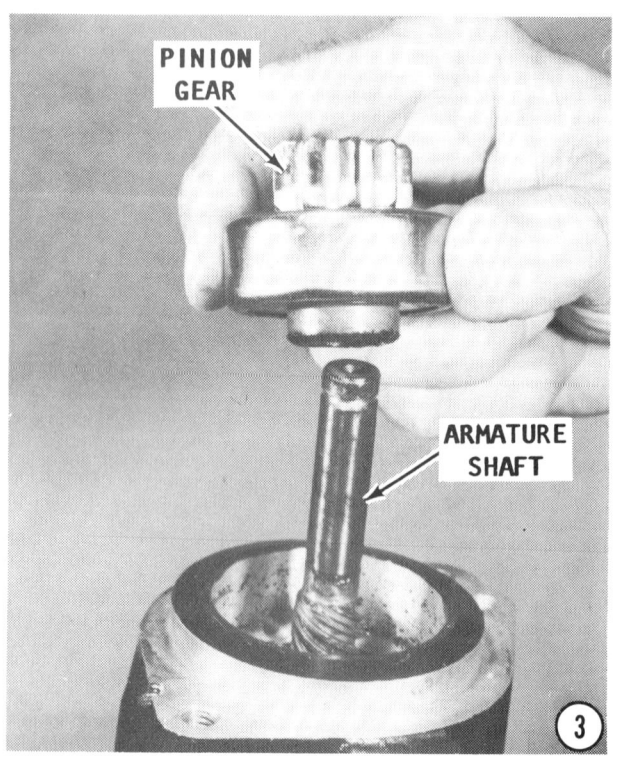

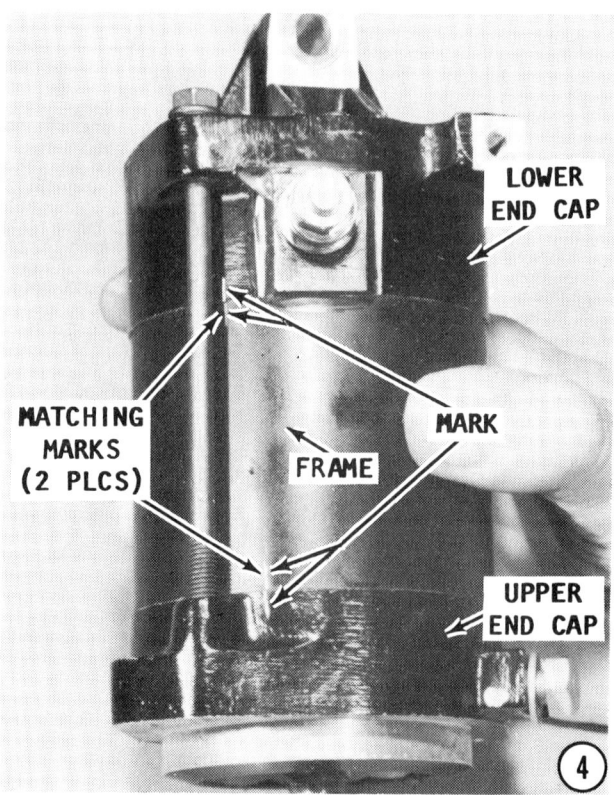

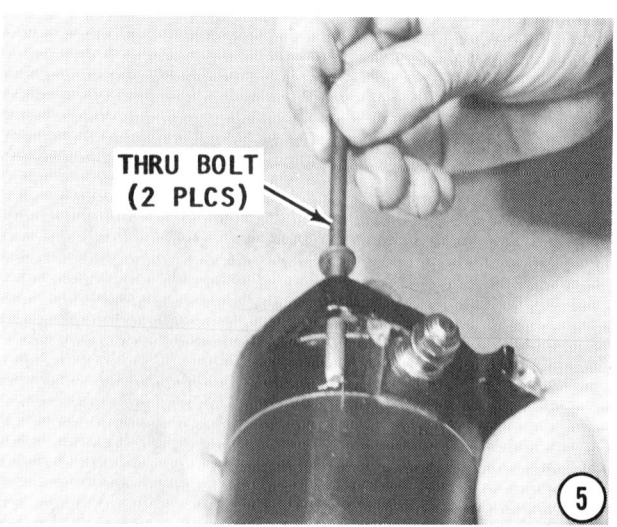

ing the snap ring in the next step. To prevent loosing the ring, hold a small container (a one pound coffee can would be ideal), over the armature shaft while removing the snap ring. The container will not catch the ring, but will deflect it downward, at least making the task of locating the ring a little easier.

1- With the cranking motor on a work surface, push the outer collar down and at the same time, pry the snap ring out of its groove and free of the armature shaft.

2- After the snap ring is free, slide the collar and spring free of the armature shaft.

3- Slide the pinion gear from the armature shaft.

4- Make a mark on both end caps and matching marks on the field frame assembly as an aid to alignment and installation of the thru bolts.

5- Remove the thru bolts and remove the upper end cap. Note the number and the

CRANKING MOTOR 9-17

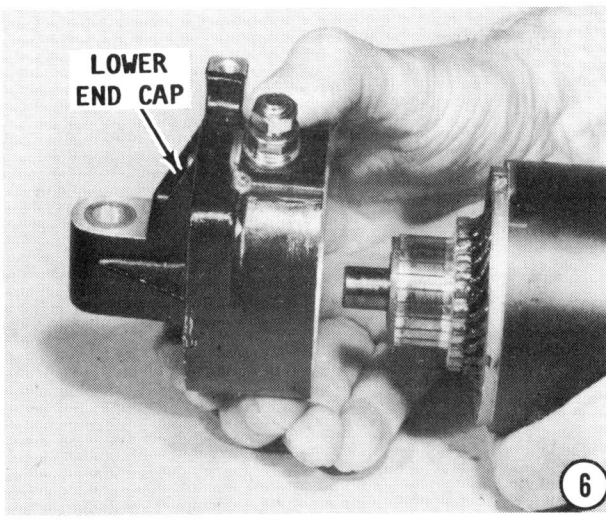

material of the washers on the end of the armature shaft. Usually there are two steel washers in front of a resin washer. Incorrect placement of these washers will cause the armature to be misaligned with the brushes and the field magnets during assembling.

6- Gently pull the lower end cap containing the brush plate from the armature shaft.

7- Pull the armature shaft from the motor frame.

Special Words on Brushes

All the positive and negative brushes are mounted in the lower cap. The two positive brushes are attached to the positive terminal. The single negative brush is grounded to the case.

If any of the three brushes are no longer fit for service, the entire brush plate is replaced, because each brush is riveted in place onto the plate.

8- Make a determination on the condition of the brushes. If the brush length is

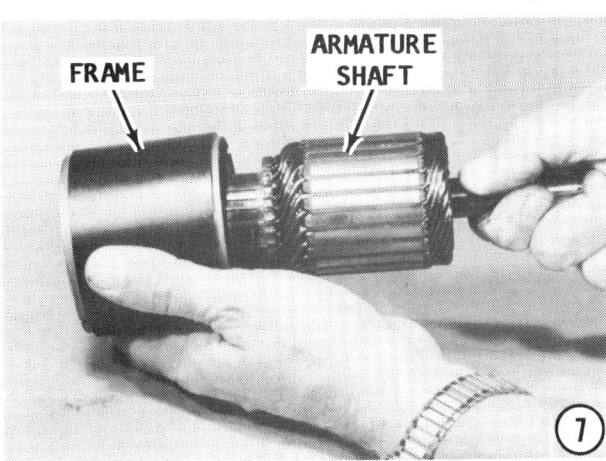

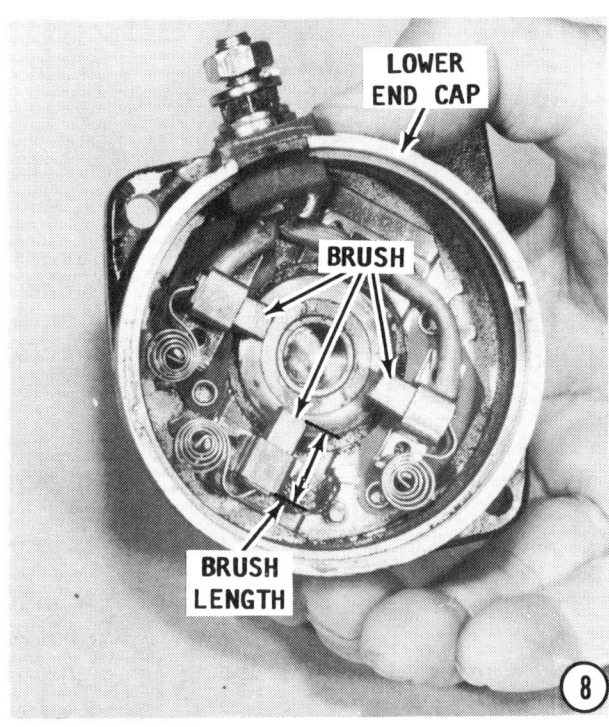

less then 1/2" (12.5mm), the brush plate must be removed and replaced.

9- Remove the hardware on the positive terminal stud. Remove the two small Phillips head screws from the underside of the lower end cap.

10- Lift out the brush plate and at the same time, pull the terminal stud from the

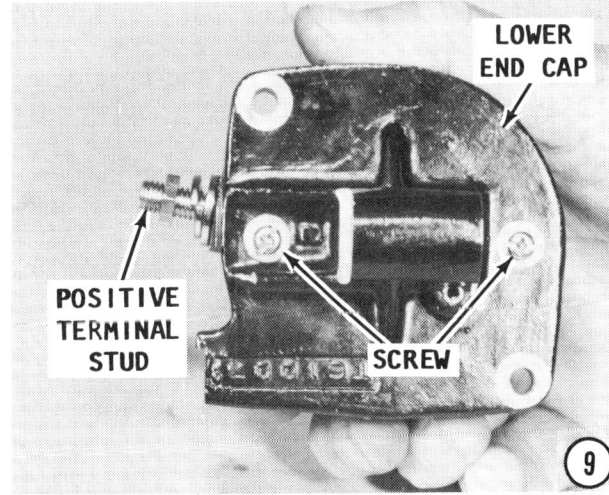

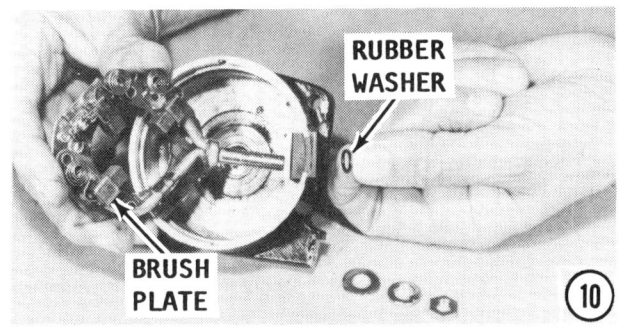

end cap. Take care to **SAVE** the small rubber insulating washer installed in the crevice of the large square grommet.

9-12 CLEANING AND INSPECTING

Pinion Gear Assembly

Inspect the pinion gear teeth for chips, cracks, or a broken tooth. Check the splines inside the pinion gear for burrs and to be sure the gear moves freely on the armature shaft.

BAD NEWS

This type of cranking motor pinion gear assembly cannot be repaired if the unit is defective. Replacement of the pinion gear assembly as a unit is the only answer.

Check to be sure the return spring is flexible and has not become distorted. Check both ends of the spring for signs of damage.

Clean the armature shaft and check to be sure the shaft is free of any burrs. If burrs are discovered, they may be removed with crocus cloth.

Frame Assembly

Clean the field coils, armature, commutator, armature shaft, brush end plate and drive end housing with a brush or compressed air. Wash all other parts in solvent and blow them dry with compressed air.

Inspect the insulation and the unsoldered connections of the armature windings for breaks or signs of burns.

Perform electrical tests on any suspected defective part, according to the procedures outlined in Section 9-13.

Check the commutator for runout. Inspect the armature shaft and both bearings for scoring.

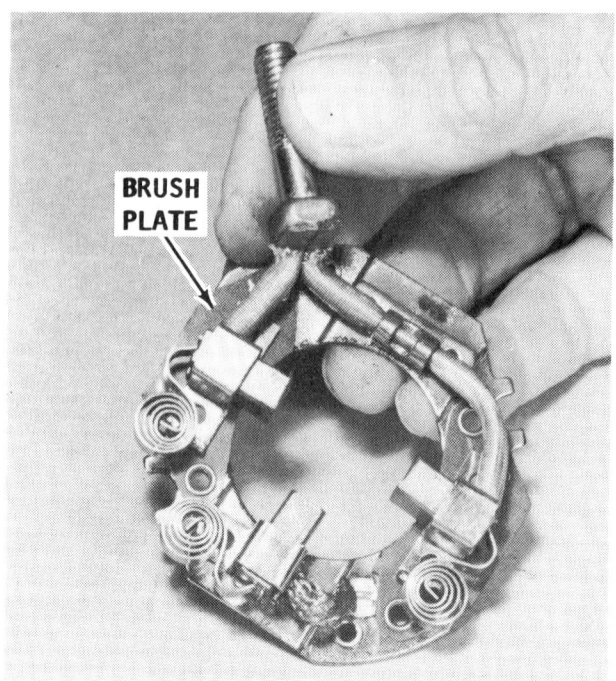

If any brush length is less than 0.4" (10mm), the entire brush plate should be replaced. The brush leads are riveted to the plate.

Turn the commutator in a lathe if it is out-of-round by more than 0.005" (0.13mm).

Check the insulated brush holders for shorts to ground. If the brushes are worn down to 0.4" (10mm) or less, they must be replaced.

The armature, fields, and brush holders must be checked before assembling the starter motor. For detailed procedures to test cranking motor parts, see Section 9-13.

Lower End Cap

The lower armature shaft rides in a special bronze bushing, prelubricated at the factory. Some evidence of lubricant must be present when the cranking motor is disassembled.

Inspect the pinion gear teeth for chips, cranks, or broken teeth.

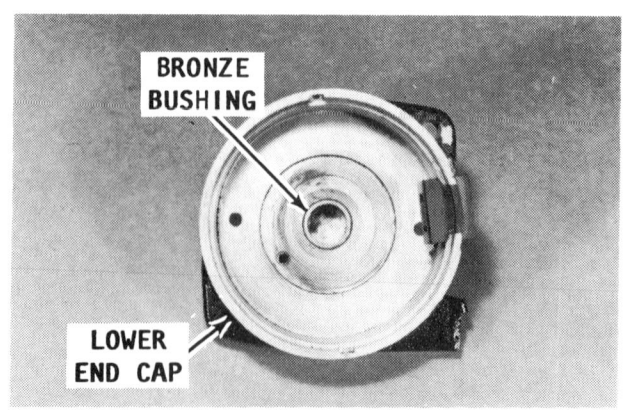

The lower armature shaft rides in a bronze bushing at the base of the lower end cap.

Complete absence of lubricant will soon lead to metal-to-metal friction and cause either a seizure of the armature shaft or spinning of the bronze bushing inside the end cap.

Normally, it is not necessary to add lubricant to this bushing, but if the bushing is completely dry, fill the base of the bushing with approximately 1/8" (3mm) of Multi Purpose Marine Lubricant. **DO NOT** use more lubricant than the 1/8" (3mm) because excessive lubricant would be squeezed out onto the commutator and brushes.

9-13 TESTING CRANKING MOTOR PARTS

SPECIAL WORDS

Most marine shops and all electrical motor rebuild shops will test an armature for a modest charge. If the armature has a short, it **MUST** be replaced.

Check the armature for a short circuit by placing it on a growler and holding a hack saw blade over the armature core while the armature is rotated. If the saw blade vibrates, the armature is shorted. Clean between the armature bars, and then check again on the growler. If the saw blade still vibrates, the armature must be replaced. Occasionally carbon dust from the brushes will short the armature. Therefore, blow the slots in the armature clean with compressed air.

1- Make contact with one probe of the test light on the armature core or shaft. Make contact with the other probe on the commutator. If the light comes on, the armature is grounded and must be replaced. This test may also be performed with an ohmmeter. Select the Rx1000 ohm scale. Make contact with one meter lead on the armature core or shaft. Make contact with

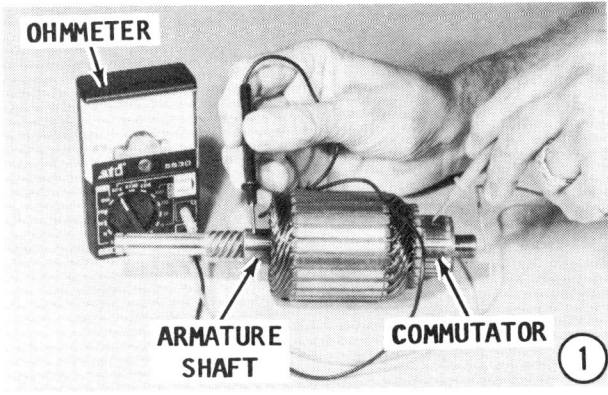

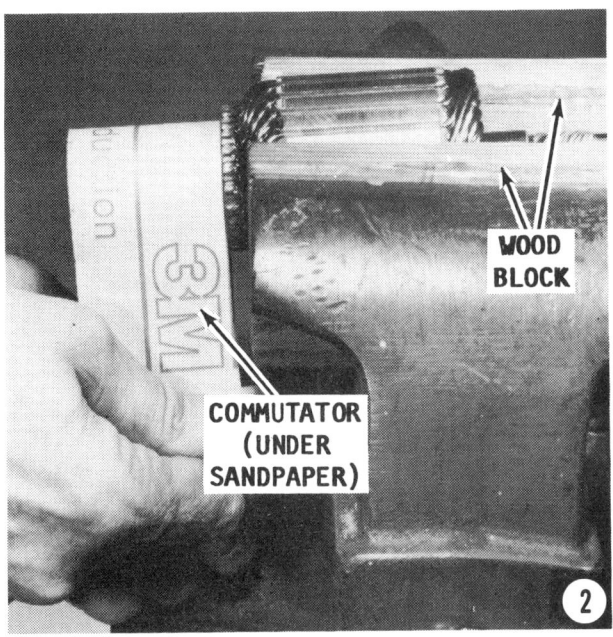

the other meter lead on the commutator. If the meter reading indicated continuity, the armature is grounded and must be replaced.

Turning the Commutator

2- True the commutator, if necessary, in a lathe. **NEVER** undercut the mica because the brushes are harder than the insulation. Undercut the insulation between the commutator bars 1/32" (0.80mm) to the full width of the insulation and flat at the bottom. A triangular groove is not satisfactory. After the undercutting work is completed, clean out the slots carefully to remove dirt and copper dust. Sand the commutator lightly with No. 500 or No. 600 sandpaper to remove any burrs left from the undercutting.

3- Test light probes, placed on any two commutator bars, should light and indicate

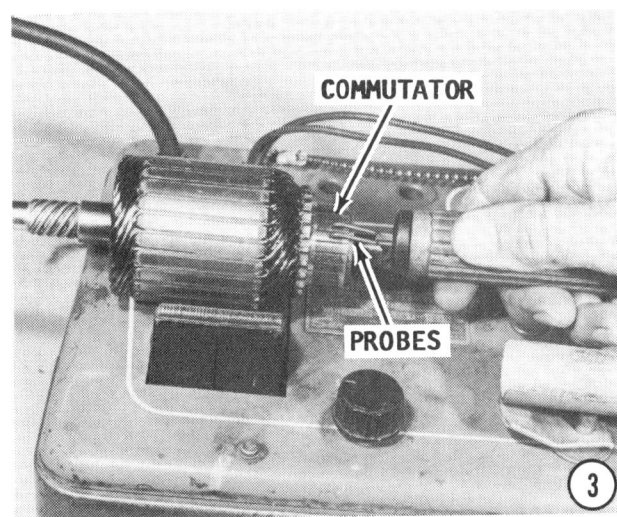

9-20 ELECTRICAL

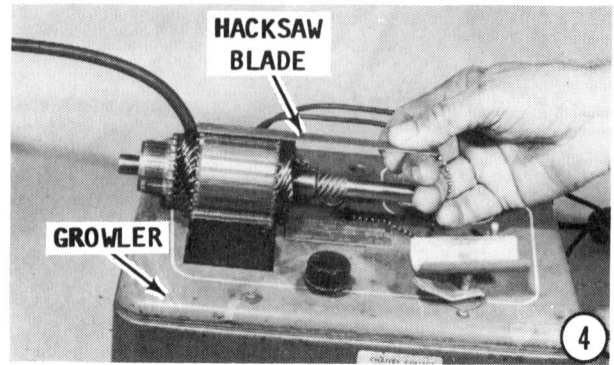

continuity. Again, this test may be performed using an ohmmeter. Select the Rx1000 ohm scale. The meter leads placed on any two commutator bars should indicate continuity.

4- Check the armature a second time on the growler for possible short circuits.

Brush Insulation Tests

5- Obtain an ohmmeter. Make contact with one lead of the ohmmeter to one of the positive brushes. Make contact with the other test lead to the other positive brush. The meter should indicate continuity. If the meter indicates no continuity, there is an

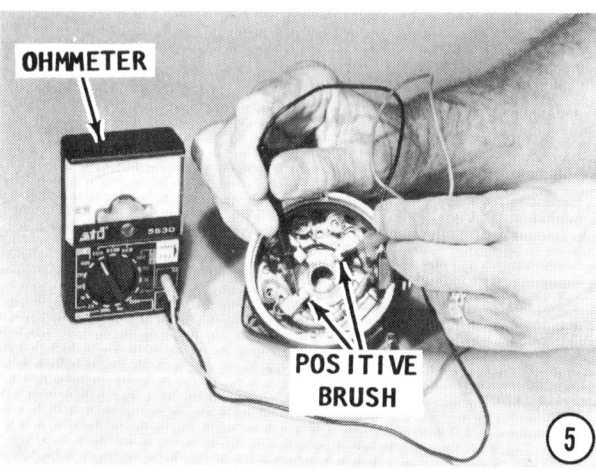

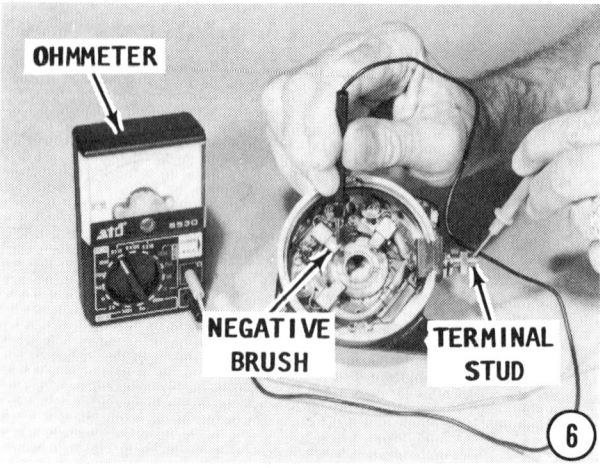

"open" between the brushes. The brush plate should be replaced. Move one of the meter leads to make contact with the case. The ohmmeter should indicate **NO** continuity. If the meter indicates continuity, one or both of the positive leads is shorted to the case and must be replaced.

6- Make contact with one lead of the ohmmeter to the negative brush. Make contact with the other test lead to the terminal stud. The ohmmeter should indicate **NO** continuity. If the meter indicates continuity, there is a short in the positive lead and the positive lead is grounded to the case.

Move the Red meter lead to make contact with the case. The ohmmeter should indicate continuity. If the meter indicates no continuity, there is an "open" in the negative lead or the lead is not properly grounded to the end cap.

9-14 CRANKING MOTOR ASSEMBLING

1- Slide the terminal stud through the large square grommet in the end cap. Place the small rubber insulating washer over the stud and seat it into the recess in the grommet. Slide the flat washer and the lockwasher over the stud. Thread on the narrow nut. Tighten the nut securely.

2- Install and tighten the two small Phillips head screws through the back of the end cap to secure the brush plate.

3- Hold the brushes back with the blades of three screwdrivers wedged between the springs and the brush holders. Insert the small end of the armature into the end cap bushing. **TAKE CARE** not to disturb the brush springs. Once assembled, rotate the shaft inside the bushing and check to be sure there is no binding and the brushes sweep across the commutator smoothly.

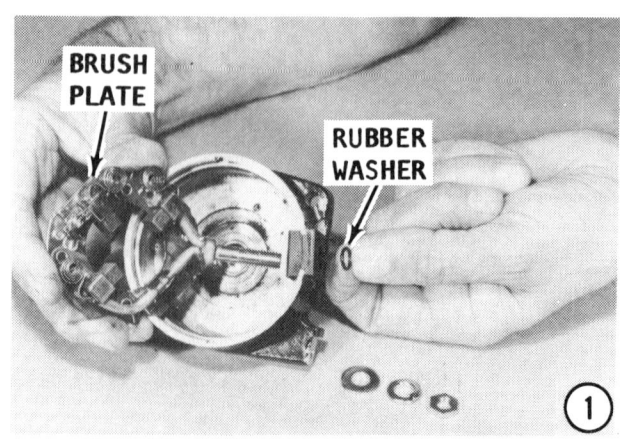

CRANKING MOTOR 9-21

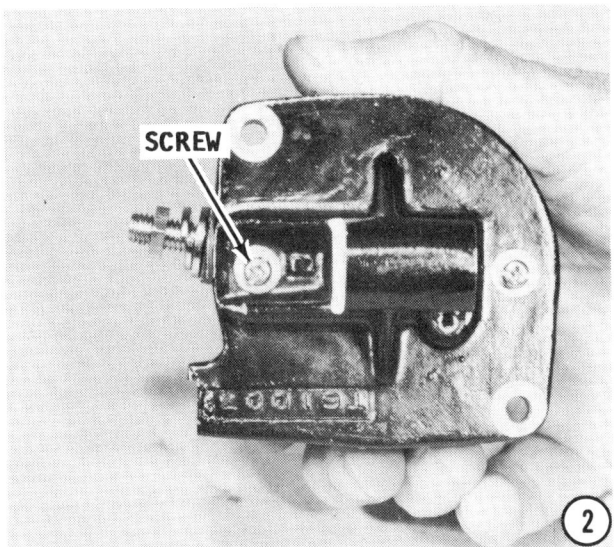

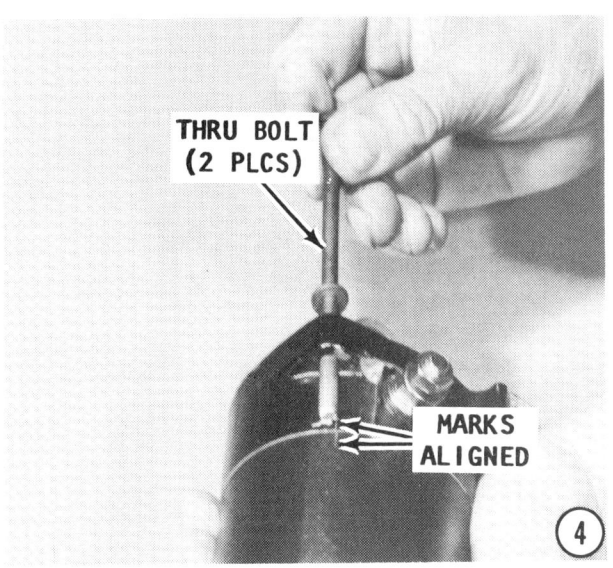

4- Slide the free end of the armature shaft into the field frame of the motor. **TAKE CARE** not to lose the brushes from the commutator. Seat the end cap into the field frame with the marks made prior to disassembly (Step 4), aligned. Apply a thin coating of Yamaha All Purpose Lubricant, or equivalent waterproof anti-seize lubricant, onto the bushing in the end cap.

Slide the washers over the commutator end of the armature shaft. Usually there are two steel washers in front of a resin washer. Incorrect placement of these washers will cause the armature to be misaligned with the brushes and the field magnets during assembling. Install the lower end cap over the armature shaft. Seat the upper end cap with the marks made during disassembly (Step 4), aligned.

Now, hold it all together and slide the two thru bolts through the lower end cap, and then thread them into the upper end cap.

5- Slide the pinion gear over the armature shaft.

6- Place the spring and the collar over the shaft. The collar must be installed with the internal groove for the snap ring facing upwards.

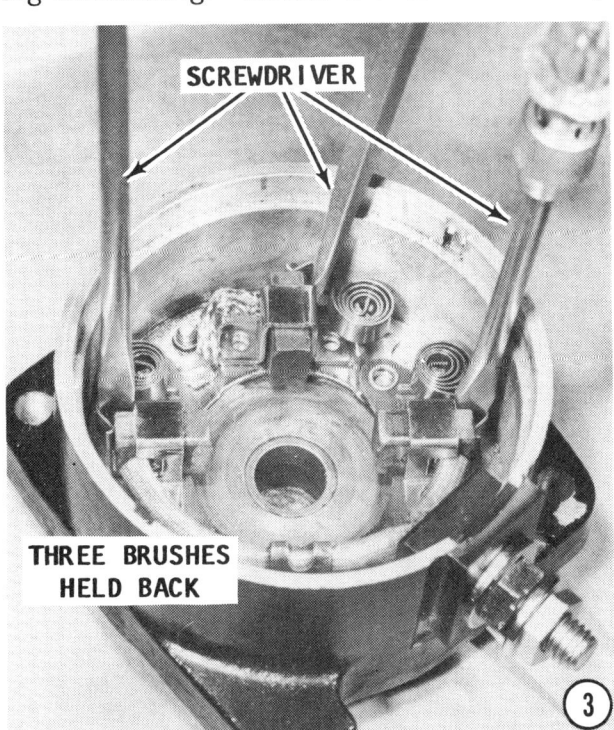

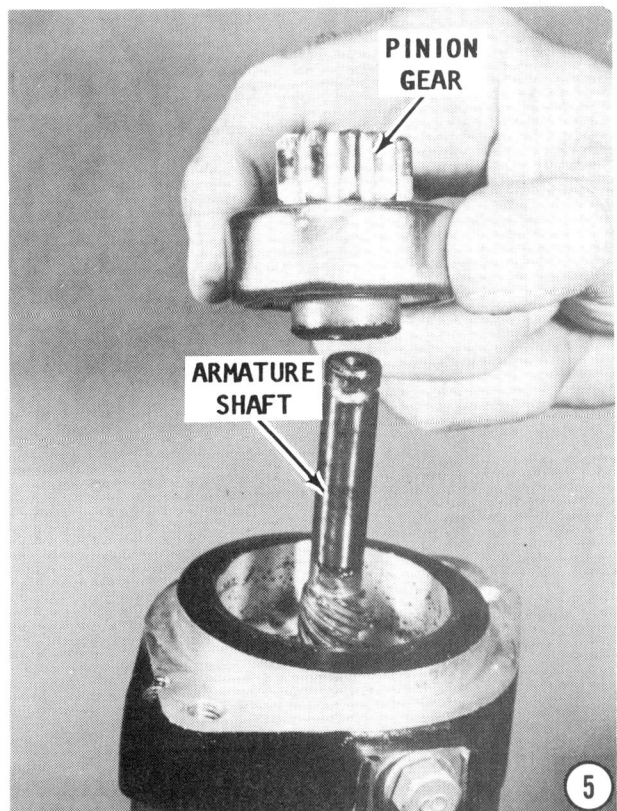

9-22 ELECTRICAL

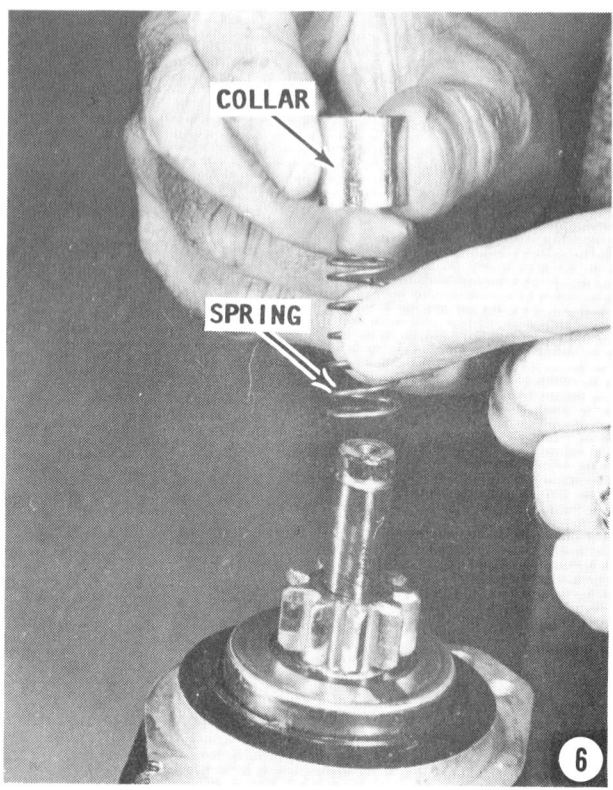

7- Hold the collar down tight against the spring while the snap ring is installed. The ring must be crimped around the groove in the shaft with a pair of pliers. When properly installed the collar will completely hide the snap ring and the top of the collar will be flush with the armature shaft.

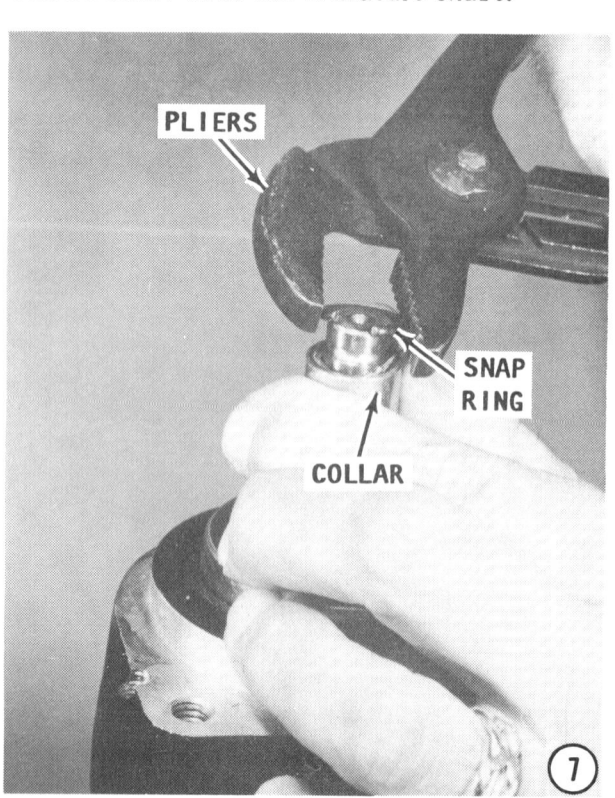

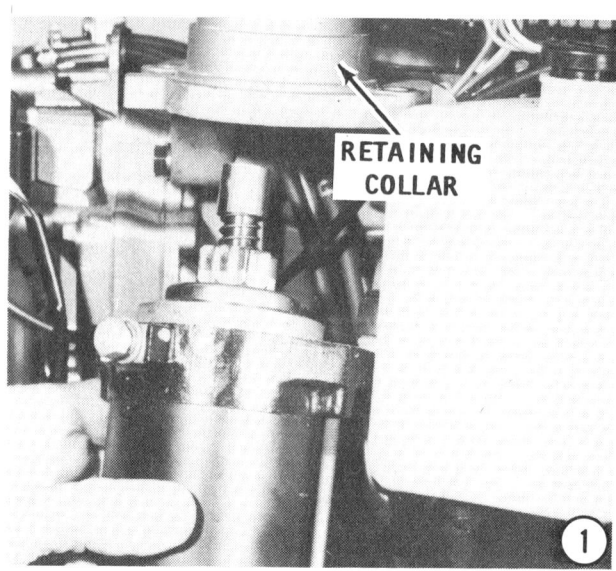

9-15 CRANKING MOTOR INSTALLATION

The following procedures pickup the work after the cranking motor has been disassembled, tested, serviced, and assembled.

1- Slide the cranking motor up into the retaining collar.

2- Start the three mounting bolts. Tighten these bolts to a torque value of 22 ft lb (30 Nm) in the following order. Tighten the lower bolt first, then the upper aft bolt and finally the upper forward bolt.

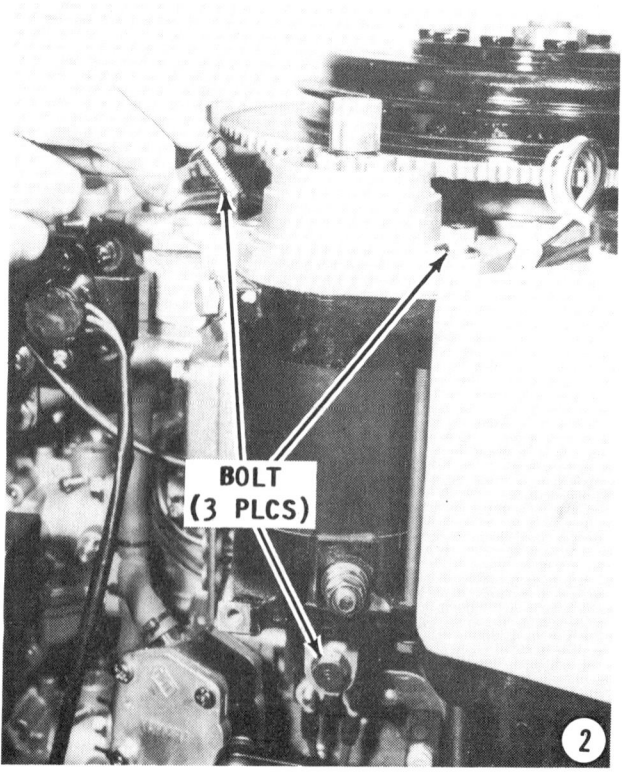

TESTING COMPONENTS 9-23

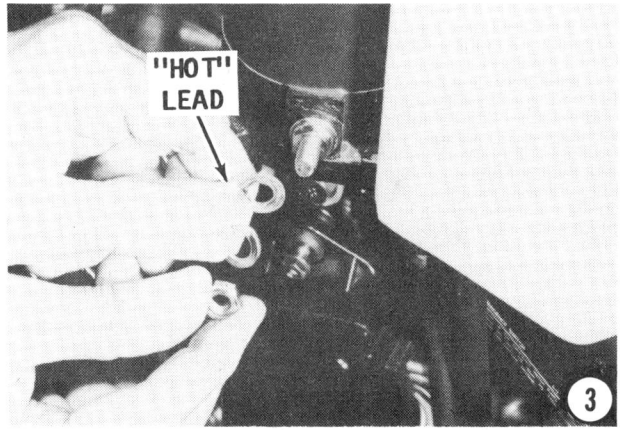

3- Connect the lower "hot" lead to the terminal of the motor.

4- Connect the upper "ground" lead to the case of the cranking motor. Slip the elbow "boots" over the installed leads.

5- Connect the two large Red leads to the front of the cranking motor relay. The lead on the right is the one from the "hot" terminal on the cranking motor. The lead on the left is also "hot" coming from the positive battery terminal. This lead also has a small Red lead attached to the relay terminal.

Install the elbow "boots" onto both relay leads. Connect the electrical leads to the battery.

Check Completed Work

Mount the outboard unit in a test tank, on the boat in a body of water, or connect a flush attachment and hose to the lower unit.

CAUTION

Water must circulate through the lower unit to the powerhead anytime the power-

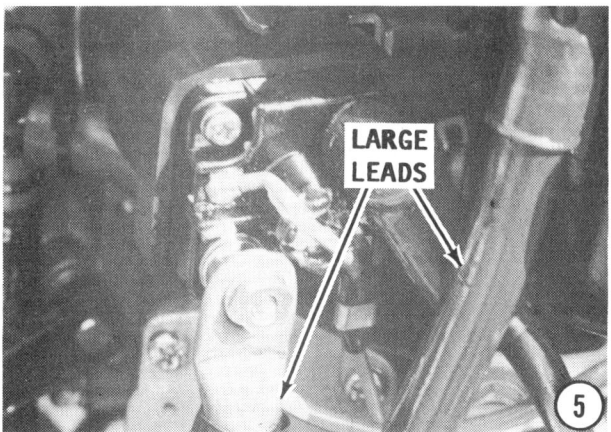

head is operating to prevent damage to the water pump in the lower unit. Just five seconds without water will damage the water pump impeller.

NEVER, AGAIN, NEVER operate the engine at high speed with a flush device attached. The engine, operating at high speed with such a device attached, would **RUNAWAY** from lack of a load on the propeller, causing extensive damage.

Crank the powerhead with the cranking motor and start the unit. Shut the powerhead down and start it several times to check operation of the cranking motor.

9-16 TESTING OTHER ELECTRICAL COMPONENTS

This short section provides testing procedures for other electrical parts installed on the powerhead. If a unit fails the testing, the faulty part must be replaced. In most cases, removal and installation is through attaching hardware.

Thermoswitch Testing

The two thermoswitches are located on the powerhead, either on or close to the exterior of the combustion chamber, in the area of highest temperatures. The thermoswitch resembles a condenser, with Pink and Black leads. The thermoswitch can be easily removed from the powerhead, by pulling it out of its recess. The thermoswitch is a very fragile sensor and should never be dropped or handled in a rough manner.

Remove the thermoswitch from the block and disconnect the two leads at their quick disconnect fittings. There are no resistance tests to be performed on this sensor while it is installed.

The resistance of the sensor is monitored at different temperatures while heating the "sensing" end of the thermoswitch in a body of water. Obtain a thermometer, an ohmmeter, and a suitable pan in which to boil water.

Place the container of water, at room temperature, on a stove. Secure the thermometer in the water in such a manner to prevent the thermometer bulb from contacting the bottom or sides of the pan.

Immerse the "sensing" end of the thermoswitch into the water up to the shoulder.

Make contact with the Red meter lead to the Pink thermoswitch lead, and the Black meter lead to the Black thermoswitch lead. The meter should indicate no continuity, as long as the water temperature is below 187°F (86°C).

Heat the water. When the temperature rises above 187°F (86°C), the meter should indicate continuity. A variation of 40°F (7°C) is allowable above or below the stated temperature. If the meter reading is acceptable within the limits given, the thermoswitch is functioning correctly. If the meter readings are not acceptable, the thermoswitch can be easily replaced with a new unit. Repeat the test for the other switch.

Testing Trim/Tilt Relay

Two identical solenoids are used, an **UP** solenoid and a **DOWN** solenoid. These units are similar in appearance to those used as electric cranking motor solenoids. Testing this type solenoid/relay is exactly the same as for testing solenoids used on cranking motors. This testing procedure is outlined in this chapter beginning on Page 9-12.

When testing these two relays, the Sky Blue and Light Green leads take the place of the Brown lead in the tests outlined for the cranking motor relay.

If the relay should fail one or both resistance tests, the unit should be replaced. No service or adjustment is possible.

Testing Power Trim/Tilt Switch

The power trim/tilt switch is located at the top of the remote control handle. The harness from the switch is routed down the handle to the base of the control box, and then into the box through a hole. The control box cover must be removed to gain access to the quick disconnect fittings and permit testing of the switch.

An auxiliary trim/tilt switch is mounted on the exterior surface of the starboard side lower cowling pan. This switch is most convenient, while performing tests on the trim/tilt unit, when the need to observe while operating the system is required.

Disconnect the three leads from the switch: the Red, Light Green, and Sky Blue, leads at their quick disconnect fittings. Obtain an ohmmeter and select the Rx1000 ohm scale. Connect the Red meter lead to the Red switch lead, and keep this connection for the following two resistance tests. Connect the Black meter lead to the Sky

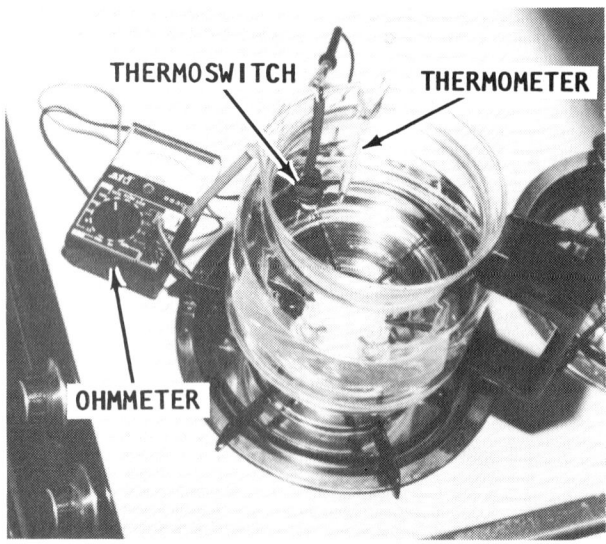

Heating a container of hot water on a stove to check performance of the thermoswitch. The resistance of the thermoswitch is monitored through a change in temperature.

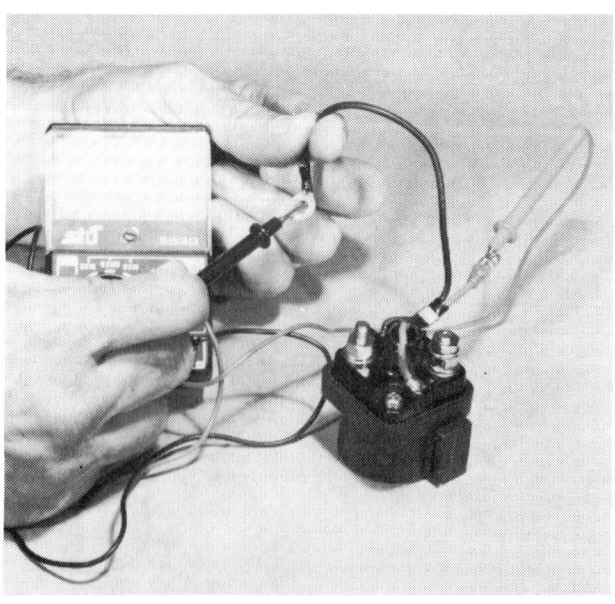

Procedures for testing the trim/tilt relay are identical to procedures for the cranking motor relay.

Blue switch lead. Depress the upper portion of the toggle switch. The meter should indicate continuity. Release the switch to the neutral position. The meter should indicate no continuity. Depress the lower portion of the switch. The meter should still indicate no continuity.

With the Red meter lead still connected to the Red switch lead, move the Black meter lead to make contact with the Light Green switch lead. Depress the lower portion of the toggle switch. The meter should indicate continuity. Release the switch to the neutral position. The meter should indicate no continuity. Depress the upper portion of the switch. The meter should still indicate no continuity.

If the switch fails one or both resistance tests, replace the switch. To remove the switch from the remote control handle, see appropriate procedures in Chapter 12.

Testing Trim/Tilt Motor

Check to be sure the manual release valve is in the **MANUAL TILT** position. This means rotated approximately three full turns from the **POWER TILT** position, as evidenced by the embossed words and directional arrow on the housing.

Disconnect the Black lead and the White lead from the tilt motor at their quick disconnect fittings. **MOMENTARILY** make contact with the two disconnected leads to the posts of a fully charged battery. Make the contact **ONLY** as long as necessary to hear the electric motor rotating.

Reverse the leads on the battery posts and again listen for the sound of the motor rotating. The motor should rotate with the leads making contact with the battery in either direction.

If the motor fails to operate in one or both directions, remove the electric motor. First, place a suitable container under the unit to catch the hydraulic fluid as it drains. Next, disconnect the two lower hydraulic lines from the bottom of the housing and remove the fill plug from the reservoir. Permit the fluid to drain into the container. After the fluid has drained, disconnect the electrical leads at the harness plug, and then remove the electric motor through the attaching hardware.

This motor is very similar in construction and operation to a cranking motor. The arrangement of the brushes differ to allow the trim/tilt motor to operate in opposite directions, but otherwise, it is almost identical to the cranking motor. Proceed to the cranking motor servicing section in this chapter for detailed procedures to return the motor to satisfactory operation.

Testing Main Switch

The main switch is located inside the control box. The box must be opened to gain access to the main switch leads.

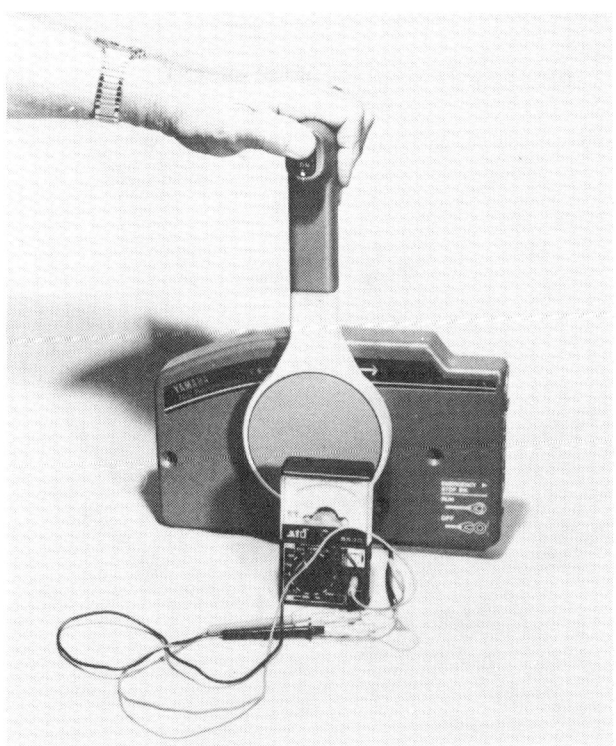

Using an ohmmeter to test the power trim/tilt switch.

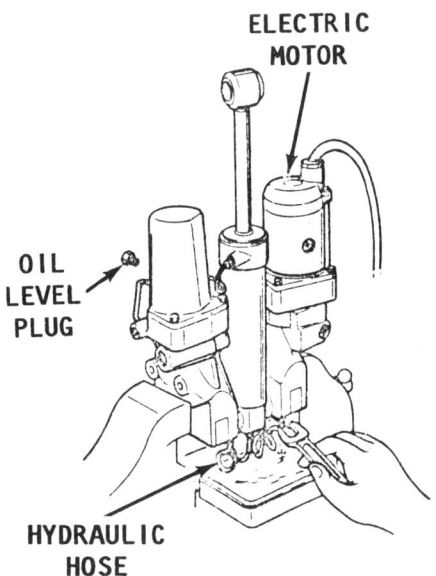

Draining the hydraulic system prior to removing the electric motor for testing and/or service work.

Disconnect the five leads from the main switch at their quick disconnect fittings: the White, Black, Red, Yellow, and Brown leads. Obtain an ohmmeter and select the Rx1000 ohm scale. Check to be sure the switch is in the OFF position. Make contact with the Red meter lead to the White switch lead, and the Black meter lead to the Black switch lead. The meter should indicate continuity. Keep both meter leads in place. Rotate the switch to the ON position, and then to the START position. If the meter indicates continuity in either or both switch positions, the switch is defective and must be replaced.

Move the Red meter lead to make contact with the Red switch lead, and the Black meter lead to make contact with the Yellow switch lead. Rotate the switch to the ON position. The meter should indicate continuity. Turn the switch back to the OFF position and then on to the START position. If the meter indicates continuity in either position, the switch is defective and must be replaced.

Keep the Red meter lead in contact with the Red switch lead for the following two resistance tests. Move the Black meter lead to the Yellow switch lead. Rotate the switch to the START position. The meter should indicate continuity. Turn the switch to the OFF position, and then on the ON position. If the meter indicates continuity in either or both positions, the switch is defective and must be replaced.

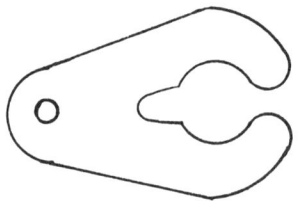

Pattern to be used to fabricate a "homemade" emergency tether, as explained in the text.

Repeat the above test with the Black meter lead in contact with the Brown switch lead.

If the main switch should fail any one of these resistance tests, even though the switch appears to function normally, the switch should be replaced.

A faulty main switch could cause the powerhead to start accidentally, while being serviced, and cause **PERSONAL INJURY**. Once testing has been satisfactorily completed connect the five leads, color to color, at their quick connect fittings. Tuck the leads neatly inside the control box, away from moving parts, and replace the outer cover.

Testing "Kill" Switch

The "kill" switch is mounted on the forward side of the box. The "kill" switch **MUST** have the emergency tether in place before testing.

Temporary Tether Replacement

If the boat owner loses the emergency tether and is unable to obtain one imme-

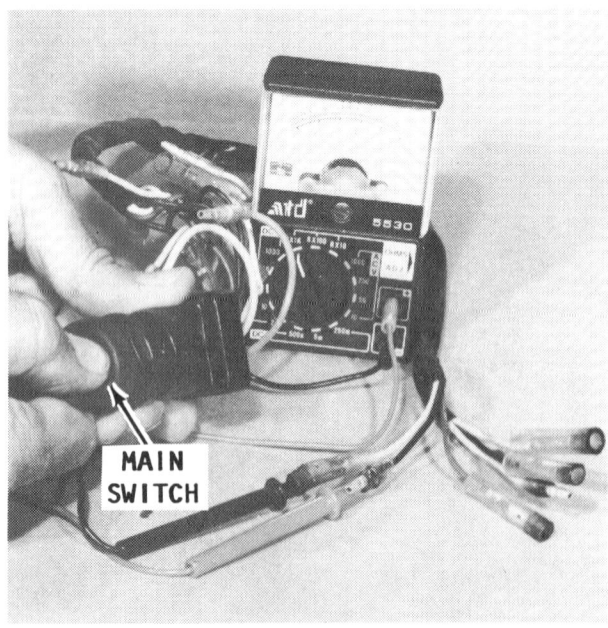

Using an ohmmeter to test the main switch, as explained in the text.

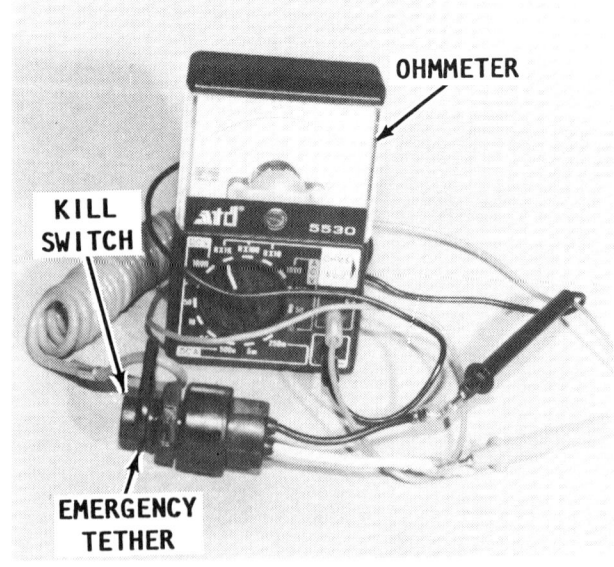

Using an ohmmeter to test the "kill" switch with the emergency tether in place.

diately from the local Yamaha dealer, an emergency substitute tether may be made using only a common knife and a couple pieces of plastic.

First, obtain a piece of plastic such as the cover of a container of margarine, whipped topping, or similar product.

Next, using the pattern shown on this page, cut out about four shapes, as shown. Stack the four cutouts together, secure them with a paper clip, or similar object, and then insert them behind the "kill" switch.

SPECIAL WORDS

If the material described is not available, obtain some other pliable material and cut the shape indicated. The thickness of the substitute tether device should be approximately 1/8" (3mm). About 100 pages (50 sheets) of this manual is approximately the proper thickness.

REMEMBER, use this device only in an emergency situation and purchase the proper tether from the local Yamaha dealer at the first opportunity. By substituting this "home made" tether, both the safety and the security features intended by the manufacturer have been lost.

TESTING

Trace the "kill" switch button harness containing two wires from the switch to their nearest quick disconnect fitting. The colors may vary for different models, but the "hot" wire is usually White/Black. Refer to the wiring diagram in the Appendix for proper color identification.

Disconnect the two leads and connect an ohmmeter across the disconnected leads. Verify the emergency tether is in place behind the "kill" switch button. Select the Rx1000 scale on the meter. Depress the "kill" button. The meter should indicate continuity. Release the button. The meter should now indicate no continuity. Both tests must be successful. If the switch fails either test, the switch is defective and **MUST** be replaced. The switch is a one piece sealed unit and cannot be serviced.

Testing Choke Solenoid

Disconnect the Blue lead from the choke solenoid, and leave the Black lead with the eyelet connected at the grounding bolt. Select the Rx1000 scale on the meter. Make contact with the Red meter lead to the Blue disconnected wire from the solenoid. Make contact with the Black meter lead to the eyelet grounding bolt. The meter should indicate continuity. If the reading is within specifications, but the choke still fails to function, then proceed to the testing procedures on the choke switch, later in this section. If the meter reading is not within specifications, but the choke still performs satisfactorily, then keep an eye on this component as a possible source of trouble in the future. If the choke solenoid failed the resistance test and fails to function, then the solenoid **MUST** be replaced. The solenoid **CANNOT** be serviced or adjusted.

Testing Choke Switch

The choke switch is a spring loaded toggle type switch located in the forward side of the control box. The control box must be opened to gain access to the switch leads. Disconnect the Blue and Yellow leads from the choke switch at the nearest quick disconnect fitting. Select the Rx1000 ohm scale on the meter. Make contact with the meter leads, one to each of the disconnected leads. In the normal **OFF** position the meter should indicate no continuity. Move the switch to the **ON** position. The meter should indicate continuity. If the switch fails either of these two tests, it must be replaced. The switch **CANNOT** be serviced or adjusted.

Testing Neutral Safety Switch

The neutral safety switch is located inside the control box. Trace the two Brown neutral safety switch leads from the switch to their nearest quick disconnect fitting.

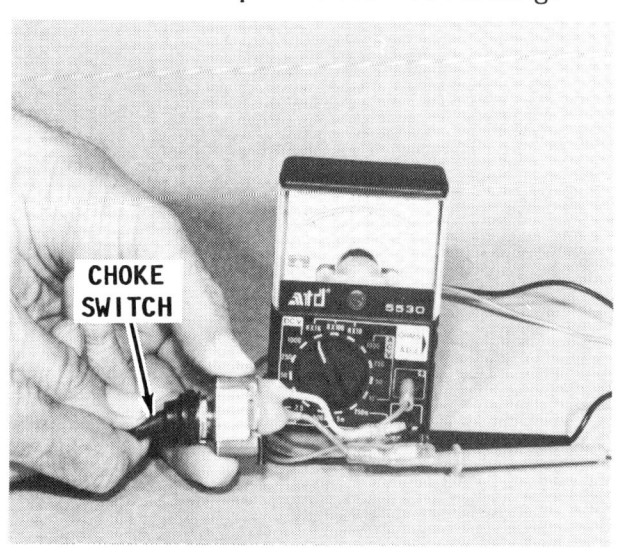

Using an ohmmeter to test the choke switch.

Disconnect the two leads and connect an ohmmeter across the two disconnected leads. Select the Rx1000 scale on the meter. When the shift lever is in the **NEUTRAL** position, the meter should indicate continuity.

When the lower unit is shifted into either **FORWARD** or **REVERSE** gear, the meter should indicate no continuity.

The switch must pass all three tests to verify the safety aspect of the switch is functioning properly. If the switch fails any one or more of the tests, the switch **MUST** be replaced.

REMEMBER this is a safety switch. A faulty switch may allow the powerhead to be started with the lower unit in gear -- an extremely dangerous situation for the boat, crew, and passengers.

Testing Buzzer

The buzzer is a warning device to indicate low oil in the reservoir, an over "rev." condition, or overheating of the powerhead. The buzzer is located inside the control box.

Remove the control box cover and identify the two leads from the buzzer, one is Yellow and the other is Pink. Disconnect these two leads at their quick disconnect fittings, and ease the buzzer out from between the four posts which anchor it in place.

Obtain a 12 volt battery. Connect the Yellow buzzer lead to the negative battery terminal. Momentarily make contact with the Pink buzzer lead to the positive battery terminal. As soon as the Pink buzzer lead makes contact with the positive battery terminal, the buzzer should sound. If the

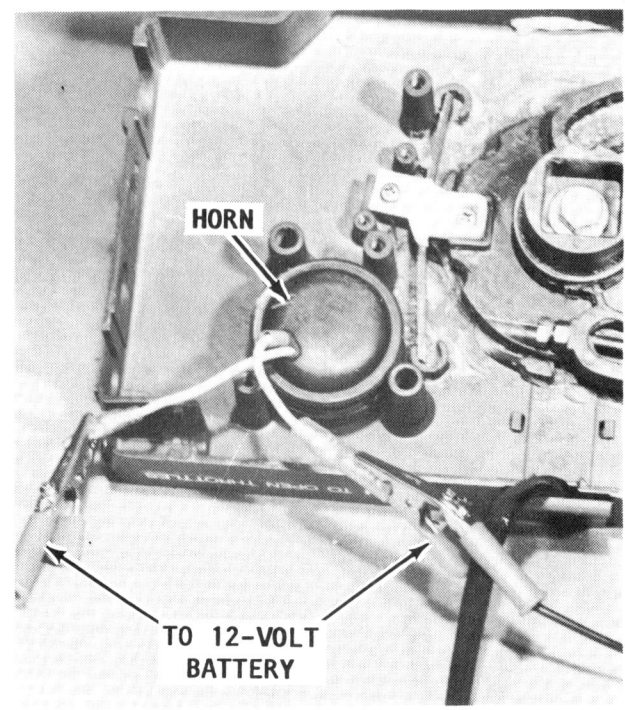

When testing the warning horn, the leads of the horn are connected to a 12-volt battery.

buzzer is silent, or the sound emitted does not capture the helmperson's attention immediately, the buzzer should be replaced. Service or adjustment is not possible. If the sound is satisfactory and immediate, install the buzzer between the four posts and connect the two leads matching color to color. Tuck the leads to prevent them from making contact with any moving parts inside the control box. Replace the cover.

Testing Oil Injection Electrical Components

Detailed testing procedures for all electrical components in the oil injection system are presented in Chapter 5.

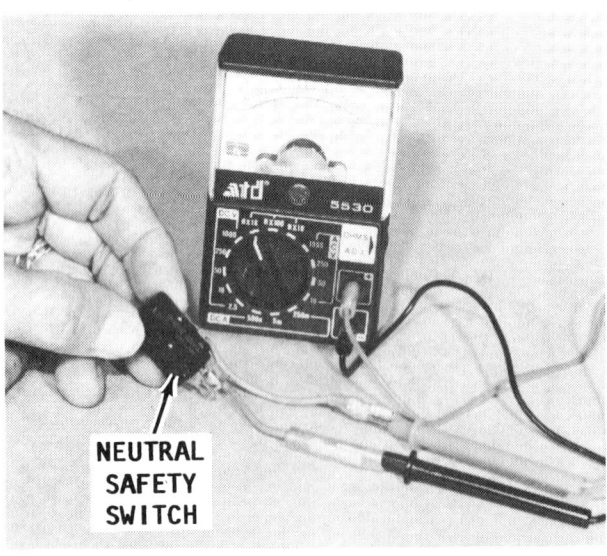

Using an ohmmeter to test a neutral safety switch.

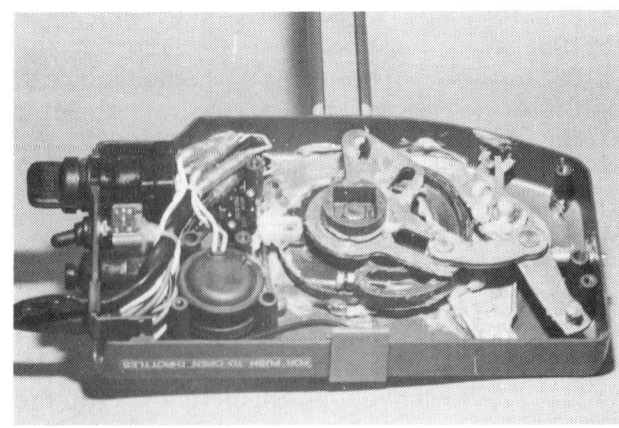

The most difficult task involving tests of the remote control unit is "stuffing" the electrical leads back into the control box in an ORDERLY manner.

10
TRIM/TILT

10-1 INTRODUCTION AND CHAPTER COVERAGE

All outboard installations are equipped with some means of raising or lowering (pivoting), the complete unit for efficient operation under various load, boat design, water conditions, and for trailering to and from the water.

The correct trim angle ensures maximum performance and fuel economy as well as more comfortable ride for crew and passengers.

The most simple form is a mechanical tilt adjustment consisting of a series of holes in the transom mounting bracket through which an adjustment pin passes to secure the outboard unit at the desired angle.

Such a mechanical arrangement works quite well for the smaller units, but with the size outboard units covered in this manual, a power system is required. The power system for these larger outboards is hydraulically operated and electrically controlled from the helmperson's position.

The trim and tilt system covered in this manual is installed between the two large clamp brackets. Two power trim/tilt relays are mounted between the air silencer and the cranking motor on the portside of the powerhead.

The trim/tilt relays are identical in shape and size to the cranking motor relay. All three relays are mounted in a vertical line. The top relay is the tilt **UP** relay, the middle relay is the tilt **DOWN** relay and the bottom relay is the cranking motor relay.

The power trim/tilt system is provided with a manual release valve to permit movement of the outboard unit manually in the remote event the system develops a malfunction, either hydraulic or electrical, preventing use of power.

10-2 DESCRIPTION AND OPERATION

The trim/tilt system consists of a housing with an electric motor, gear driven hydraulic pump, hydraulic reservoir, two trim cylinders, and one tilt cylinder attached. The tilt cylinder performs a double function as tilt cylinder and also as a shock absorber, should the lower unit strike an underwater object while the boat is underway.

The necessary valves, check valves, relief valves, and hydraulic passageways are incorporated internally and externally for

Proper operation of the trim/tilt system is essential to efficient outboard performance on the water.

efficient operation. A manual release valve is provided, on the starboard side of the housing, to permit the outboard unit to be raised or lowered should the battery fail to provide the necessary current to the electric motor or if a malfunction should occur in the hydraulic system.

The gear driven pump operates in much the same manner as an oil circulation pump installed on motor vehicles. The gears may revolve in either direction, depending on the desired cylinder movement, up or down. One side of the pump is considered the "suction" side, and the other the "pressure" side, when the gears rotate in a given direction. These sides are reversed, the "suction" side becomes the "pressure" side and the "pressure" side becomes the "suction" side when gear movement is changed to the opposite direction.

SPECIAL WORDS

Yamaha engineers have been constantly working to improve the operational performance of the trim and tilt system installed on their outboard units. Therefore, many of the units will appear to be quite similar but internal hydraulic passages and check valves have been changed as well as the external routing of hydraulic lines.

One of many changes was the addition of hydraulic circuits to connect the upper chambers in the port trim cylinder with the starboard trim cylinder. Another change involved the main valve to permit the trim pistons to be retracted when the engine is tilted up. This action reduced operating noises and provided protection for the trim cylinders against corrosion and marine growth.

Therefore, the basic principles described in this section apply to all units, but the specific location and number of components, may vary, and the routing of hydraulic lines may not be exactly as viewed on the unit being serviced.

OPERATION

Trim Up

When the UP portion of the trim/tilt switch on the remote control handle is depressed, the UP circuit, through the relay, is closed and the electric motor rotates in a **CLOCKWISE** direction. As a convenience, an auxiliary trim/tilt switch is installed on the exterior of the starboard side lower cowling on some models

The pump "sucks" in fluid from the reservoir through a check valve and forces the fluid out the "pressure" side of the pump.

From the pump, the pressurized fluid passes through a series of valves to the lower chamber of both trim cylinders and the pistons are extended. The outboard unit raises. The fluid in the upper chamber of

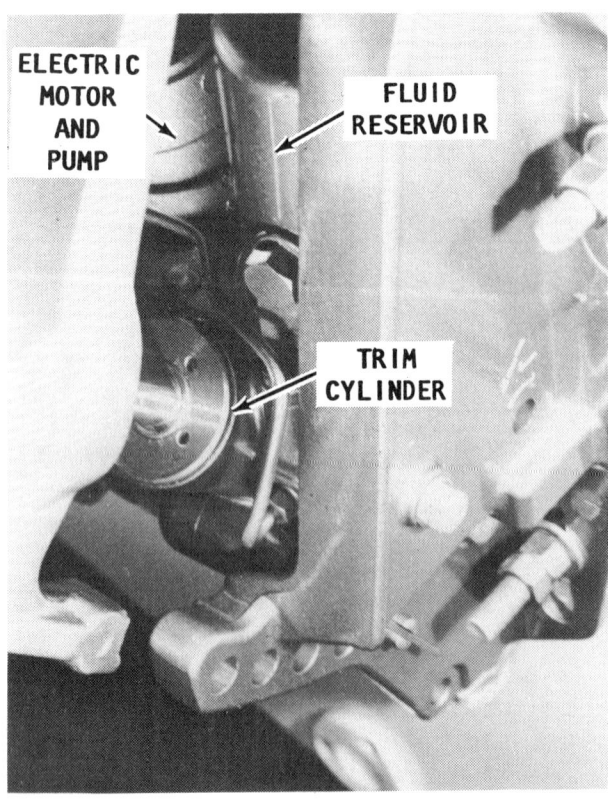

Major components of the power trim/tilt system installed between the clamp brackets.

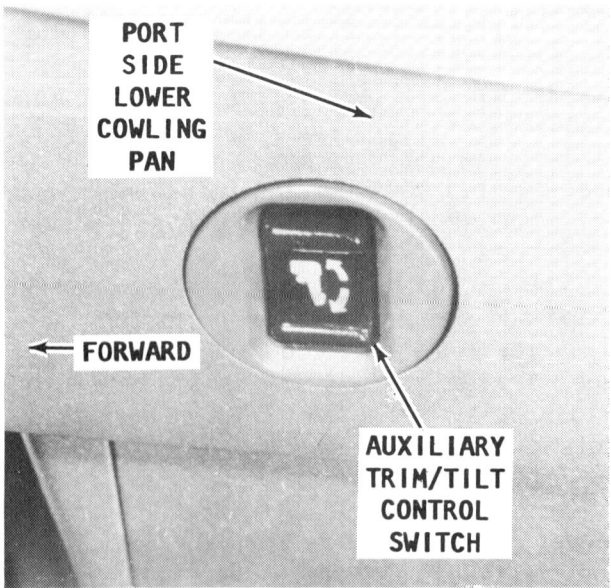

Close view of the auxiliary trim/tilt control switch on the exterior of the Port side lower cowling.

OPERATION 10-3

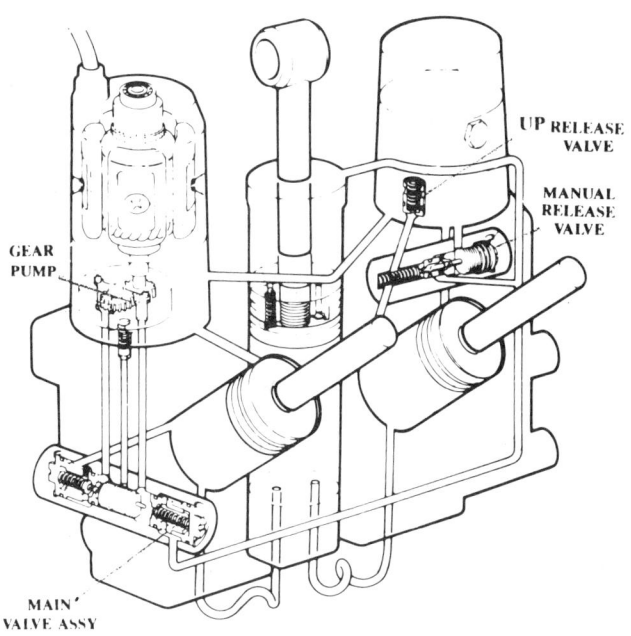

Cutaway drawing of a 6H1 trim/tilt system with major parts identified.

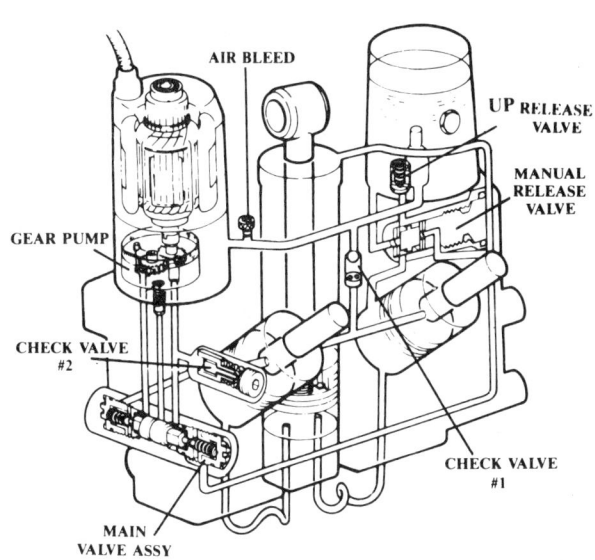

Cutaway drawing of a Type 03 trim/tilt system with major parts identified.

the pistons is routed back to the reservoir as the piston is extended. When the desired position for trim is obtained, the switch on the control handle is released and the outboard is held stationary.

A trim sender unit is installed on the port side clamp bracket. The unit sends a signal to an indicator on the control panel to advise the helmperson of the relative position of the outboard unit.

If the switch is not released when the trim cylinders are fully extended, the tilt cylinder will continue to move the outboard unit upward, as described later in this section.

Trim Down

When the **DOWN** portion of the trim/tilt switch on the remote control handle, or the auxiliary switch, is depressed, the **DOWN** circuit, through the relay, is closed and the electric motor rotates **COUNTERCLOCKWISE**. The "pressure" side of the pump now becomes the "suction" side and the original "suction" side becomes the "pressure" side.

Fluid is forced through a series of check valves to the upper chamber of each trim cylinder and the pistons begin to retract, moving the outboard unit downward. Fluid from the lower chamber of each trim cylinder is routed back through the pump and a relief valve to the reservoir.

Tilt Up

The first phase of the tilt up movement is the same as for the trim up function.

When the pistons of the trim cylinders are fully extended, fluid is forced through the system to the lower chamber of the tilt cylinder. Pressure increases in the lower chamber and the piston is extended, raising the outboard unit.

As fluid pressure in the upper chamber of the tilt cylinder increases, the fluid is routed through check valves back to the pump and the reservoir.

When the tilt piston is fully extended, fluid pressure in the lower chamber of the trim cylinders increases. This increase in pressure opens an **UP** relief valve and the fluid is routed to the reservoir.

SPECIAL WORDS

When the tilt piston becomes fully extended, the outboard is in the full **UP** position. The sound of the electric motor and the pump will have a noticeable change. The switch on the remote control handle should be released immediately.

If the switch is not released, the motor will continue to rotate, the pump will continue to pump, but the **UP** relief valve will open and the pressurized fluid will be routed to the reservoir.

If the boat is underway when the tilt cylinder is extended and powerhead rpm is increased beyond a very slow speed, the forward thrust of the propeller will increase the pressure on the tilt piston. This increase in pressure will cause the UP relief valve to open and the outboard unit will begin a downward movement.

Tilt Down

When the DOWN portion of the trim/tilt switch on the remote control handle, or the auxiliary trim/tilt switch, is depressed, the down circuit is closed through the relay and the electric motor rotates COUNTER-CLOCKWISE, as in the case of Trim Down.

The hydraulic pump "sucks" fluid from the reservoir. Fluid, under pressure is then routed to the upper chamber of the tilt cylinder and the piston begins to retract and the outboard unit moves downward.

Fluid in the lower chamber of the tilt cylinder is routed through the lower chamber of each trim cylinder and then back to the pump. When the outboard unit makes physical contact with the ends of the trim cylinders, the trim cylinders also retract until the outboard is in the full DOWN position.

SHOCK ABSORBER ACTION

The lower end of the tilt piston is "capped" with a "free" piston. This "free" piston normally moves up and down with the tilt piston -- "goes along for the ride".

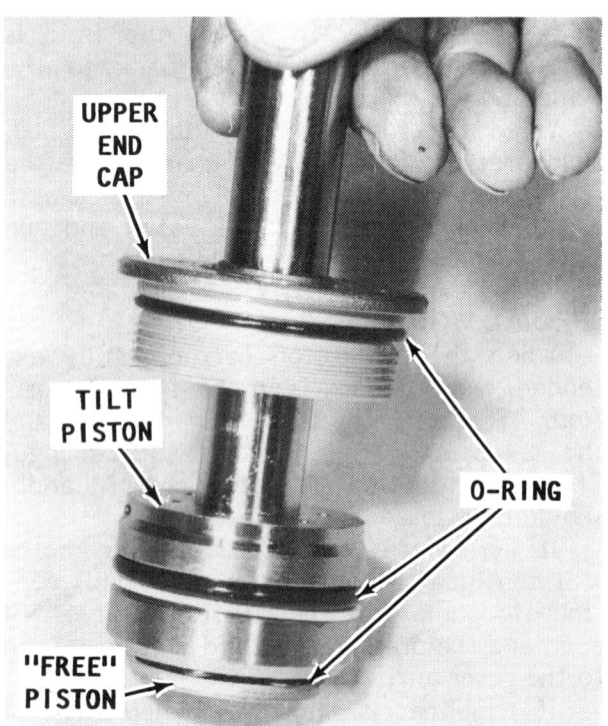

The tilt piston removed for inspection, cleaning, and necessary service work.

In the event the outboard lower unit should strike an underwater object while the boat is underway, the tilt piston would be suddenly and forcibly extended, moved UPWARD.

The "free" piston also moves upward but at a much slower rate than the tilt piston. The action of the tilt piston separating from the "free" piston causes two actions. First, the hydraulic fluid in the upper chamber above the piston is compressed and pressure builds in this area. Second, a vacuum is formed in the area between the tilt piston and the "free" piston.

This vacuum in the area between the two pistons "sucks" fluid from the upper chamber. The fluid fills the area slowly and the shock of the lower unit striking the object is absorbed. After the object has been passed the weight of the outboard unit tends to retract the piston. The fluid between the tilt piston and the "free" piston is compressed and forced through check valves to the reservoir until the "free" piston reaches its original "neutral" position.

10-3 PURGING AIR FROM THE SYSTEM UNIT INSTALLED IN CLAMP BRACKET

FIRST, THESE WORDS

Air must be purged (bled) from the system if any of the following conditions exist.

a- Erratic motion is felt during operation of the system.

b- Fluid brands have been mixed.

c- Fluid has been contaminated with moisture or other foreign material.

d- A component in the system has been replaced.

e- The system has been opened for any reason.

f- The outboard unit has been left in the full UP position and lowers slowly over a period of time.

SYSTEM REMOVED

If the unit has been removed, disassembled, and serviced on the work bench, a separate set of purging procedures must be performed, as outlined in Section 10-6, this chapter.

1- Operate the system until the outboard unit is in the full UP position.

2- Open the manual release valve to the MANUAL TILT position by rotating the screw approximately three full turns

CLOCKWISE from the POWER TILT position.

SPECIAL WORDS

The words: **POWER TILT** and **MANUAL TILT** in addition to a directional arrow for the manual position are embossed on the housing.

3- Permit the outboard unit to move to the full **DOWN** position under its own weight.

4- Close the manual release valve to the **POWER TILT** position by rotating the head **COUNTERCLOCKWISE** approximately three full turns.

5- Perform steps 1 thru 4 two or three times.

6- Move the outboard to the full **UP** position.

7- Clean the area around the fill plug to prevent contaminants entering the system when the plug is removed. Remove the fill plug. Check the fluid level in the reservoir. The fluid should reach the lower edge of the fill opening. Replenish the system using Yamalube Power Trim/Tilt Fluid, or any brand name Automatic Transmission Fluid (Dexron or Type F). **NEVER** mix brand name fluids.

8- Install and tighten the fill plug.

10-4 TROUBLESHOOTING
SPECIAL WORDS

When moving through the listed troubleshooting procedures, **ALWAYS** stop and check the system after each task. The problem may have been corrected, intentionally, or not.

The manufacturer's line of recommended products includes fluid to be used in the power trim/tilt systems.

PRELIMINARY CHECKS

1- Check to be sure the manual release head is tightened snugly **CLOCKWISE**, approximately three full turns from the power tilt position.

2- Verify the hydraulic reservoir is filled with fluid. The level of fluid should reach to the lower edge of the fill opening. Replenish as required with Yamalube Power Trim/Tilt Fluid, or any brand name Automatic Transmission Fluid (Dexron or Type F). **NEVER** mix brand name fluids.

WARNING
TRIM SYSTEM IS PRESSURIZED! DO NOT REMOVE FILL SCREW UNLESS OUTBOARD UNIT IS RAISED TO FULL UP POSITION. TIGHTEN FILL SCREW SECURELY BEFORE LOWERING OUTBOARD.

3- Inspect the hydraulic system, lines and fittings, for leaks. If an external leak is discovered, correct the condition.

4- Purge air from the system, if there is any indication the hydraulic fluid contains air.

To check for air in the system, first verify the manual release valve is snug in the **POWER TILT** position. Next, activate the **UP** circuit and raise the outboard slightly with the trim cylinders. Now, exert a heavy, steady, downward force on the lower unit.

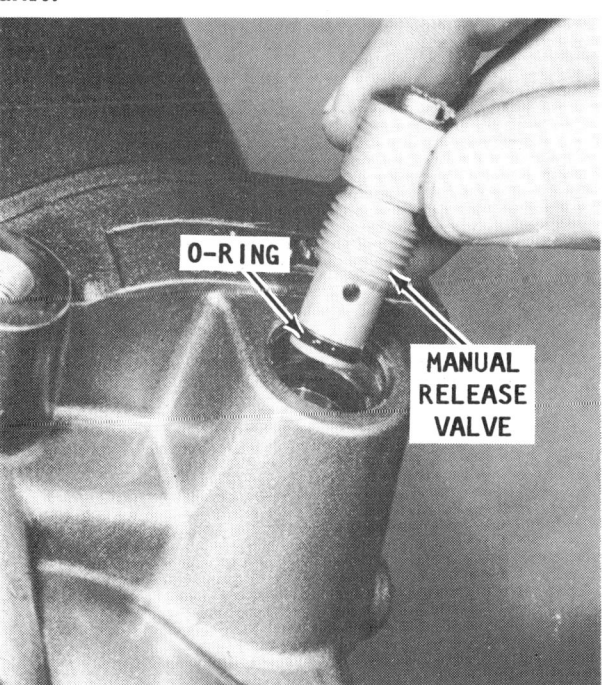

*The manual release valve on some trim/tilt systems covered in this manual has **LEFT HAND** threads. Check the arrow embossed on the housing for direction of release.*

10-6 TRIM/TILT

If the trim pistons retract into the trim cylinders more than 1/8" (3.2mm) there is air in the system.

Purge (bleed) air from the system according to the procedures beginning on Page 10-4.

5- If the system fails to hold the outboard unit in the full tilted (trailering) position, service the tilt cylinder.

6- If the system fails to hold the outboard unit in the desired trim position, service the trim cylinders.

7- If the hydraulic pump "whines" during operation, there is probably air in the system. Purge air from the system according to the procedures beginning in Section 10-3.

8- If the electric motor makes "strange" sounds or seems to be "laboring", the electric motor may require service. See Chapter 9 for service procedures.

SPECIAL WORDS

If a problem is encountered with the trim/tilt system, it is important to determine, if possible, whether the malfunction is in the hydraulic system or in an electrical circuit.

SYMPTOMS OF HYDRAULIC PROBLEM

The most common problem in the hydraulic system is failure of an O-ring to hold pressure. Therefore, if any of the following problems are encountered, see the appropriate service instructions in Section 10-5.

Outboard Behaves Abnormally

Check to be sure: The battery is adequately charged; the fluid level in the reservoir reaches the lower edge of the fill opening, with the outboard fully raised; the hydraulic fluid is not contaminated (murky color); the system does not contain air; the unit does not physically bind somewhere due to an accident.

Outboard Fails to Trim Up or Down

Check to be sure: Adequate fluid in the reservoir; manual release valve is tightened COUNTERCLOCKWISE snugly to the POWER TILT position, approximately three full turns from the manual tilt position.

Cause may be hydraulic pump failure; O-rings in trim cylinders failing to hold pressure; trim cylinders damaged due to accident.

Outboard Trims/Tilts Up, Fails to Trim/Tilt Down

Manual release valve is leaking; O-rings in trim cylinder or in the tilt cylinder failing to hold pressure; main valve has sticky or damaged shuttle piston; sticky or contaminated check valves; down relief valve has weak spring; damaged check ball or seat; or contamination is holding a valve open.

Outboard Trims/Tilts Down, Fails to Trim/Tilt Up

Manual release valve is leaking; O-rings in trim cylinders or tilt cylinder failing to hold pressure; shuttle piston in main valve assembly is sticking; contaminated check valves; UP relief valve has damaged seat or contamination holds the valve open.

Outboard "Shudders" When Shifted From One Gear to Another

Hydraulic system contaminated with air or foreign matter; internal cylinder leaks -- O-rings failing to hold pressure.

Outboard Fails to Hold Set Trim or Tilt Position

O-rings in trim and/or tilt cylinder failing to hold pressure; check valves in tilt piston contaminated requiring cleaning; external leak — fitting or part; manual release valve damaged; shuttle piston in main valve assembly sticking or contaminated check valve; UP relief valve damaged or contaminated causing a slow leak.

Outboard Tilts Up When Unit In Reverse Gear

Tilt piston has leaky absorber valve or metering valve; main valve assembly has sticky or damaged shuttle piston or leaky check valves; manual release valve is leaking; O-rings in tilt piston failing to hold pressure.

Outboard Makes Excessive Noise

Hydraulic fluid level is low; fluid is contaminated with air.

Outboard Begins To Trail Out When Throttle Backed Off at High Speed

Manual release valve not tightened snugly COUNTERCLOCKWISE to POWER TILT position; O-rings in tilt cylinder failing to hold pressure.

DISASSEMBLING 10-7

**Outboard Fails to Hold Trim Position
When Unit Operating In Reverse**

Manual release valve not tightened snugly **COUNTERCLOCKWISE** to **POWER TILT** position; O-rings in trim cylinders failing to hold pressure.

**Outboard Moves With Jerky Motion
Outboard Fails to Reach Full Down Position**

System contaminated with air; internal leaks in cylinders.

SYMPTOMS OF ELECTRICAL PROBLEM

If any of the following problems are encountered, troubleshoot the electrical system as outlined in this section.

a- Outboard trims up and down, but the electrical motor "grinds".

b- Outboard will not trim up or down.

c- Outboard trims up, but will not trim down.

d- Outboard trims down, but will not trim up.

Manual Operation

If the battery is dead, or sufficient power cannot be supplied to the electric motor to drive the hydraulic pump for any number of reasons, the outboard unit may be raised or lowered manually, by opening the manual release valve. This is accomplished by rotating the manual release valve approximately three full turns **CLOCKWISE** from the power tilt position. Words and a directional arrow embossed on the housing indicate proper direction.

WARNING

IF OUTBOARD UNIT IS IN THE UP POSITION WHEN THE MANUAL RELEASE VALVE IS OPENED, THE OUTBOARD WILL DROP TO THE FULL DOWN POSITION RAPIDLY. THEREFORE, ENSURE ALL PERSONS STAND CLEAR.

10-5 SERVICING

The following procedures provide detailed instructions to service most accessible parts of the power trim/tilt system. Many of the instructions require the end cap of the tilt cylinder and/or the end caps of the trim cylinders to be removed. As explained in the previous paragraphs, this is not an easy task.

If the attempt to remove the end cap is not successful, the unit must be taken to a shop with the proper test equipment and trained personnel with the expertise to service a high pressure hydraulic unit. If the holes provided in the end cap for the special tool are damaged, the trim/tilt unit will probably have to be replaced.

AUTHORS' WORDS

Some of the accompanying illustrations were made with the trim/tilt unit on the work bench for photographic clarity. However, the work described, with the exception of the tilt cylinder removal, may be performed without removing the unit from the clamp bracket.

Preliminary Tasks

With power or manually, raise the outboard unit to the full **UP** position and lock it in place.

Obtain a suitable container to receive the hydraulic fluid from the trim/tilt system.

Remove the two lines from the bottom of the trim/tilt housing.

Remove the fill plug from the reservoir and allow the hydraulic fluid to drain into the container.

After the fluid has drained, replace the drain plug as an assist in preventing contamination from entering the reservoir.

DISASSEMBLING

FIRST, THESE WORDS

Make an attempt to keep parts identified as they are removed. Many parts may appear similar, but components should be installed into the same location from which they are removed.

1- Obtain Yamaha special tool P/N YB6175. Observe the pattern of pins and

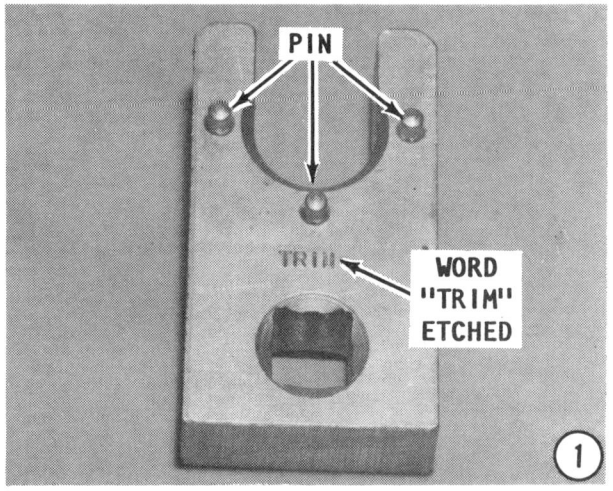

10-8 TRIM/TILT

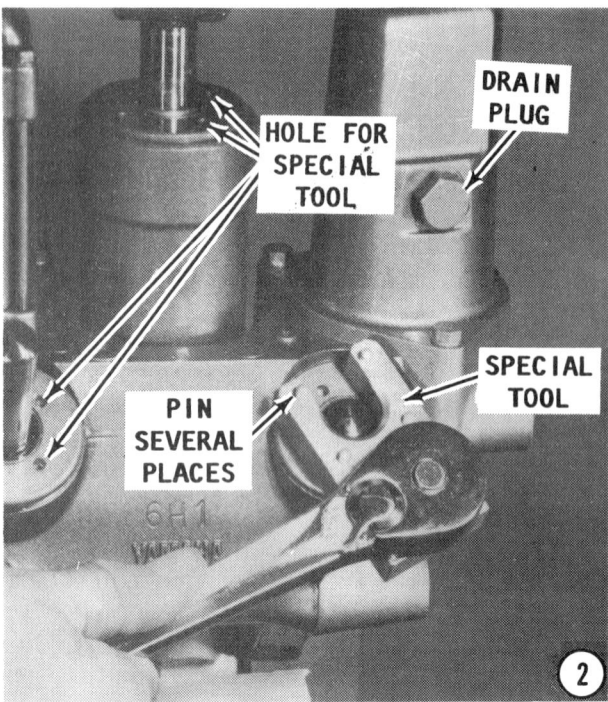

the words etched on each side of the special tool. The side with the three pins and the word "TRIM" is used to remove/install the end cap on the trim cylinders. The side with the four pins and the word "TILT" is used for the end cap on the tilt cylinder.

2- Using the trim side of the special tool, index the pins into the recess holes provided in the end cap. Index a socket wrench adaptor into the square hole in the tool and remove the end cap from the trim cylinder to be serviced. As mentioned several times in this section, this is not an easy task, but under certain favorable conditions it can be accomplished.

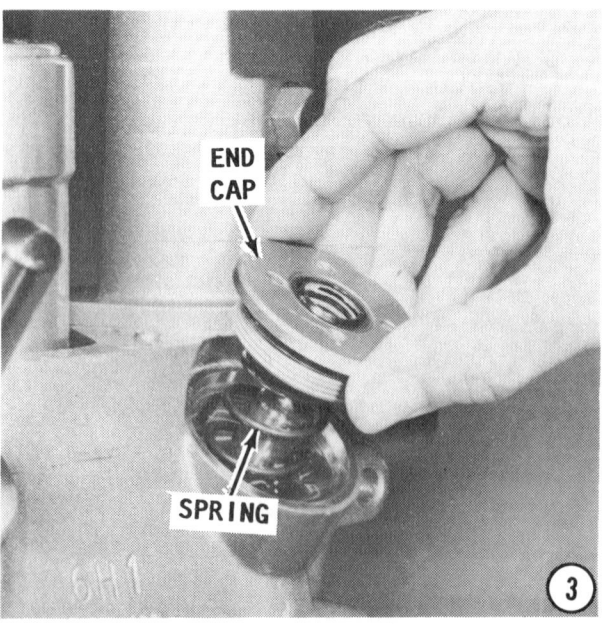

3- Remove the end cap and spring. Early models before 1987 may not have the spring. The spring is partially attached to the end cap and will come free with the cap.

4- Withdraw the trim piston straight up and out of the cylinder. Repeat Steps 2 thru 4 to remove the other trim piston.

SPECIAL WORDS

If the tilt piston must be removed from the cylinder for servicing, the trim/tilt unit must be removed from the clamp bracket assembly. This task is accomplished by first removing the end caps from both ends of the pivot pin, and then removing the pin. The pin can **ONLY** be removed from the port side out the starboard side. After the pin is free, remove the attaching hardware securing the trim/tilt assembly in the clamp bracket, and remove the assembly. Clamp the unit in a vise equipped with soft jaws or pad the jaws with a couple pieces of wood. The vise will provide stability and access to almost all parts.

5- After the trim/tilt unit has been removed from the clamp bracket and clamped in a vise, as described in the previous paragraph, the tilt end cap may be removed. Using the tilt side of the same tool as for the trim cylinder end cap, index the pins into the recess holes provided in the end cap. Remove the end cap in the same

manner with a socket adaptor and wrench as was used for the end cap of the trim cylinders.

6- Withdraw the tilt piston straight up and out of the cylinder.

7- Remove the circlip or Truarc snap ring, depending on the model being serviced, and then remove the manual release head. On early model units, prior to 1987, the threads of the release head were standard right hand. However, in 1987 the threads were changed to **LEFT HAND**.

Therefore, as the valve head is rotated, watch to see in which direction the head "lifts" as it is rotated to determine left or right hand threads. The inner parts of the manual release valve, the ball, release rod, seat, spring, and pin, are all secured by the valve seat screw.

Removal of this screw is extremely difficult. Without good cause, an attempt to remove the seat screw should not be made. Individual replacement parts are not available for the items behind the screw. Therefore, the only gain in removing them would be cleaning. In most cases replacement of the O-ring in the lower groove of the manual release head will solve a problem in this area.

NOW, THESE WORDS

In the majority of cases, service of the hydraulic items removed thus far will solve any rare problems encountered with the trim/tilt system.

The reservoir can be removed through the attaching hardware and cleaned if the

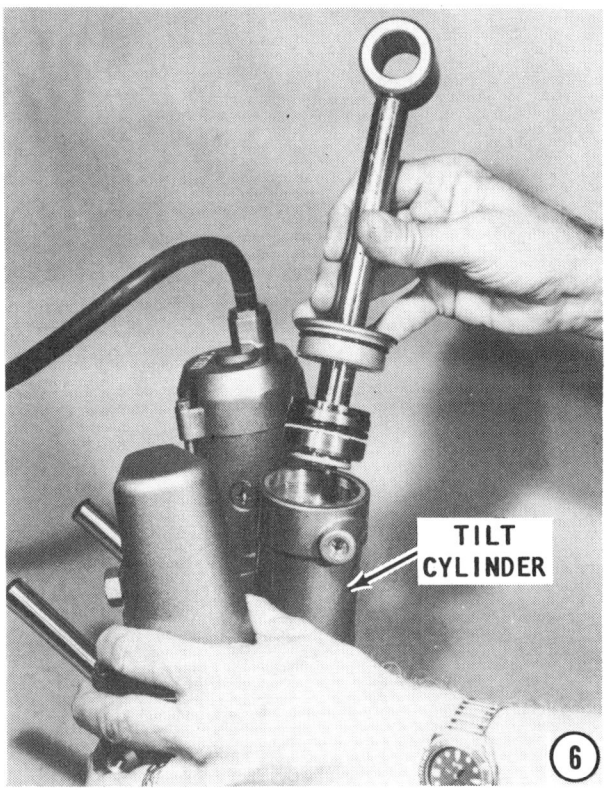

system was considered contaminated with foreign material, which is highly unlikely, because the system is a "closed" system. The only route for entry of foreign material would be through the fill opening.

Further disassembly and service to the system would best be left to a shop properly equipped with the proper test equipment and trained personnel with the expertise to work with high pressure hydraulic systems.

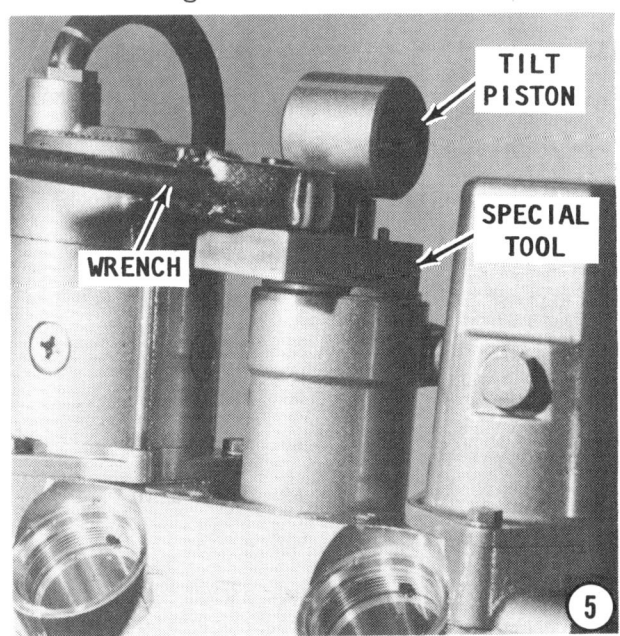

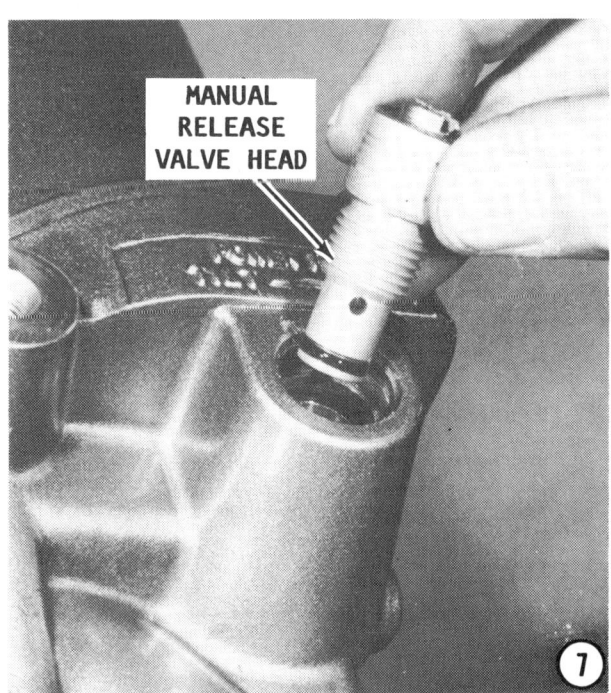

10-10 TRIM/TILT

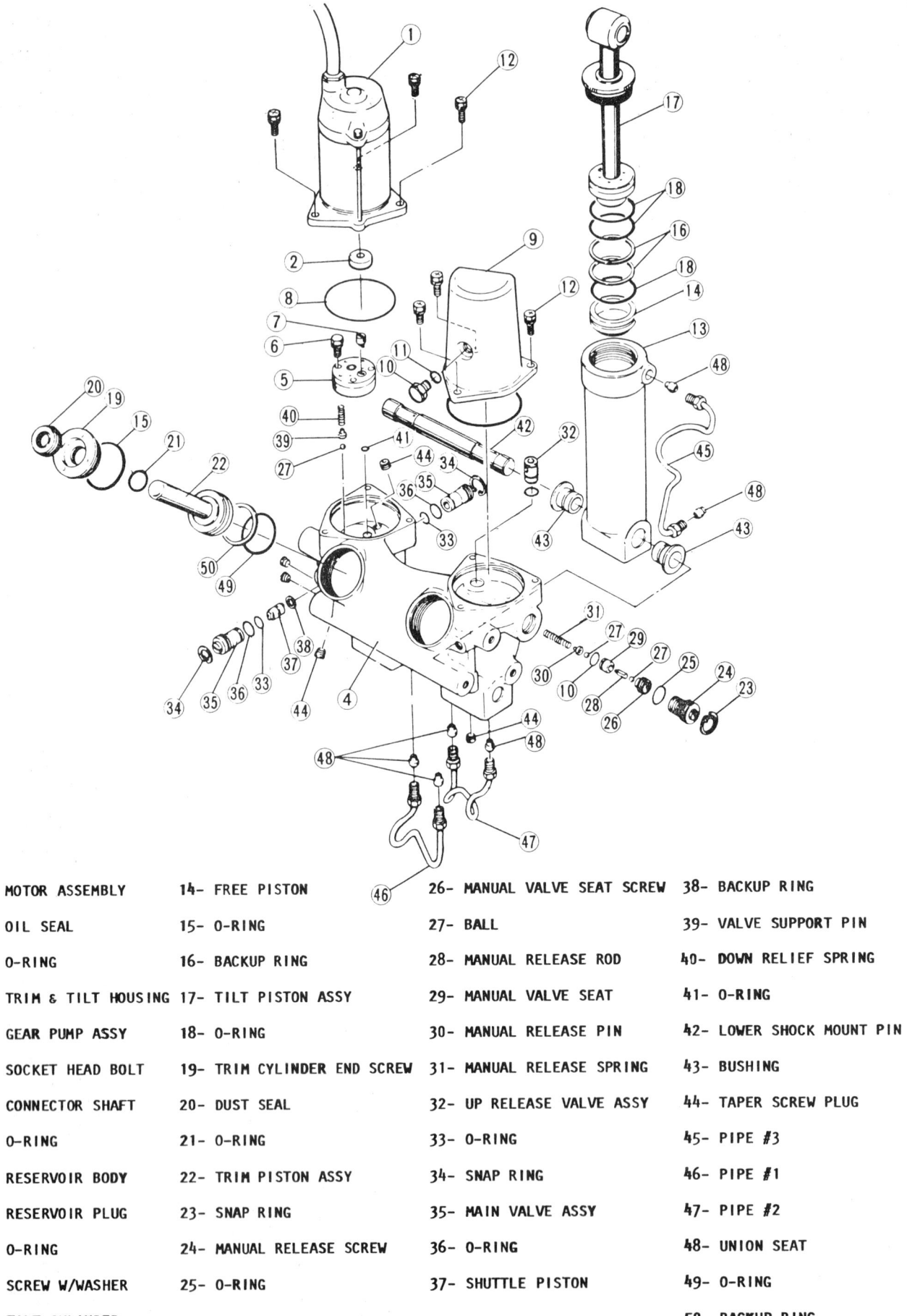

1- MOTOR ASSEMBLY
2- OIL SEAL
3- O-RING
4- TRIM & TILT HOUSING
5- GEAR PUMP ASSY
6- SOCKET HEAD BOLT
7- CONNECTOR SHAFT
8- O-RING
9- RESERVOIR BODY
10- RESERVOIR PLUG
11- O-RING
12- SCREW W/WASHER
13- TILT CYLINDER
14- FREE PISTON
15- O-RING
16- BACKUP RING
17- TILT PISTON ASSY
18- O-RING
19- TRIM CYLINDER END SCREW
20- DUST SEAL
21- O-RING
22- TRIM PISTON ASSY
23- SNAP RING
24- MANUAL RELEASE SCREW
25- O-RING
26- MANUAL VALVE SEAT SCREW
27- BALL
28- MANUAL RELEASE ROD
29- MANUAL VALVE SEAT
30- MANUAL RELEASE PIN
31- MANUAL RELEASE SPRING
32- UP RELEASE VALVE ASSY
33- O-RING
34- SNAP RING
35- MAIN VALVE ASSY
36- O-RING
37- SHUTTLE PISTON
38- BACKUP RING
39- VALVE SUPPORT PIN
40- DOWN RELIEF SPRING
41- O-RING
42- LOWER SHOCK MOUNT PIN
43- BUSHING
44- TAPER SCREW PLUG
45- PIPE #3
46- PIPE #1
47- PIPE #2
48- UNION SEAT
49- O-RING
50- BACKUP RING

Exploded drawing of a typical power trim/tilt system, with major parts identified. Engineering changes have been made to the internal routing of hydraulic fluid and the number and location of internal valves on various models. However the procedures listed in this chapter are valid and any differences are clearly indicated.

ASSEMBLING 10-11

CLEANING AND INSPECTING

Keep the work area as clean as possible to prevent contamination through foreign material entering the system on the parts to be installed.

Clean all parts thoroughly with solvent and blow them dry with compressed air.

Carefully inspect the trim pistons and the tilt piston for any sign of damage.

Purchase, if available, new O-rings and discard the old items, but **ONLY** after the replacement O-ring is verified as correct for the intended installation.

ASSEMBLING

The following procedures outline the steps required to install the parts removed in the disassembling section. Make every effort to keep the parts as clean as possible during the installation work to prevent contaminating the system.

Good shop practice dictates new O-rings be installed anytime the unit is disassembled and the rings are exposed.

1- Coat a new O-ring with Yamalube Power Trim and Tilt Fluid or a good grade of automatic transmission fluid, and then install the O-ring into the groove in the manual release head. The accompanying illustration shows two different type heads used on the trim/tilt units installed on the outboard units covered in this manual. As mentioned several times, the manual release valve is one area of constant engineering changes.

2- Insert and thread the head into the manual release valve opening. **REMEMBER** some units are standard right hand threads and others have **LEFT HAND** threads. Tighten the head just snug because it will be rotated for power tilt and manual tilt operation.

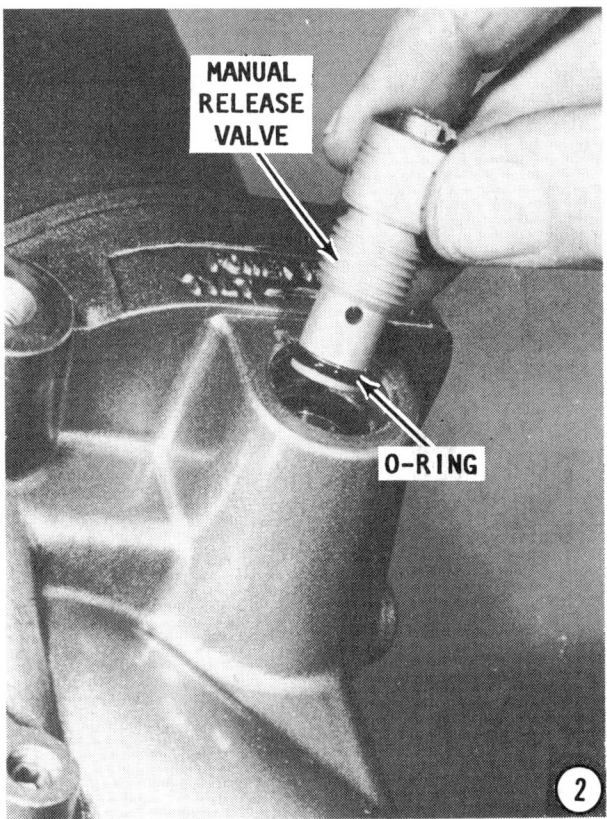

3- Secure the head in the trim/tilt housing with a circlip or Truarc snap ring, depending on the trim/tilt unit being serviced.

4- Check to be sure the back up rings are in place in the groove of the tilt piston and the free piston, or install the rings if they were removed. Apply a coating of Yamalube Power Trim and Tilt Fluid to new O-rings, and then install the O-rings into the grooves of the tilt piston, free piston, and the tilt piston end cap.

With the piston rod facing **UP** the O-ring of the free piston **MUST** be installed under the back up ring. The O-ring of the tilt piston **MUST** be installed on top of the back up ring.

Insert the tilt piston into the piston and thread the end cap into the housing.

5- Clamp the trim/tilt housing in a vise

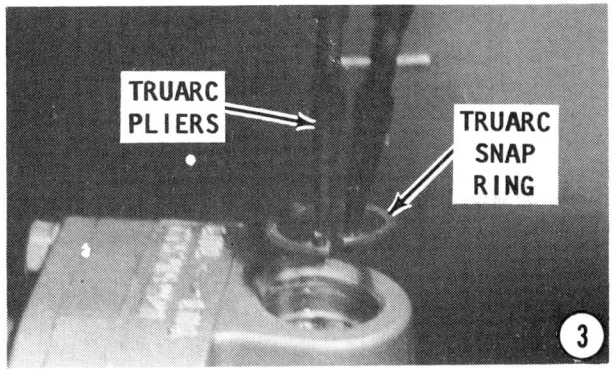

10-12 TRIM/TILT

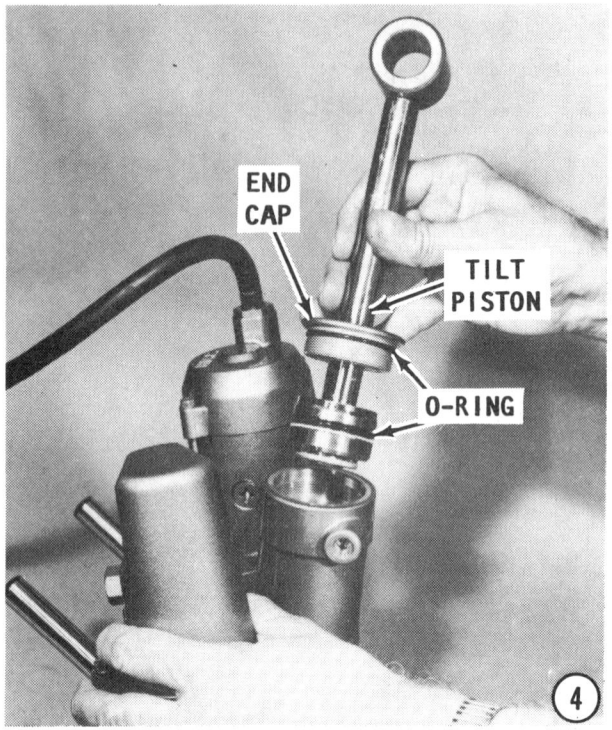

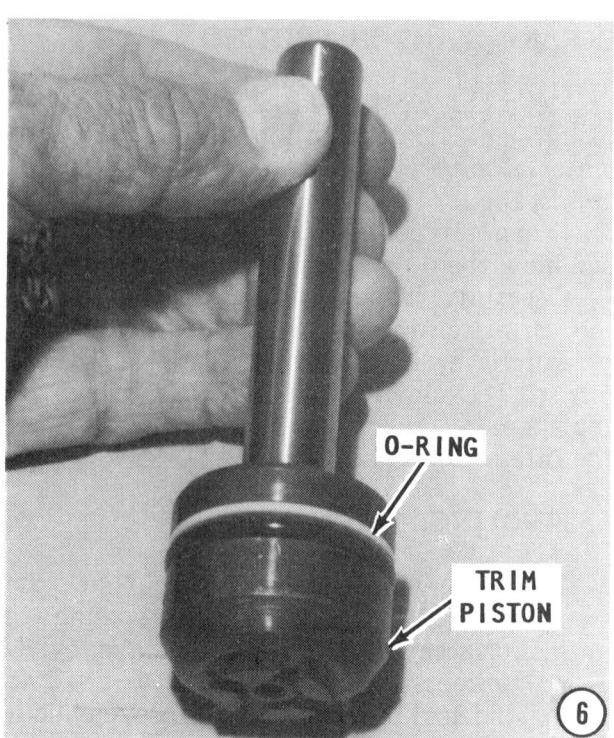

equipped with soft jaws or pad the jaws with a couple pieces of wood. Using the tilt side of Yamaha special tool YB6175 with the pins indexed into the recess holes of the end cap, tighten the end cap to a torque value of 61 ft lbs (85Nm).

6- Check to be sure the back up ring is properly installed in the trim piston groove, or install the ring if it was removed. Coat a new O-ring with Yamalube Power Trim and Tilt Fluid, and then install the O-ring into the groove of the trim piston. With the trim piston rod facing **UP**, the O-ring **MUST** be installed **UNDER** the back up ring.

7- Insert the trim piston straight down into the cylinder. Push the piston as far down as possible. Repeat Steps 6 and 7 for the other trim piston.

8- Coat a new O-ring with Yamalube Power Trim and Tilt Fluid, and then install the O-ring into the groove of the trim end cap. On models since early 1987, check to be sure the spring is properly seated. If the two seals in the end cap were removed, coat

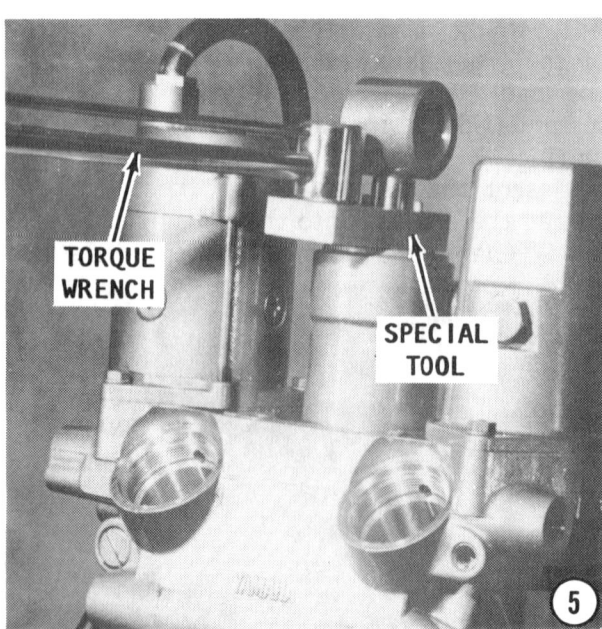

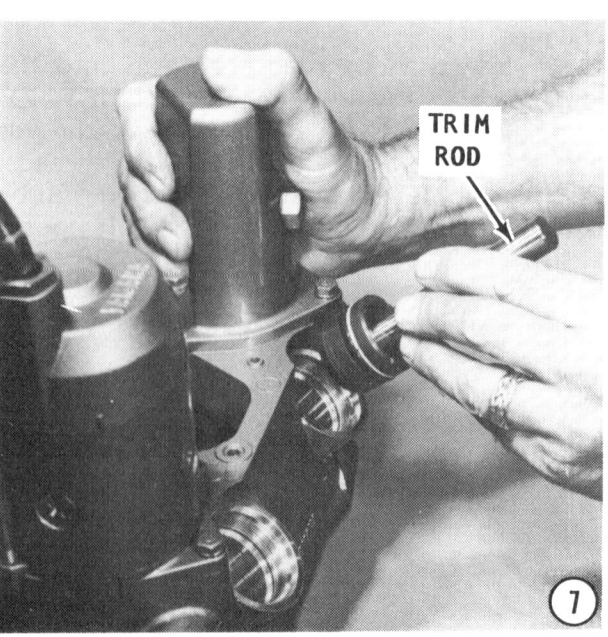

PURGING SYSTEM 10-13

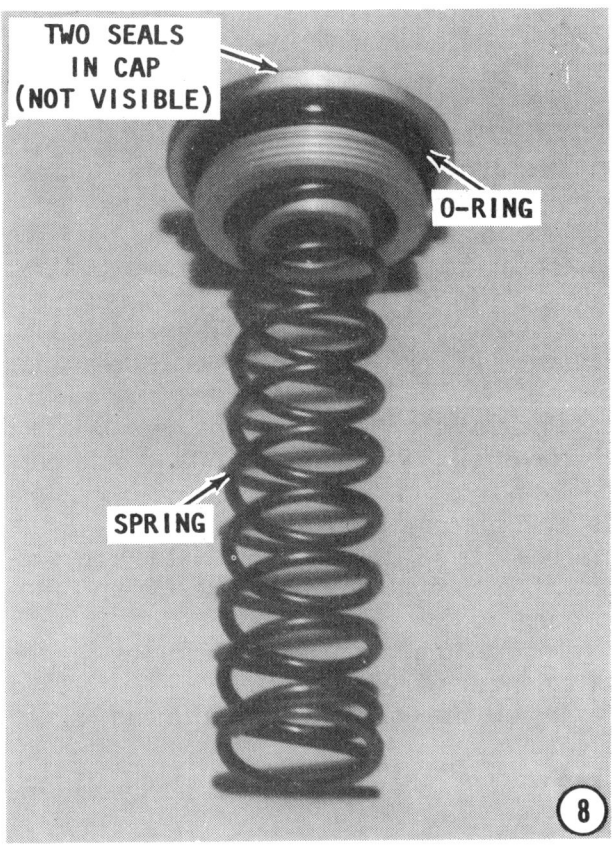

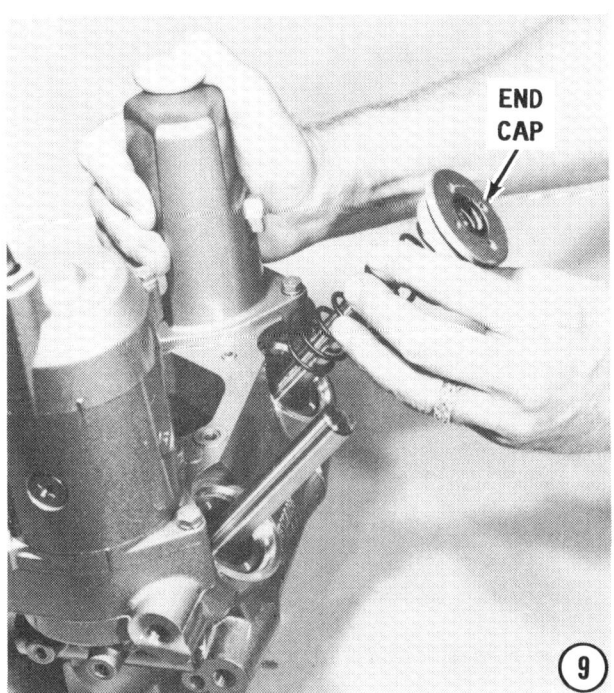

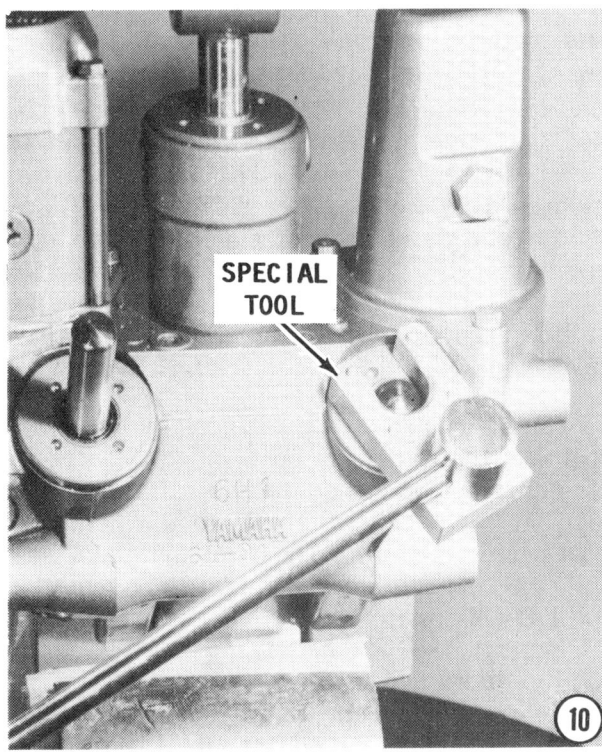

new seals with hydraulic fluid, and then insert them into the individual grooves of the end cap.

9- Slide the spring down over the trim piston rod, and then thread the end cap into the trim/tilt housing.

10- Using the trim side of Yamaha special tool YB6175 with the pins indexed into the recess holes of the end cap, tighten the end cap to a torque value of 61 ft lbs (85Nm).

Repeat Steps 8 thru 10 for the other trim cylinder.

CLOSING TASKS

Unit NOT Removed

If the trim/tilt unit was **NOT** removed from the clamp brackets, fill the system with Yamalube Power Trim/Tilt Fluid or a good grade of Automatic Transmission Fluid according to the procedures outline in Section 10-3, beginning on Page 10-4.

Purge the system of air. See purging procedures beginning in the following paragraph.

10-6 PURGING AIR FROM THE SYSTEM UNIT REMOVED

If the trim/tilt unit **WAS** removed from the clamp bracket, the system is filled and purged with the unit on the bench. Fill the system with Yamalube Power Trim/Tilt Fluid or a good grade of Automatic Transmission Fluid.

Purge air from the system by first pulling the tilt cylinder to its fully extended position. Check the fluid level and add fluid as required to bring the level up to the bottom of the fill opening.

Next, rotate the manual release valve to the **POWER TILT** position, approximately three full turns from the manual position.

Now, obtain a 12 volt battery and connect the Blue lead from the trim/tilt unit to the positive lead from the battery. Now, momentarily make contact with the Black lead from the unit to the negative post of the battery.

The trim cylinders should move to the fully extended position. Check the fluid level and add fluid, as required.

Finally, connect the Green lead from the unit to the positive post on the battery. Momentarily make contact with the Black lead from the trim/tilt unit to the negative post of the battery. The trim cylinders and and the tilt cylinder should retract to the full **DOWN** position.

Repeat the extension and retraction of the cylinders two or three times.

To check for air in the system, extend the tilt cylinder, and then apply a downward heavy force manually to the end of the piston. The piston should feel solid. If the piston retracts more than about 1/8" (3.2mm), the system still contains air. Perform the purging sequence until the tilt cylinder is solid.

Install the unit between the clamp bracket. Begin by checking to be sure the sleeve for the piston rod end is in place, and the insert in both port and starboard bracket arms are in place. Coat the inside surfaces of the sleeve and inserts with Yamalube Grease "A" or equivalent water resistant grease.

Position the trim/tilt in place and insert the pivot pin. Remember, the pin could only be removed in one direction. Now, the reverse is true, it must be installed from the starboard side, through the piston rod end, and then through the port side clamp bracket arm.

Secure the trim/tilt housing with the attaching hardware.

Install the end caps onto both ends of the pivot pin.

Perform an operational check of the system.

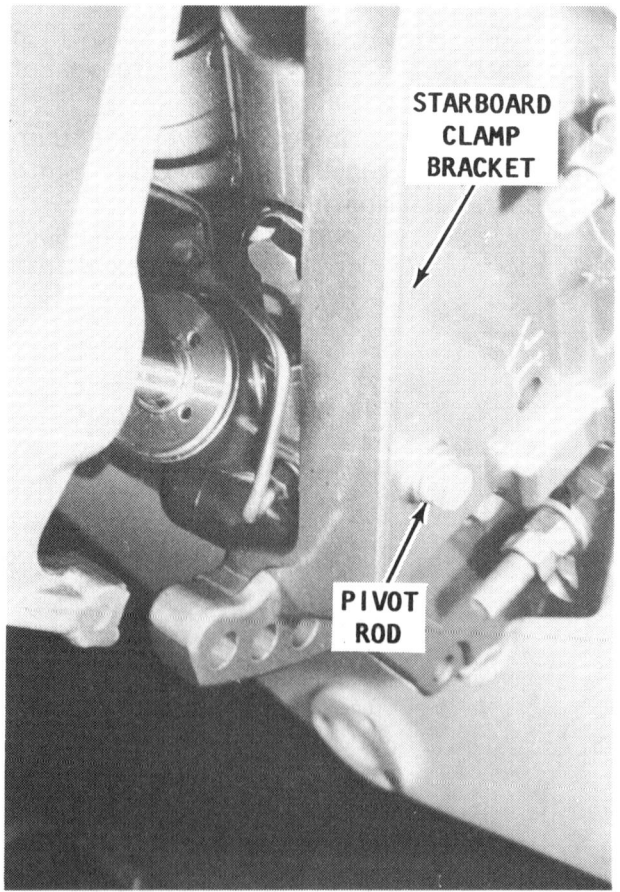

Port side view of a typical trim/tilt system installation.

Starboard side view of a typical trim/tilt system installation.

11
LOWER UNIT

11-1 DESCRIPTION

The lower unit is considered as the part of the outboard below the exhaust housing.

Two distinctly different lower units, identified as Type "A" and Type "B" are covered in this chapter with separate sections for each.

The Type "A" is a conventional propeller driven lower unit which includes both the standard model and the counterrotating model installed on some 150hp and 200hp outboards. This lower unit is covered in Section 11-2 thru 11-8 beginning on the next page.

The Type "B" lower unit is a jet drive propulsion system. Water is drawn in from the forward edge of the lower unit and forced out under pressure to propel the boat forward. This type lower unit is installed on some 115hp and 200hp units. Detailed procedures for this Type "B" unit are covered in Section 11-9, beginning on page 11-47.

Each section is presented with complete detailed instructions for removal, disassem-

A dual engine installation. The portside unit has a standard lower unit, and the starboard unit has the counterrotating unit. Special procedures for the counterrotating unit are given and clearly indicated throughout this chapter.

11-2 LOWER UNIT

bly, cleaning and inspecting, assembling, adjusting, and installation of only one unit. There are no cross-references. Each section is complete from removal of the first item to final test operation.

Shifting

Shifting for the Type "A", both standard and counterrotating lower units, is accomplished through a cable arrangement from a shift box, installed near the helm, to the engine. The cable hookup involves two individual cables, one for shifting and the other for throttle control.

Shifting for the Type "B" lower unit is accomplished through a small lever mounted on the inside of the transom. Movement can be controlled from the control box through a variety of arrangements at the owner's discretion.

The lower unit may be removed and serviced without disturbing the remainder of the outboard motor.

11-2 TYPE A STANDARD AND COUNTER-ROTATING LOWER UNIT

DESCRIPTION AND OPERATION

A single design shifting mechanism is employed on both the standard and counterrotating units, with the counterrotating shift mechanism being a "mirror image" of the standard shift mechanism.

On the standard lower unit, the propeller rotates **CLOCKWISE** and employs a standard propeller. On the counterrotating lower unit, the propeller rotates in the opposite direction -- **COUNTERCLOCKWISE**. These units use a left-hand propeller. The counterrotating lower unit is identified with a decal on the lower unit housing and the letter "L" embossed on the aft side just above the trim tab.

Actually, the lower unit housing casting is identical for both the standard and the counterrotational units. The main physical differences are in the shifting mechanism (mirror image) and the propeller shaft. The counterrotating unit has a two piece propeller shaft. Another difference regards nomenclature: what would be the forward gear on a standard unit becomes the reverse gear on a counterrotating unit and what would normally be the reverse gear on a standard unit, becomes the forward gear on a counterrotating unit. Thankfully, the pinion gear remains the same and driveshaft rotation remains the same.

SPECIAL WORDS

All threaded parts are right-hand unless otherwise indicated.

If there is any water in the standard or counterrotating lower unit or metal particles are discovered in the gear lubricant,

Standard lower unit serviced and ready for a day's work on the water for its owner.

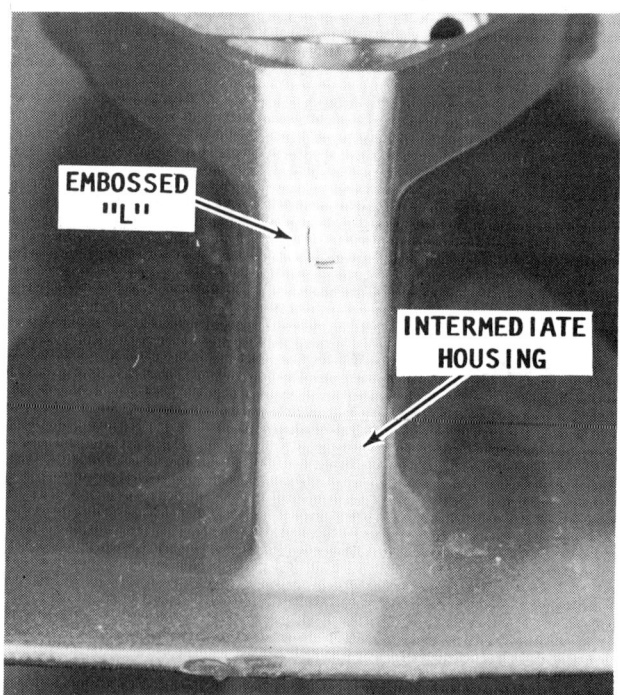

A counterrotating lower unit has the letter "L" embossed on the leading edge of the intermediate housing.

the lower unit should be completely disassembled, cleaned and inspected.

11-3 STANDARD LOWER UNIT SHIFTING PRINCIPLES

The standard lower unit is equipped with a clutch "dog" permitting operation in **NEUTRAL, FORWARD,** and **REVERSE.**

When the unit is shifted from neutral to reverse, the shift rod is rotated. This action moves the plunger and clutch "dog" toward the reverse gear.

When the unit is shifted from reverse to neutral and then to forward gear, the rotation of the shift rod is reversed and the clutch "dog" is moved toward the forward gear.

The shift mechanism design consists of a cam on the shift rod, a shifter, a shift slide, a compression spring, and six steel balls: two small, two medium, and two large. The shift slide is a hollow tube, closed at one end. The closed end of this tube is "necked", and the "head" or knob of the tube fits into the shifter. The shifter has a vertical hole all the way through it. The shift rod passes through this hole in the shifter. The cam in the shift rod engages the shifter at the side notch, so any rotating movement in the shift rod will translate to a back and forth movement in the shifter, and the shift slide.

The shift slide rests inside the propeller shaft. If the shift slide were to be inserted into the propeller shaft without any of the six balls, it would be a snug fit. But, there are grooves in the internal recess of the propeller shaft and there are holes drilled into the shift slide at strategic locations.

The six balls inside the shift slide are held under tension by the compression spring and arranged to protrude from the holes drilled in the slide. As the shift slide moves back and forth within the propeller shaft, four of the balls seat themselves in the grooves of the propeller shaft, to hold a certain gear position.

NEUTRAL

In the **NEUTRAL** position, the cam is centered in the shifter and the two medium steel balls are settled in internal grooves in the recess of the propeller shaft.

FORWARD GEAR

When the lower unit is shifted into **FORWARD** gear, the shift cam rotates in a counterclockwise direction and pulls on the shifter and shift slide. Because the shifter and shift slide are connected, they are moved out of the propeller shaft -- moving the clutch dog toward the **FORWARD** gear. The steel balls are forced out of their grooves and apply a torsional force on the shift slide. The shift into **FORWARD** gear is complete.

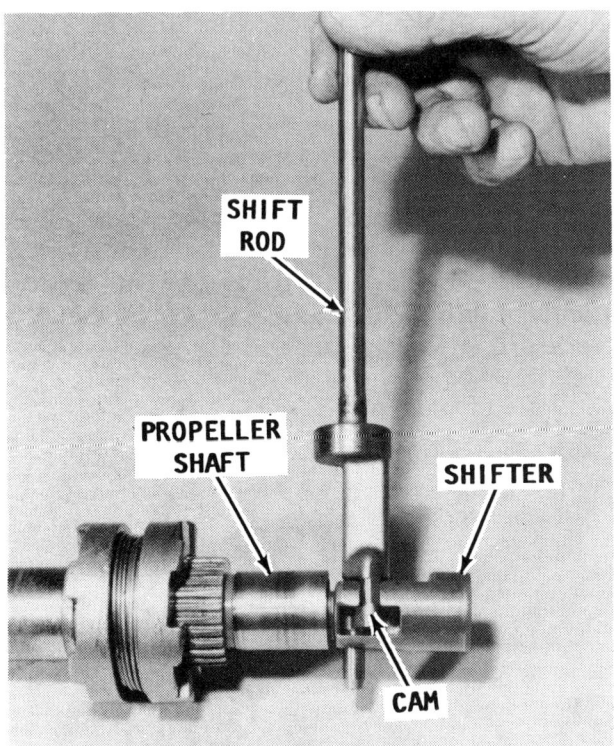

The shifting mechanism removed from a standard lower unit with major parts visible.

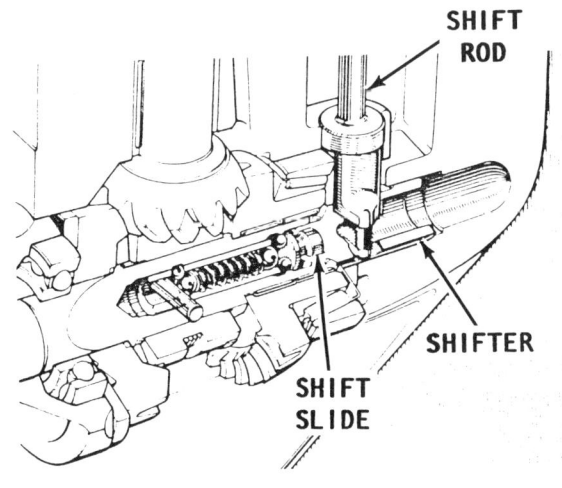

Cutaway drawing of a standard lower unit. This type illustration assists in understanding the relationship of the internal working parts.

REVERSE GEAR

When the lower unit is shifted into **REVERSE** gear, the unit first moves into the **NEUTRAL** position. The shift cam rotates in a clockwise direction. The cam pushes the shifter and the shift slide in toward the propeller shaft. The clutch "dog" moves toward the **REVERSE** gear and two steel balls are moved past their detent groove and on to the other side. This action applies a torsional force on the shift slide. The shift movement to **REVERSE** gear is complete.

GEAR OPERATION

The pinion gear on the lower end of the driveshaft is in constant mesh with both the forward and reverse gear. These three gears constantly rotate anytime the powerhead is operating.

A sliding clutch "dog" is mounted on the propeller shaft. A shifting motion at the control box will translate into a back and forth motion at the clutch "dog" via a series of shift mechanism components. When the clutch "dog" is moved forward, it engages with the forward gear. The clutch "dog" then picks up the rotation of the forward gear. Because the clutch is secured to the propeller shaft with a pin, the shaft rotates at the same speed as the clutch. The propeller is thereby rotated to move the boat forward.

When the clutch "dog" is moved aft, the the clutch engages only the reverse gear. The propeller shaft and the propeller are thus moved in the opposite direction to move the boat sternward.

When the clutch "dog" is in the neutral position, neither the forward nor reverse gear is engaged with the clutch and the propeller shaft does not rotate.

From this explanation, an understanding of wear characteristics can be appreciated. The pinion gear and the clutch "dog" receive the most wear, followed by the forward gear, with the reverse gear receiving the least wear.

All three gears, the forward, reverse and pinion, are spiral bevel type gears.

A mixture of ball bearings, tapered roller bearings, and caged or loose needle bearings is used in each unit. The type bearing used is clearly indicated in the procedures.

Cutaway photograph of a counterrotating lower unit with the clutch "dog" in the NEUTRAL position.

11-4 COUNTERROTATING LOWER UNIT SHIFTING PRINCIPLES

As mentioned earlier in this chapter a single design shifting mechanism is employed on both the standard and counterrotating units, with the counterrotating shift mechanism being a "mirror image" of the standard shift mechanism.

The main physical differences lie in the shifting mechanism (mirror image) and the propeller shaft. The counterrotating unit has a two piece propeller shaft. Another difference regards nomenclature: what would be the forward gear on a standard unit becomes the reverse gear on a counterrotating unit and what would normally be the reverse gear on a standard unit, becomes the forward gear on a counterrotating unit. The pinion gear remains the same and driveshaft rotation remains the same as on a standard lower unit.

If Section 11-3 is read and understood, then it is easy to see how a "mirror image" shifting mechanism would produce counter rotation of the propeller shaft under these conditions. This type lower unit consists of the same major identical components as the standard unit (with the exception of the two piece propeller shaft).

Cutaway photograph of a counterrotating lower unit with the clutch "dog" in the FORWARD position.

Cutaway photograph of a counterrotating lower unit with the clutch "dog" in the REVERSE position.

On a standard lower unit, the cam on the shift shaft is located on the starboard side of the shifter. Therefore, when the rod is rotated counterclockwise, the clutch dog is pulled forward and the forward gear is engaged.

On a counterrotating lower unit, the cam on the shift rod is located on the port side of the shifter. Therefore, when the rod is rotated counterclockwise, the clutch dog is pushed back and the gear in the aft end of the housing (which normally is the reverse gear) is engaged. In this manner, the rotation of the propeller shaft is reversed. The same logic applies to the selection of reverse gear.

CRITICAL WORDS

Counterrotational shifting is accomplished **WITHOUT** modification to the shift cable at the shift box. The normal setup is essential for correct shifting. The only special equipment the counterrotating unit requires is the installation of a left hand propeller.

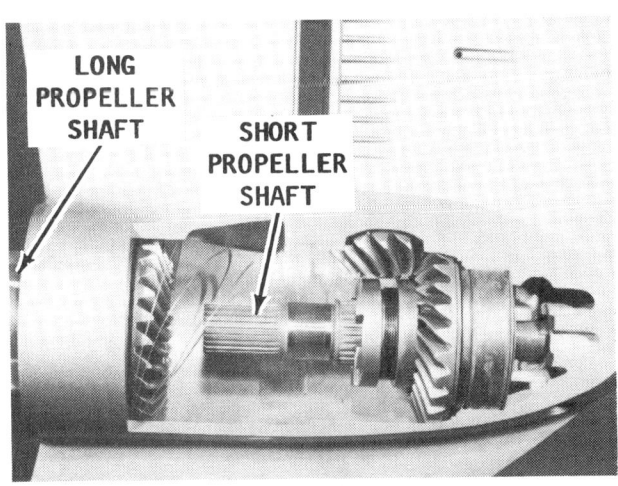

A counterrotating lower unit has a two piece propeller shaft -- a long section with the propeller attached, and a short internal section.

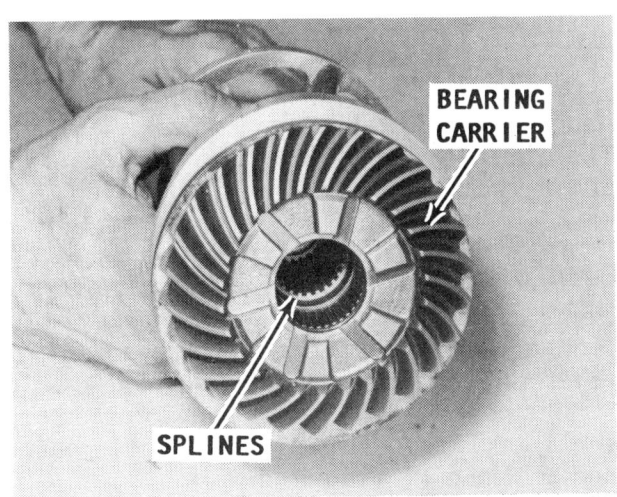

Bearing carrier of a counterrotating lower unit with the internal splines for the short section of propeller shaft clearly visible.

11-5 EXHAUST GASES

At low powerhead speed, the exhaust gases from the powerhead escape from an idle hole in the intermediate housing. As powerhead rpm increases to normal cruising rpm or high speed rpm, these gases are forced down through the intermediate housing and lower unit, then out with cycled water through the propeller.

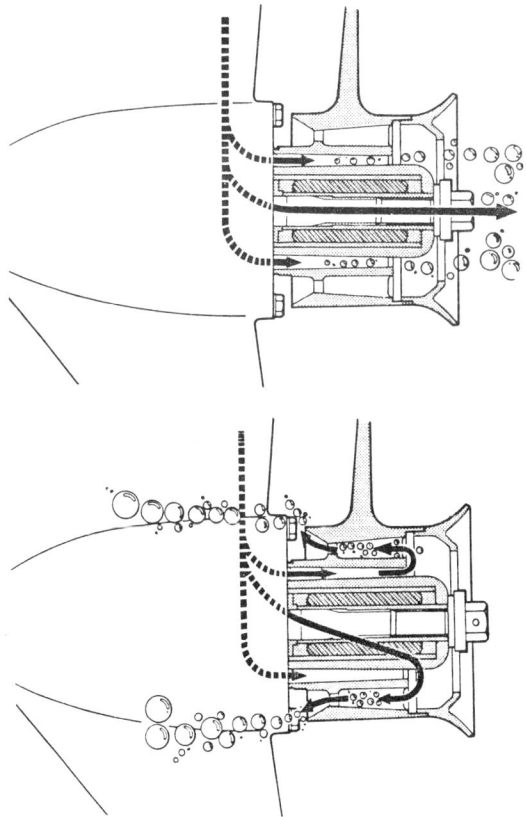

Cross-section drawing of the lower unit showing route of the exhaust gases with the unit in forward gear (top), and in reverse gear (bottom).

11-6 LOWER UNIT

Water for cooling is pumped to the powerhead by the water pump and is expelled with the exhaust gases. The water pump impeller is installed on the driveshaft. Therefore, the water output of the pump is directly proportional to powerhead rpm.

WORDS OF WISDOM

Procedural steps are given to remove all items in the lower unit. However, do **NOT** remove bearings, bushings, or seals, if a determination can be made the item is fit for further service. As the work progresses, simply skip the steps involving these parts and continue with required work.

Before beginning work on the lower unit, take time to **READ** and **UNDERSTAND** the information presented in Section 11-2 or 11-3, whichever applies to the unit being serviced.

Disconnect the high tension spark plug leads and remove the spark plugs before working on the lower unit.

11-6 TROUBLESHOOTING TYPE "A" STANDARD AND COUNTERROTATING PROPELLER DRIVE LOWER UNIT

Troubleshooting **MUST** be done **BEFORE** the unit is removed from the powerhead to permit isolating the problem to one area. Always attempt to proceed with troubleshooting in an orderly manner. The shot-in-the-dark approach will only result in wasted time, incorrect diagnosis, frustration, and unnecessary replacement of parts.

The following procedures are presented in a logical sequence with the most prevalent, easiest, and less costly items to be checked listed first.

1- Check the propeller and the rubber hub. See if the hub is shredded. If the propeller has been subjected to many strikes against underwater objects, it could slip on its hub. If the hub appears to be damaged, replace it with a **NEW** hub. Replacement of the hub must be done by a propeller rebuilding shop equipped with the proper tools and experience for such work.

Shift Mechanism Check

2- Verify the ignition switch is **OFF**, to prevent possible personal injury, should the engine start. Shift the unit into **REVERSE** gear and at the same time have an assistant turn the propeller shaft to ensure the clutch is fully engaged. If the shift handle is hard to move, the trouble may be in the lower unit shift rod, requiring an adjustment, or in the shift box.

Isolate the Problem

3- Disconnect the remote control cable at the engine and then lift off the remote control shift cable. Operate the shift lever. If shifting is still hard, the problem is in the shift cable or control box, see Chapter 12. If the shifting feels normal with the remote control cable disconnected, the problem

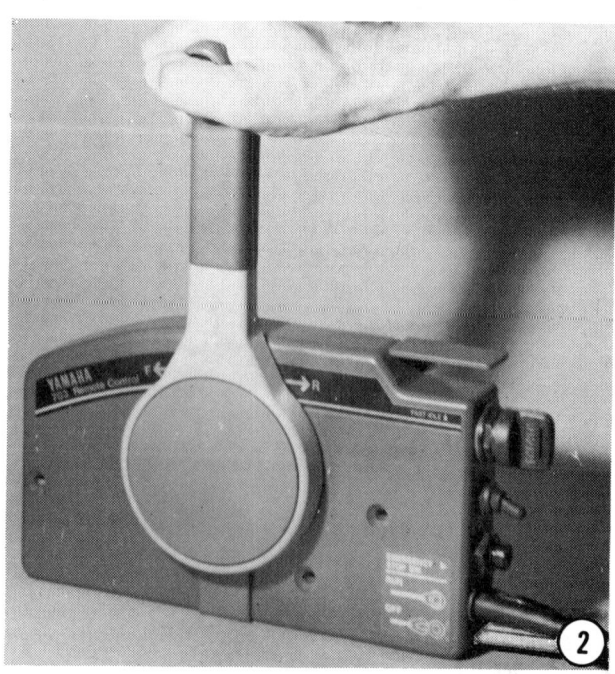

TROUBLESHOOTING 11-7

must be in the lower unit. To verify the problem is in the lower unit, have an assistant turn the propeller and at the same time move the shift cable back and forth. Determine if the clutch engages properly.

11-7 PROPELLER SERVICE

REMOVAL

1- Straighten the cotter pin and pull it free of the propeller shaft. Remove the castellated nut, washer, and outer spacer. **NEVER** pry on the edge of the propeller. Any small distortion will affect propeller performance.

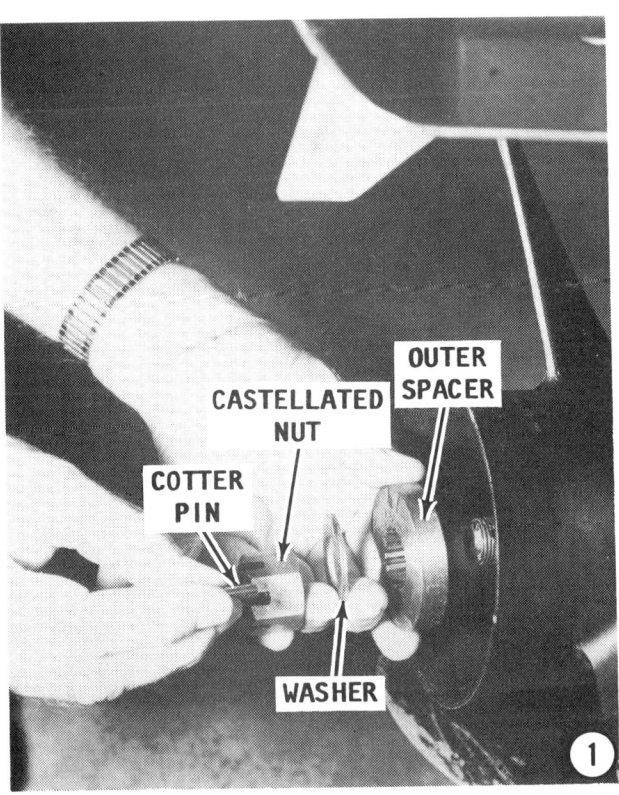

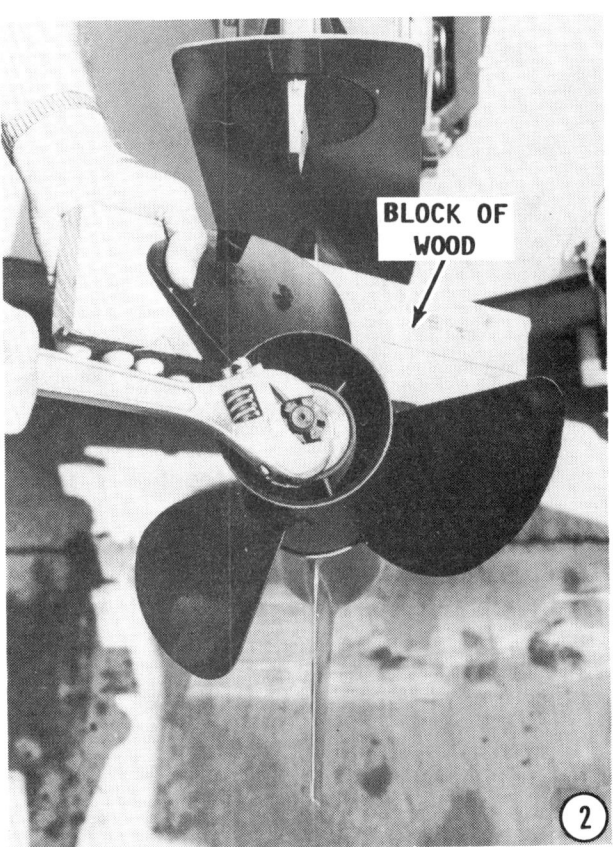

2- If the nut is "frozen", place a block of wood between one blade of the propeller and the anti-cavitation plate to keep the shaft from turning. Use a socket and breaker bar to loosen the castellated nut. Remove the nut, washer, outer spacer, and then the propeller and inner spacer.

3- If the propeller is "frozen" to the shaft, heat must be applied to the shaft to melt out the rubber inside the hub. Using heat will destroy the hub, but there is no other way. As heat is applied, the rubber will expand and the propeller will actually be blown from the shaft. Therefore, **STAND CLEAR** to avoid personal injury.

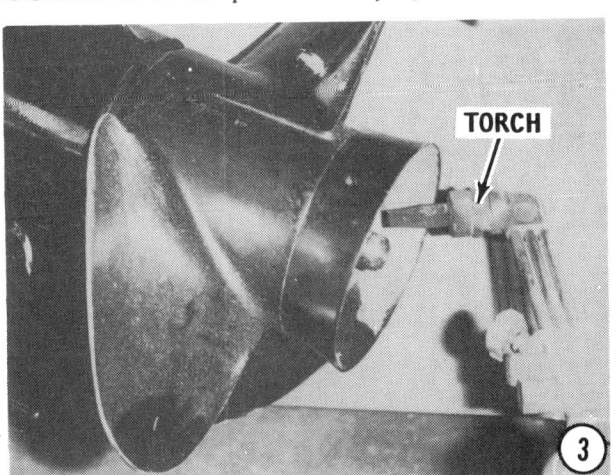

11-8 LOWER UNIT

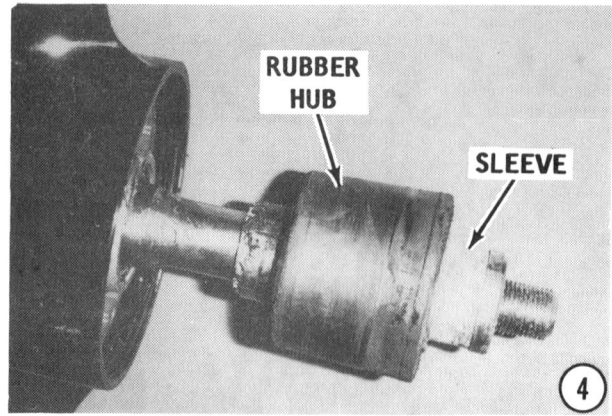

4- Use a knife and cut the hub off the inner sleeve.

5- The sleeve can be removed by cutting it with a hacksaw, or it can be removed with a puller. Again, if the sleeve is frozen, it may be necessary to apply heat. Remove the thrust hub from the propeller shaft.

INSTALLATION

GOOD WORDS

An anti-seizing compound will prevent the propeller from "freezing" to the shaft and permit propeller removal, without difficulty, the next time the propeller needs to be "pulled".

Apply Yamalube Grease "A", or equivalent anti-seizing compound, to the propeller shaft.

1- Install the inner spacer onto the propeller shaft, and then the propeller.

2- Install the outer spacer, washer, and the castellated nut. Place a block of wood between one of the propeller blades and the anti-cavitation plate to prevent the propel-

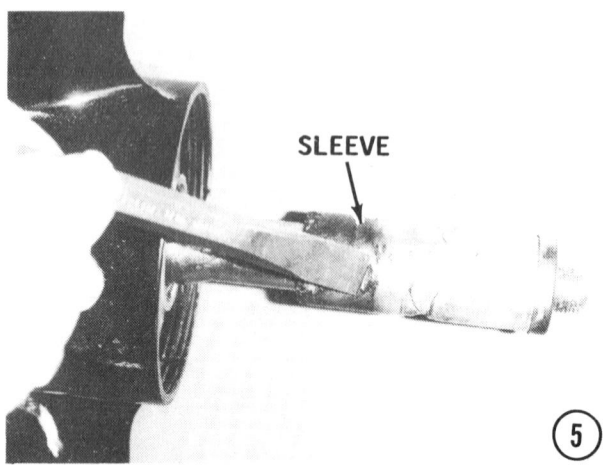

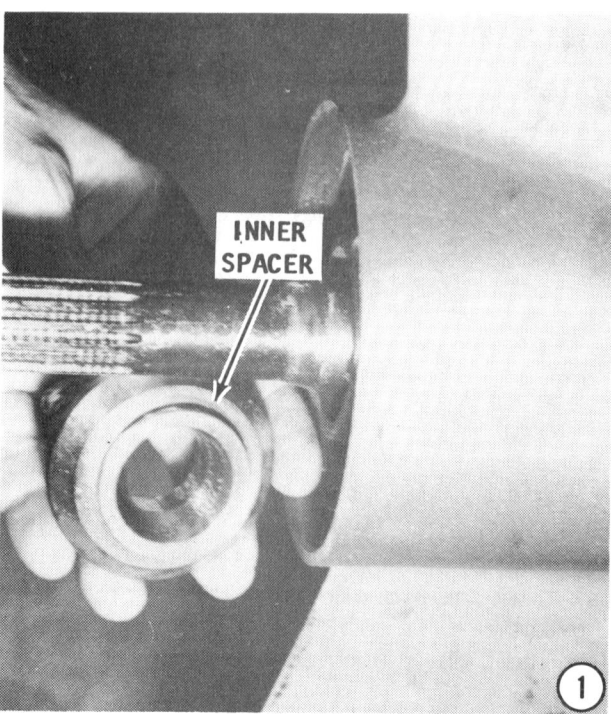

ler from rotating. Tighten the nut to a torque value of 40 ft lbs (55Nm), then back off the nut until the cotter pin can be inserted through the nut and the hole in the propeller shaft. Bend the arms of the cotter pin around the nut to secure it in place.

Remove the block of wood. Connect the spark plug wires to the spark plugs. Connect the electrical lead to the battery terminal.

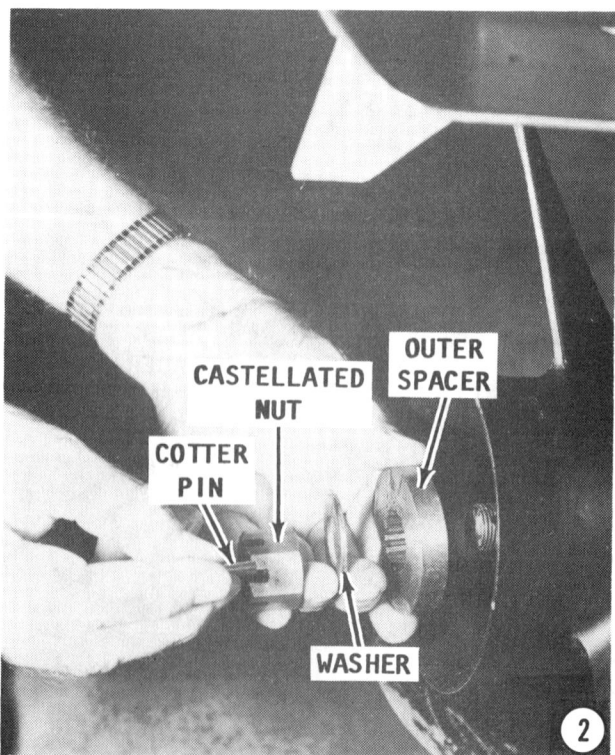

REMOVAL   11-9

TRIM TAB ADJUSTMENT

The trim tab should be positioned to enable the helmsperson to handle the boat with equal ease to starboard and port at normal cruising speed. If the boat seems to turn more easily to starboard, loosen the mounting bolt and move the trim tab trailing edge to the right. Move the trailing edge of the trim tab to the left if the boat tends to turn more easily to port.

11-8 LOWER UNIT SERVICE
PROPELLER DRIVE

LOWER UNIT REMOVAL

The following procedures present complete instructions to remove, disassemble, assemble, and adjust the lower unit of **ALL** Type "A" units covered in this manual. When the standard and counterrotating units differ, special instructions are given for each unit.

AUTHORS' WORDS

Because so many different models are covered in this manual, it would not be feasible to provide an illustration of each and every unit.

Therefore, the accompanying illustrations are of a "typical" unit. The unit being serviced may differ slightly in appearance due to engineering or cosmetic changes but the procedures are valid. If a difference should occur, the models affected will be clearly identified.

1- Mark the position of the trim tab, as an aid to installing it back in its original location.

2- Remove the plastic cap above the trim tab, and then using the correct size socket and a long extension, remove the bolt, and the trim tab.

3- Position a suitable container under the lower unit, and then remove the **OIL** screw and the **OIL LEVEL** screw. Allow the gear lubricant to drain into the container. As the lubricant drains, catch some with your fingers from time to time, and rub it between your thumb and finger to determine if any metal particles are present. If metal

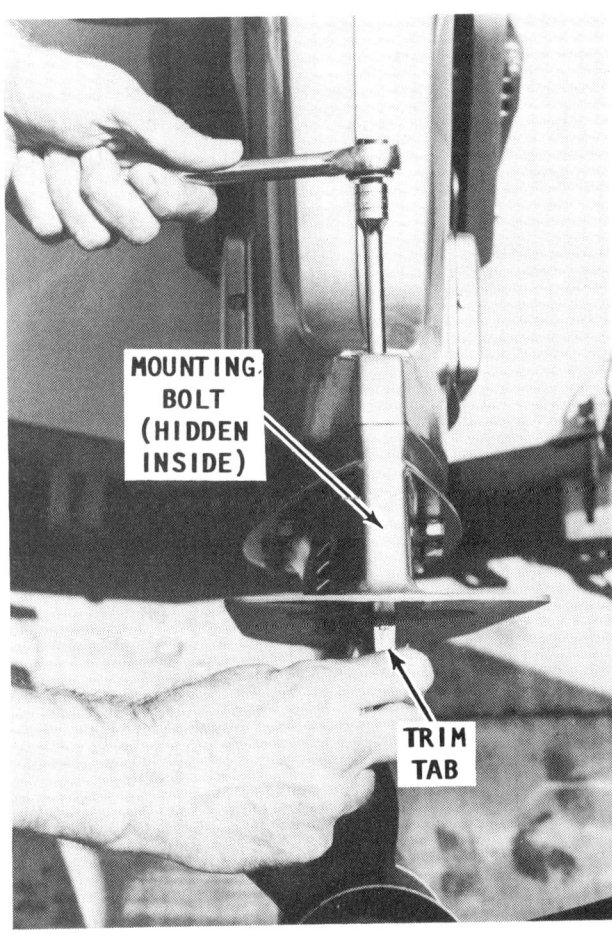

Making a trim tab adjustment, as explained in the text. The bolt is hidden inside the housing.

11-10 LOWER UNIT

separation from the lubricant. The presence of any water in the gear lubricant is "bad news". The unit must be completely disassembled, inspected, the cause of the problem determined, and then corrected.

After the lubricant has drained, temporarily install both the drain and oil level screws.

Special Words for Counterrotating Models

The first few times lower unit lubricant is drained, a discoloration may be noticed. This condition is quite normal and is no cause for alarm unless accompanied by the presence of metal chips. The discoloration is due to the use of molybdenum-disulfide assembly grease at the factory.

4- Straighten the cotter pin, and then pull it free of the castellated nut with a pair of pliers or a cotter pin removal tool. Remove the castellated propeller nut by first placing a block of wood between one of the propeller blades and the anti-cavitation plate to prevent the propeller from rotating, and then remove the nut. Remove the washer and the outer spacer. Remove the outer thrust hub from the propeller shaft. If the thrust hub is stubborn and refuses to budge, use two **PADDED** pry bars on opposite sides of the hub and work the hub loose. **TAKE CARE** not to damage the lower unit.

Remove the propeller. All models have a spacer between the bearing carrier and the propeller. Slide this spacer free of the propeller shaft.

If the propeller is "frozen" to the shaft, see Section 11-6 for special procedures to "break" it loose.

is detected in the lubricant, the unit must be completely disassembled, inspected, and the damaged parts replaced.

Check the color of the lubricant as it drains. A whitish or creamy color indicates the presence of water in the lubricant. Check the drain pan for signs of water

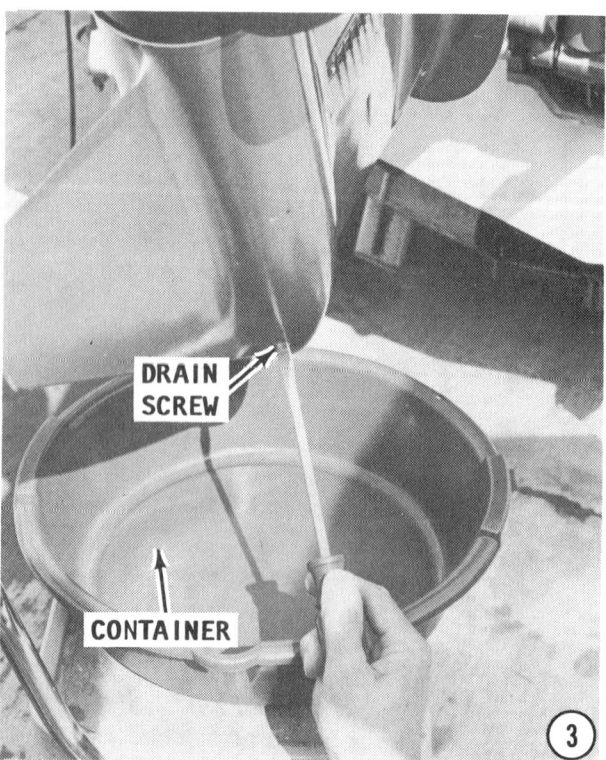

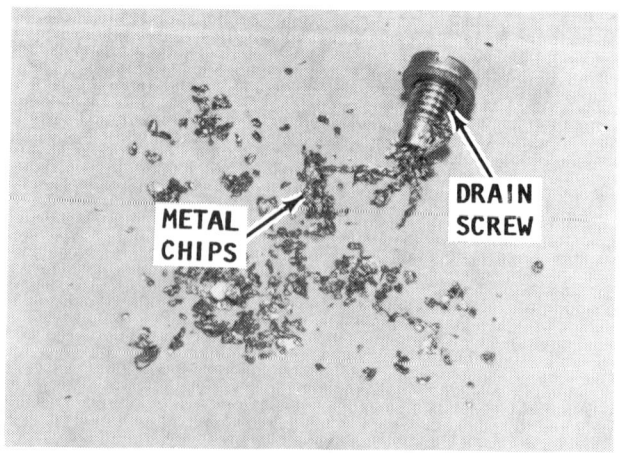

The drain screw is magnatized to pickup metal particles in the lubricating fluid. The presence of metal particles on the screw or in the fluid is BAD NEWS! The lower unit must be disassembled and the damaged parts replaced.

REMOVAL 11-11

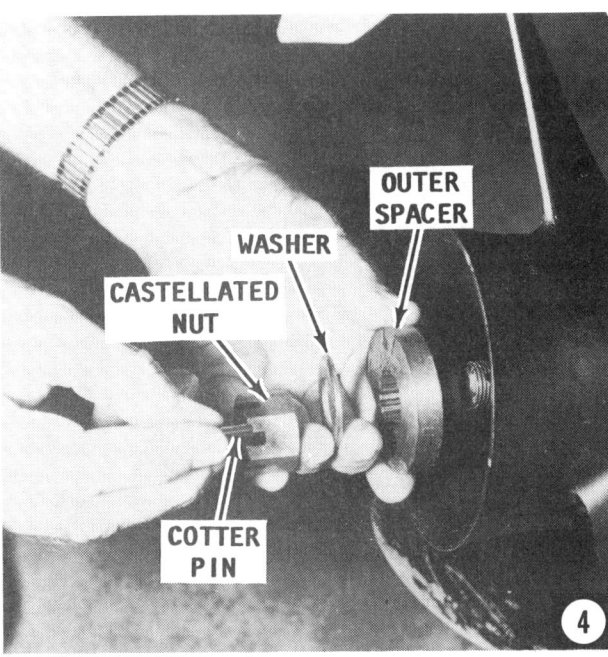

Remove the inner thrust hub. If this hub is also stubborn, use padded pry bars and work the hub loose. Again, **TAKE CARE** not to damage the lower unit.

5- Snip the band retaining the pilot tube to the swivel bracket and gently pull the two ends of the tube apart, at their connector.

Tilt and lock the outboard unit in the raised position. Check to be sure the lower unit is in neutral gear.

Separating Lower Unit
From Intermediate Housing

6- Remove the six external bolts securing the lower unit to the intermediate housing.

7- One bolt securing the lower unit to the intermediate housing is located **UNDER** the trim tab.

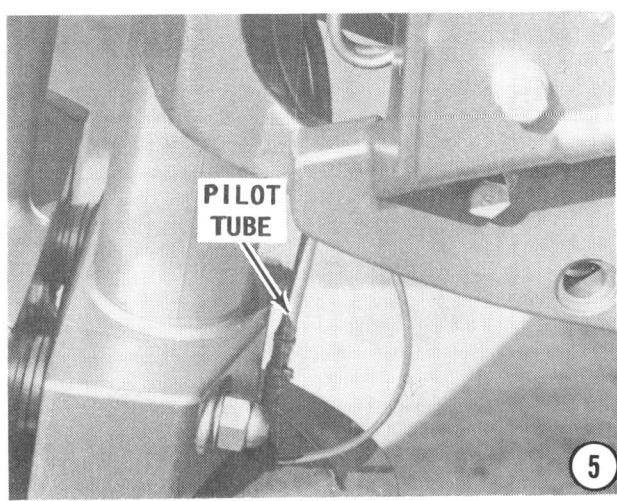

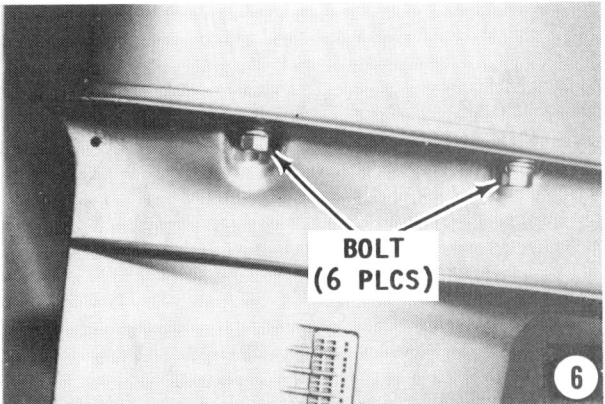

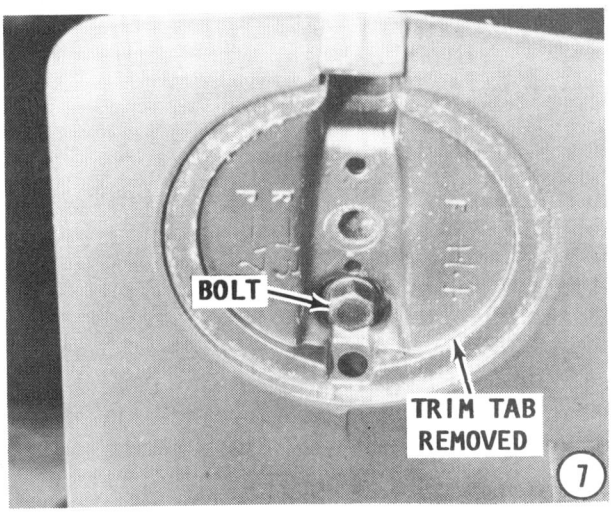

8- Separate the lower unit from the intermediate housing. **WATCH FOR** and **SAVE** the two dowel pins when the two units are separated. The water tube will come out of the grommet and remain with the intermediate housing. The driveshaft will remain with the lower unit.

The upper and lower driveshafts will automatically disengage as the unit separates.

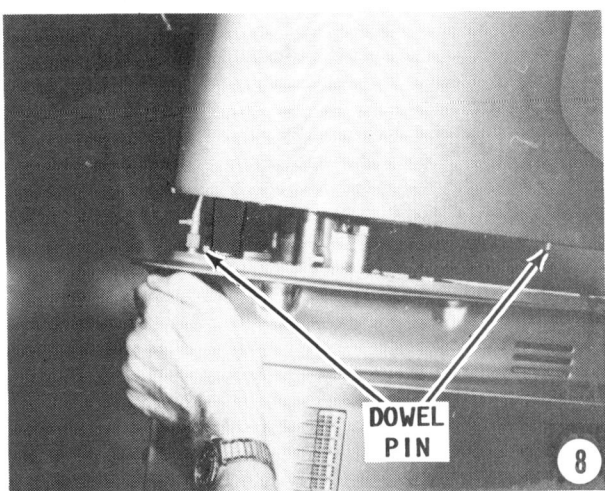

11-12 LOWER UNIT

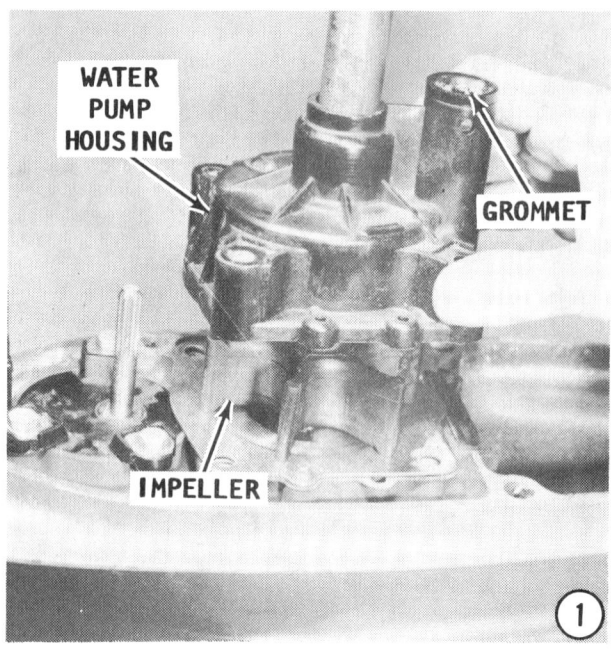

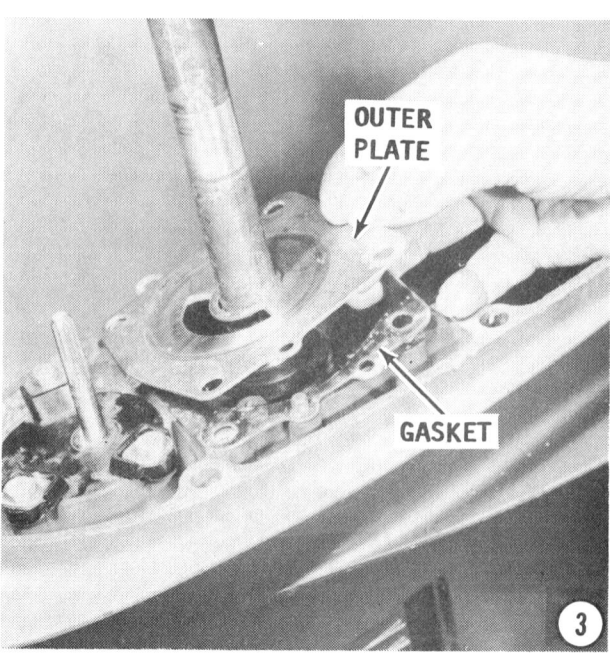

DISASSEMBLING

Water Pump Removal

1- Remove the four bolts and washers securing the water pump housing. Slide the water pump housing upward and free of the driveshaft. Pull out the water tube grommet from atop the water pump housing. Slide the impeller upward and free of the driveshaft. Remove and **SAVE** the small Woodruff key.

2- Pull out the insert cartridge from the water pump housing. Remove and discard the O-ring.

3- Lift off the outer plate, and then the gasket beneath the plate. Slide both pieces upward and off the driveshaft.

VERY SPECIAL WORDS

If the only work to be performed on the lower unit is servicing the water pump, proceed directly to Step 40, Assembling.

Shift Rod Removal

4- Obtain Yamaha special tool P/N YB6052. Place the tool over the splines of the shift rod. Check to ensure the lower unit is in neutral gear. The shift rod cannot be removed if the unit is in forward or reverse.

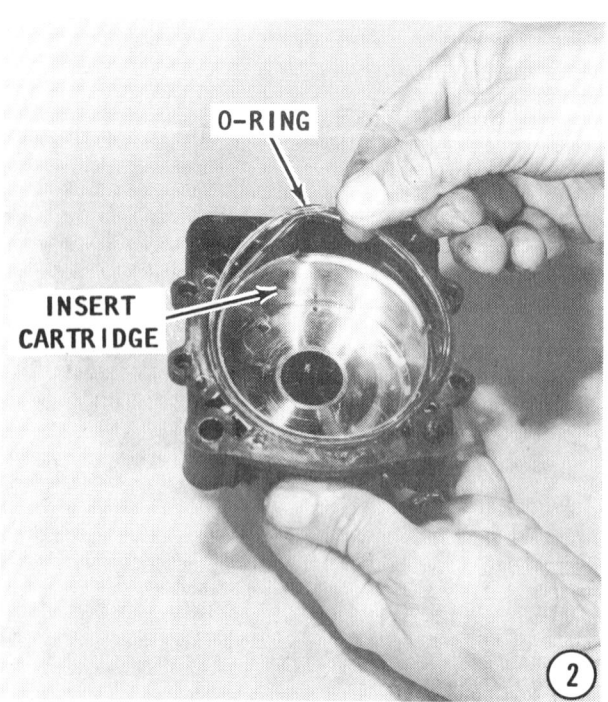

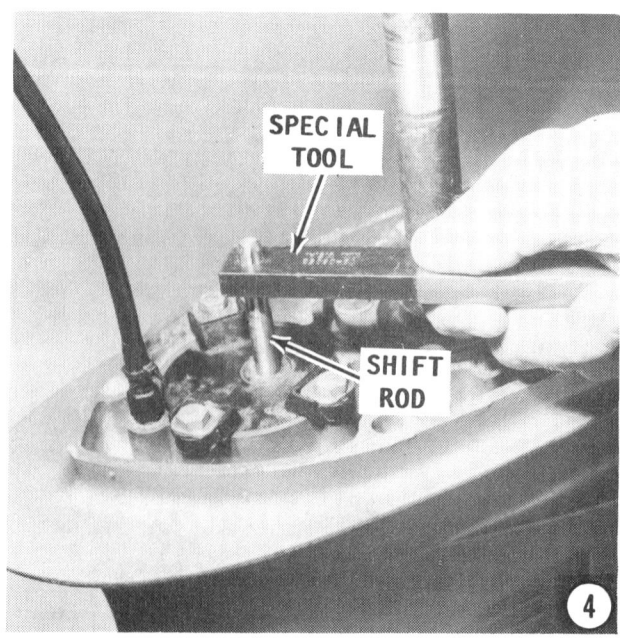

DISASSEMBLING 11-13

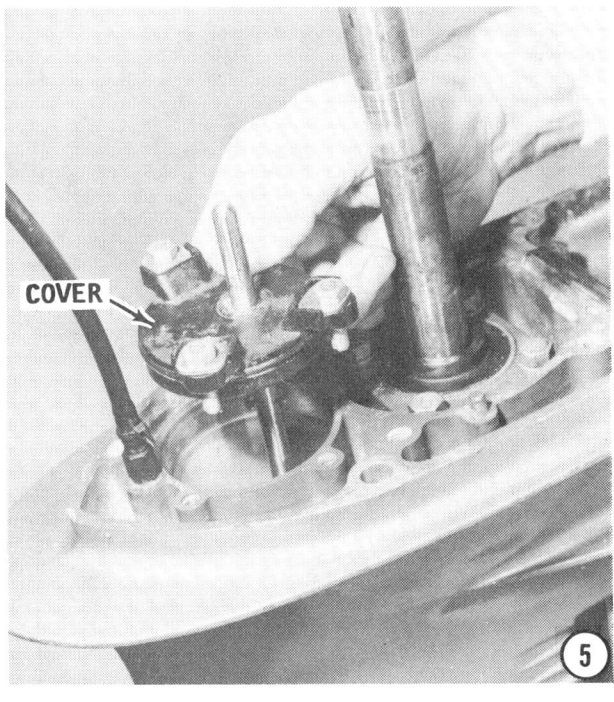

5- Remove the three bolts and washers securing the shift rod cover, and then pull the rod and cover off the lower unit.

Oil Seal Housing Removal

6- Remove the four bolts securing the oil seal housing.

7- Using two screwdrivers, gently pry up on both sides of the oil seal housing. Lift the housing up and free of the driveshaft. If the housing is stubborn and cannot be removed at this time, proceed with the work. The housing will be removed in a later step when the driveshaft is pulled from the lower unit and then Steps 8 and 9 may be completed.

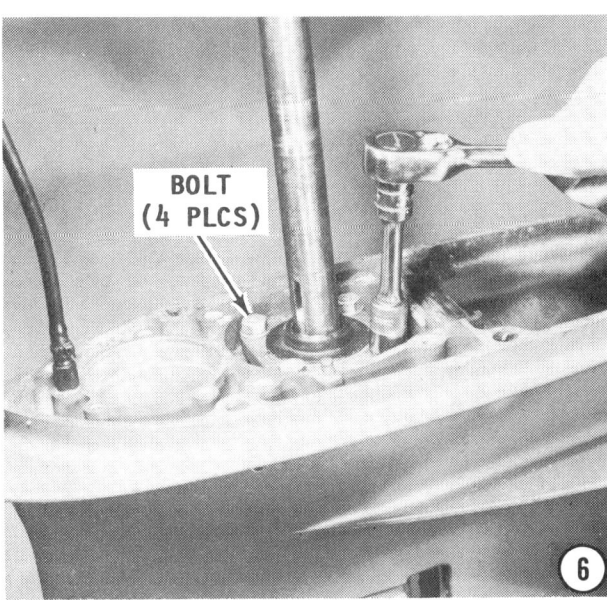

8- Remove and discard the O-ring around the oil seal housing.

SPECIAL WORDS

Removal of the seals and caged needle bearing, as described in the next paragraph, destroys the sealing qualities of the seals and distorts the needle bearing cage. Therefore, once removed, these parts are unfit for further service and cannot be installed a second time. Remove the seals and bearing **ONLY** if they need to be replaced **AND** be absolutely certain new seals and bearing are available before removing the old parts.

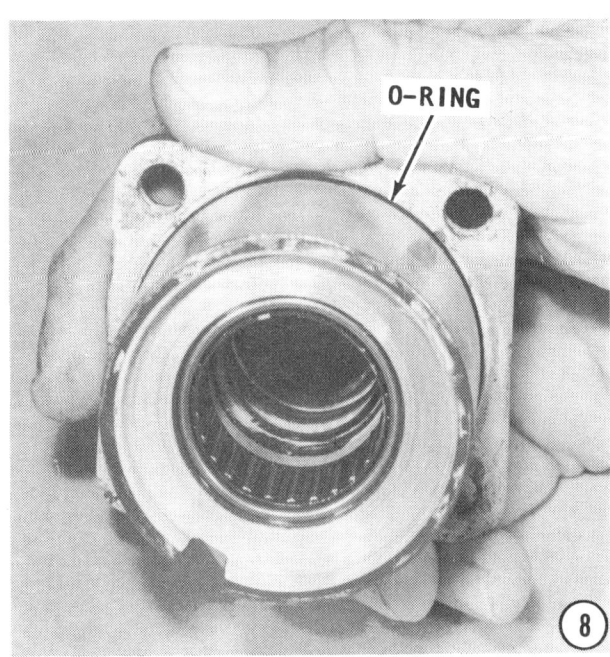

11-14 LOWER UNIT

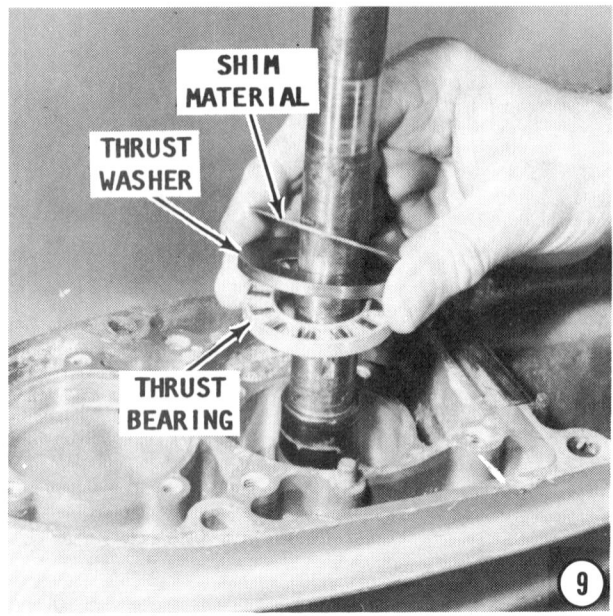

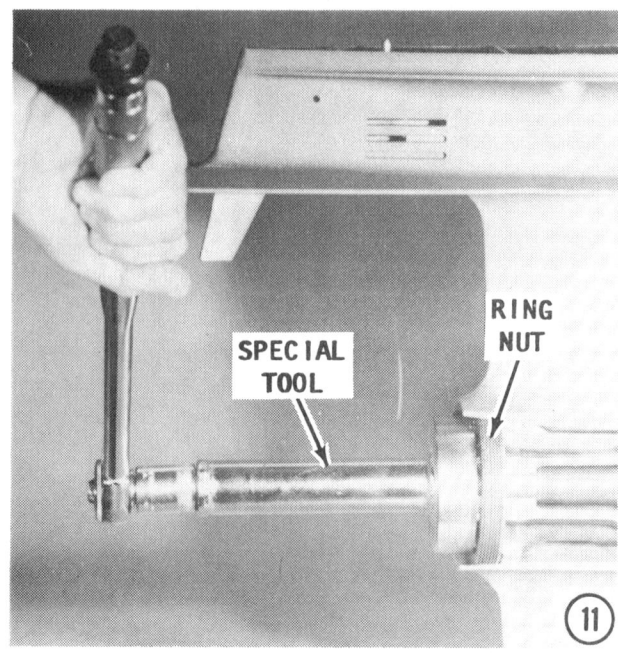

Step 8 continues: Obtain special tool, Yamaha P/N YB6096 slide hammer with jaw expander attachment. Use the tool to remove the oil seals one by one from the oil seal housing.

Obtain special tool, Yamaha P/N YB6196 and a suitable driver. Use the tool and the driver to drive the bearing from the oil seal through the **TOP** of the housing.

9- Watch for and **SAVE** any shim material installed around the driveshaft. The shim material is critical in obtaining the correct backlash during assembling. Using the old shim material will save considerable time, especially starting with no shim material.

Lift out the shim material, thrust washer and thrust bearing. Slide all the pieces upward and off the driveshaft.

Bearing Carrier Removal

10- Straighten the tabs on the lockwasher bent over the ring nut.

11- Obtain Yamaha special tool P/N YB-34447. Note the word "OFF" embossed on the ring nut. Install the tool into the ring and rotate the ring in the direction indicated. Continue rotating the ring nut until the nut is free.

12- Remove the ring nut and the tabbed washer behind the nut.

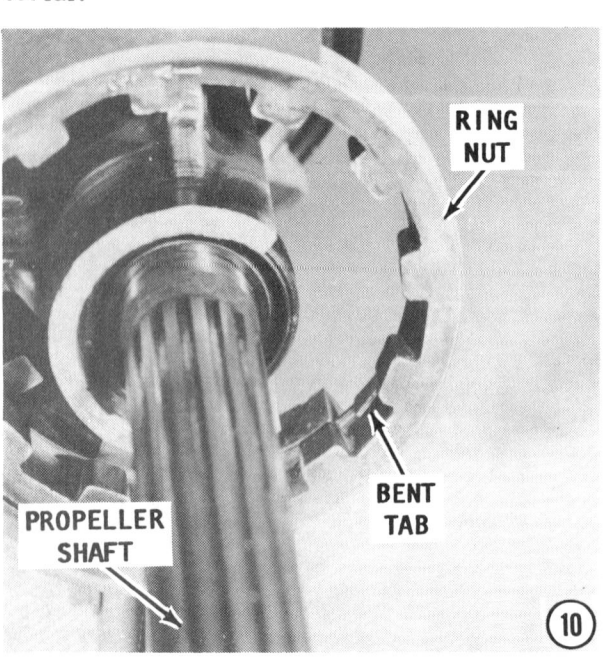

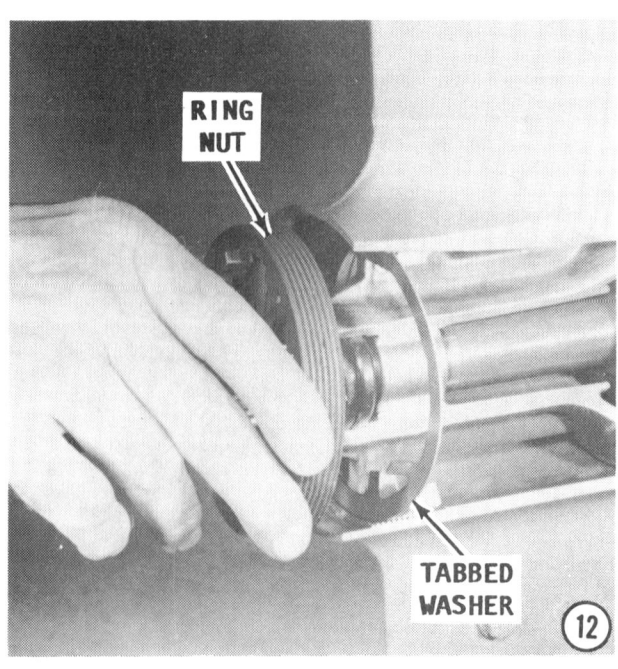

DISASSEMBLING 11-15

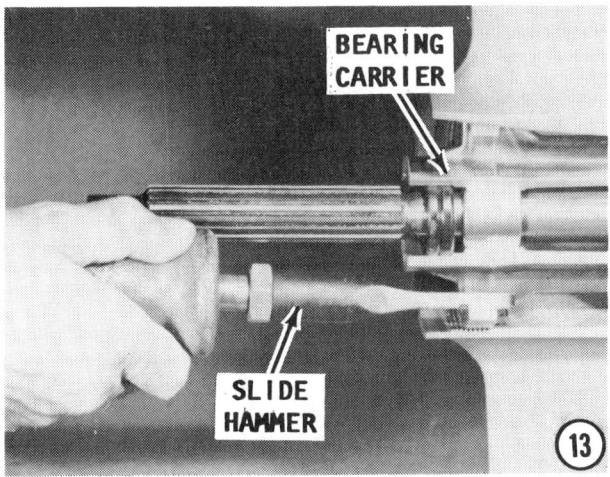

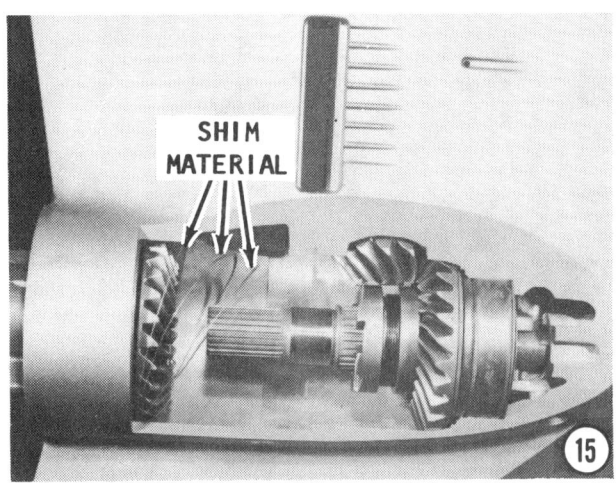

13- Obtain Yamaha special tool P/N YB6096 slide hammer with jaw expander attachment and "pull" the bearing carrier free.

14- Remove the bearing carrier. **TAKE CARE** not to lose the tiny key in the groove at the bottom of the lower unit. On counterrotating lower units, one half of the propeller shaft will come away with the bearing carrier, the other half will remain in the lower unit.

15- Watch for and **SAVE** any shim material from the back side of the reverse gear. **BE ADVISED,** on counterrotating units, this gear is the forward gear. The shim material is critical in obtaining the correct backlash during assembling. Using the old shim material will save considerable time, especially starting with no shim material.

16- Remove the spacer washer from the front of the reverse gear (forward gear on counterrotating units).

17- Remove and discard the O-ring around the bearing carrier.

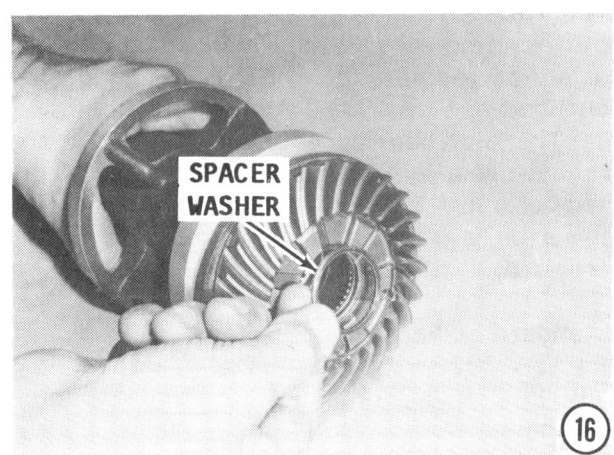

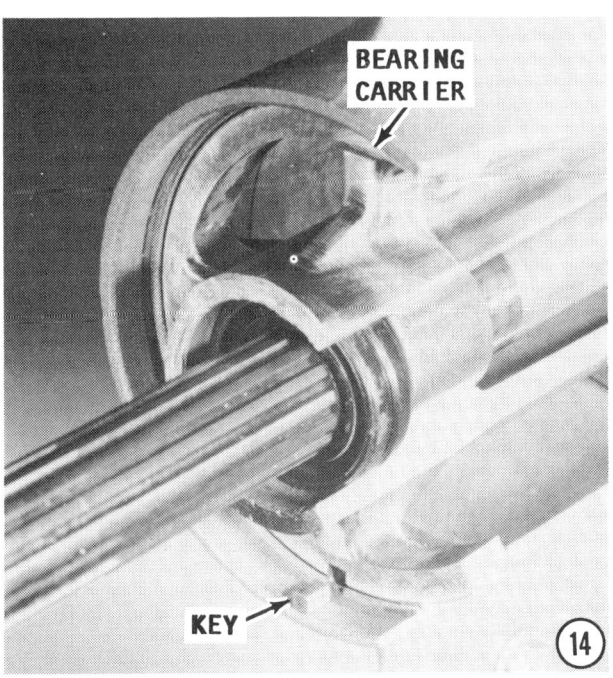

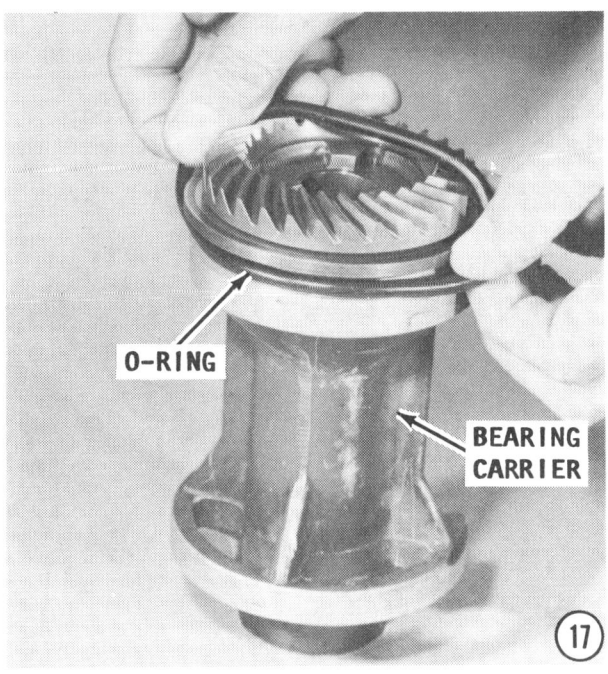

11-16 LOWER UNIT

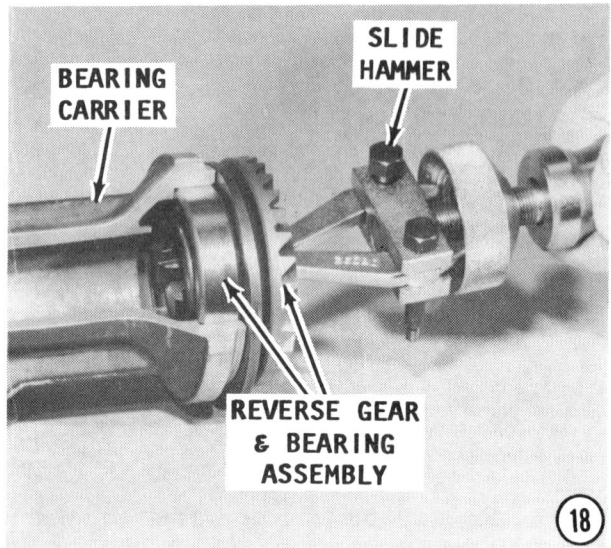

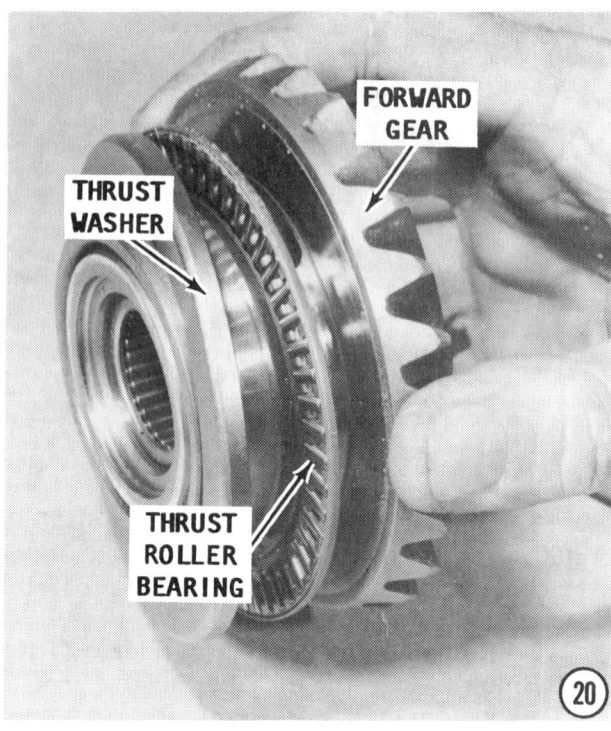

SPECIAL WORDS

The reverse gear and ball bearing assembly is pressed into the bearing carrier. The ball bearing is pressed onto the back of the reverse gear and must be separated using a special tool. This procedure is outlined in Step 19.

The ball bearing or the ball bearing and reverse gear together must be "pulled" from the bearing carrier. For counterrotating lower units, these special words apply to the forward, not the reverse gear.

Standard Lower Units

18- Obtain special tool, Yamaha P/N YB6096 slide hammer with jaw expander attachment and "pull" the bearing or the bearing and reverse gear from the bearing carrier. When using the tool, check to be **SURE** the jaws are hooked onto the inner race.

19- The ball bearing is pressed onto the back of the reverse gear. To remove this bearing, first obtain a universal bearing separator tool, and then press on the inner race to separate the bearing from the gear.

Counterrotating Lower Units

20- Obtain a universal bearing separator. Separate the forward gear from the bearing carrier. Remove the large thrust washer and the thrust roller bearing.

SPECIAL WORDS

Perform the following half of this step **ONLY** if the needle bearing in question is no longer fit for service.

The caged needle bearing set is pressed into the bearing carrier. To remove the bearing set, first obtain Yamaha Handle P/N

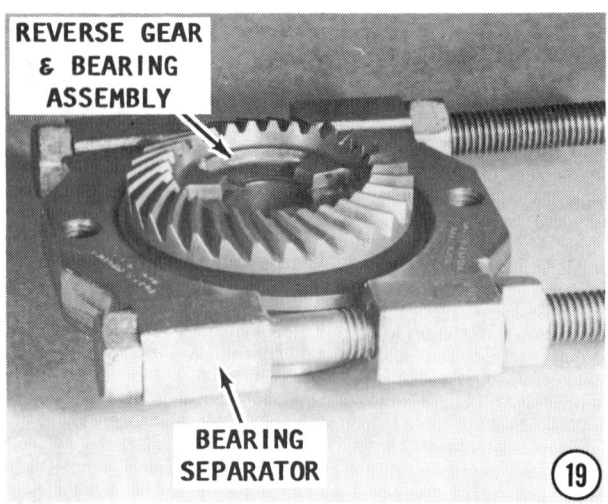

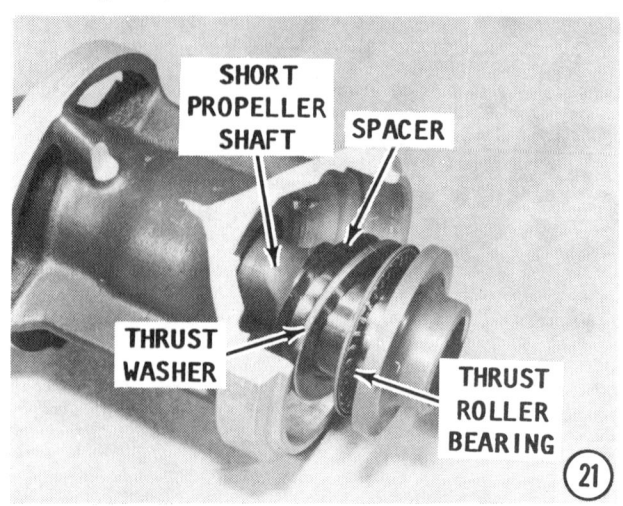

DISASSEMBLING 11-17

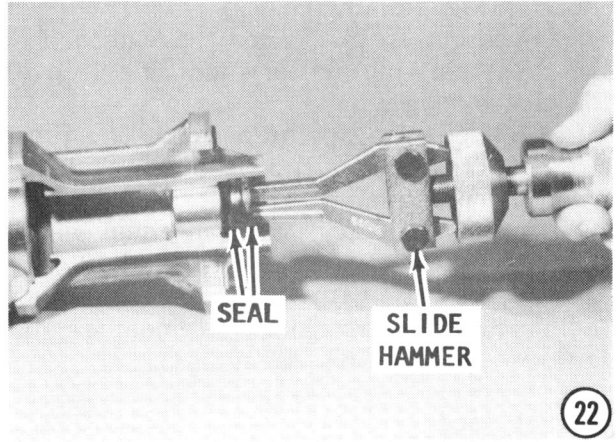

YB6071 and Driver P/N YB6196. Drive the needle bearing set free of the bearing carrier.

21- Pull the short propeller shaft from the bearing carrier. Remove the spacer, large thrust washer and thrust bearing.

All Lower Units

CRITICAL WORDS

Perform Step 22 only if the seals have been damaged and are no longer fit for service. Removing the seals destroys their sealing qualities. Therefore, they cannot be installed a second time. Be absolutely sure a new seal is available **BEFORE** removing the old seal in the next step.

22- Inspect the condition of the two seals in the bearing carrier. If the seals

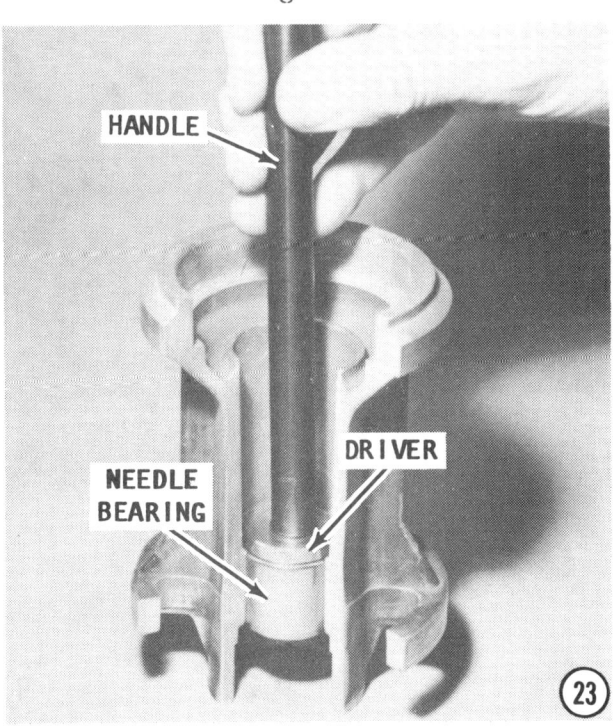

appear to be damaged and replacement is required, obtain the same slide hammer and jaw attachments as used in Step 18. Use these tools to remove the seals.

SPECIAL WORDS

Perform Step 23 **ONLY** if the needle bearing in question is no longer fit for service.

23- The caged needle bearing set is pressed into the bearing carrier. To remove the bearing set, first obtain Yamaha Handle P/N YB6071 and Driver P/N YB6196. Drive the needle bearing set free of the bearing carrier.

Propeller Shaft Removal and Disassembling

24- Pull the propeller shaft, clutch, and shifter assembly straight back and free of the lower unit.

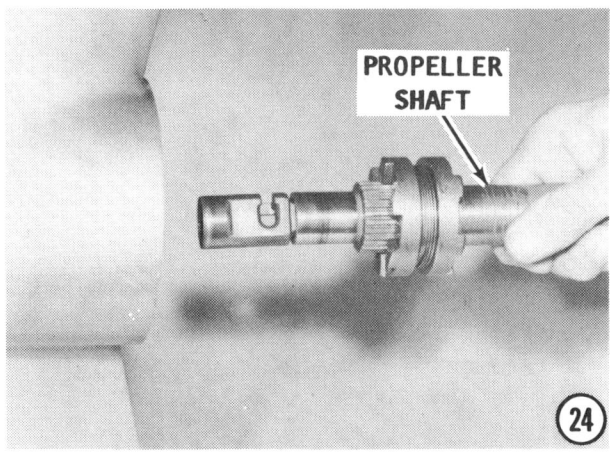

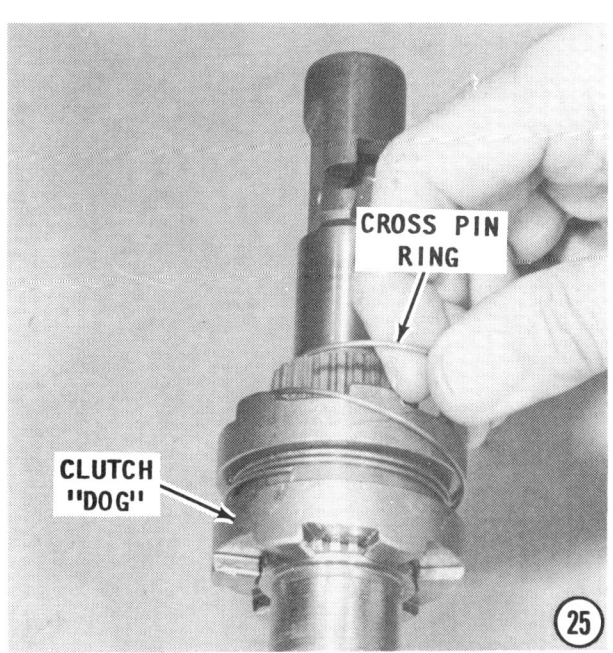

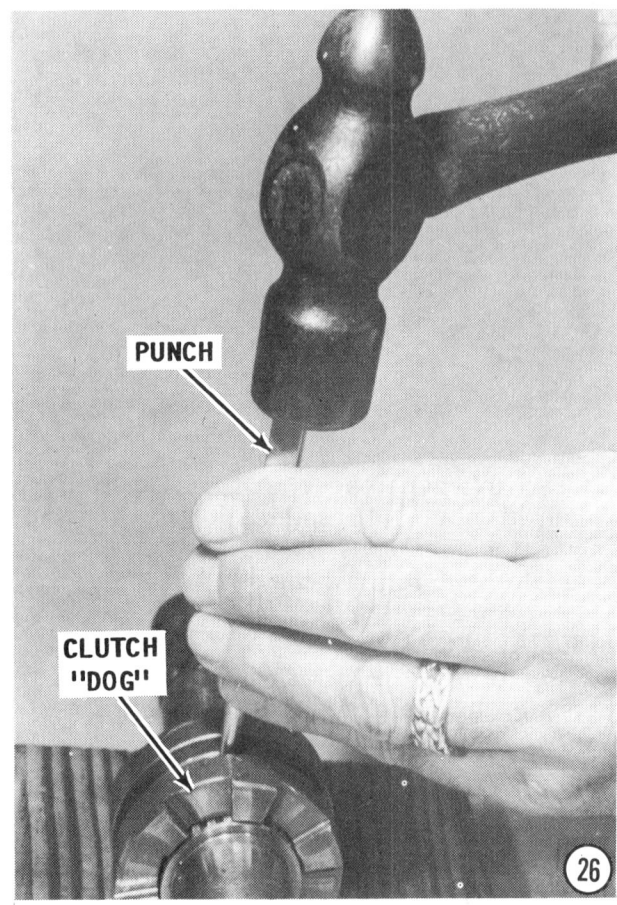

Clutch "Dog" Removal

25- Insert an awl under the end loop of the cross pin ring and pry the ring free of the clutch "dog".

26- Use a long pointed punch and press out the cross pin. Slide the clutch "dog" from the propeller shaft. Notice how both shoulders of the "dog" are an equal width, therefore the "dog" may be reinstalled either way, usually with the least worn side facing the forward gear. Remember this fact, as an aid during assembling.

Special Words

As the work proceedes in the disassembly of the shift slide, six ball bearings will be found. A set of two small balls, two medium balls and two large balls. Take care to store these balls separately to avoid confusion during assembly.

27- Unhook the shifter from the "head" or knob on the end of the shift slide. Pull out the shift slide. Take care not to lose the two medium ball bearings wedged between the shift slide and the propeller shaft.

28- Pry out the two medium balls from the shift slide.

29- Continue to pull out the shift slide until two more smaller balls appear. Pry out these balls from the shift slide.

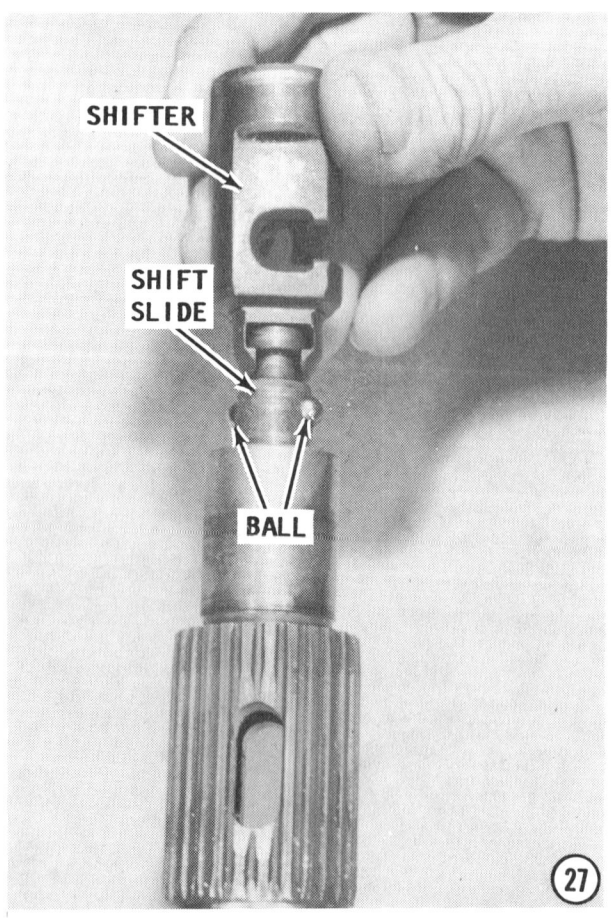
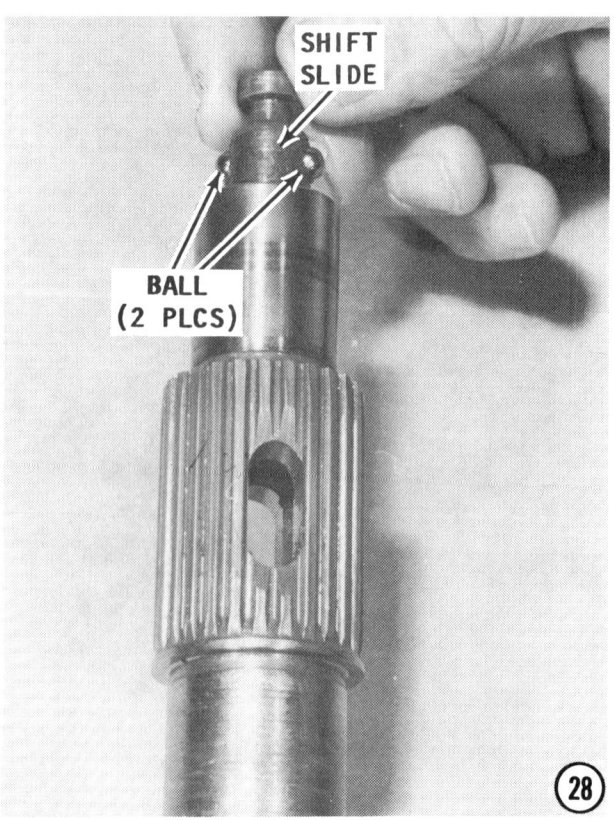

DISASSEMBLING 11-19

30- Remove one large ball, a compression spring and another large ball from the shift slide.

Driveshaft Removal

Before the driveshaft can be removed, the pinion gear at the lower end of the shaft must be removed. The pinion gear is secured to the driveshaft with a nut.

SPECIAL WORDS

In most cases, when working with tools, a nut is rotated to remove or install it to a particular bolt, shaft, etc. In the next two steps, the reverse is required because there is no room to move a wrench inside the lower unit cavity. The nut on the lower end of the driveshaft is held steady and the shaft is rotated until the nut is free.

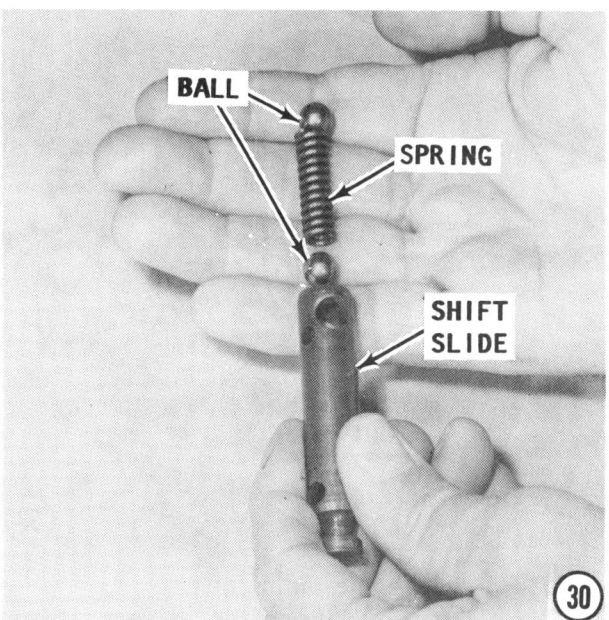

31- Obtain driveshaft tool Yamaha P/N YB6201 and a suitable size wrench with which to rotate the tool.

32- Obtain a 22mm size socket. This socket will be used to prevent the nut from rotating while the driveshaft is rotated.

Now, hold the pinion nut steady with the socket, and at the same time install the driveshaft tool on top of the driveshaft. With both tools in place, one holding the nut and the other on the driveshaft, use muscle to rotate the driveshaft **COUNTERCLOCKWISE** and "break" the nut free.

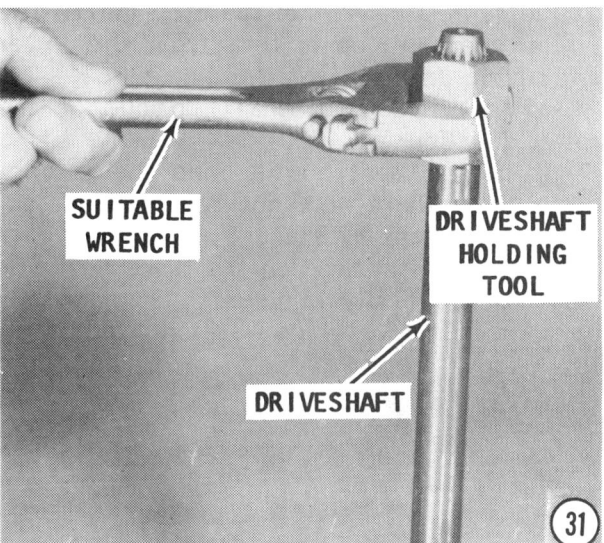

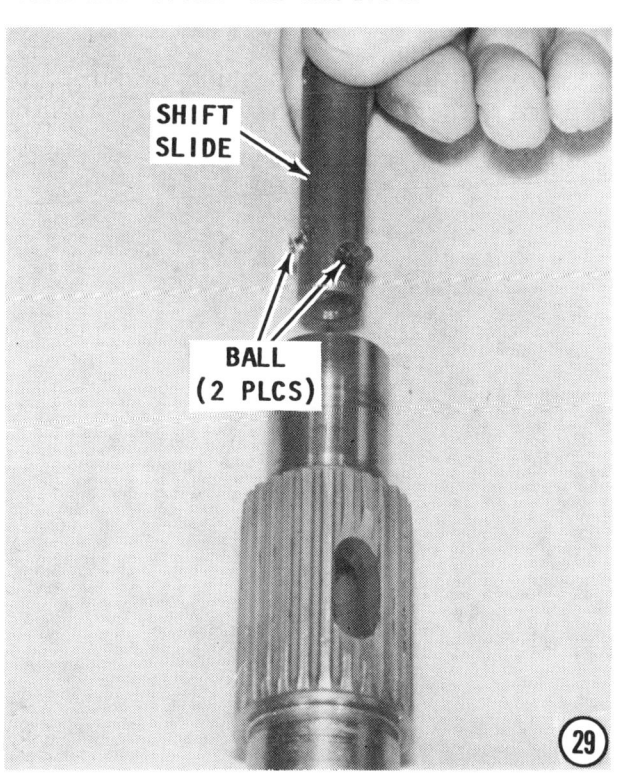

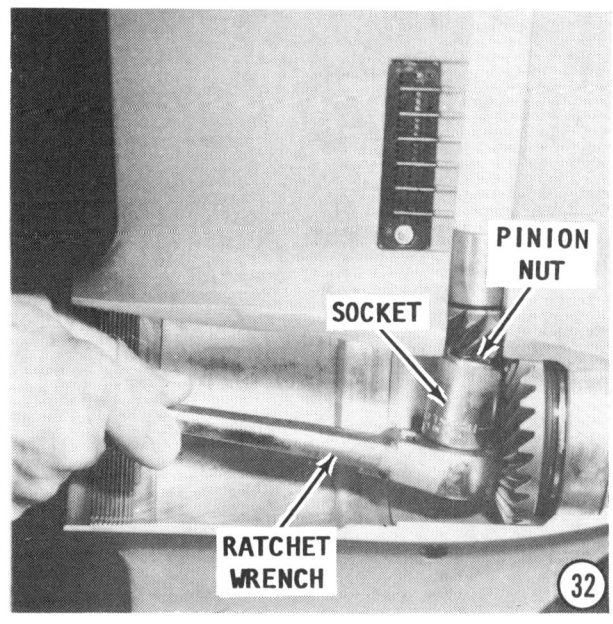

11-20 LOWER UNIT

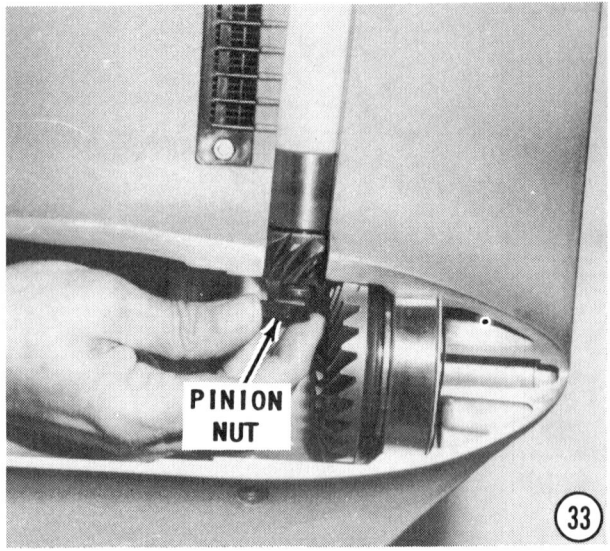

33- Remove the pinion nut.

34- Gently pull up on the driveshaft and at the same time rotate the driveshaft. The pinion gear will come free from the lower end of the driveshaft.

35- Pull the driveshaft up out of the lower unit housing.

If the oil seal housing and the components beneath it were not removed as directed in Steps 7, 8 and 9, the housing at this point will still be on the driveshaft and can now be easily removed. Go back and perform the steps which were omitted.

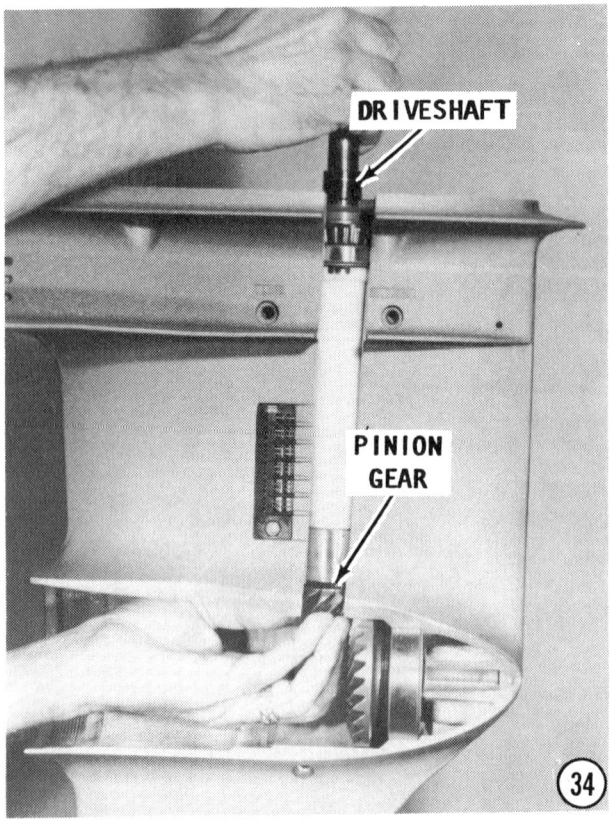

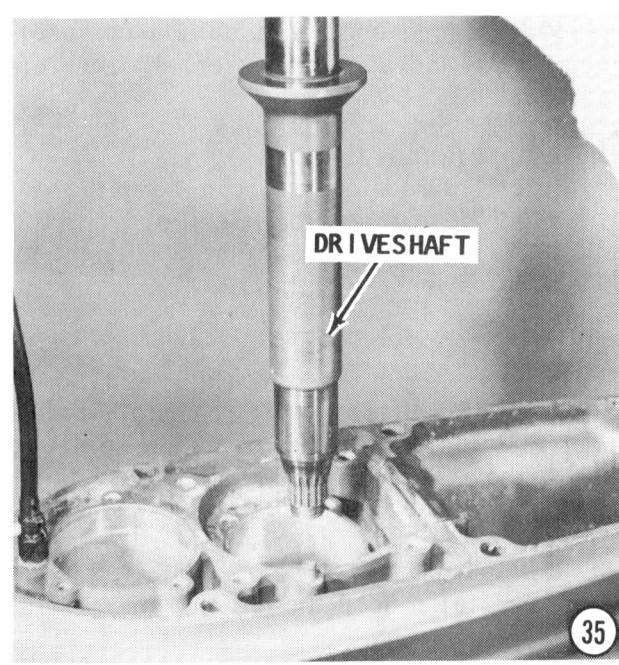

36- Lift out the driveshaft sleeve.

SPECIAL WORDS

Perform Step 37 **ONLY** if the needle bearing in question is no longer fit for service.

Driveshaft Needle Bearing Removal

37- The loose needle bearing set is pressed into the driveshaft housing. To remove the bearing set, first obtain Yamaha Handle P/N YB6071 and Driver P/N YB6194. Drive the needle bearing set downward free of the bearing carrier.

Even if the entire bearing is not to be replaced, there is a good chance, when the

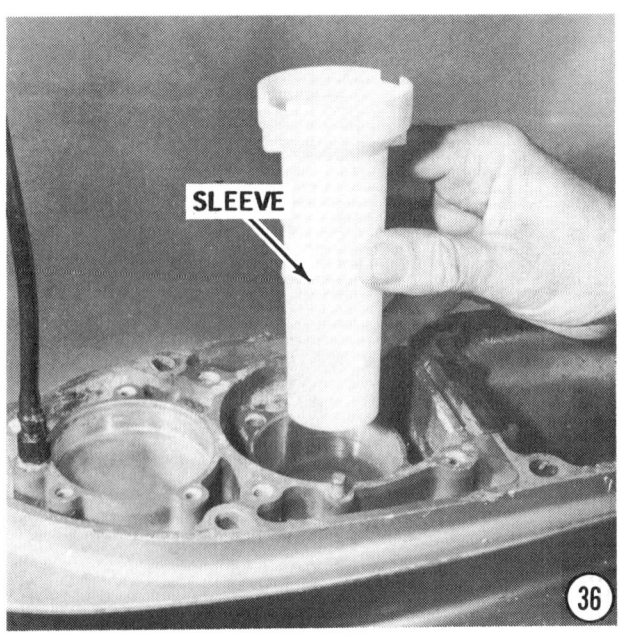

DISASSEMBLING 11-21

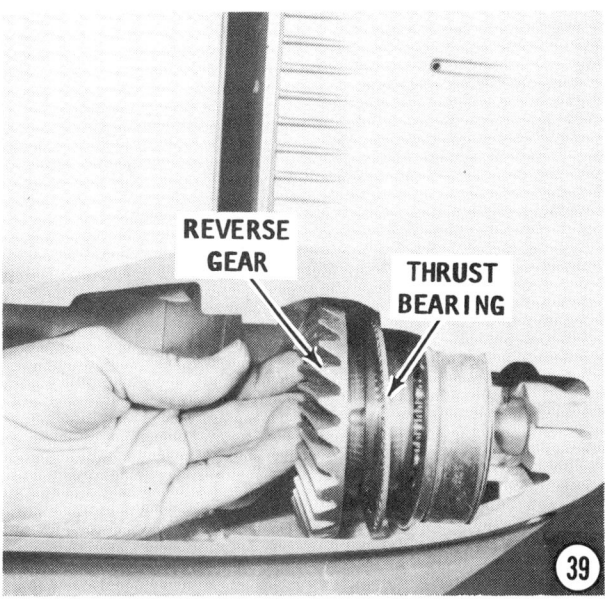

driveshaft was removed some of the needle bearings fell away from the cage. This is not an indication the entire cage and needle set must be replaced. Save the needle bearings. They can be held in place with grease while the driveshaft is installed later. If any of the needle bearings are lost --- **BAD NEWS**, the cage must be driven out and an entire new bearing set installed.

Forward Gear Removal
Standard Lower Unit

38- Lift out the forward gear and the tapered roller bearing.

Counterrotating Lower Unit

39- Lift out the reverse gear and tapered roller bearing assembly, then lift out the thrust bearing. Take care not to confuse this thrust bearing with the one removed in Step 20. If these thrust bearing are to be reused, they must be installed in their original locations, because each develops a unique wear pattern. If a bearing is installed in a different location, the wear on the individual tiny roller bearings would be greatly accelerated and lead to premature failure of the bearing.

40- Lift out the large thrust washer and spacer.

All Lower Units

41- Obtain a slide hammer and jaw attachment Yamaha P/N YB6096. Use the

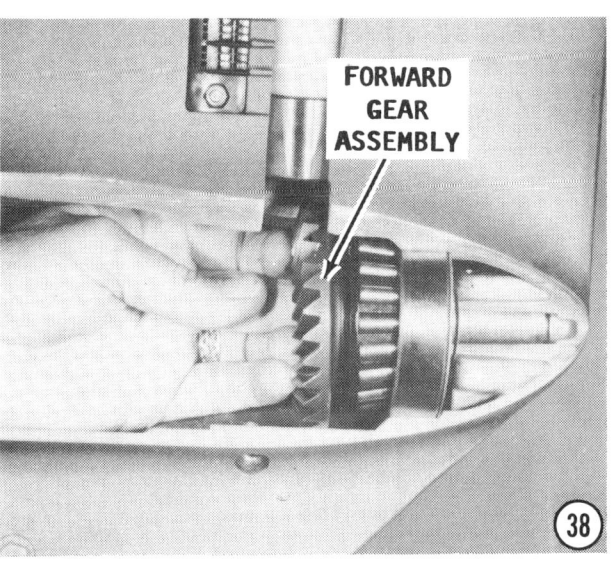

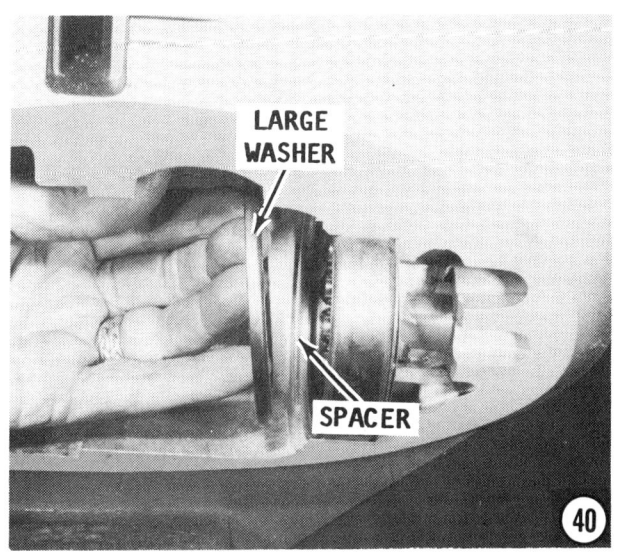

11-22 LOWER UNIT

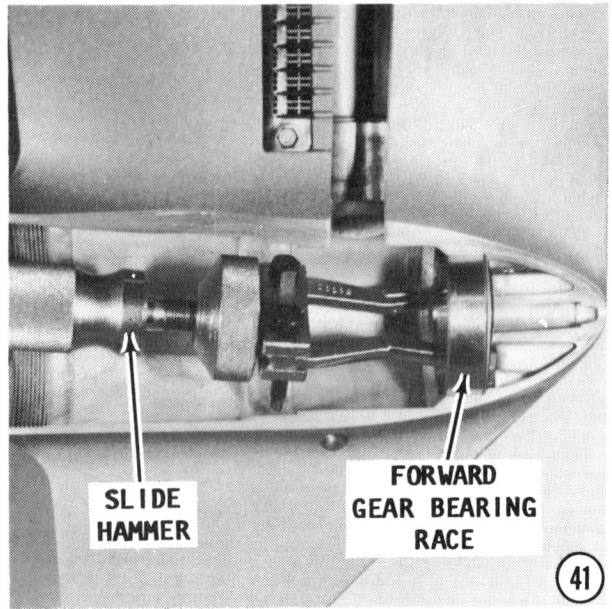

slide hammer and special tool to pull the bearing race from the lower unit housing. Be sure to hold the slide hammer at right angles to the driveshaft while working the slide hammer.

42- **WATCH** for and **SAVE** any shim material found behind the forward gear (reverse gear on counterrotating units). The shim material is critical to obtaining the correct backlash during installation. Using the old shim material will save considerable time, especially starting with no shim material.

CRITICAL WORDS

If a two piece bearing is to be replaced, the bearing and the race must be replaced as a matched set. Remove the bearing **ONLY** if it is unfit for further service.

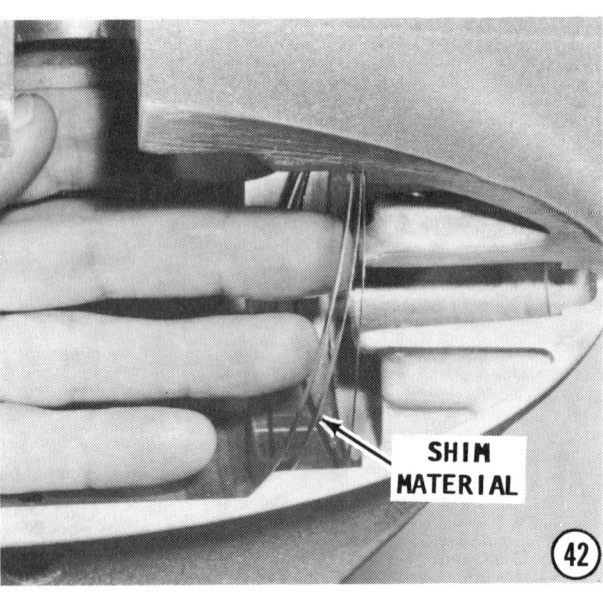

43- Position a bearing separator between the forward gear (reverse gear on counterrotating units), and the tapered roller bearing. Using a hydraulic press, separate the gear from the bearing.

CLEANING AND INSPECTING

Good shop practice requires installation of new O-rings and oil seals **REGARDLESS** of their appearance.

Clean all water pump parts with solvent, and then dry them with compressed air. Inspect the water pump housing and oil seal housing for cracks and distortion, possibly caused from overheating. Inspect the plate and water pump cartridge for grooves and/or rough surfaces. If possible, **ALWAYS** install a new water pump impeller while the lower unit is disassembled. A new impeller will ensure extended satisfactory service and give "peace of mind" to the owner. If the old impeller must be returned to service, **NEVER** install it in reverse to the original direction of rotation. Installation in reverse will cause premature impeller failure.

If installation of a new impeller is not possible, check the seal surfaces. All must be in good condition to ensure proper pump operation. Check the upper, lower, and ends of the impeller vanes for grooves, cracking, and wear. Check to be sure the indexing notch of the impeller hub is intact and will not allow the impeller to slip.

Clean around the Woodruff key. Clean all bearings with solvent, dry them with compressed air, and inspect them carefully.

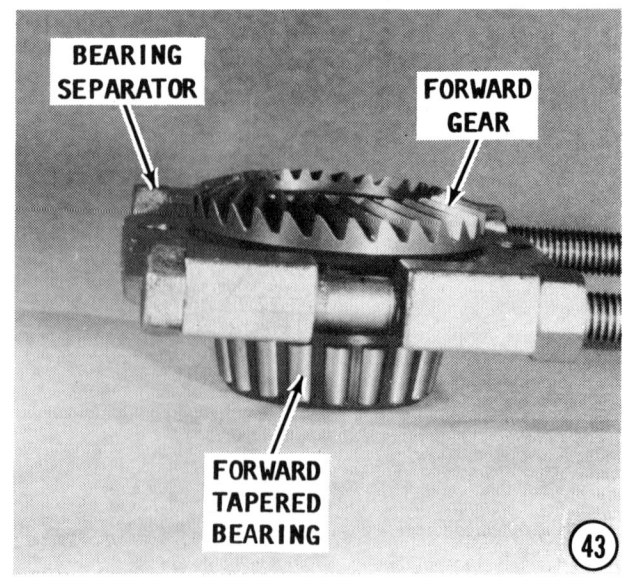

CLEANING & INSPECTING 11-23

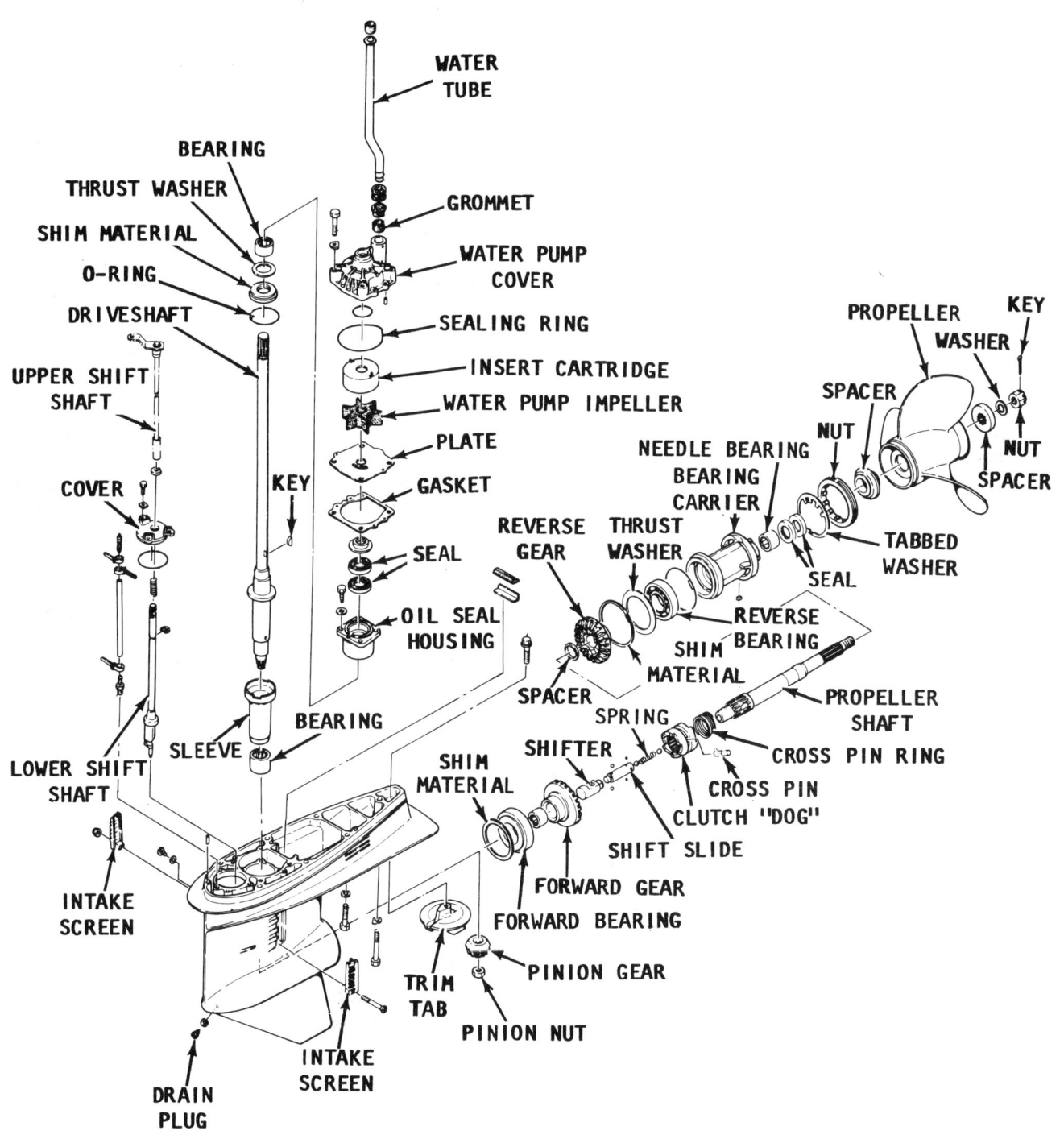

Exploded drawing of a standard lower unit, with major parts identified. The internal mechanism of the counterrotating lower unit is shown vividly with photographs throughout this chapter.

After 60 seconds at 1500 rpm.

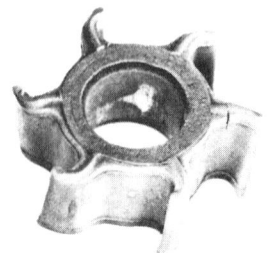

After 90 seconds at 1500 rpm.

After 30 seconds at 2000 rpm.

After 45 seconds at 2000 rpm.

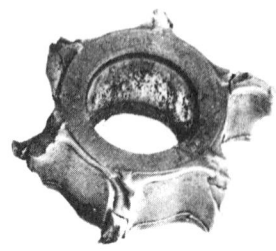

After 60 seconds at 2000 rpm

Cautions throughout this manual point out the danger of operating the powerhead without water passing through the water pump. The above photographs are self evident.

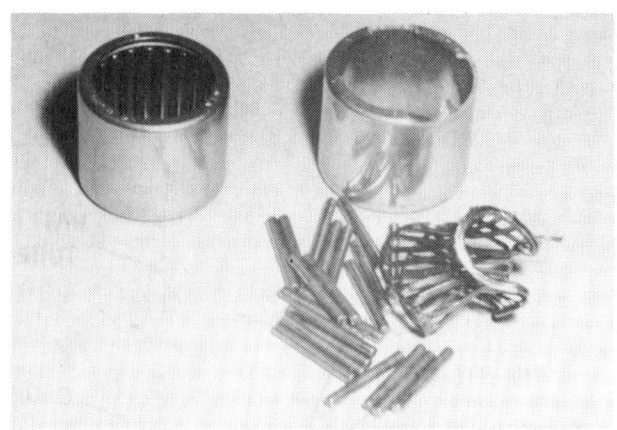

A new needle bearing (left), alongside a used needle bearing after the bearing was removed using the "bigger hammer" method.

Be sure there is no water in the air line. Direct the air stream through the bearing. **NEVER** spin a bearing with compressed air. Such action is highly dangerous and may cause the bearing to score from lack of lubrication. After the bearings are clean and dry, lubricate them with Formula 50 oil, or equivalent. Do not lubricate tapered bearing cups until after they have been inspected.

Inspect all ball bearings for roughness, scratches and bearing race side wear. Hold the outer race, and work the inner bearing race in and out, to check for side wear.

Determine the condition of tapered bearing rollers and inner bearing race, by inspecting the bearing cup for pitting, scoring, grooves, uneven wear, imbedded particles, and discoloration caused from overheating. **ALWAYS** replace tapered roller bearings as a set.

Clean the forward gear with solvent, and then dry it with compressed air. Inspect the gear teeth for wear. Under normal conditions the gear will show signs of wear but it will be smooth and even.

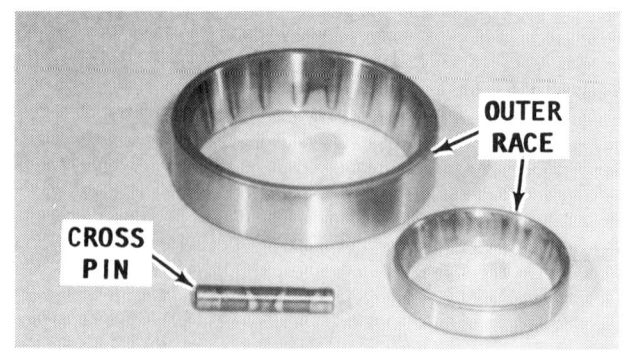

Grooves on the cross pin or marked wear patterns on the outer roller bearing races are evidence of premature failure of these or associated parts.

CLEANING & INSPECTING 11-25

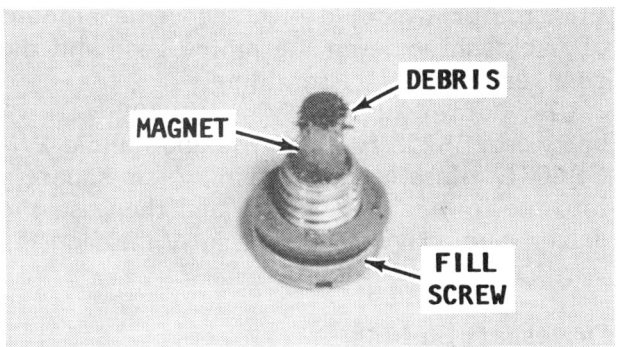

A magnet is an integral part of the fill screw. This magnet will catch small metallic parts in the lower unit lubricant.

Clean the bearing carrier with solvent, and then dry it with compressed air. **NEVER** spin bearings with compressed air. Such action is highly dangerous and may cause the bearing to score from lack of lubrication. Check the gear teeth of the reverse gear for wear. The wear should be smooth and even.

Check the clutch "dogs" to be sure they are not rounded-off, or chipped. Such damage is usually the result of poor operator habits and is caused by shifting too slowly or shifting while the engine is operating at high rpm. Such damage might also be caused by improper shift rod adjustments.

Rotate the reverse gear and check for catches and roughness. Check the bearing for side wear of the bearing races.

Clean the driveshaft with solvent, and then dry it with compressed air.

Inspect the driveshaft splines for excessive wear. Check the oil seal surfaces above and below the water pump drive pin or Woodruff key area for grooves. Replace the shaft if grooves are discovered.

Inspect the driveshaft bearing surface above the pinion gear splines for pitting, grooves, scoring, uneven wear, embedded

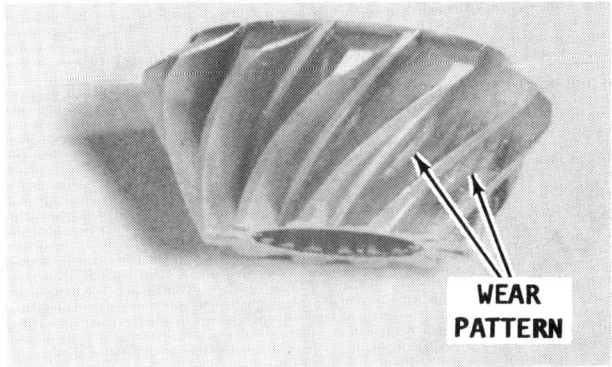

Unacceptable pinion gear wear pattern, probably caused by inadequate lubrication in the lower unit.

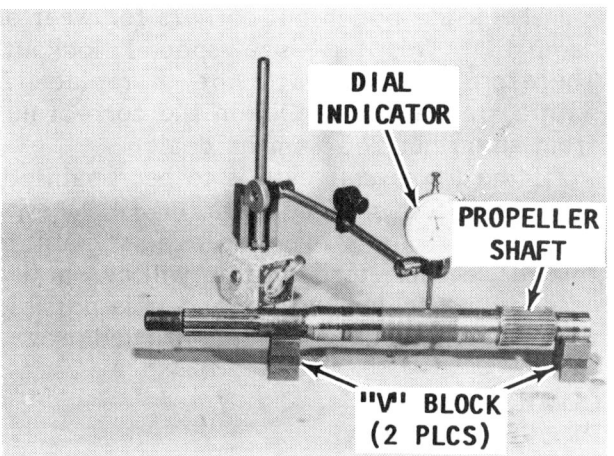

Using a dial indicator and two "V" blocks to measure the propeller shaft runout, as described in the text.

metal particles and discoloration caused by overheating.

Inspect the propeller shaft oil seal surface to be sure it is not pitted, grooved, or scratched. Inspect the roller bearing contact surface on the propeller shaft for pitting, grooves, scoring, uneven wear, embedded metal particles, and discoloration caused from overheating.

Inspect the propeller shaft splines for wear and corrosion damage. Check the propeller shaft and driveshaft for runout.

Place each shaft on V blocks and measure the runout with a dial indicator gauge at a point midway between the V blocks. The maximum acceptable runout for the propeller shaft is 0.0008" (0.02mm). The maximum acceptable runout for the driveshaft is 0.02" (0.5mm).

Inspect:

All bearing bores for loose fitting bearings.

Gear housing for impact damage.

Cover nut threads for cross-threading and corrosion damage.

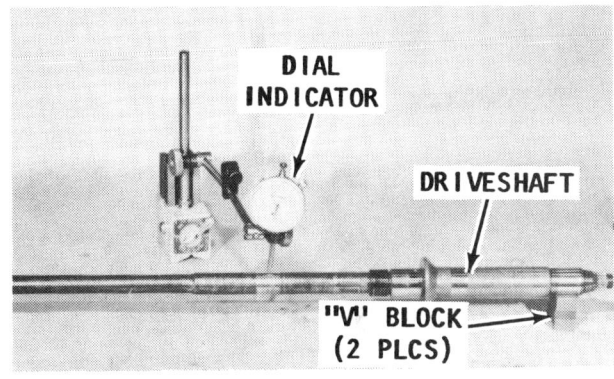

Using a dial indicator and two "V" blocks to measure the driveshaft runout.

Check the pinion nut corners for wear or damage. This nut is a special locknut. Therefore, do **NOT** attempt to replace it with a standard nut. Obtain the correct nut from an authorized Yamaha dealer.

If the lower unit case is to be repainted. Mask off the threads engaging the cover nut. If this nut is installed against painted threads, a false torque value will be obtained upon tightening the nut and it is possible the nut could back off with continued use.

LOWER UNIT
PROPELLER DRIVE
ASSEMBLING

FIRST, THESE WORDS

Procedural steps are given to assemble and install virtually all items in the lower unit. However, if certain items, i.e. bearings, bushings, seals, etc. were found fit for further service and were not removed, simply skip the assembly steps involved. Proceed with the required tasks to assemble and install the necessary components.

Forward Gear — Standard Lower Unit
Reverse Gear — Counterrotating Unit
Tapered Bearing Installation

The next two steps apply **ONLY** if the ball bearing race was removed. If the bearing was not disturbed, proceed directly to Step 3. Do not pass go. Do not collect $200.00.

1- Insert the same amount of shim material saved during disassembling, Step 42, into the lower unit bearing race cavity. The shim material should give the same amount of backlash between the pinion gear and the gear, as before disassembling.

2- Obtain driver P/N YB6199 for V4 units or YB6258 for V6 units and handle P/N YB6071. Insert the bearing race squarely into the lower unit housing, and then use the driver and drive the race into the housing until it is fully seated.

Driveshaft Bushing
and Bearing Installation

Perform the next step **ONLY** if the needle bearing set was removed during disassembling Step 37. If the needle bearing set was not disturbed, apply a coat of Yamalube to the inner bearing surfaces and check to be sure all the needle bearings are still in place, proceed directly to Step 4.

3- A single needle bearing is used at the lower end of the driveshaft just above the pinion gear. The installation procedure differs for V4 and V6 units.

V4 Units

Obtain driver P/N YB6194, handle P/N YB6071, and guide plate P/N YB6213.

Apply a coat of Yamalube around the inside of the bearing to help keep the needle bearings in place during installation.

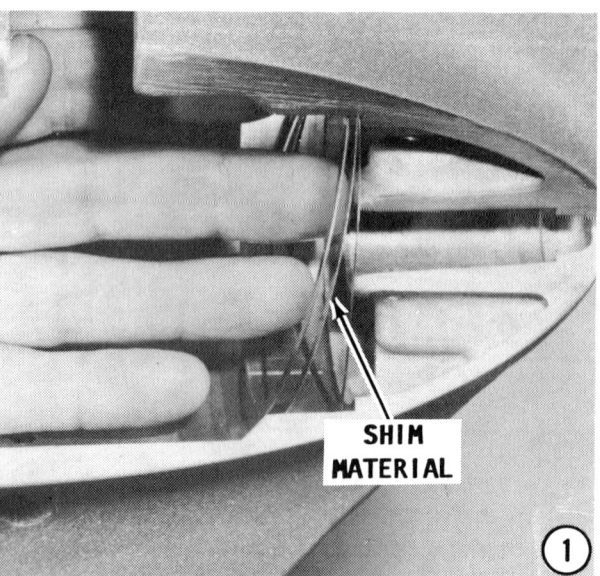

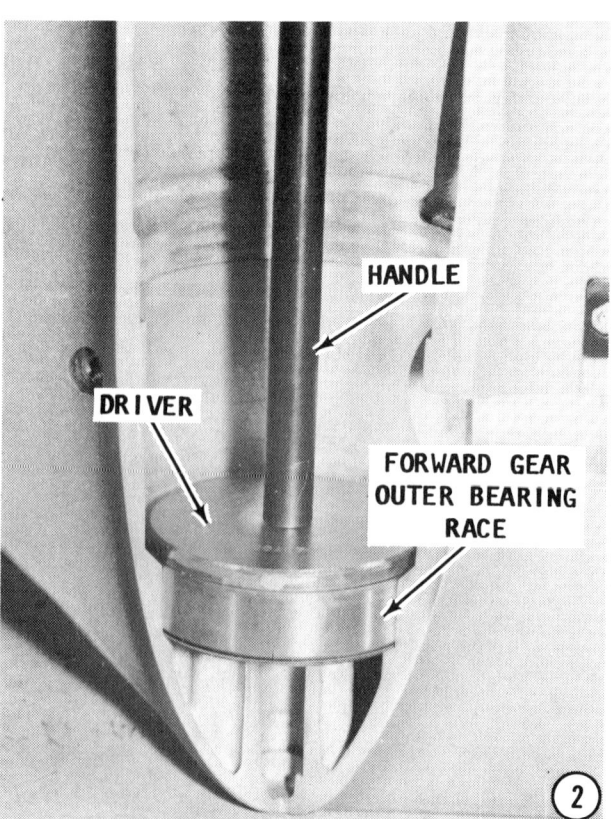

ASSEMBLING 11-27

Slide the guide plate onto the driver with the stamped mark facing **UPWARD**. Install the driver onto the threaded end of the handle. Use the guide plate to center the handle in the hole, and then install the needle bearing with the stamped mark on the bearing facing **UPWARD** toward the driver until the shoulder of the handle seats against the guide plate. This position places the needle bearing at a predetermined location in the lower unit.

V6 Units

Obtain bearing installation tool P/N YB6029, driver P/N YB6246 and plate P/N YB6169.

Apply a coat of Yamalube around the inside of the bearing to help keep the needle bearings in place during installation. Hold the bearing with the embossed numbers facing **UPWARD**. Insert the driver into the bearing from the bottom and move the bearing into position into the "torpedo" housing of the lower unit. The bearing will be "pulled" up into place.

Insert the bearing installation tool down into the driveshaft cavity, with the centering plate (supplied with the tool), resting on the cavitation plate. Lower the long bolt through the bearing and driver. Slide the special plate up over the end of the bolt and install and tighten the lockwasher and nut. Place a wrench on the bolt head of the installing tool and tighten the bolt to seat the bearing. This action will draw the bearing upward to a predetermined location in the lower unit.

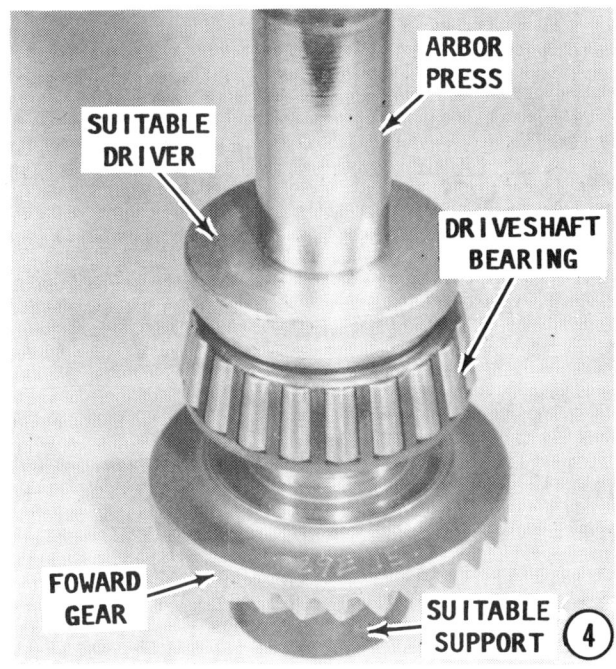

All Units

Slide the driveshaft sleeve into the upper end of the lower unit with the notch in the upper flange facing **AFT**

4- Position the forward gear (reverse gear on counterrotating models), tapered bearing over the gear. Use a suitable mandrel and press the bearing flush against the shoulder of the gear. **ALWAYS** press on the **INNER** race, never on the cage or the rollers.

Commercial product for foot problems and also very useful in determining the wear pattern of meshing gears, as explained in the text.

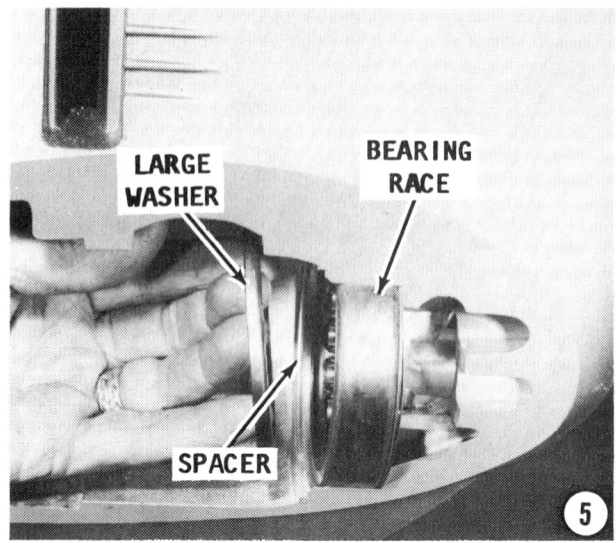

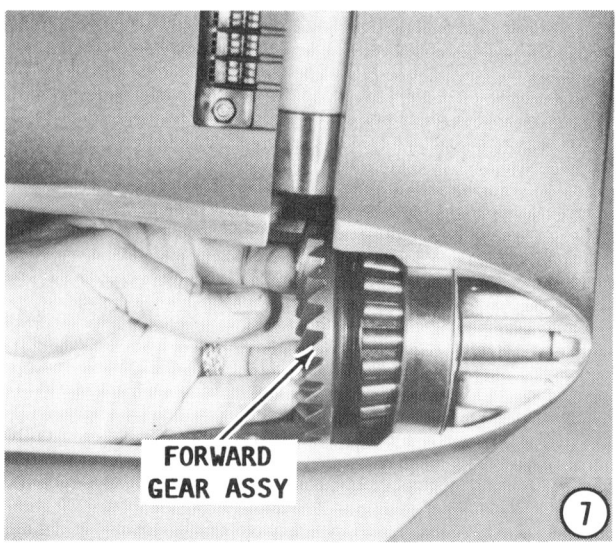

PLAN AHEAD

Obtain a suitable substance which can be used to indicate a wear pattern on the forward and pinion gears as they mesh. Machine dye may be used and if this material is not available, Desenex Foot Powder (obtainable at the local Drug Store/Pharmacy), or equivalent may be substituted. Desenex is a white powder available in an aerosol container. Before assembling either gear, apply a light film of the dye, Desenex, or equivalent, to the driven side of the gear. After the gears are assembled and rotated several times, they will be disassembled and the wear pattern can be examined. The substance will be removed from the gears prior to final assembly.

Counterrotating Lower Unit

5- Insert the spacer and large thrust washer to rest up against the reverse gear bearing race.

6- Insert the thrust bearing and reverse gear assembly.

Standard Lower Unit

7- Position the forward gear assembly into the forward bearing race.

Pinion Gear
and Driveshaft Installation

8- Lower the driveshaft down through the sleeve in the upper end of the lower unit.

Coat the pinion gear with a fine spray of Desenex, as instructed in the "Plan Ahead" paragraph prior to assembling Step 5. Handle the gear carefully to prevent disturbing the powder.

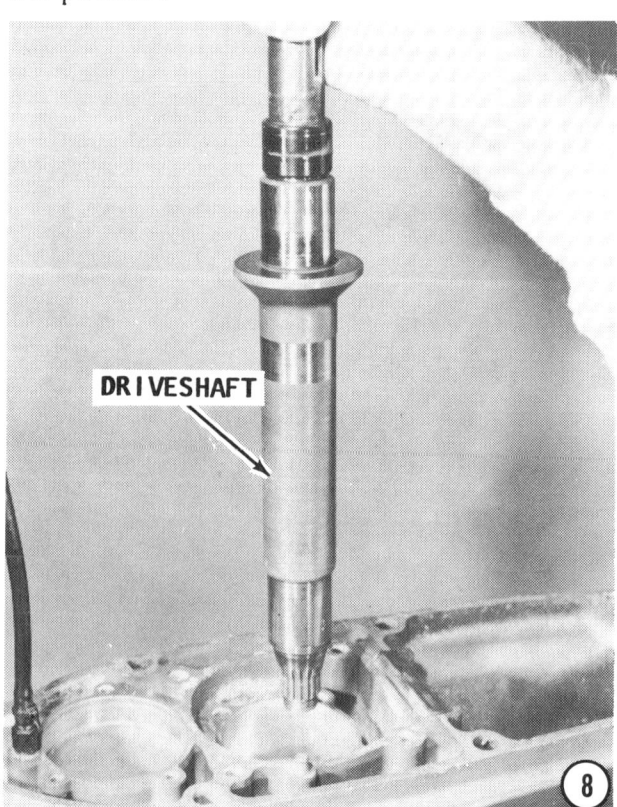

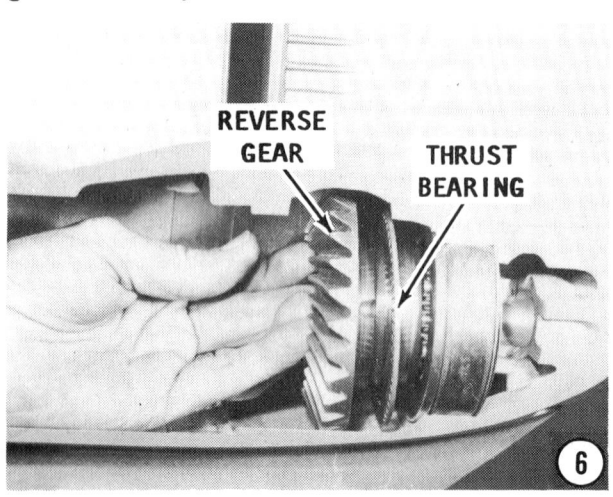

ASSEMBLING 11-29

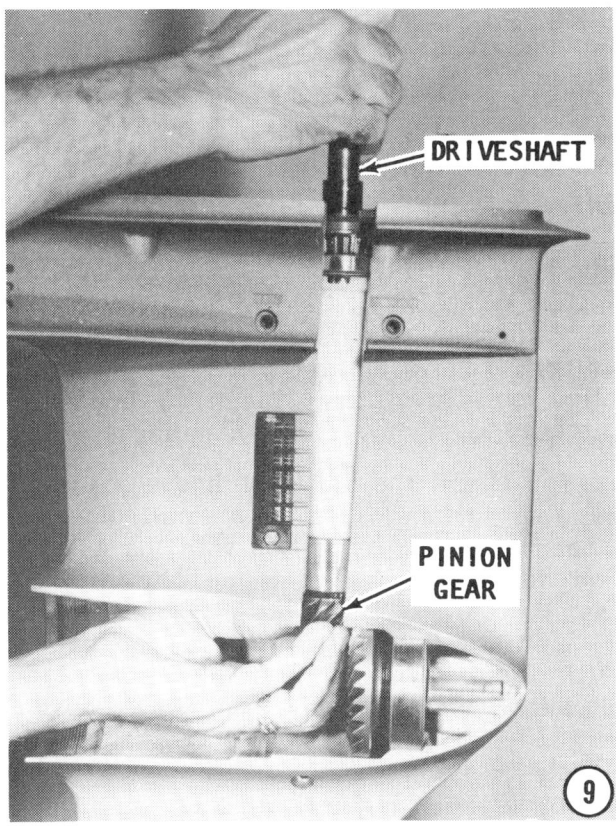

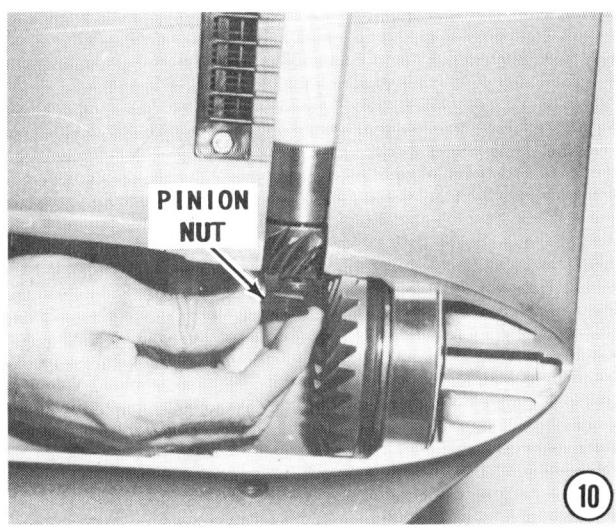

9- Raise the pinion gear to allow the driveshaft to pass through the pinion gear. It may be necessary to rotate the driveshaft slightly to allow the splines on the driveshaft to index with the internal splines of the pinion gear. The teeth of the pinion gear will index with the teeth of the forward gear.

10- Once the pinion gear is in place, start the threads of the pinion gear nut. Tighten the nut as much as possible by rotating the driveshaft with one hand and holding the nut with the other hand. The next step tightens the nut.

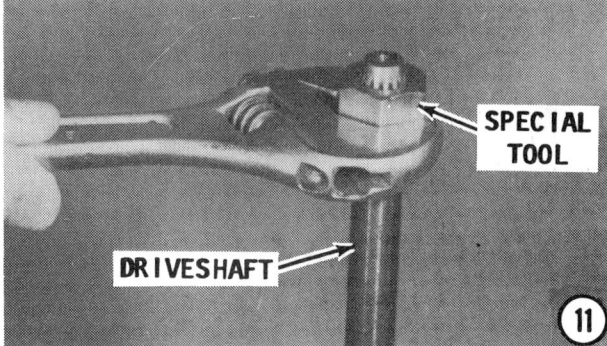

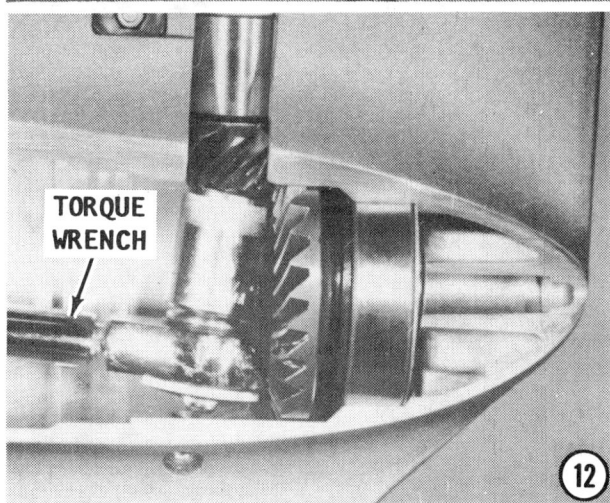

11- Tighten the pinion gear nut as follows: Obtain driveshaft tool P/N YB6151 and a socket the same size as the pinion gear nut. Now, place the tool over the splines on the upper end of the driveshaft.

12- Position another socket and torque wrench on the tool. Reach into the lower unit cavity and hold the pinion gear nut steady with the proper size socket. Rotate the driveshaft **CLOCKWISE** until the torque wrench indicates a torque value of 68 ft lbs (95Nm).

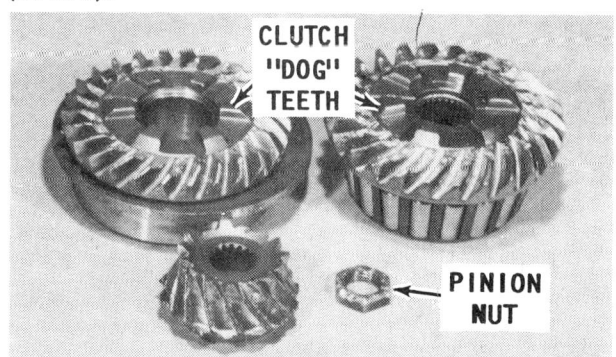

The damage to this gear set could possibly have been caused by failure to tighten the pinion gear nut to the required torque value. The pinion nut may have shaken loose, allowing the pinion gear to lower, drastically changing the backlash and gear mesh pattern, resulting in the damaged gears. Notice the clutch "dog" engaging teeth were not affected, indicating the gear damage was not caused by a "speed shift".

11-30 LOWER UNIT

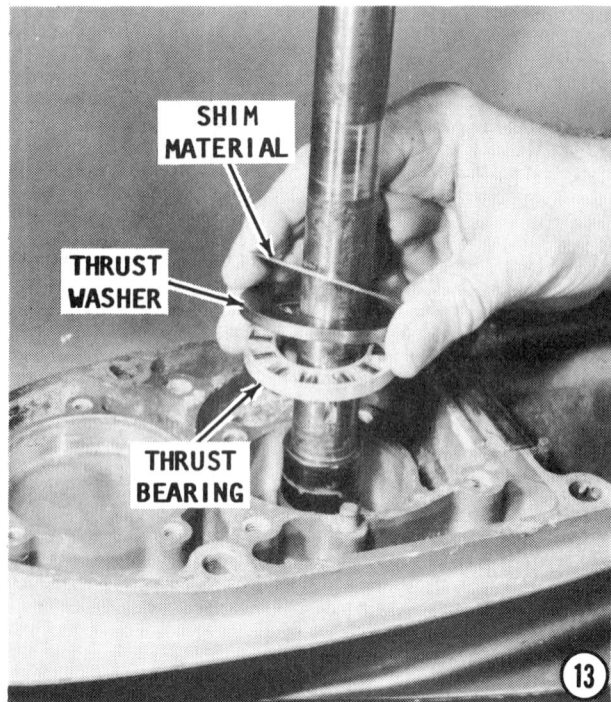

13- Slide the thrust bearing and then the thrust washer over the driveshaft. Next, install the same amount of shim material saved during disassembling, Step 9. The shim material should give the same backlash between the pinion gear and the other two gears as was obtained prior to disassembling.

SPECIAL WORDS

Perform the next step **ONLY** if the needle bearing set or oil seals were removed from the oil seal housing during disassembling Step 8. If the needle bearing set was not disturbed: apply a coat of Yamalube to the inner bearing surfaces; check to be sure all the needle bearings are still in place; and then proceed directly to Step 15.

Driveshaft Oil Seals

SPECIAL WORDS FROM YAMAHA ENGINEERS

Yamaha engineers have concluded it is better to have both seals face the same direction rather than "back to back" as directed by other outboard manufacturers.

14- Place the oil seal housing on a suitable surface, such as a small block of wood, in the same position with the correct end facing up (as it is positioned after installation).

Obtain driver P/N YB6196 and handle P/N YB6071. Insert the driver into the top of the bearing, with the numbers embossed on the bearing facing **UPWARD**. Attach the handle to the driver and tap the bearing into the oil seal housing until it seats.

If servicing a V4 unit, obtain handle P/N YB6071 and seal installer P/N YB6016. If servicing a V6 unit, obtain P/N YB6196.

Lower the first seal **SQUARELY** into the top of oil seal housing recess, with the lip of the seal facing **UPWARD**. Place the seal installer over the seal and attach the handle. Tap the end of the handle with a hammer until the seal is fully seated. After installation, pack the seal with Yamaha

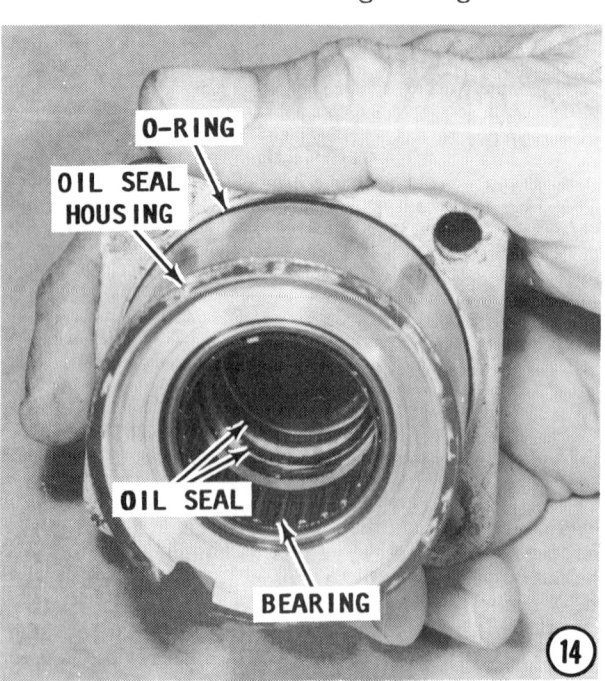

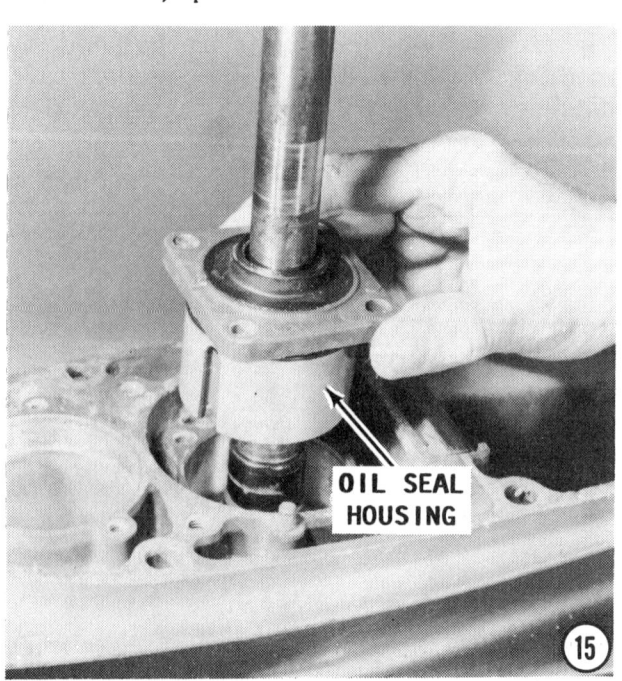

ASSEMBLING 11-31

Grease "A" or equivalent water resistant lubricant.

Install the second seal in the same manner, with the seal lip facing **UPWARD**. Pack the second seal with grease.

Install a new O-ring around the oil seal housing. Coat the O-ring with Yamalube All Purpose Grease.

15- Lower the oil seal housing down over the driveshaft. Tap it lightly to seat it properly in the lower unit. Check to be sure the bolt holes align.

16- Install and tighten the four bolts securely.

Pinion Gear Depth

SPECIAL WORDS

The proper amount of pinion gear depth (pinion gear engagement with the forward gear), is critical for proper operation of the lower unit and long life of the internal parts.

17- Grasp the driveshaft and **PULL** up. At the same time, check the pinion gear tooth engagement with the forward gear teeth (reverse gear on a counterrotating lower unit), to be sure contact is made the full length of the tooth. This can be accomplished by using a flashlight and looking into the lower unit opening. If the pinion gear depth is **NOT** correct, the amount of shim material behind the pinion gear must be adjusted. The addition or removal of shim material behind the pinion gear will not affect the gear backlash procedures, beginning on Page 11-38.

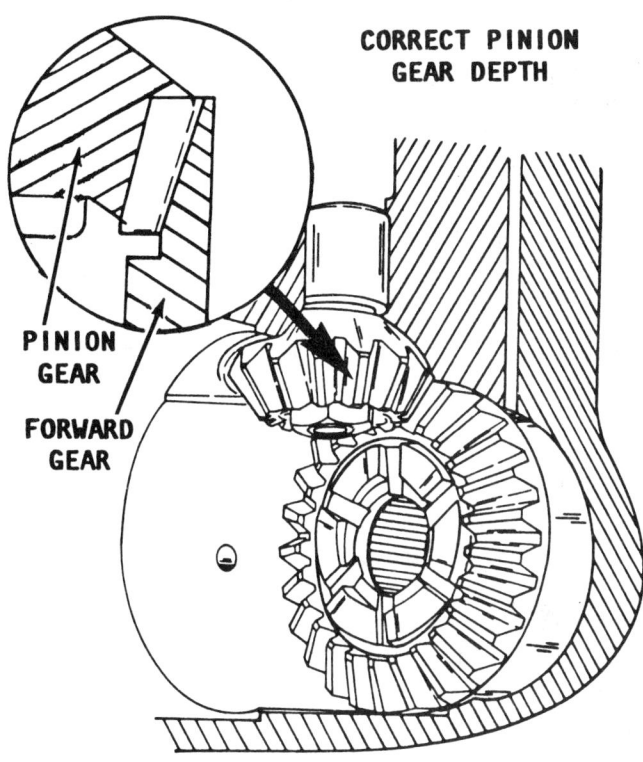

CORRECT PINION GEAR DEPTH

PINION GEAR
FORWARD GEAR

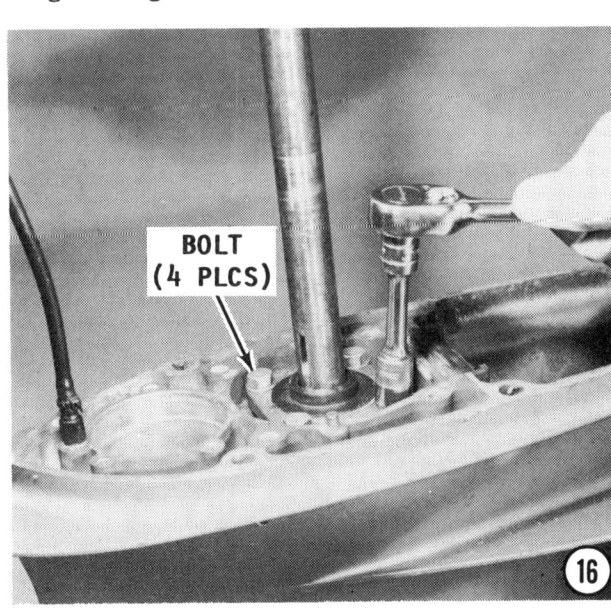

BOLT (4 PLCS)

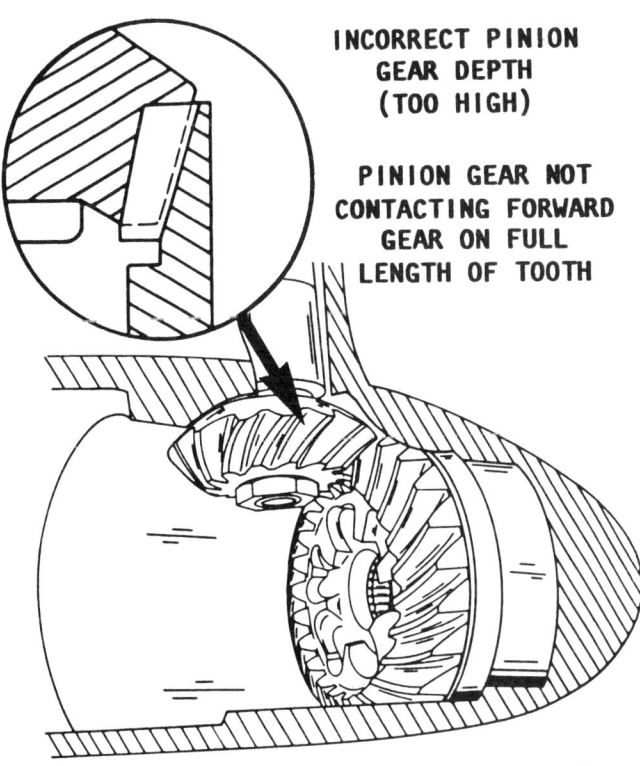

INCORRECT PINION GEAR DEPTH (TOO HIGH)

PINION GEAR NOT CONTACTING FORWARD GEAR ON FULL LENGTH OF TOOTH

11-32 LOWER UNIT

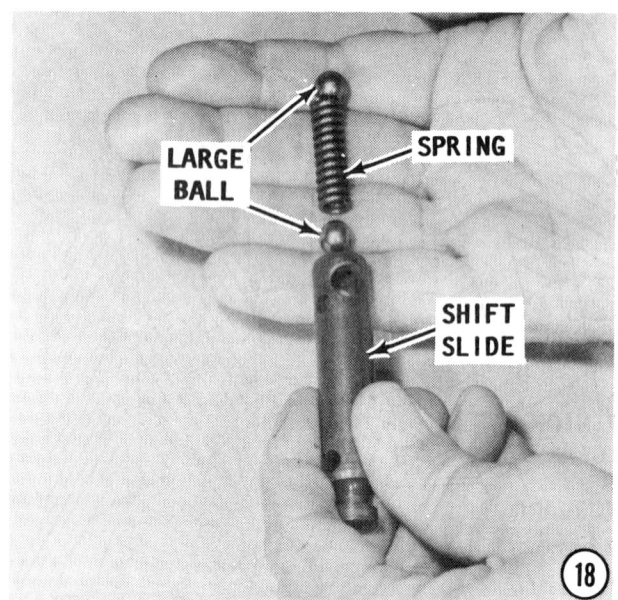

Clutch "Dog" Installation

18- Insert one large ball bearing, the compression spring, and the other large ball bearing into the shift slide.

19- Apply just a dab of grease to the two smallest balls to help keep them in place, and then insert them into the holes nearest the open end of the shift slide. Insert the shift slide into the propeller shaft with the large hole in the slide aligned with the slot in the propeller shaft. Continue to insert the shift slide into the propeller shaft until both small balls enter the shaft -- then stop.

20- Apply a dab of grease to the medium size balls to help keep them in place, and then insert them into the holes nearest the

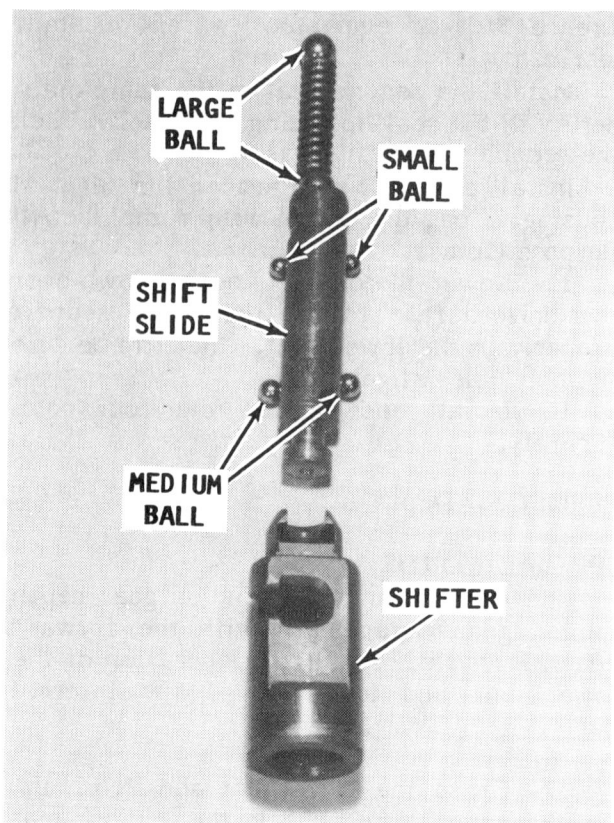

Arrangement of the three different size ball bearings around the shift slide.

necked end of shift slide. Continue to insert the shift slide into the propeller shaft until both medium balls are about to enter the shaft -- then stop.

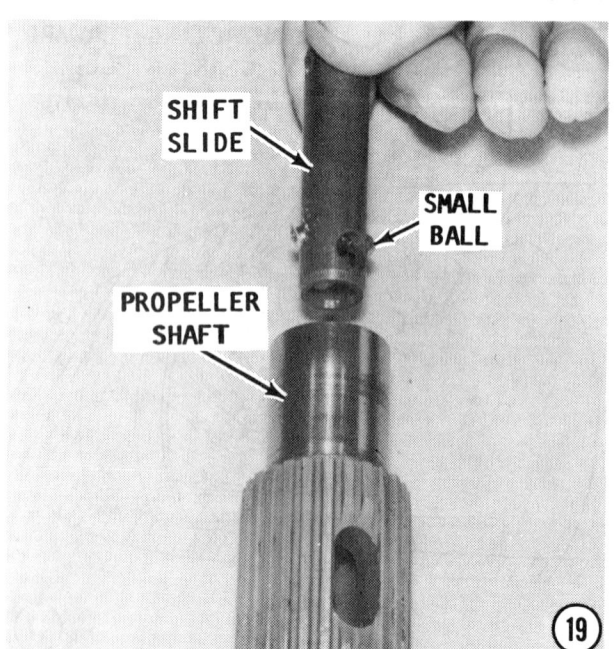

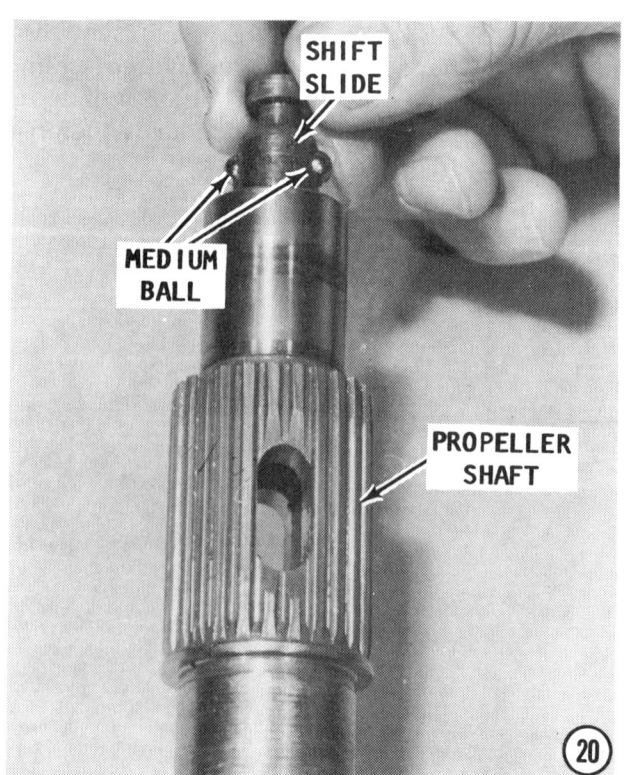

ASSEMBLING 11-33

21- Hook the shifter onto the knob on the end of the shift slide. Push in the shift slide until the medium balls enter the propeller shaft and seat themselves into internal grooves inside the shaft. Stop the action when a "click" is heard, indicating the two small balls have moved into the neutral groove in the propeller shaft. There is only one groove, so wiggle the slide back and forth until the ball is definitely seated in the groove. Refer to the three illustrations on Page 11-3 for a sectional view of this installation. This illustration will be a helpful guide in determining if the balls are seated properly.

22- Slide the clutch "dog" onto the propeller shaft, with the hole in the "dog" aligned with the two holes (hopefully already aligned), of the shaft and shift slide. Notice how both shoulders of the "dog" are an equal width, therefore the "dog" may be installed either way, usually with the least worn side facing the forward gear.

Insert the cross pin into the hole and push it through until the pin ends are flush with the clutch "dog" groove surface.

23- Wrap the cross pin retaining ring around the groove to retain the cross pin.

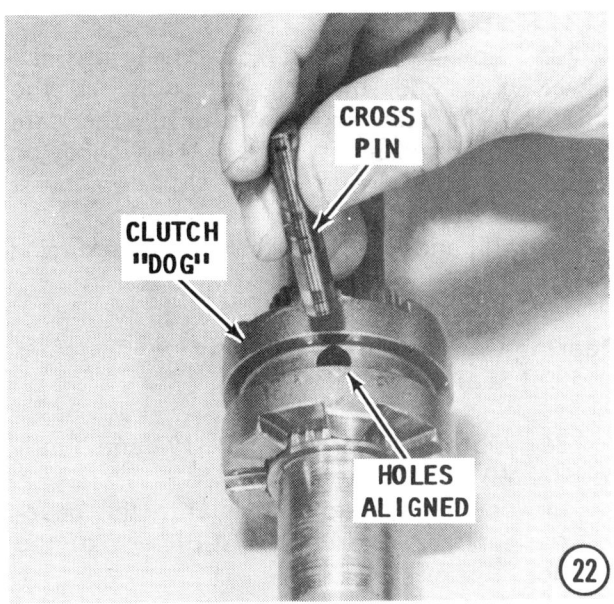

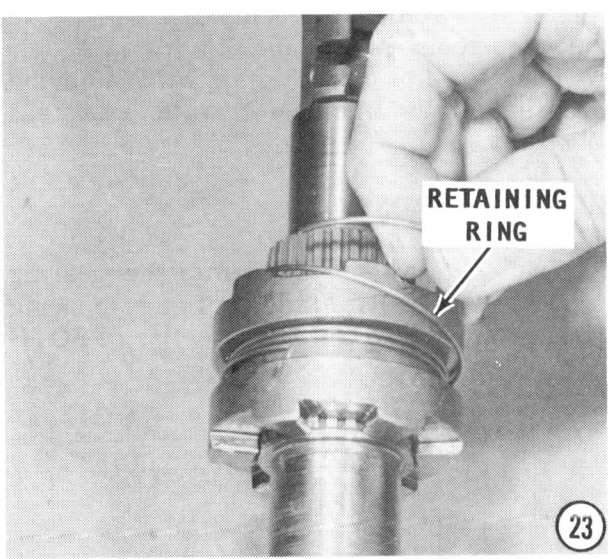

The ring may be wrapped in either direction. Check to be sure the ring "turns" do not overlap one another.

24- Guide the assembled propeller shaft into the lower unit.

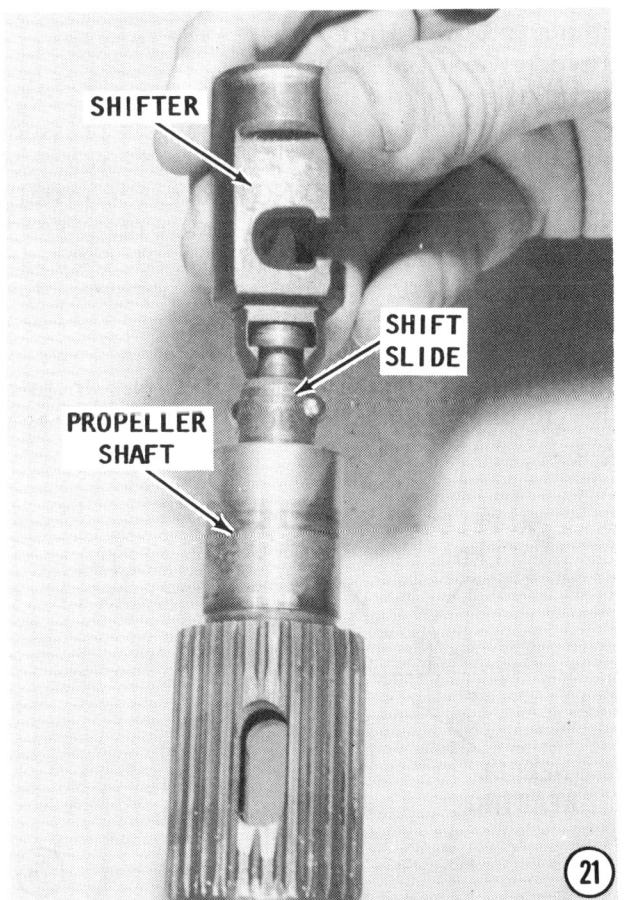

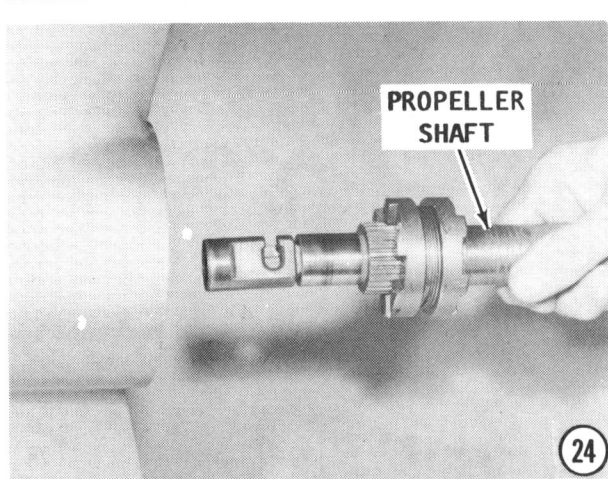

Shift Rod Installation

25- Lower the shift rod down into the lower unit and through the hole in the shifter. On standard lower units, the cam on the shift rod must face to starboard, on counterrotating lower units, this cam must face to port.

Install and tighten the three securing bolts.

Bearing Carrier Needle Bearing and Oil Seal Installation

26- Obtain driver P/N YB6196 and handle P/N YB6071.

Lower the needle bearing into the bearing carrier from the **PROPELLER** end of the carrier.

In all cases, install the bearing with the stamped mark on the bearing facing **UPWARD** toward the driver. Seat the bearing squarely into the bore, until the shoulder of the driver seats against the shoulder of the bearing housing. The bearing will then be correctly seated in the bearing carrier. Use the same tools and install both oil seals.

Pack the lips of both seals wth Yamalube Grease "A", or equivalent water resistant lubricant.

SPECIAL WORDS FROM YAMAHA ENGINEERS

Two seals are installed into the bearing carrier. After installation of the bearing carrier into the lower unit, the lips of **BOTH** seals face toward the propeller. Yamaha engineers have concluded it is better to have both seals face the same direction rather than "back to back" as directed by other outboard manufacturers. In this position, with both lips facing outward toward the water after bearing carrier installation, the engineers feel the seals will be more effective in keeping water out of the lower unit. Any lubricant lost will be negligible.

Obtain driver P/N YB6195 and handle P/N YB6071.

Both seals are driven into the bearing carrier from the **PROPELLER** end of the carrier with the lips of the seals facing **DOWNWARD** toward the driver.

Lower the first seal **SQUARELY** into the bearing carrier with the lip of the seal facing **DOWNWARD**. Place the seal installer over the seal and attach the handle. Tap the end of the handle with a hammer until the seal is fully seated. After installation, pack the seal with Yamaha Grease "A" or equivalent water resistant lubricant.

Install the second seal in the same manner, with the seal lip facing **DOWNWARD**. Pack the second seal with grease.

Standard Lower Units Reverse Gear Ball Bearing Installation

The reverse gear ball bearing is pressed onto the reverse gear, because the reverse gear is a press fit. **DO NOT** forget to install the thrust washer, between the reverse gear and the ball bearing assembly. This thrust washer will ensure proper mesh between the reverse gear and the pinion gear.

27- Position the reverse gear on a press with the gear teeth facing **DOWN**. Install the thrust washer on top of the reverse gear.

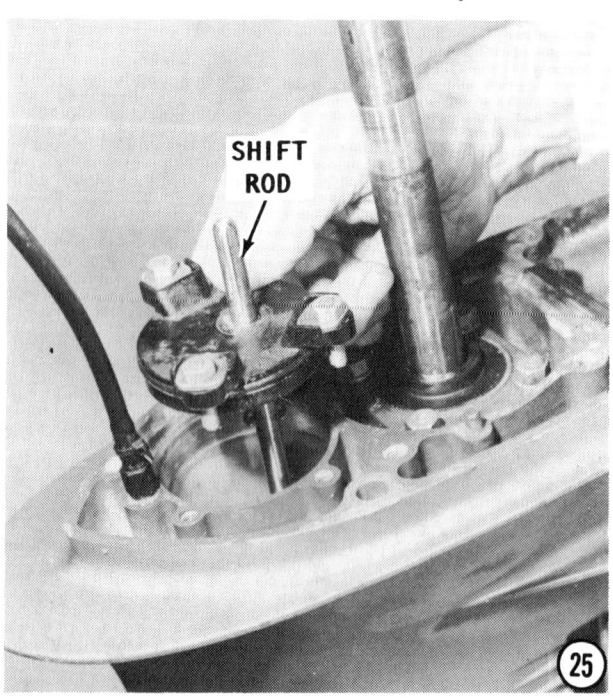

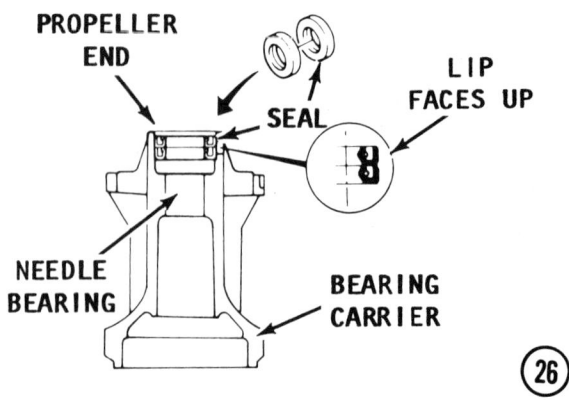

ASSEMBLING 11-35

Position the ball bearing assembly on top of the thrust washer, with the embossed marks on the bearing facing **UPWARD** toward the press shaft.

Now, use a suitable mandrel and press against the inner bearing race.

CRITICAL WORDS

Take care to ensure the mandrel is pressing on the inner race and not on the outer race or the ball bearings. Such action would destroy the bearing.

Continue to press the bearing into place until the bearing, thrust washer, and back of the reverse gear are all seated against each other.

28- Place the bearing carrier on the press with the lower unit end facing up, as shown. Fit the reverse gear ball bearing into the top of the bearing carrier. Obtain a suitable mandrel which will rest **ONLY** on the inner hub of the reverse gear. **DO NOT** use a mandrel which would apply pressure on the gear teeth.

Continue to press until the outer race of the bearing seats against the bearing carrier surface.

Counterrotating Lower Units- Short Propeller Shaft and Forward Gear Installation

29- Slide the thrust bearing, large thrust washer and spacer onto the short propeller

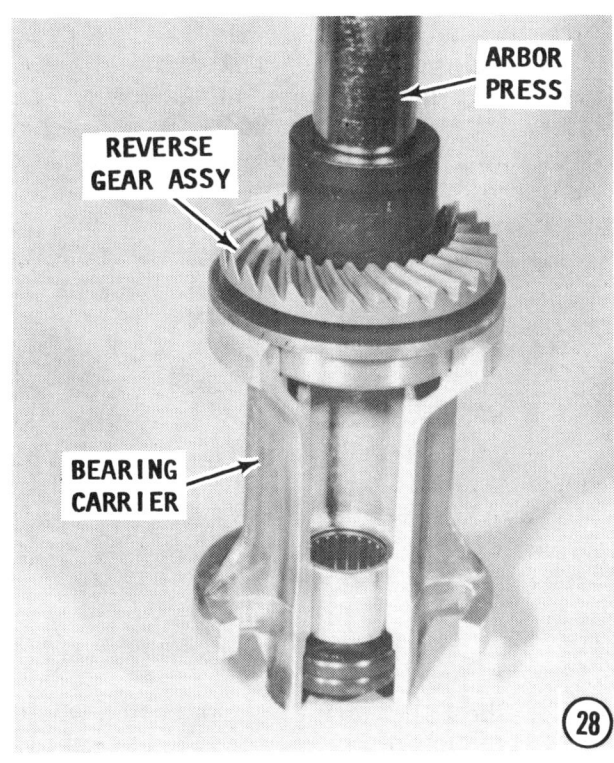

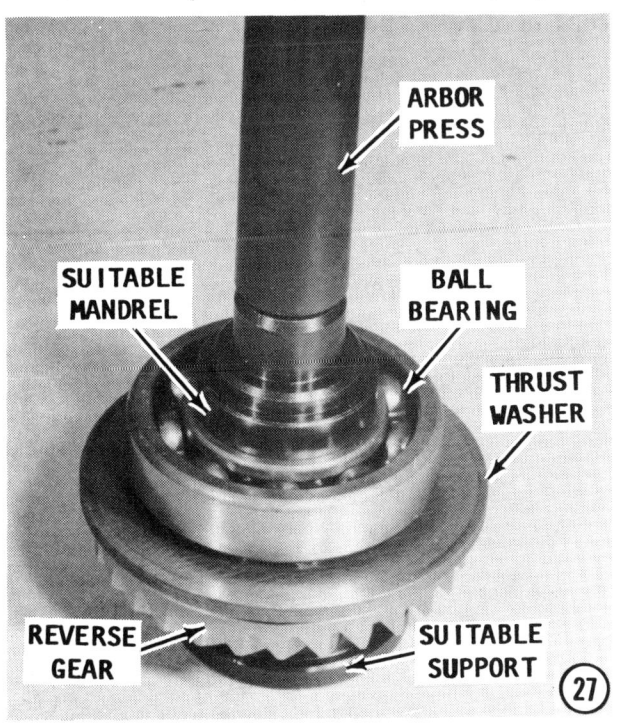

shaft. Insert the shaft into the bearing carrier.

30- Position the forward gear on a press with the gear teeth facing **DOWN**. Install the thrust bearing and then the large thrust washer on top of the reverse gear.

Position the ball bearing assembly on top of the thrust washer, with the embossed marks on the bearing facing **UPWARD** -- toward the press shaft.

Now, use a suitable mandrel and press against the inner bearing race.

CRITICAL WORDS

Take care to ensure the mandrel is pressing on the inner race and not on the outer

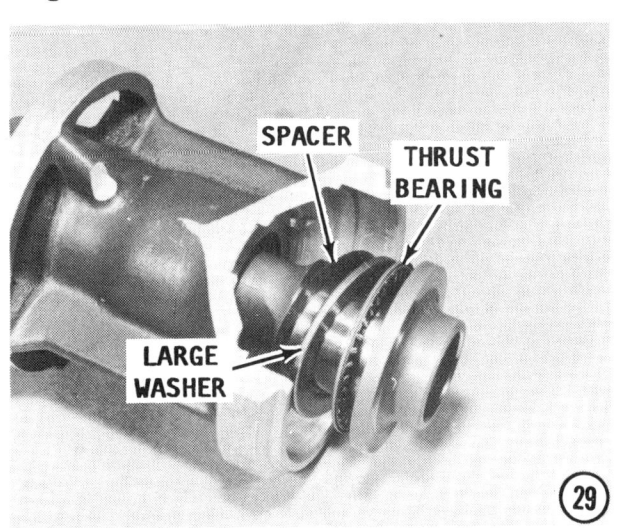

11-36 LOWER UNIT

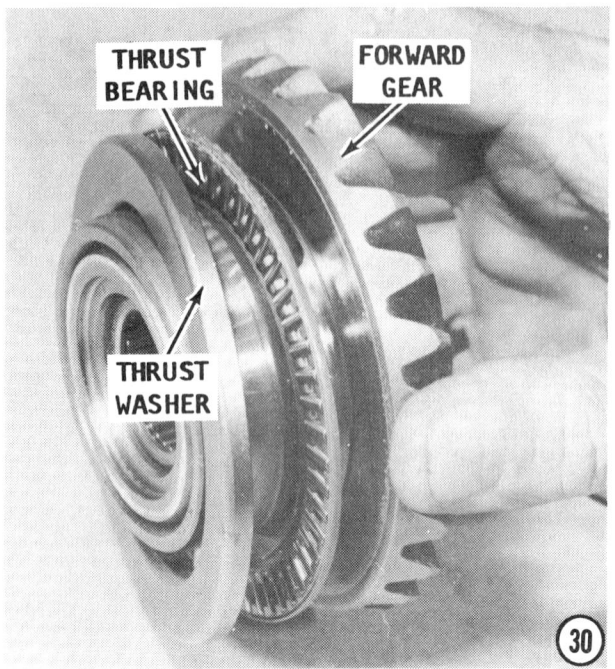

race or the ball bearings. Such action would destroy the bearing.

Continue to press the bearing into place until the ball bearing, thrust bearing, thrust washer, and back of the forward gear are all seated against each other.

Place the bearing carrier on the press with the lower unit end facing up, with the short propeller shaft inserted. Fit the for-

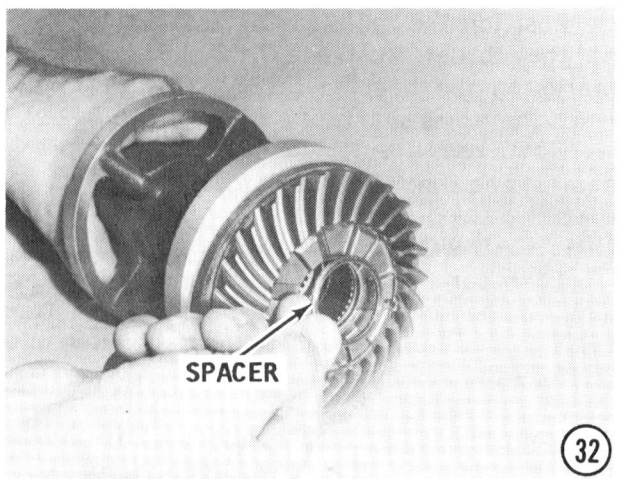

ward gear ball bearing into the top of the bearing carrier. Obtain a suitable mandrel which will rest **ONLY** on the inner hub of the forward gear. **DO NOT** use a mandrel which would apply pressure on the gear teeth.

Continue to press until the outer race of the bearing seats against the bearing carrier surface.

All Lower Units

31- Install a new O-ring into the outer groove of the bearing carrier. Apply a coating of Yamabond #4 to the O-ring and to its mating surface.

Coat the teeth of the reverse gear with a fine spray of Desenex, as instructed in the "Plan Ahead" paragraph prior to assembling Step 8. Handle the gear carefully to prevent disturbing the powder.

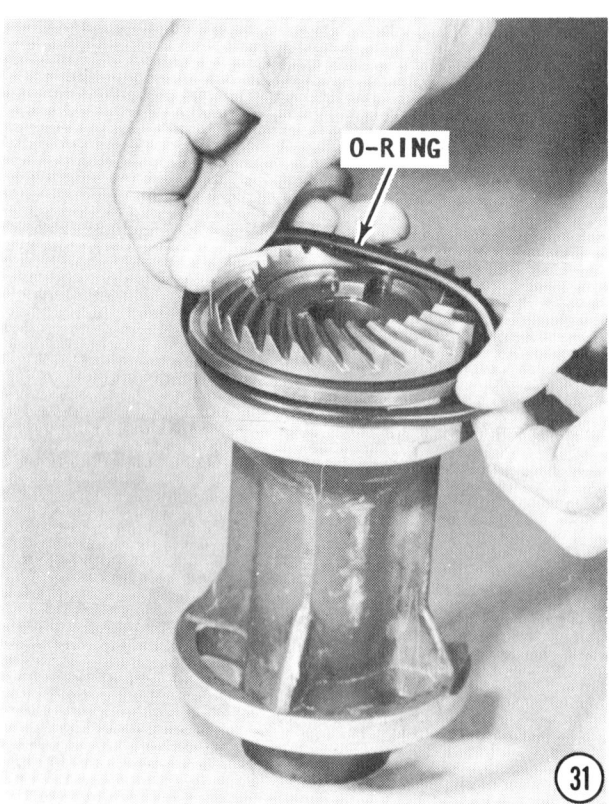

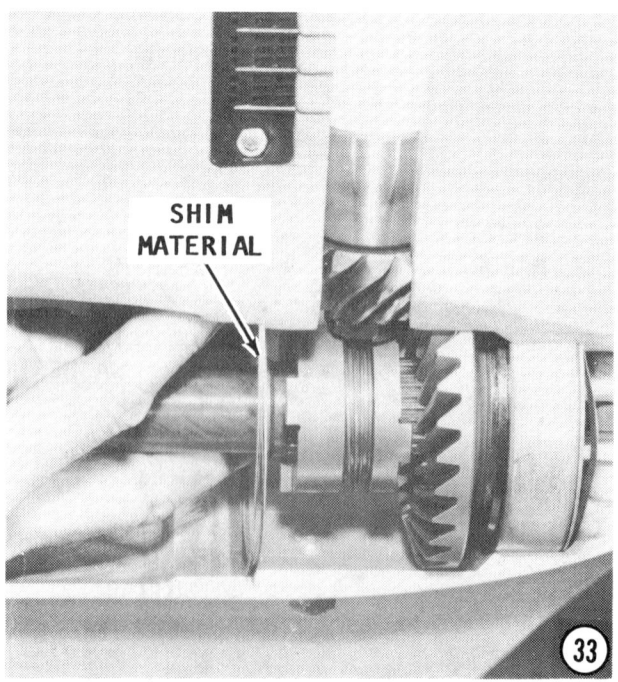

ASSEMBLING 11-37

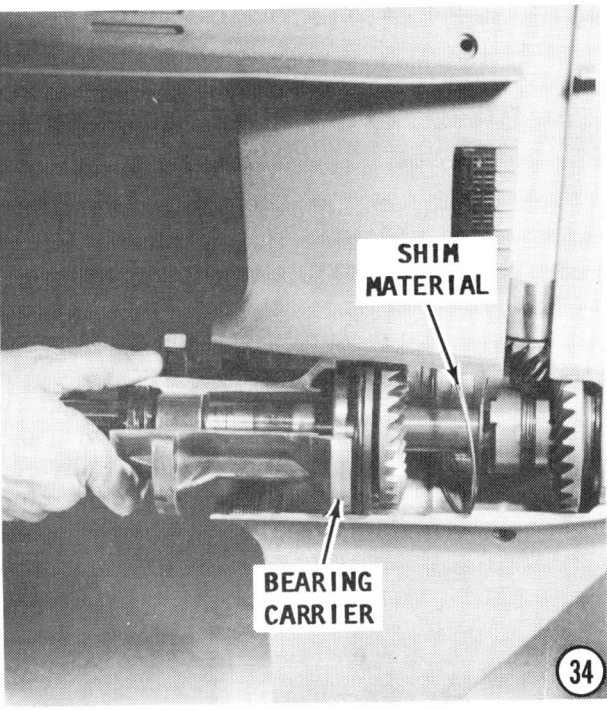

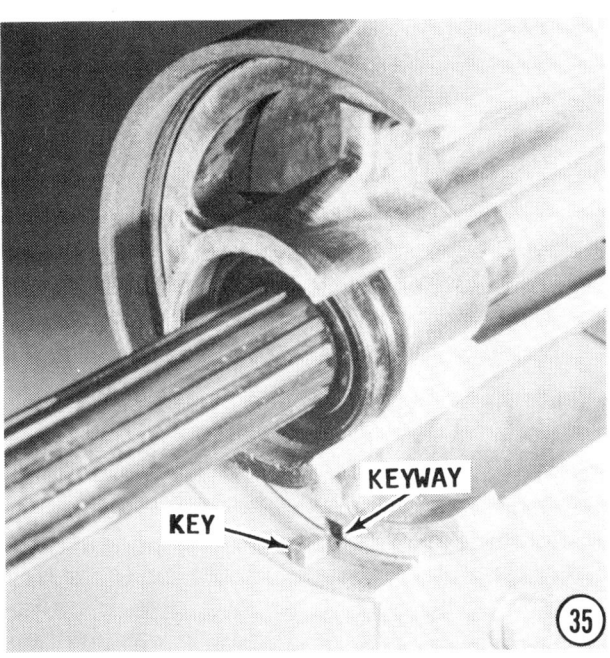

32- Install the spacer in front of the reverse gear (forward gear on counterrotating units).

33- Insert the same amount of shim material saved during disassembling, Step 15, over the installed propeller shaft and into the lower unit. The shim material should give the same backlash between the pinion gear and the reverse gear (forward gear on counterrotating units) as was obtained prior to disassembling.

34- Install the bearing carrier into the lower unit.

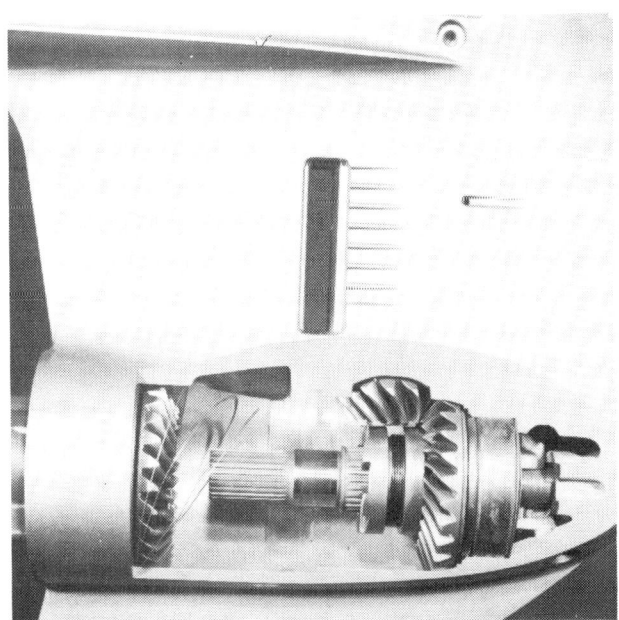

Installation of the bearing carrier into a counterrotating lower unit.

VERY SPECIAL WORDS
Assembling of parts at this time is **NOT** to be considered as final. The three gears are coated with the Desenex powder, or equivalent, to determine a gear pattern. Therefore, the assemblies will be separated to check the pattern. During final installation the two mounting bolts will be coated with Loctite, or equivalent.

If the assembler has omitted the application of the Desenex powder and does not have plans to check the gear pattern, then this step may be considered as the final assembly of the bearing carrier. Loctite, or equivalent, should be applied to the threads of the bearing carrier attaching bolts.

35- Align the keyway in the lower unit housing with the keyway in the bearing carrier. Insert the key into both grooves, and then push the bearing carrier into place in the lower unit housing.

11-38 LOWER UNIT

36- Install a tabbed washer against the bearing carrier, and then install the ring nut with the embossed marks facing **OUTWARD**, away from the bearing carrier.

37- Obtain special bearing carrier holding tool P/N YB34447. Tighten the locknut, in the direction indicated by the embossed **ON** mark to a torque value of 105 ft lb (145Nm) for V4 units and 140 ft lb (190Nm) for V6 units.

Remove the special tool.

GEAR PATTERN WORDS

Assembling of parts at this time is **NOT** to be considered as final. The three gears are coated with the Desenex powder, or equivalent, to determine a gear pattern. Therefore, the assemblies will be separated to check the pattern. During final installation, one or more of the lock washer tabs will be bent down over the locknut.

If the assembler has omitted the application of Desenex powder and does not have plans to check the gear pattern, then this step may be considered as the final assembly of the bearing carrier. If such is the case, bend one or more of the tabbed washer tabs down over the locknut to secure it in place.

PROPELLER

The propeller will be installed after the gear backlash measurements have been made, the water pump installed, and the lower unit attached to the intermediate housing.

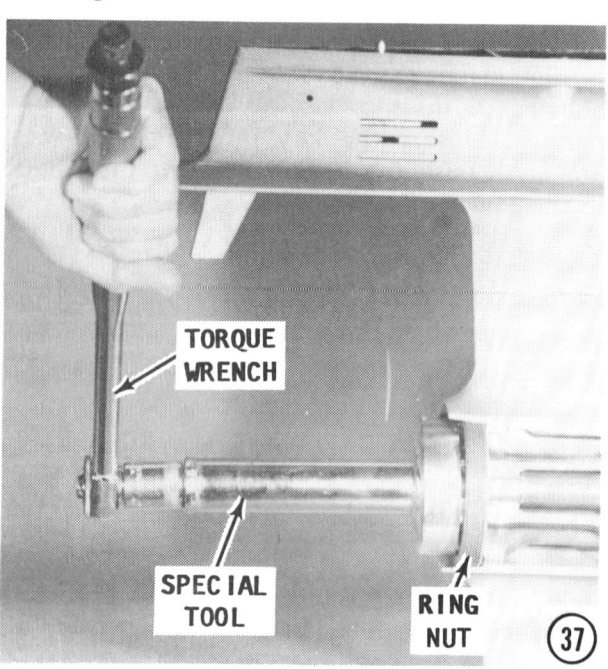

GEAR MESH PATTERN

This step is only necessary if Desenex, or similar material, was applied to the three gears prior to assembling.

38- Grasp the driveshaft and pull **UPWARD**. At the same time, rotate the propeller shaft **COUNTERCLOCKWISE** through about six or eight complete revolutions. This action will establish a wear pattern on the gears with the Desenex powder.

Now, disassemble the unit and compare the pattern made on the gear teeth with the accompanying illustrations. The pattern should almost be oval on the drive side and be positioned about halfway up the gear teeth.

If the pattern appears to be satisfactory, clean the dye or powder from the gear teeth and assemble the unit one final time.

If the pattern does **NOT** appear to be satisfactory, add or remove shim material, behind the pinion gear, as required. Adding or removing shim material will move the gear pattern towards or away from the center of the teeth.

After the gear mesh pattern is determined to be satisfactory, assemble the bearing carrier again to check for backlash, as described in the following paragraphs.

BACKLASH MEASUREMENT

Backlash Education

Backlash is the acceptable clearance between two meshing gears, in order to take

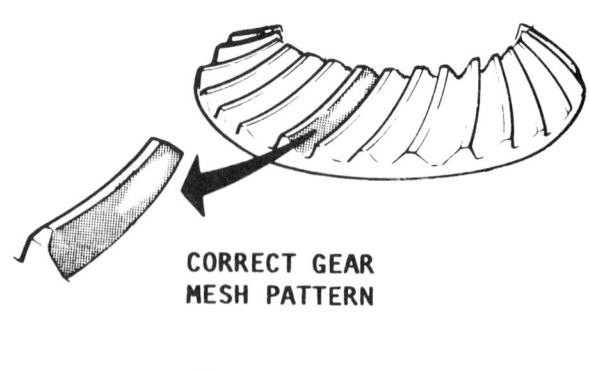

CORRECT GEAR MESH PATTERN

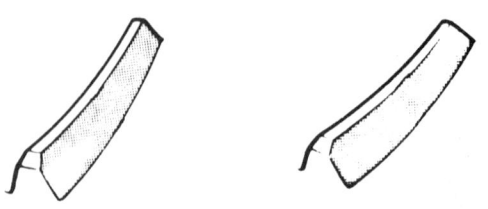

BOTH INCORRECT PATTERNS

BACKLASH MEASUREMENT   11-39

into account possible errors in machining, deformation due to load, expansion due to heat generated in the lower unit, and center-to-center distance tolerances. A "no backlash" condition is unacceptable, as such a condition would mean the gears are locked together or are too tight against each other which would cause phenomenal wear and generate excessive heat from the resulting friction.

Excessive backlash which cannot be corrected with shim material adjustment indicates worn gears. Such worn gears must be replaced. Excessive backlash is usually accompanied by a loud whine when the lower unit is operating in **NEUTRAL** gear.

The backlash is measured at this time before the water pump is installed. If the amount of backlash needs to be adjusted, the lower unit must be disassembled to change the amount of shim material behind one of the gears.

As a general rule, if the lower unit was merely disassembled, cleaned and then assembled with only a new water pump impeller, new gaskets, seals and O-rings, there is no reason to believe the backlash would have changed. Therefore, it is safe to say this next section may be skipped.

HOWEVER, if any one or more of the following components were replaced, the gear backlash should be checked for possible shim adjustment:

New lower unit housing -- check forward and reverse gear shim material and pinion gear depth.

New forward gear tapered roller bearing -- check forward gear backlash (on standard lower units).

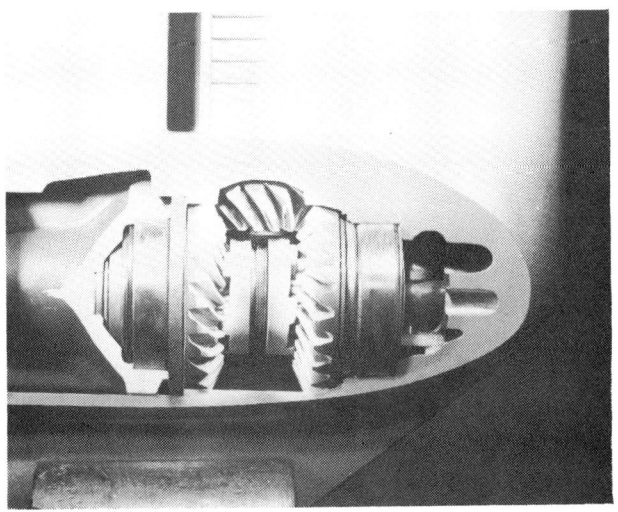

Wear pattern on a gear set sprayed with Desenex foot powder, as described in the text.

New reverse gear tapered roller bearing --check reverse gear backlash (on counterrotating units).

New reverse gear ball bearing -- check reverse gear backlash (on standard lower units).

New forward gear ball bearing -- check forward gear backlash (on counterrotating lower units).

New pinion gear -- check pinion gear depth.

New forward gear -- check forward gear backlash.

New reverse gear -- check reverse gear backlash.

New bearing carrier -- check reverse gear backlash (on standard lower units), check forward gear backlash (on counterrotating lower units).

New drive shaft -- check pinion gear depth.

New driveshaft thrust bearing, thrust washer or spacer - check pinion gear depth.

New driveshaft oil seal housing -- check pinion gear depth.

New forward gear thrust bearing, thrust washer or spacer - check forward gear backlash.

New reverse gear thrust bearing, thrust washer or spacer - check reverse gear backlash.

**Standard Lower Unit --
Forward Gear Backlash Measurement
Counterrotating Lower Unit --
Reverse Gear Backlash Measurement**

39- Install the bearing carrier puller and "J" bolts onto the ribs of the carrier. With the bearing carrier attaching bolts or the bearing carrier locknut secured, the puller locks the shaft and prevents rotation in any direction. The puller will also lock the lower unit in **FORWARD** gear (on standard lower units) or **REVERSE** gear (on counterrotating lower units).

Obtain backlash adjusting plate P/N YB7003. Secure the plate over the top of the lower unit with the nut and bolt provided with the tool.

Secure a dial indicator to the plate.

Obtain and install backlash indicator gauge P/N YB6265 onto the driveshaft. Adjust the end of the dial indicator to rest on the mark on the gauge.

CRITICAL WORDS

After the dial indicator is in place and all is secure, leave the lower unit in the **UPSIDEDOWN** position. The forward gear and reverse gear backlash is measured for these models with the lower unit in this position.

Backlash Measurement

Slowly rotate the driveshaft, "rocking" the shaft back and forth through about a 25° to 30° arc, without any rotation of the propeller shaft. At the outer limits of the arc, a "click" will be heard. This "click" sound occurs when the pinion gear tooth contacts one face of the driven gear tooth. Another "click" is heard when the pinion gear swings back, and the pinion tooth contacts the face of the adjacent driven gear tooth. The arc between the two "clicks" represents the backlash, or free play, between the gears.

Zero the dial indicator gauge at the first "click". As soon as the second "click" is heard, stop all motion and observe the maximum deflection of the dial indicator needle. Acceptable backlash measurement is as follows for the models listed:

V4 units   0.013-0.018" (0.32-0.45mm)
V6 units   0.011-0.015" (0.28-0.39mm)
  w/standard lower unit.
V4 and V6   0.038-0.049" (0.97-1.25mm)
  w/counterrotating lower unit.

Remove or add shim material behind the forward gear (on standard lower units) or reverse gear (on counterrotating lower units), to bring the backlash within specifications.

On Standard Lower Units

Forward gear -- adding shim material **DECREASES** backlash.

Forward gear -- removing shim material **INCREASES** backlash.

On Counterrotating Lower Units

Reverse gear -- adding shim material **DECREASES** backlash.

Reverse gear -- removing shim material **INCREASES** backlash.

HELPFUL WORDS

The actual measurement and the amount of shim material to be added or removed are related but **NOT** equal. Therefore, some very simple arithmetic must be performed.

To determine how much shim material must be moved to obtain the desired backlash, the manufacturer gives a very simple formula:

A constant (a number depending on the unit being serviced) minus the backlash measurement, and then the answer multiplied by a multiplying factor. This may sound complicated but it is **NOT**.

Formula:
(Measurement - constant) x factor.

If the constant has a greater value than the measurement, the formula becomes:
(Constant - measurement) x factor.

The constant and the multiplying factor vary, depending on the model hp unit being serviced. **NO PROBLEM!** The following list indicates the exact constant and multiplying factor for each of the powerheads covered in this manual.

Model	Constant Inches (mm)	Multiplying Factor
V4	0.015 (0.39)	1.25
V6	0.013 (0.34)	1.45

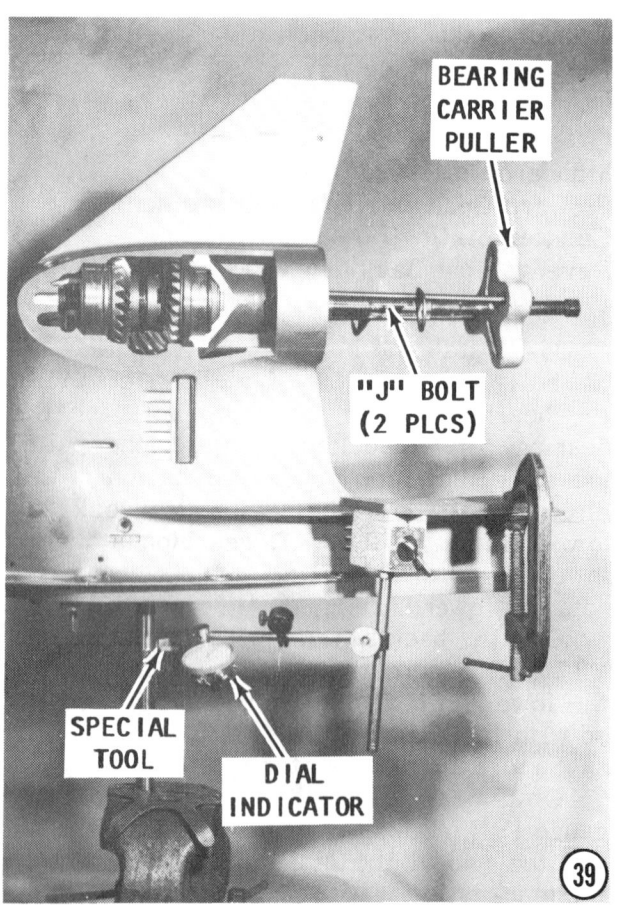

NOTE:
The millimeter answer should be the same if the original measurement was made in millimeters and the millimeter constant used, instead of the measurement being made in inches.

EXAMPLE
Model 115hp
w/standard lower unit

From the above listing, observe the constant for these units is 0.015, if measuring in inches and 0.39, if measuring in millimeters. Also, the multiplying factor is 1.25 for both inches and millimeters.

Assume the measurement to be 0.020" (0.5mm). The amount of shim material required is determined as follows:

Working in inches:
(Measurement - constant) x factor = answer.
(0.020 - 0.015) x 1.25 = 0.006" (0.14mm)

Working in millimeters:
(Measurement - constant) x factor = answer.
(0.5 - 0.39) x 1.25 = 0.14mm (0.006")

Now, the measurement was 0.020" (0.5mm) and the maximum permitted is 0.018" (0.45mm), as listed at the beginning of this gear backlash section. To decrease backlash **ADD** 0.006" (0.14mm) shim material.

Use this procedure, measurement, and formula, in the same sequence for other models covered in this manual.

CRITICAL WORDS
If the gear backlash specification cannot be reached by adding and removing shim material, all three gears may have to be replaced.

Remove the bearing carrier puller and "J" bolts from the propeller shaft.

Standard Lower Unit --
Reverse Gear Backlash Measurement
Counterrotating Lower Unit --
Forward Gear Backlash Meaurement

40- Obtain Yamaha special tool P/N YB6052. Place the tool over the installed shift rod and shift the lower unit into reverse gear. Rotate the driveshaft **CLOCKWISE** and check to be sure the propeller shaft rotates in the correct direction, **COUNTERCLOCKWISE** on standard lower units and **CLOCKWISE** on counter rotating lower units. This is a test to see if the clutch mechanism was properly assembled and functions correctly. Operate the shift rod to shift the lower unit back into **NEUTRAL**. Rotate the driveshaft **CLOCKWISE** 20° to 30° with the other hand. This action will preload the reverse gear.

Hold the propeller to prevent the propeller shaft from turning. At the same time, "rock" the driveshaft back and forth through about a 25° to 30° arc, without any rotation of the propeller shaft. At the outer limits of the arc, a "click" sound will be heard. This "click" sound occurs when the pinion gear tooth contacts one face of the driven gear tooth. Another "click" is heard when the pinion gear swings back, and the pinion tooth contacts the face of the adjacent driven gear tooth. The arc between the two "clicks" represents the backlash, or free play, between the gears.

Zero the dial indicator gauge at the first "click". As soon as the second "click" is heard, stop all motion and observe the maximum deflection of the dial indicator needle. Acceptable backlash measurement is as follows for the models listed:

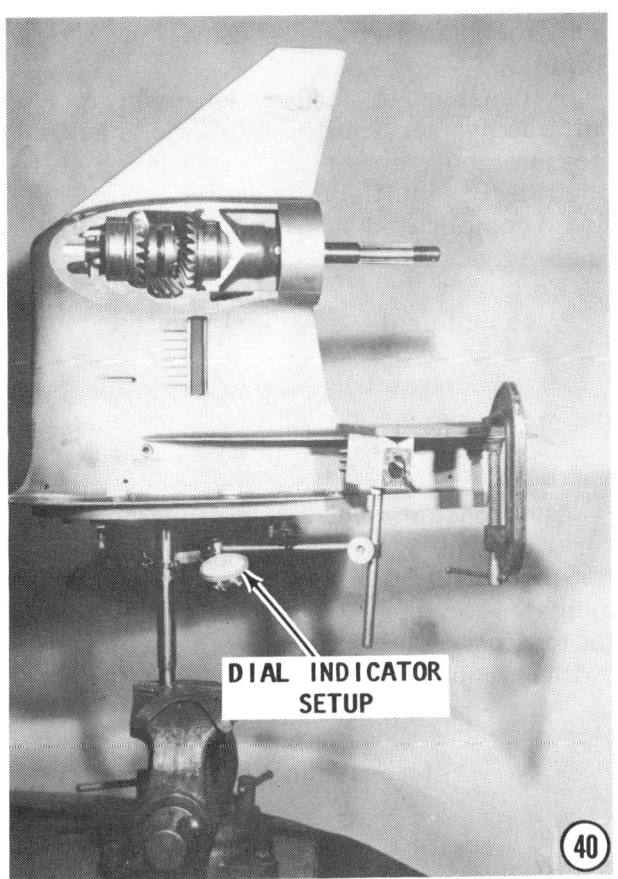

DIAL INDICATOR SETUP

40

V4 units   0.031-0.044" (0.80-1.12mm)
V6 units   0.038-0.049" (0.97-1.25mm)
w/standard lower unit.

V4 and V6   0.011-0.015" (0.28-0.39mm) units w/counterrotating lower unit.

Remove or add shim material behind the gear to bring the backlash to within specifications.

On Standard Lower Units
Reverse gear -- adding shim material **DECREASES** backlash.

Reverse gear -- removing shim material **INCREASES** backlash.

On Counterrotating Lower Units
Forward gear -- adding shim material **DECREASES** backlash.

Forward gear -- removing shim material **INCREASES** backlash.

HELPFUL WORDS
The actual measurement and the amount of shim material to be added or removed are related but **NOT** equal. Therefore, some very simple arithmetic must be performed.

To determine how much shim material must be moved to obtain the desired backlash, the manufacturer gives a very simple formula:

A constant (a number depending on the unit being serviced) minus the backlash measurement, and then the answer multiplied by a multiplying factor. This may sound complicated but it is **NOT**.

Formula:
(Measurement - constant) x factor.

If the constant has a greater value than the measurement, the formula becomes:
(Constant - measurement) x factor.

The constant and the multiplying factor vary, depending on the model hp unit being serviced. **NO PROBLEM!** The following list indicates the exact constant and multiplying factor for each of the powerheads covered in this manual.

Model	Constant Inches (mm)	Multiplying Factor
V4	0.043 (1.11)	1.45
V6	0.038 (0.96)	1.25

NOTE:
The millimeter answer should be the same if the original measurement was made in millimeters and the millimeter constant used, instead of the measurement being made in inches.

EXAMPLE
Model 200hp
From the above listing, observe the constant for these units is 0.038, if measuring in inches and 0.96, if measuring in millimeters. Also, the multiplying factor is 1.25 for both inches and millimeters.

Assume the measurement to be 0.034" (0.86mm). The amount of shim material required is determined as follows:

Working in inches:
(Constant - measurement) x factor = answer.
(0.038 - 0.034) x 1.25 = 0.005" (0.13mm)

Working in millimeters: (Constant - measurement) x factor = answer.
(0.96 - 0.86) x 1.25 = 0.13mm (0.005")

The measurement was 0.034" (0.86mm). The minimum permitted is 0.038" (0.96mm), as listed at the beginning of this gear backlash section. To increase backlash **REMOVE** 0.005" (0.13mm) shim material.

Use this procedure, measurement, and formula, in the same sequence for other models covered in this manual.

CRITICAL WORDS
If the gear backlash specification cannot be reached by adding and removing shim material, all three gears may have to be replaced.

After the proper amount of backlash and pinion gear depth (gear pattern) has been obtained, install the bearing carrier using the special instruction outlined in Step 34.

41- Bend down one or more lockwasher tabs over the locknut.

WATER PUMP INSTALLATION

42- Place a new gasket and the outer plate over the driveshaft and onto the oil seal housing with the holes in the plate and gasket indexed over the two pins in the housing.

BACKLASH MEASUREMENT 11-43

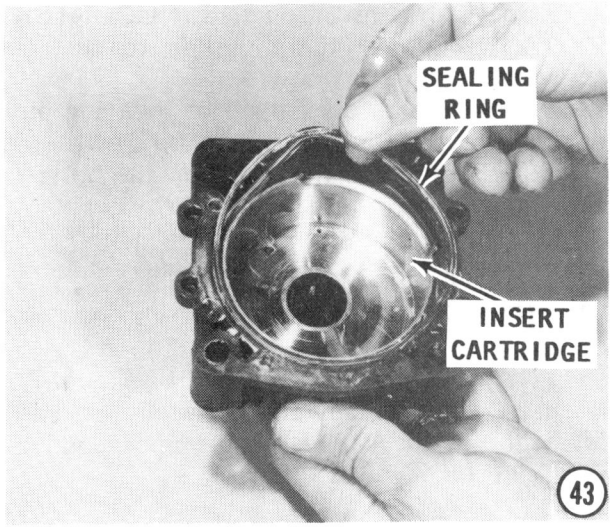

43- Fit the insert cartridge into the water pump housing. Push the water tube grommet into place in the housing. Apply a coat of Yamalube to the sealing ring of the water pump housing and place the ring in the groove of the housing.

44- Fit the Woodruff key into the driveshaft. Just a dab of grease on the key will help to hold the key in place.

Slide the water pump impeller over the driveshaft with the rubber membrane on the top side and the keyway in the impeller indexed over the Woodruff key. **TAKE CARE** not to damage the membrane. Coat the impeller blades with Yamalube Grease "A" or equivalent water resistant lubricant.

The water pump housing is manufactured for use on either a V4 or V6 unit. Therefore, one hole on each side of the housing will not be used for each installation.

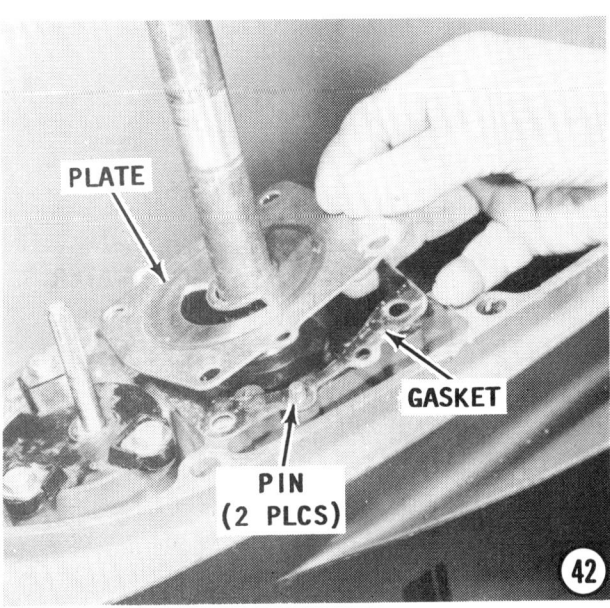

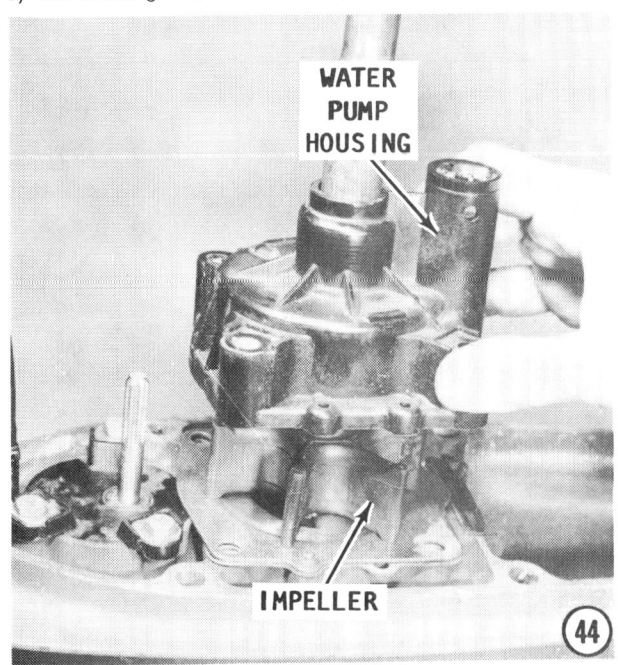

11-44 LOWER UNIT

Rotate the insert cartridge **COUNTERCLOCKWISE** over the impeller to tuck in the vanes.

Seat all parts over the two locating pins and secure the water pump housing with the four washers and bolts. Tighten the bolts to a torque value of 5.8 ft lbs (8Nm).

INSTALLATION — LOWER UNIT TO INTERMEDIATE HOUSING

1- Apply just a **"Dab"** of Yamalube Grease "A", or equivalent water resistant lubricant to the splines at the upper end of the driveshaft.

BAD NEWS

An excessive amount of lubricant on top of the driveshaft to crankshaft splines will be trapped in the clearance space. This trapped lubricant will not allow the driveshaft to fully engage with the crankshaft.

Apply some of the same lubricant to the end of the water tube in the intermediate housing.

Apply just a **"Dab"** of Yamalube Grease "A" or equivalent water resistant lubricant, to the indexing pin on the mating surface of the lower unit. Obtain Yamaha special tool P/N YB6052 and check to see if the lower unit is in **NEUTRAL** gear position, and then push the shift lever on the powerhead into the neutral gear position.

Begin to bring the intermediate housing and lower gear housing together.

SPECIAL WORDS

The next step takes time and patience. Success will probably not be achieved on the first attempt. **THREE** items must mate at the same time before the lower unit can be seated against the intermediate housing.

a- The top of the driveshaft on the lower unit indexes with the lower end of the crankshaft.

b- The water tube in the intermediate housing slides into the grommet on the water pump housing.

c- The top splines of the lower shift rod in the lower unit slide into the internal splines of the upper shift rod in the intermediate housing.

2- As the two units come closer, rotate the propeller shaft slightly to index the upper end of the lower driveshaft tube with the crankshaft. At the same time, feed the water tube into the water tube grommet, and feed the lower shift rod into the upper shift rod.

Push the lower unit housing and the intermediate housing together. The pin on the upper surface of the lower unit will index with a matching hole in the lower surface of the anti-cavitation plate.

WORDS FROM EXPERIENCE

If all three items appear to mate properly, but the lower unit seems locked in position about 4 inches (10cm) away from the intermediate housing and it is not possible, with ease, to bring the two housings closer, the driveshaft has missed the cylindrical lower oil seal housing leading to the crankshaft.

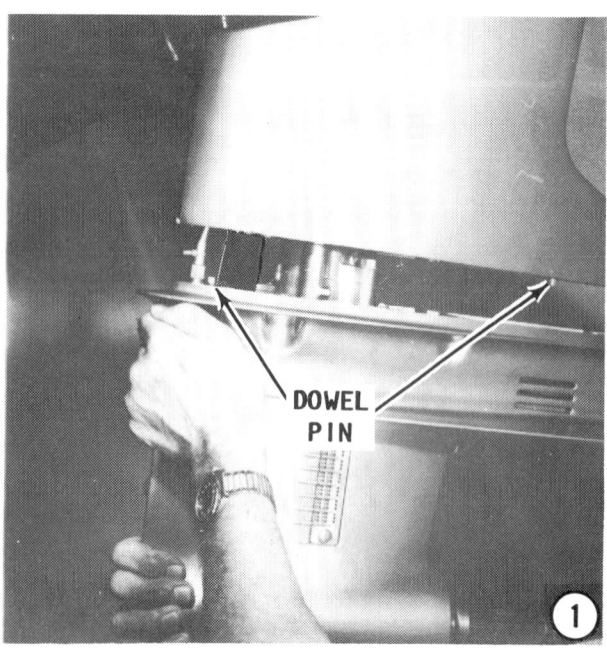

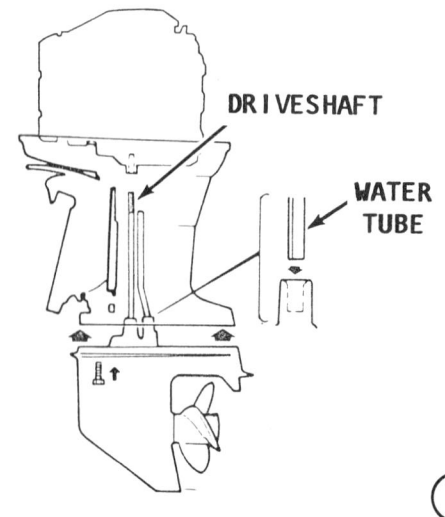

Move the lower unit out of the way. Shine a flashlight up into the intermediate housing and find the oil seal housing. Now, on the next attempt, "find" the edges of the oil seal housing with the driveshaft before trying to mate anything else -- the water tube and the top of the shift rod. If the driveshaft can be made to enter the oil seal housing, the driveshaft can then be easily indexed with the crankshaft.

3- Apply Loctite to the threads of the six bolts used to secure the lower unit to the intermediate housing. Remember, one bolt passes through the area covered by the trim tab. Install and tighten the bolts to a torque value of 29 ft lbs (40Nm).

4- Apply Loctite to the threads of the bolt located **UNDER** the trim tab, install and tighten this bolt to the same torque specification as given in the previous step.

Install and secure the trim tab with the attaching hardware.

5- Slide the two ends of the pilot tube together at the push-on fitting. Secure the tube to the swivel bracket using a new band retainer.

Check the Shifting

Operate the shift lever through all gears. The shifting should be smooth and the propeller should rotate in the proper direction when the flywheel is rotated by hand in a **CLOCKWISE** direction. Naturally the propeller should not rotate when the unit is in neutral.

Propeller Installation

GOOD WORDS

An anti-seizing compound will prevent the propeller from "freezing" to the shaft and permit propeller removal, without diffi-

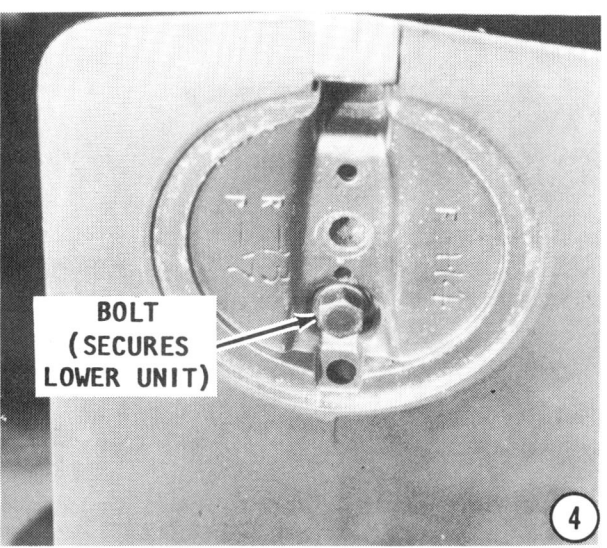

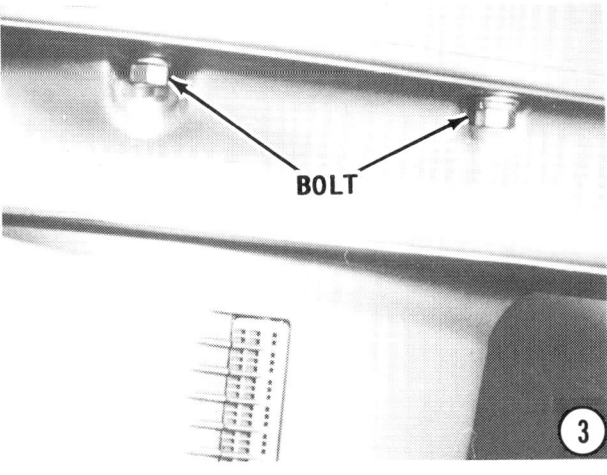

culty, the next time the propeller needs to be "pulled".

6- Apply Yamalube Grease "A", or equivalent anti-seizing compound, to the propeller shaft. Install the propeller as directed by the procedures in Section 11-7, this chapter.

Remove the block of wood. Connect the spark plug wires to the spark plugs. Connect the electrical lead to the battery terminal.

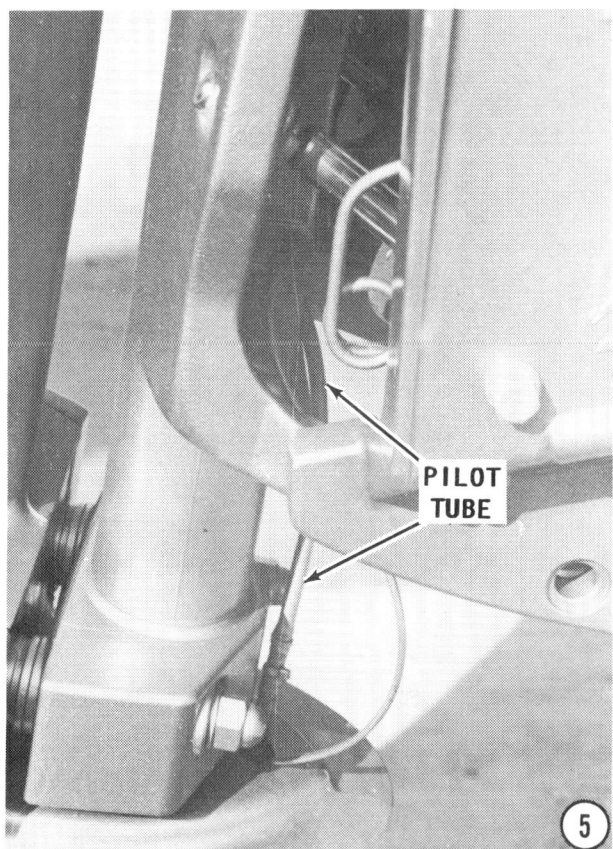

11-46 LOWER UNIT

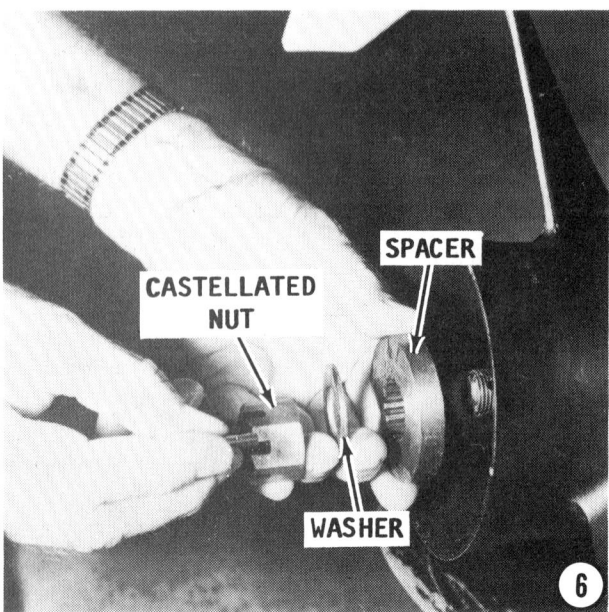

Trim Tab Adjustment

7- The trim tab should be positioned to enable the helmsperson to handle the boat with equal ease to starboard and port at normal cruising speed. If the boat seems to turn more easily to starboard, loosen the socket head screw and move the trim tab trailing edge to the right. Move the trailing edge of the trim tab to the left if the boat tends to turn more easily to port.

Closing Tasks

Shift the unit into **FORWARD** gear, release the tilt lock lever and lower the outboard to the normal operating position.

8- Remove the oil level plug and the drain plug. Fill the lower unit with Yamalube gearcase lubricant or Hypoid gear oil 90 weight until the lubricant flows from the top hole. Install both plugs and clean any excess lubricant from the lower unit.

Use the following table as a guide to lower unit capacity for the model indicated.

Note: "crttg" = counterrotating

Model	Capacity
115hp w/std lwr unit	26.7oz. (790cc)
115hp w/crttg lwr unit	25.0oz. (740cc)
130hp	26.7oz. (790cc)
150hp to 1987	30.1oz. (910cc)
150hp 1987 & on	33.1oz. (980cc)
L150hp	32.1oz. (900cc)
ProV 150	33.1oz. (980cc)
175hp to 1987	30.1oz. (910cc)
175hp 1987 & on	33.1oz. (980cc)
200hp to 1987 w/std lwr unit	30.1oz. (910cc)
200hp 1987 & on w/std lwr unit	33.1oz. (980cc)
200hp to 1987 w/crttg lwr unit	29.4oz. (870cc)
200hp 1987 & on w/crttg lwr unit	32.1oz. (900cc)
220hp to 1987	30.1oz. (910cc)
220hp 1987 only	33.1oz. (980cc)
225hp	33.1oz. (980cc)

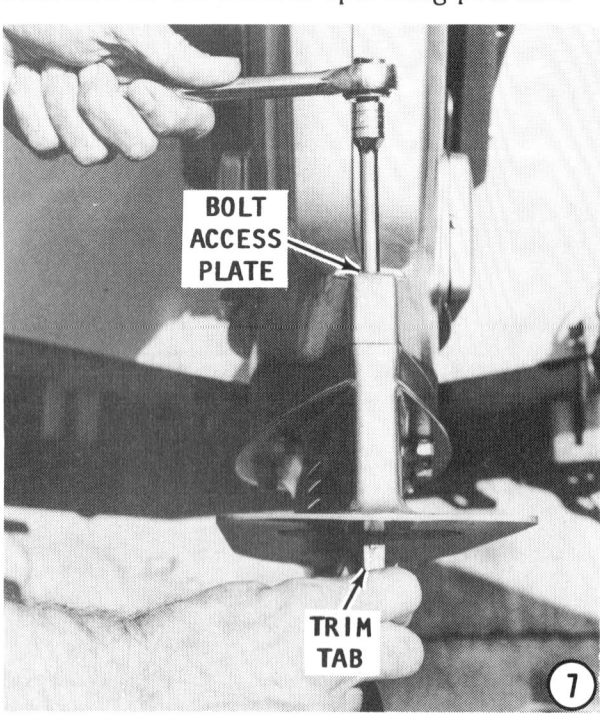

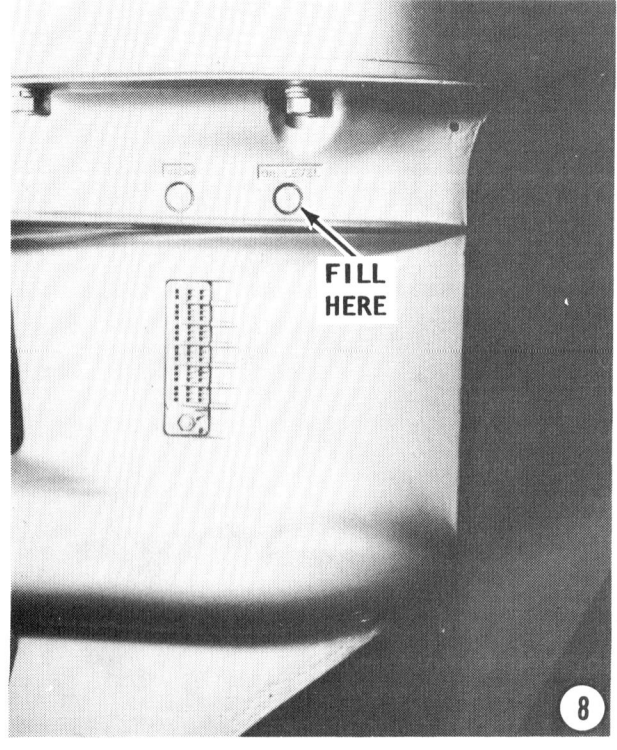

Mount the engine in a test tank or body of water.

CAUTION

Water must circulate through the lower unit to the powerhead anytime the powerhead is operating to prevent damage to the water pump in the lower unit. Just five seconds without water will damage the water pump impeller.

Start the engine and check the completed work for satisfactory operation, shifting, and **NO** leaks.

11-9 LOWER UNIT SERVICE JET DRIVE

DESCRIPTION AND OPERATION

The jet drive unit is designed to permit boating in areas prohibited to a boat equipped with a conventional propeller drive system. The housing of the jet drive barely extends below the hull of the boat allowing passage in ankle deep water, white water rapids, and over sand bars or in shoal water which would foul a propeller drive.

The jet drive provides reliable propulsion with a minimum of moving parts. Simply stated, water is drawn into the unit through an intake grille by an impeller driven by a driveshaft off the crankshaft of the powerhead. The water is immediately expelled under pressure through an outlet nozzle directed away from the stern of the boat.

As the speed of the boat increases and reaches planing speed, the jet drive discharges water freely into the air, and only the intake grille makes contact with the water.

The jet drive is provided with a gate arrangement and linkage to permit the boat to be operated in reverse. When the gate is moved downward over the exhaust nozzle, the pressure stream is reversed by the gate and the boat moves sternward.

Conventional controls are used for powerhead speed, movement of the boat, shifting and power trim and tilt.

Model Identification and Serial Numbers

A model letter identification is stamped on the rear, port side of the jet drive housing. A serial number for the unit is stamped on the starboard side of the jet drive housing, as indicated in the accompanying illustration.

Since their introduction by Yamaha in 1987, jet drive units have been installed on Models 115 and 200hp and certain 3-cylinder units.

In 1987, the units covered in this manual were identified as 115ETL-JD and 200ETL-JD. In 1988 identity of the units changed and simplified. The 115ETL-JD became the Jet Power 80, the 200ETL-JD became the

A jet drive mounted to the intermediate housing ready to give its owner a "fun day" on the water.

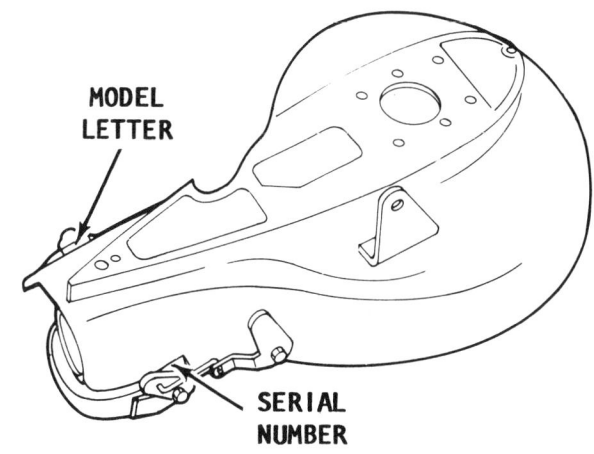

The model letter designation and the serial numbers are embossed on the jet drive housing.

11-48 LOWER UNIT

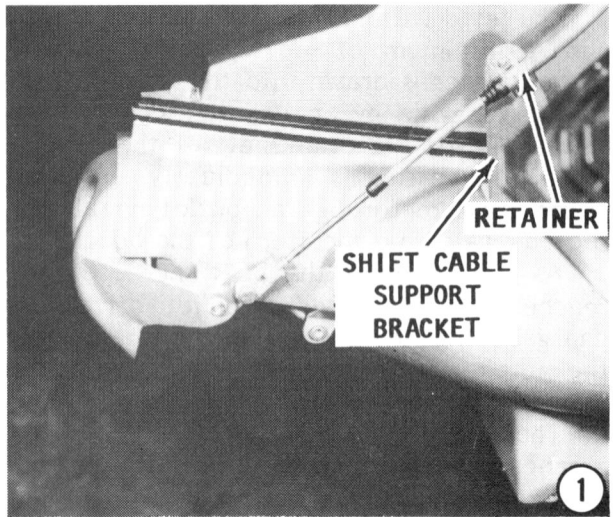

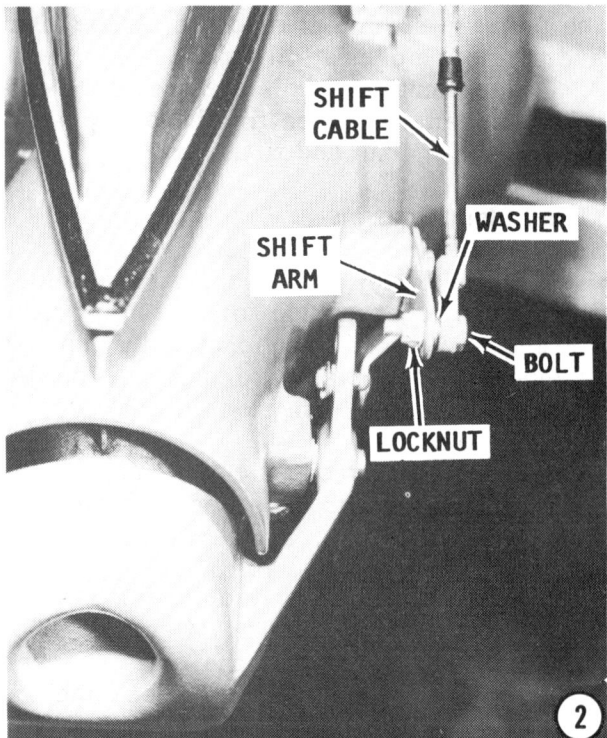

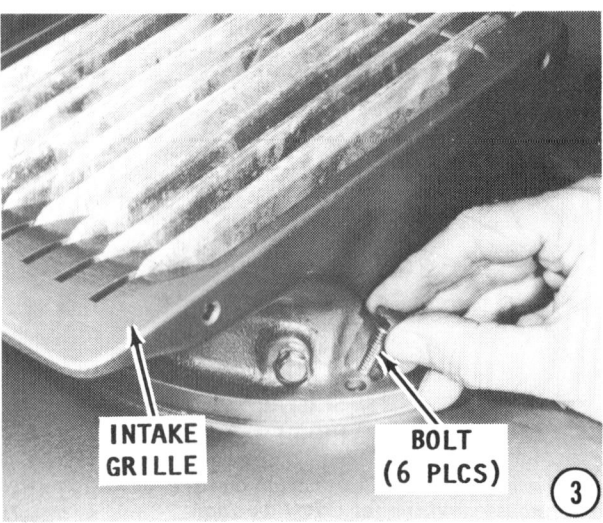

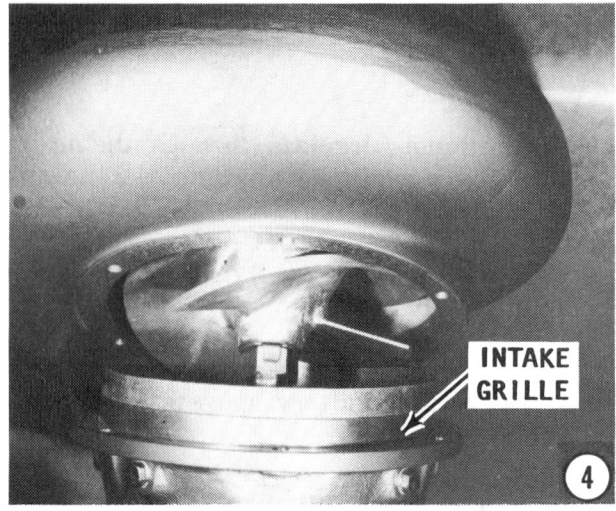

Jet Power 140. In all future references in this and other Seloc manuals, the jet drives will be referred to by their Jet Power designations.

A single size jet drive is used on both outboard units covered in this manual: The designation AA4 (installed on V4 units) or AA6 (installed on V6 units) may be found on the port side of the jet housing. Both units are identical in size and design. Two more models of the Jet Drive exist, one slightly smaller than the other. These are installed on 3 cylinder models and are covered in Seloc's Yamaha Outboard Manual Vol. II.

REMOVAL & DISASSEMBLING

1- Remove the two bolts and retainer securing the shift cable to the shift cable support bracket.

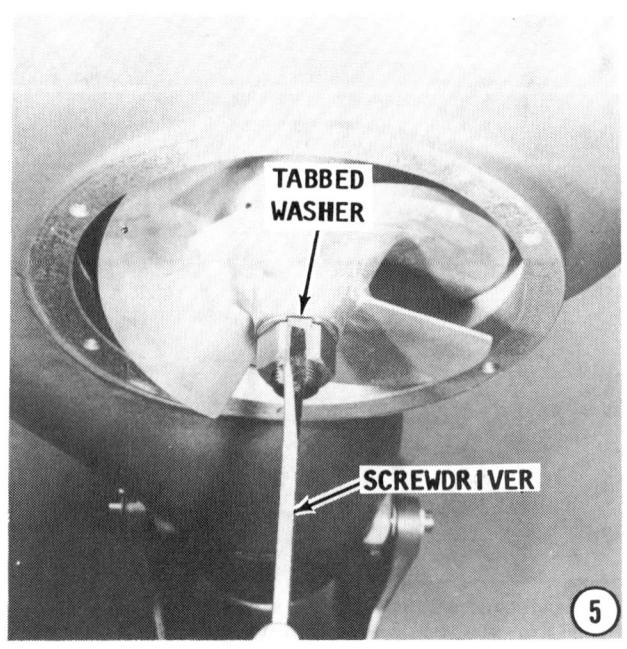

JET DRIVE DISASSEMBLING 11-49

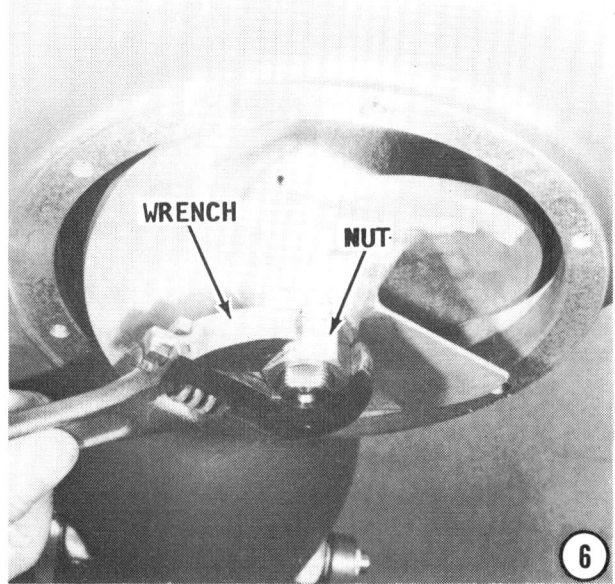

2- Remove the locknut, bolt, and washer securing the shift cable to the shift arm. Try not to disturb the length of the cable.

3- Remove the six bolts securing the intake grille to the jet casing.

4- Ease the intake grille from the jet drive housing.

5- Pry the tab or tabs of the tabbed washer away from the nut to allow the nut to be removed.

6- Loosen, and then remove the nut.

7- Remove the tabbed washer and spacers. Make a careful count of the spacers behind the washer. If the unit is relatively new, there could be as many as **EIGHT** spacers stacked together. If less than eight spacers are removed from behind the washer, the others will be found behind the jet impeller, which is removed in the following step. A **TOTAL** of **EIGHT** spacers will be found.

8- Remove the jet impeller from the shaft. If the impeller is "frozen" to the shaft, obtain a block of wood and a hammer. Tap the impeller in a **CLOCKWISE** direction to release the shear key.

9- Slide the nylon sleeve and shear key free of the driveshaft and any spacers found

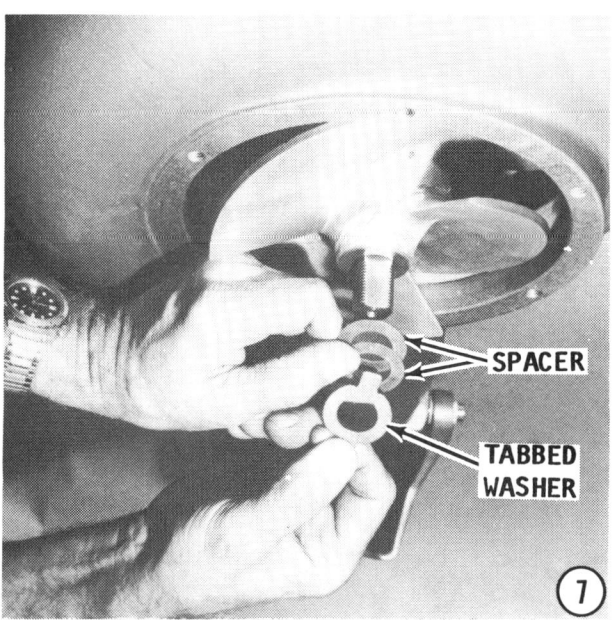

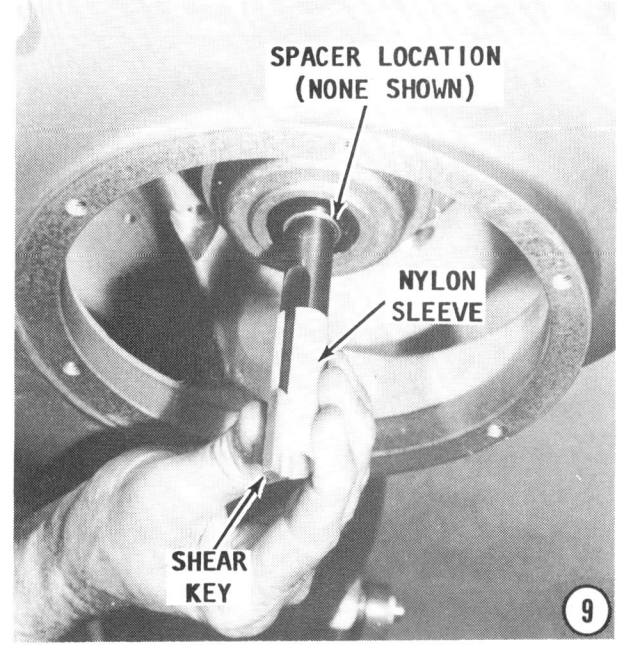

11-50　LOWER UNIT

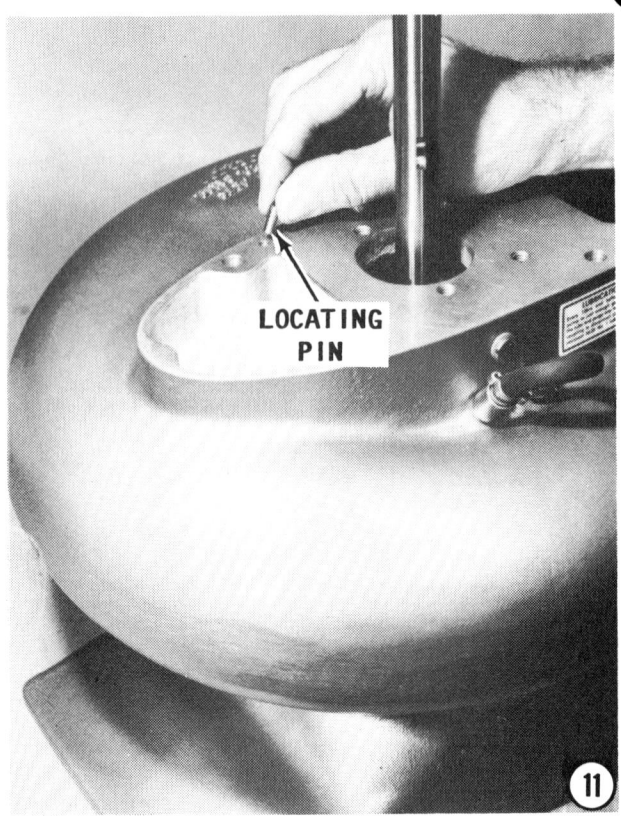

behind the impeller. Make a note of the number of spacers at both locations --behind the impeller **AND** on top of the impeller, under the nut and tabbed washer.

10- One external bolt and four internal bolts are used to secure the jet drive to the intermediate housing. The external bolt is located at the aft end of the anti-cavitation plate. The four internal bolts are located inside the jet drive housing, as indicated in the accompanying illustration. Remove the five attaching bolts.

11- Lower the jet drive from the intermediate housing. Remove the locating pin from the forward starboard side (or center forward, depending on the model being serviced) of the upper jet housing.

Location of the one exterior bolt securing the jet drive to the intermediate housing.

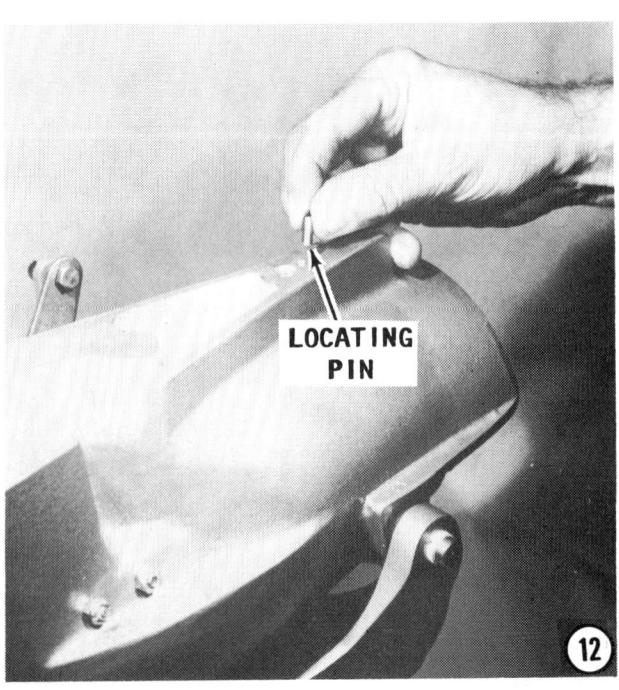

JET DRIVE DISASSEMBLING 11-51

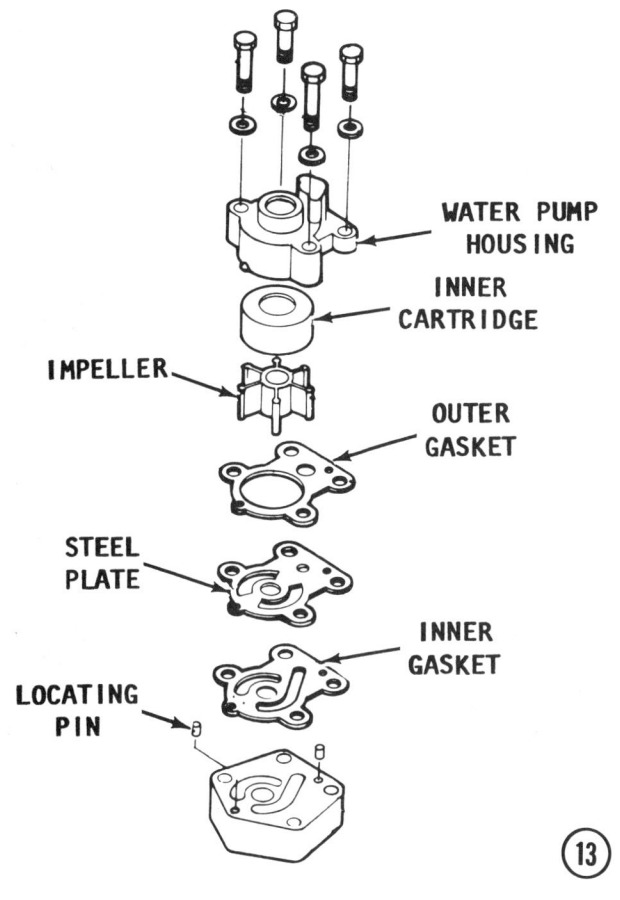

SPECIAL WORDS

There will be a total of **SIX** locating pins to be removed in the following steps. Make careful note of the size and location of each when they are removed, as an assist during assembling.

12- Remove the locating pin from the aft end of the housing. This pin and the one

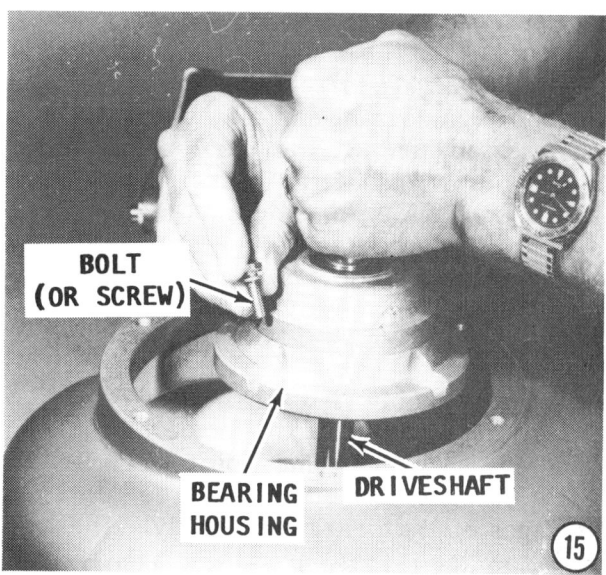

removed in the previous step should be of identical size.

13- Remove the four bolts and washers from the water pump housing.

Pull the water pump housing, the inner cartridge and the water pump impeller, up and free of the driveshaft. Remove the Woodruff key from its recess in the driveshaft. Next, remove the outer gasket, the steel plate and the inner gasket.

14- Remove the two small locating pins and lift the aluminum spacer up and free of the driveshaft.

15- Remove the four securing bolts and then remove the driveshaft and bearing assembly from the housing.

16- Remove the large thick adaptor plate from the intermediate housing. This plate is secured with seven bolts and lockwashers. Lower the adaptor plate from the intermediate housing and remove the two small locating pins, one on the forward port side and another from the last aft hole in the adaptor plate. Both pins are identical size.

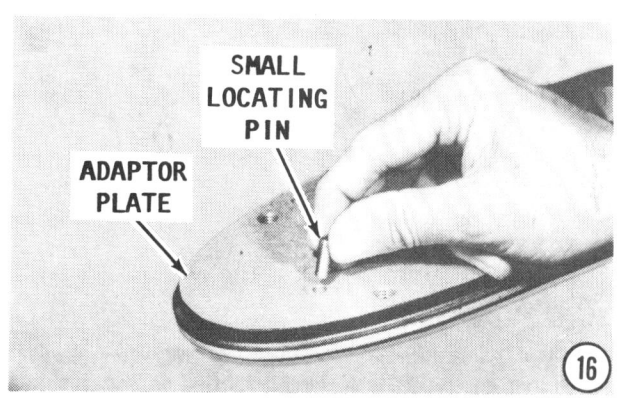

CLEANING AND INSPECTING

Wash all parts, except the driveshaft assembly, in solvent and blow them dry with compressed air. Rotate the bearing assembly on the driveshaft to inspect the bearings for "rough" spots, binding, and signs of corrosion or damage.

Saturate a shop towel with solvent and wipe both extensions of the driveshaft.

Bearing Assembly

Lightly wipe the exterior of the bearing assembly with the same shop towel. Do not allow solvent to enter the three lubricant passages of the bearing assembly. The best way to clean these passages is **NOT** with solvent - because any solvent remaining in the assembly after installation will continue to dissolve good useful lubricant and leave bearings and seals dry. This condition will cause bearings to fail through friction and seals to dry up and shrink - losing their sealing qualities.

The only way to clean and lubricate the bearing assembly is after installation to the jet drive - via the exterior lubrication fitting. This procedure is described in Section 3-12 on Page 3-12 and explains how the old lubricant may be completely replaced with new.

If the old lubricant emerging from the hose coupling is a dark, dirty, grey color, the seals have already broken down and water is attacking the bearings. If such is the case, it is recommended the entire driveshaft bearing assembly be taken to the dealer for service of the bearings and seals.

Dismantling Bearing Assembly

A complicated procedure must be followed to dismantle the bearing assembly including "torching" off the bearing housing. Naturally, excessive heat might ruin the seals and bearings. Therefore, the best recommendation is to leave this part of the service work to the experts at your local Yamaha dealership.

Driveshaft and Associated Parts

Inspect the threads and splines on the driveshaft for wear, rounded edges, corrosion and damage.

Carefully check the driveshaft to verify the shaft is straight and true without any sign of damage.

Inspect the jet drive housing for nicks, dents, corrosion, or other signs of damage. Nicks may be removed with No. 120 and No. 180 emery cloth.

Reverse Gate

Inspect the gate and its pivot points. Check the swinging action to be sure it moves freely the entire distance of travel without binding.

Inspect the slats of the water intake grille for straightness. Straighten any bent slats, if possible. Use the utmost care when prying on any slat, as they tend to break if excessive force is applied. Replace the intake grille if a slat is lost, broken, or bent, and cannot be repaired. The slats are spaced evenly and the distance between them is critical, to prevent large objects from passing through and becoming lodged between the jet impeller and the inside wall of the housing.

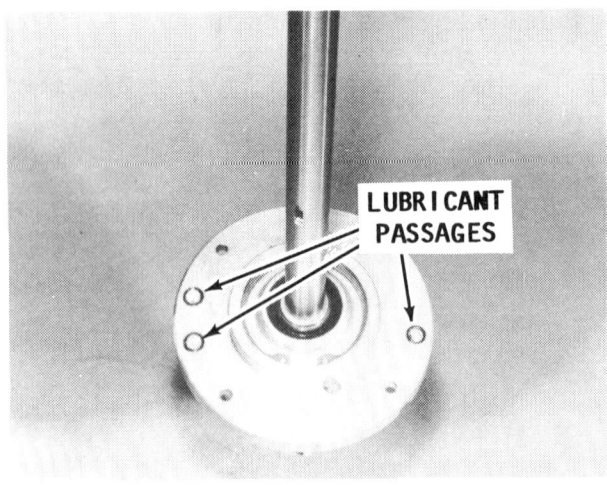

*Take extra precautions to **PREVENT** cleaning solvent from entering the three lubrication passages.*

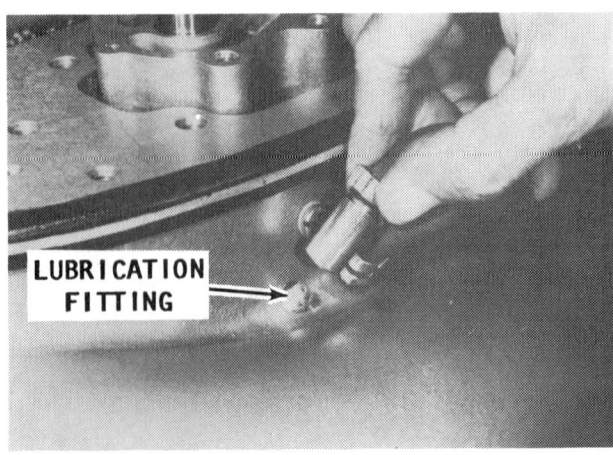

Cleaning and lubricating the bearing assembly is best accomplished by completely replacing the old lubricant with new, as described in the text.

CLEANING & INSPECTING   11-53

Jet Impeller

The jet impeller is a precisely machined and dynamically balanced aluminum spiral. Observe the drilled recesses at exact locations to achieve this delicate balancing. Some of these drilled recesses are clearly shown in the accompanying illustration.

Excessive vibration of the jet drive may be attributed to an out-of-balance condition caused by the jet impeller being struck excessively by rocks, gravel or cavitation "burn".

The term cavitation "burn" is a common expression used throughout the world among people working with pumps, impeller blades, and forceful water movement.

"Burns" on the jet impeller blades are caused by cavitation air bubbles exploding with considerable force against the impeller blades. The edges of the blades may develop small "dime size" areas resembling a porous sponge, as the aluminum is actually "eaten" by the condition just described.

Excessive rounding of the jet impeller edges will reduce efficiency and performance. Therefore, the impeller should be inspected at regular intervals.

If rounding is detected, the impeller should be placed on a work bench and the edges restored to as sharp a condition as possible, using a file. Draw the file in only one direction. A back-and-forth motion will not produce a smooth edge. **TAKE CARE** not to nick the smooth surface of the jet impeller. Excessive nicking or pitting will create water turbulence and slow the flow of water through the pump.

Inspect the shear key. A slightly distorted key may be reused although some diffi-

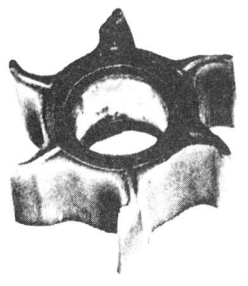
After 60 seconds at 1500 rpm.

After 90 seconds at 1500 rpm.

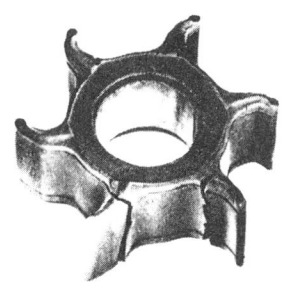
After 30 seconds at 2000 rpm.

After 45 seconds at 2000 rpm.

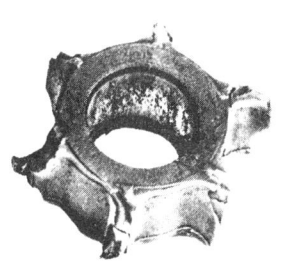
After 60 seconds at 2000 rpm

The slats of the grille must be carefully inspected and any bent slats straightened, for maximum performance of the jet drive.

Cautions throughout this manual point out the danger of operating the powerhead without water passing through the water pump. The above photographs are self evident.

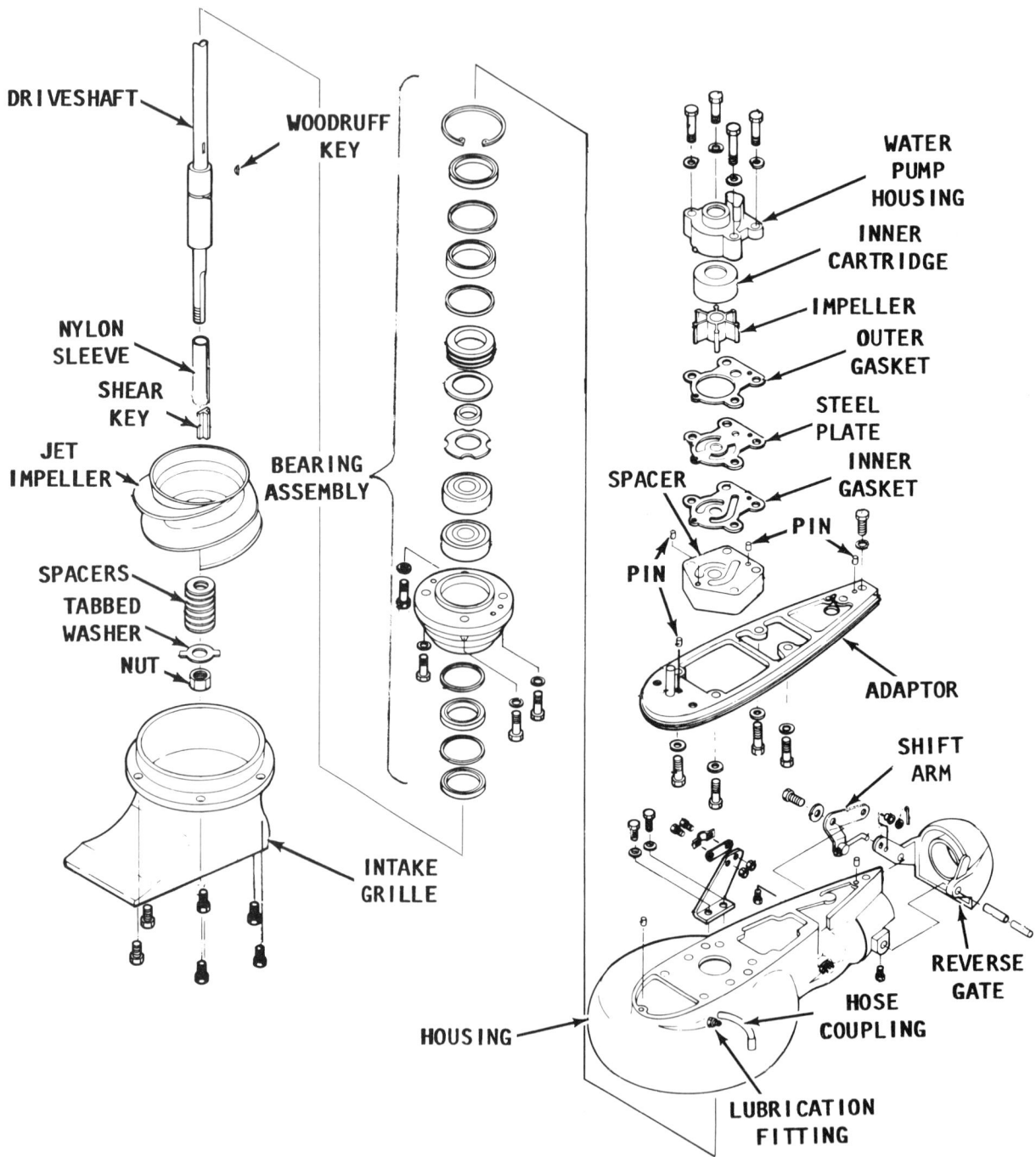

Exploded drawing of a jet drive lower unit, with major parts identified.

JET DRIVE ASSEMBLING   11-55

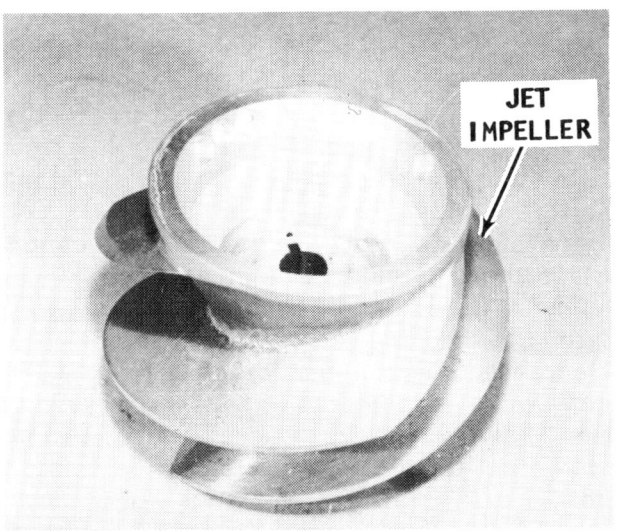

The edges of the jet impeller should be kept as sharp as possible for maximum jet drive efficiency.

culty may be encountered in assembling the jet drive. A cracked shear key should be discarded and replaced with a new key.

Water Pump

Clean all water pump parts with solvent, and then blow them dry with compressed air. Inspect the water pump housing for cracks and distortion, possibly caused from overheating. Inspect the steel plate, the thick aluminum spacer and the water pump cartridge for grooves and/or rough spots. If possible **ALWAYS** install a new water pump impeller while the jet drive is disassembled. A new water pump impeller will ensure extended satisfactory service and give "peace of mind" to the owner. If the old water pump impeller must be returned to service, **NEVER** install it in reverse of the original direction of rotation. Installation in reverse will cause premature impeller failure.

If installation of a new water pump impeller is not possible, check the sealing surfaces and be satisfied they are in good condition. Check the upper, lower, and ends of the impeller vanes for grooves, cracking, and wear. Check to be sure the indexing notch of the impeller hub is intact and will not allow the impeller to slip.

JET DRIVE ASSEMBLING

1- Identify the two small locating pins used to index the large thick adaptor plate to the intermediate housing. Insert one pin into the last hole aft on the topside of the

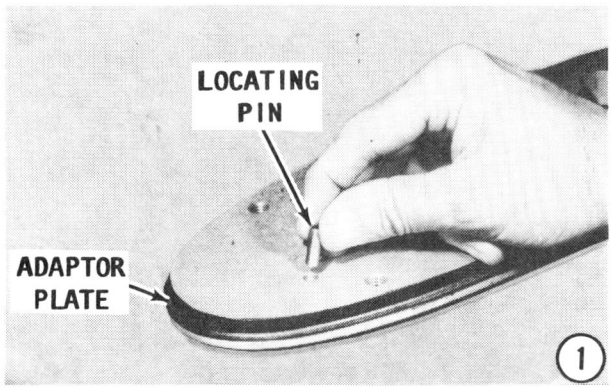

plate. Insert the other pin into the hole forward toward the port side, as shown.

Lift the plate into place against the intermediate housing with the locating pins indexing with the holes in the intermediate housing. Secure the plate with the seven bolts.

Tighten the bolts to a torque value of 22 ft lbs (30Nm).

2- Place the driveshaft bearing assembly into the jet drive housing. Rotate the bearing assembly until all bolt holes align. There is only **ONE** correct position.

Tighten the four securing bolts to a torque value of 5 ft lbs (7Nm).

SPECIAL WORDS

If installing a new jet impeller, place all eight spacers at the lower or "nut" end of the impeller, and skip the following step.

Shimming Jet Impeller

The clearance between the outer edge of the jet drive impeller and the water intake

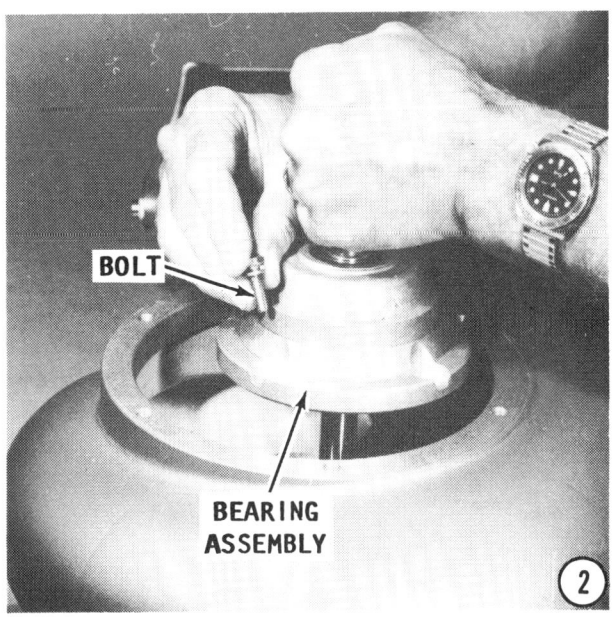

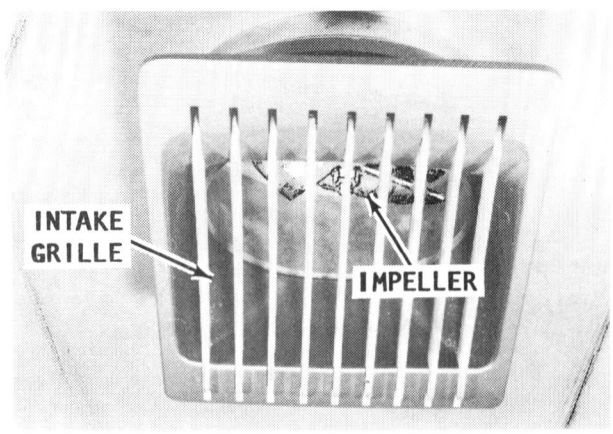

The clearance between the jet impeller and the casing cone, can be fairly well estimated by shining a flashlight up through the grille and visually checking the distance between the impeller and the cone.

housing cone wall should be maintained at approximately 1/32" (0.8mm). This distance can be visually checked by shining a flashlight up through the intake grille and estimating the distance between the impeller and the casing cone, as indicated in the accompanying illustrations. It is not humanly possible to accurately measure this clearance, but by observing closely and estimating the clearance, the results should be fairly accurate.

After continued use, the clearance will increase. The spacers removed in Steps 7 & 8 are used to position the impeller along the driveshaft with a desired clearance of 1/32" (0.8mm) between the jet impeller and the housing wall.

3- A total of **EIGHT** spacers are used. When new, all spacers are located at the tapered (or "nut") end of the impeller. As the clearance increases, the spacers are transferred from the tapered ("nut") end and

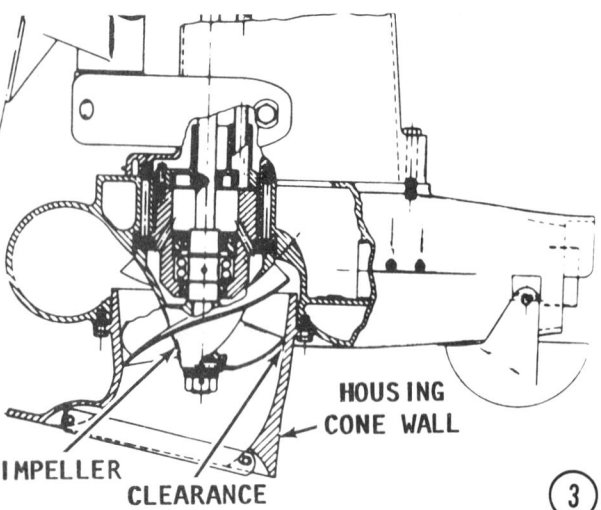

placed at the wide ("intermediate housing") end of the jet impeller.

This procedure is best accomplished while the jet drive is removed from the intermediate housing.

Secure the driveshaft with the attaching hardware. Installation of the shear key and nylon sleeve is not vital to this procedure. Place the unit on a convenient work bench. Shine a flashlight through the intake grille into the housing cone and "eyeball" the clearance between the jet impeller and the cone wall, as indicated in the accompanying line drawing. Move spacers one-at-a-time from the tapered end to the wide end to obtain a satisfactory clearance. Dismantle the driveshaft and note the exact count of spacers at both ends of the bearing assembly. This count will be recalled in Steps 9 & 11 of Assembly to properly install the jet impeller.

Water Pump Assembling

4- Place the aluminum spacer over the driveshaft with the two holes for the indexing pins facing **UPWARD**. Fit the two locating pins into the holes of the spacer.

GOOD WORDS

The manufacturer recommends **NO** sealant be used on either side of the water pump gaskets.

5- Slide the inner water pump gasket (the gasket with two curved openings) over the driveshaft. Position the gasket over the two locating pins. Slide the steel plate

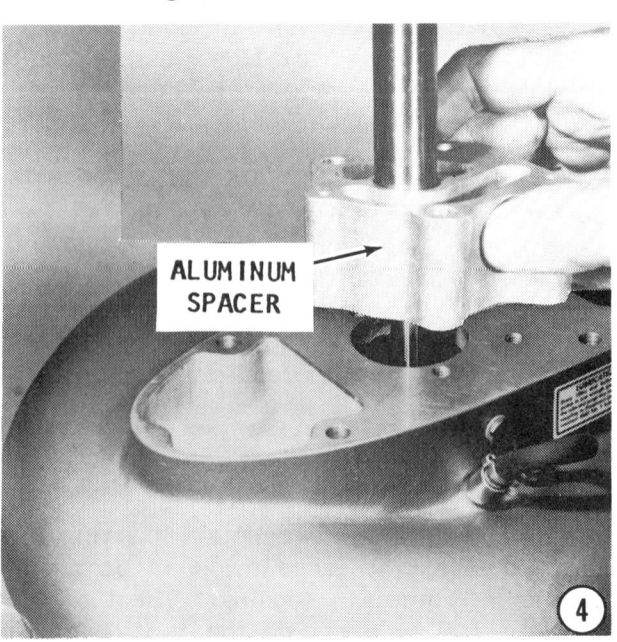

JET DRIVE ASSEMBLING   11-57

down over the driveshaft with the tangs on the plate facing **DOWNWARD** and with the holes in the plate indexed over the two locating pins.

Check to be sure the tangs on the plate fit into the two curved openings of the gasket beneath the plate. Now, slide the outer gasket (the gasket with the large center hole) over the driveshaft. Position the gasket over the two locating pins.

Fit the Woodruff key into the driveshaft. Just a dab of grease on the key will help to hold the key in place. Slide the water pump impeller over the driveshaft with the rubber membrane on the top side and the keyway in the impeller indexed over the Woodruff key. **TAKE CARE** not to damage the membrane. Coat the impeller blades with Yamalube Grease "A" or equivalent water resistant lubricant.

Install the insert cartridge, the inner plate, and finally the water pump housing over the driveshaft. Rotate the insert cartridge **COUNTERCLOCKWISE** over the impeller to tuck in the impeller vanes. Seat all parts over the two locating pins.

Tighten the four bolts to a torque value of 11 ft lbs (15Nm).

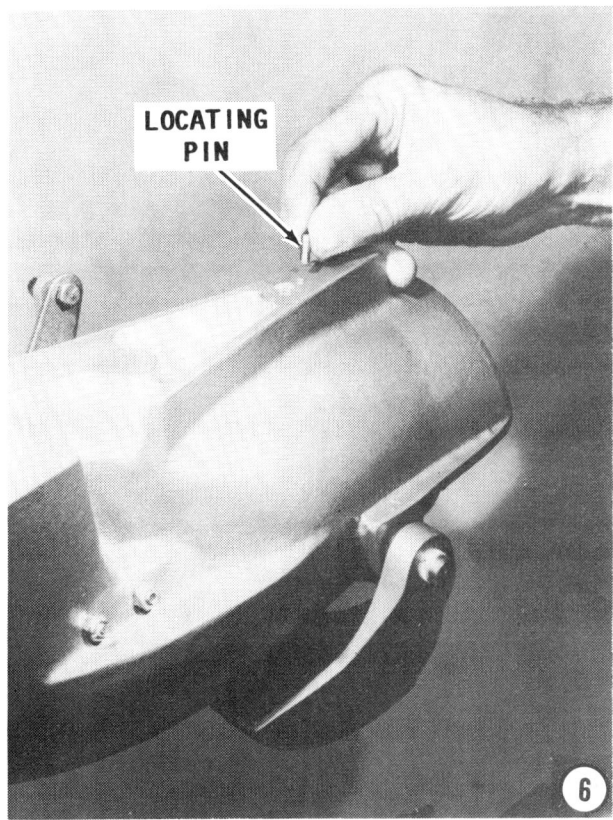

6- Install one of the small locating pins into the aft end of the jet drive housing.

Jet Drive Installation

7- Install the other small locating pin into the forward starboard side (or center forward end, depending on the model being serviced).

8- Raise the jet drive unit up and align it with the intermediate housing, with the

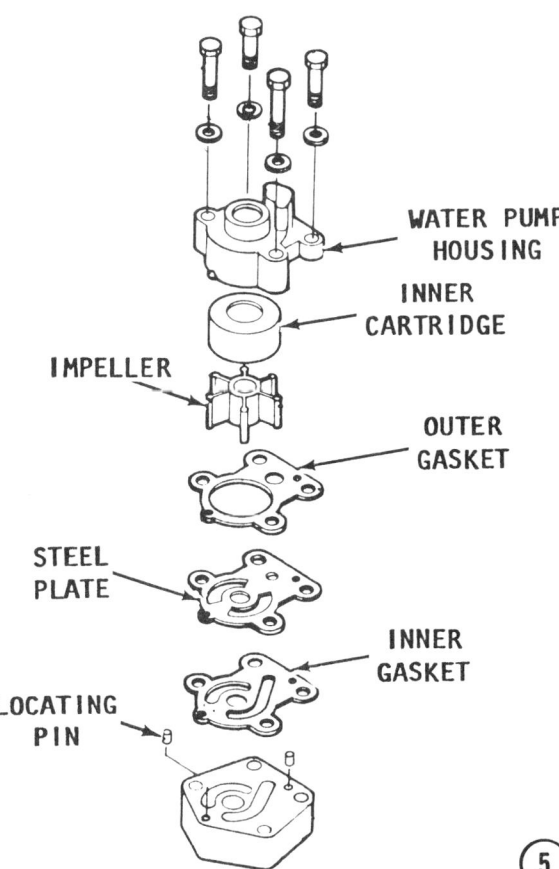

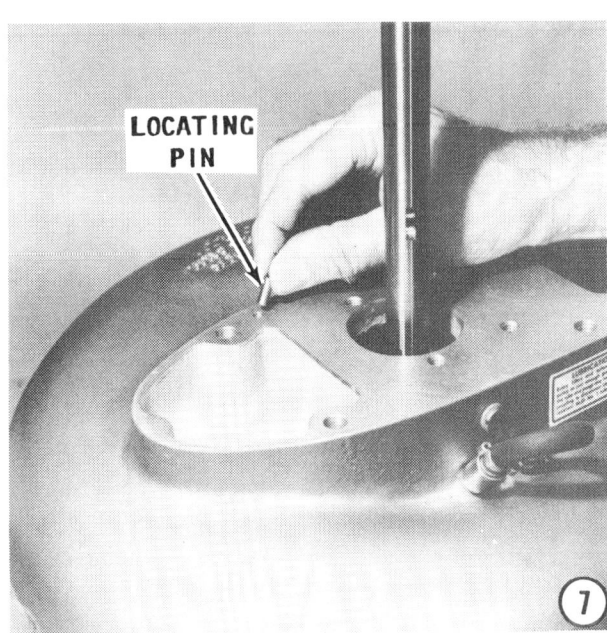

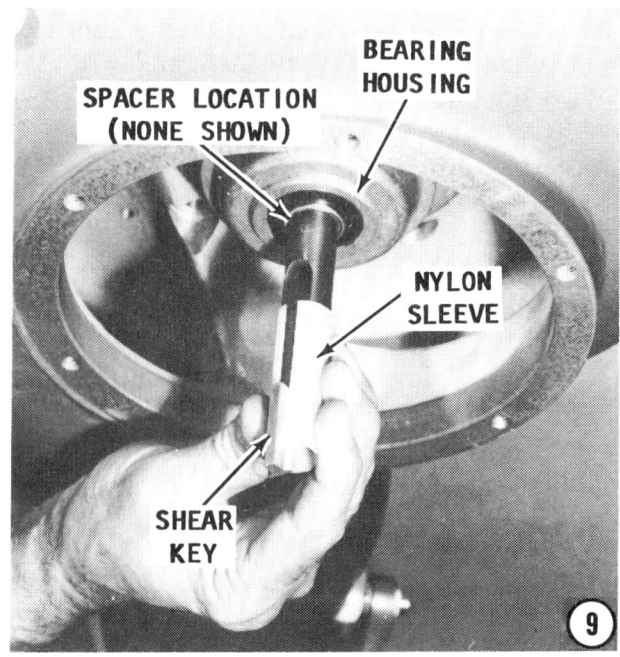

small pins indexed into matching holes in the adaptor plate. Install the four internal bolts and the one external bolt. Location of the external bolt is at the aft end of the anticavitation plate. Tighten the five bolts to a torque value of 11 ft lbs (15Nm).

9- Place the required number of spacers (if any) as determined in Step 3, **OR** in the paragraphs following "Shimming the Jet Impeller", up against the bearing housing.

Slide the nylon sleeve over the driveshaft and insert the shear key into the slot of the nylon sleeve with the key resting against the flattened portion of the driveshaft.

10- Slide the jet impeller up onto the driveshaft, with the groove in the impeller collar indexing over the shear key.

11- Place the remaining spacers over the driveshaft. The number of spacers will be eight minus the number used in Step 9.

12- Tighten the nut to a torque value of 17 ft lbs (23Nm). If neither of the two tabs on the tabbed washer aligns with the sides of the nut, remove the nut and washer. Invert the tabbed washer. Turning the washer over will change the tabs by approxi-

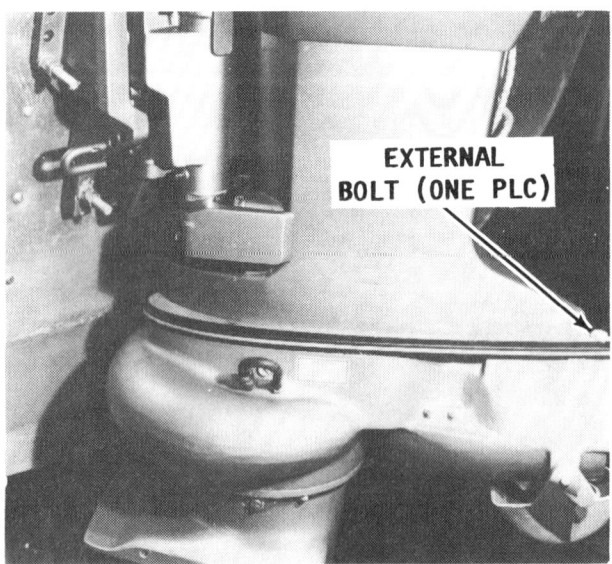

Location of the one exterior bolt securing the jet drive to the intermediate housing.

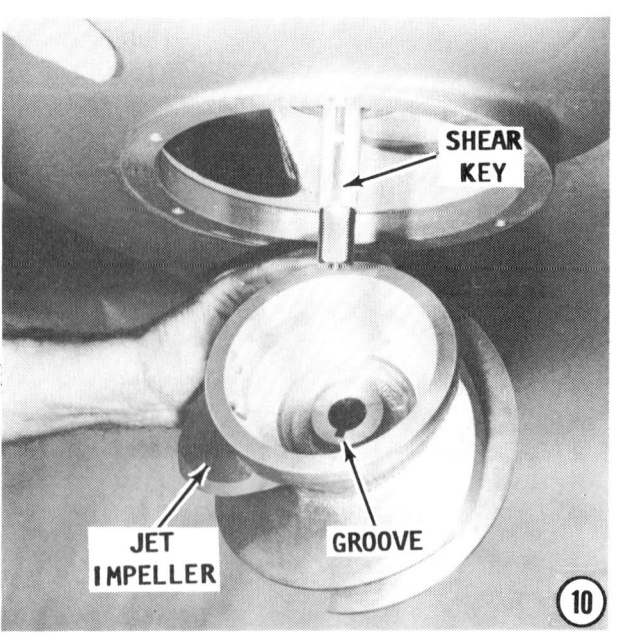

JET DRIVE ASSEMBLING 11-59

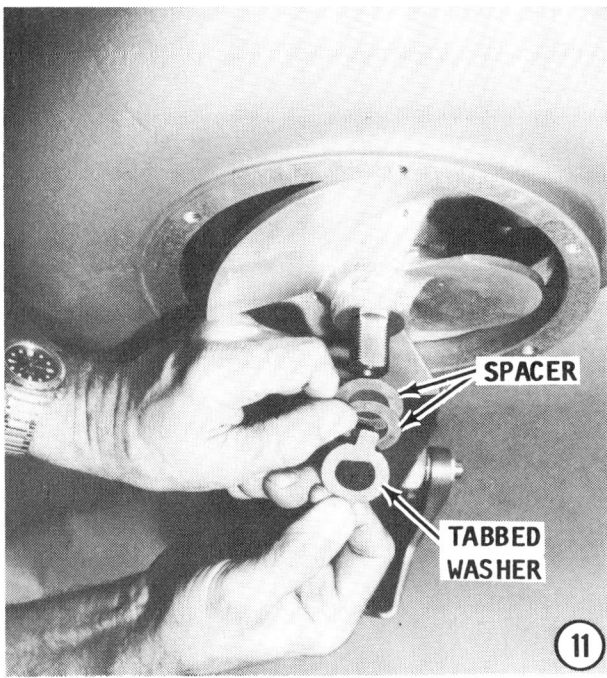

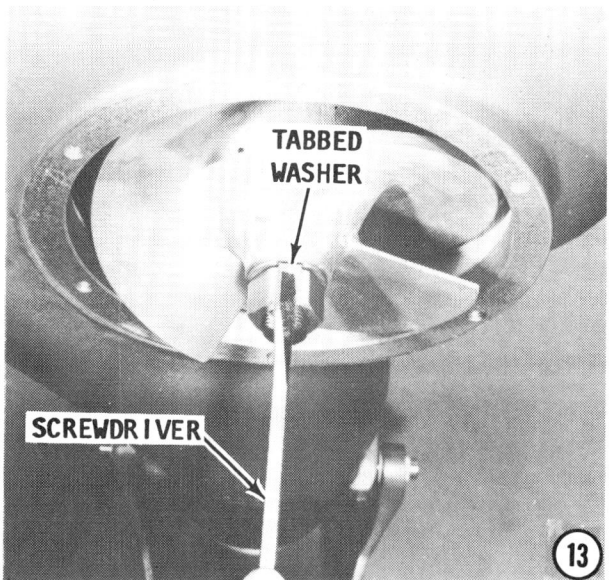

mately 15°. Install and tighten the nut to the required torque value. The tabbed washer is designed to align with the nut in one of the two positions described.

13- Bend the tabs up against the nut to prevent the nut from backing off and becoming loose.

14- Install the intake grille onto the jet drive housing with the slots facing aft. Install and tighten the six securing bolts to a torque value of 11 ft lbs (15Nm).

15- Slide the bolt through the end of the shift cable, washer, and into the shift arm. Install the locknut onto the bolt and tighten the bolt securely.

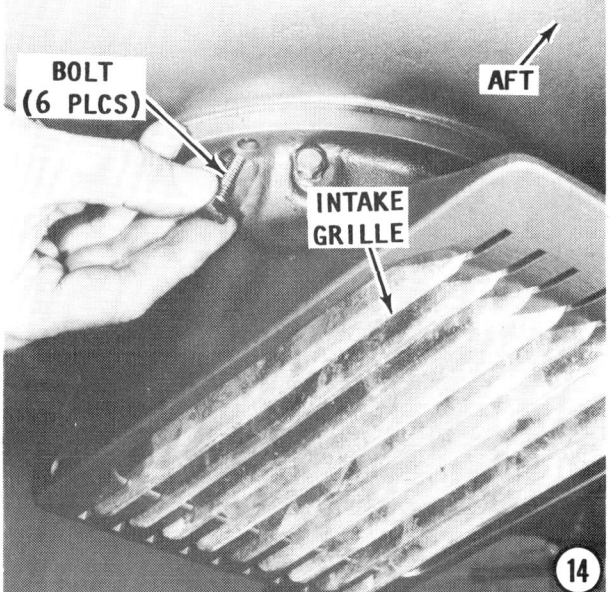

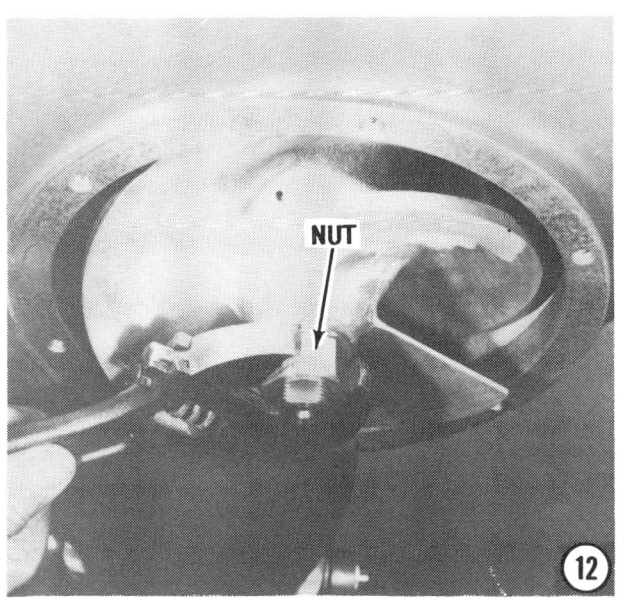

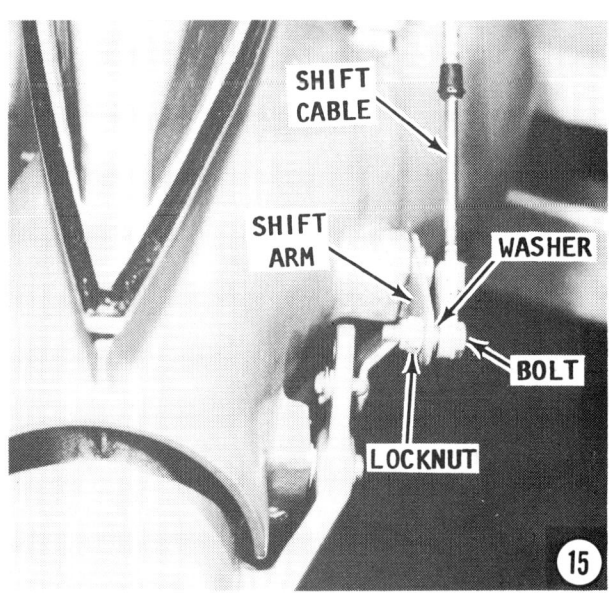

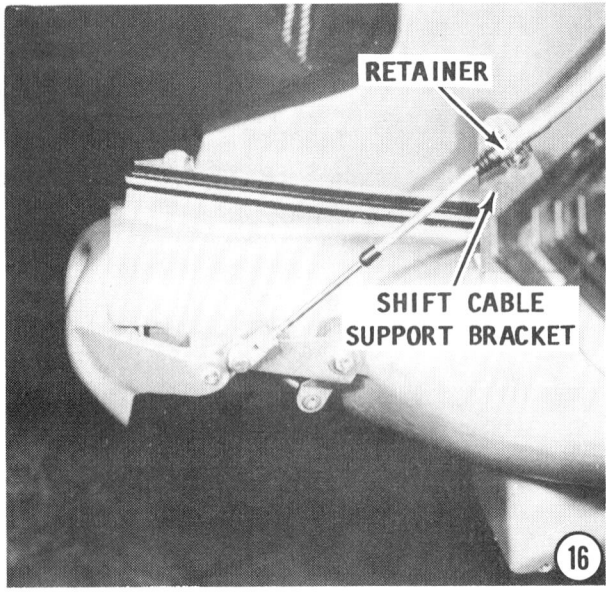

16- Install the shift cable against the shift cable support bracket and secure it in place with the two bolts.

11-10 JET DRIVE ADJUSTMENTS

Cable Alignment and Free Play Adjustment

1- Move the shift lever downward into

the **FORWARD** position. The leaf spring should snap over on top of the lever to lock it in position.

2- Remove the locknut, washer, and bolt from the threaded end of the shift cable. Push the reverse gate firmly against the rubber pad on the underside of the jet drive housing.

Check to be sure the link between the reverse gate and the shift arm is hooked into the **LOWER** hole on the gate.

Hold the shift arm up until the link rod and shift arm axis form an imaginary straight line, as indicated in the accompanying illustration. Adjust the length of the shift cable by rotating the threaded end, until the cable can be installed back onto the shift arm **WITHOUT** disturbing the imaginary line. Pass the nut through the cable end, washer, and shift arm. Install and tighten the locknut.

Neutral Stop Adjustment

FIRST, THESE WORDS

In the **FORWARD** position, the reverse gate is neatly tucked underneath and clear of the exhaust jet stream.

In the **REVERSE** position, the gate swings up and blocks the jet stream deflecting the water in a forward direction under the jet housing to move the boat sternward.

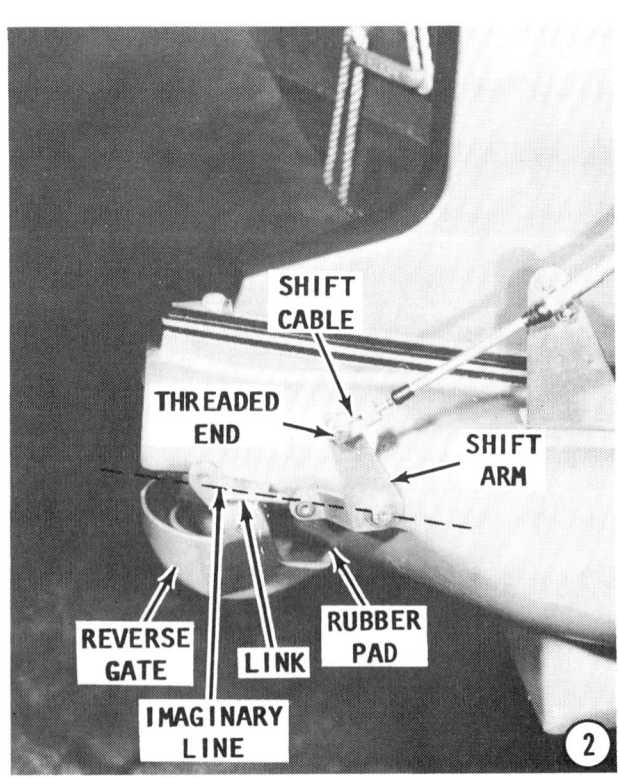

JET DRIVE ADJUSTMENTS 11-61

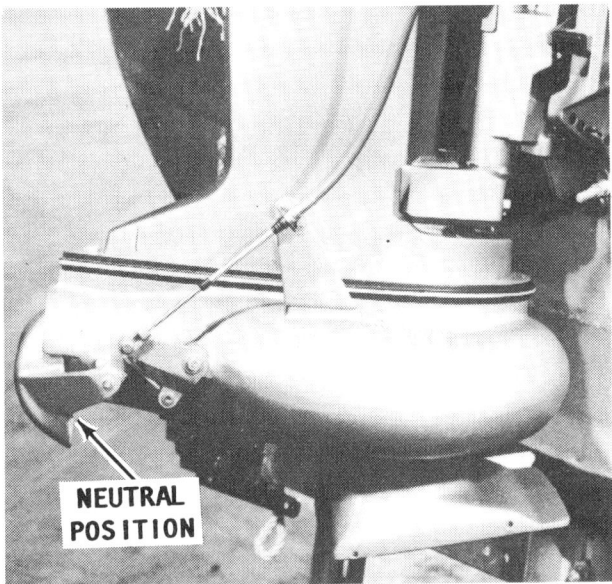

The reverse gate in the neutral position.

In the neutral position, the gate assumes a "happy medium" -- a balance between forward and reverse when the powerhead is operating at **IDLE** speed. Actually, the gate is deflecting some water to prevent the boat from moving forward, but not enough volume to move the boat sternward.

WARNING
THE GATE MUST BE PROPERLY ADJUSTED FOR SAFETY OF BOAT AND PASSENGERS. IMPROPER ADJUSTMENT COULD CAUSE THE GATE TO SWING UP TO THE REVERSE POSITION WHILE THE BOAT IS MOVING FORWARD CAUSING SERIOUS INJURY TO BOAT OR PASSENGERS.

3- Loosen, but do **NOT** remove the locknut on the neutral stop lever. Check to be sure the lever will slide up and down along the slot in the shift lever bracket.

CRITICAL WORDS
The following procedure **MUST** be performed with the boat and jet drive in a body of water. Only with the boat in the water can a proper jet stream be applied against the gate for adjustment purposes.

CAUTION
Water must circulate through the lower unit to the powerhead anytime the powerhead is operating to prevent damage to the water pump in the lower unit. Just five seconds without water will damage the water pump impeller.

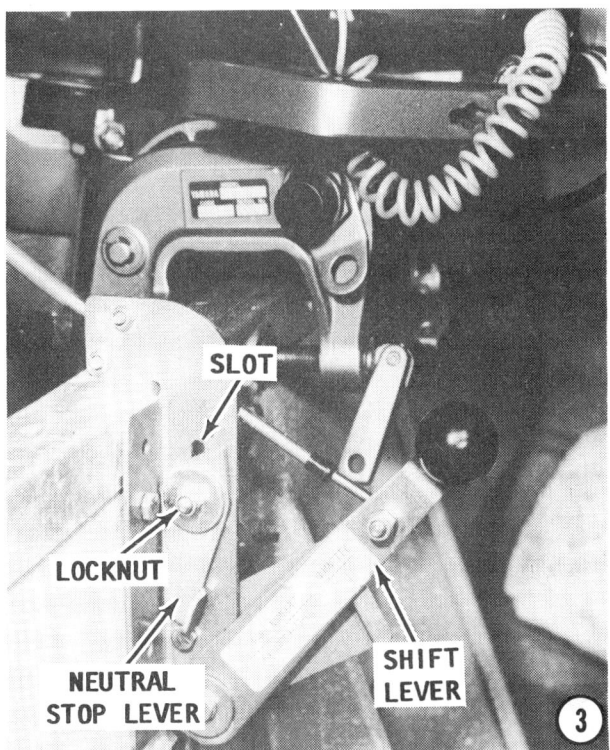

Start the powerhead and allow it to operate **ONLY** at **IDLE** speed. With the neutral stop lever in the down position, move the shift lever until the jet stream forces on the gate are "balanced". "Balanced" means the water discharged is divided in

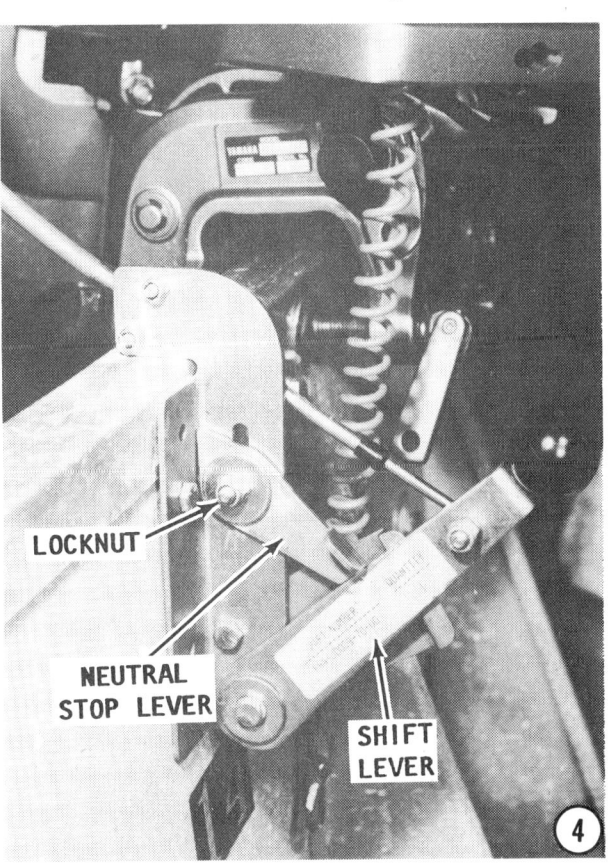

11-62 LOWER UNIT

both directions and the boat moves neither forward nor sternward. The gate is then in the neutral position with the powerhead at idle speed.

4- Move the neutral stop lever up against the shift lever until the stop lever barely makes contact with the shift lever. Tighten the locknut to maintain this new adjusted position. Shut down the powerhead.

GOOD WORDS

The reverse gate may not swing to the full **UP** position in reverse gear after Steps 1 thru 4 have been performed. Do **NOT** be concerned. This condition is acceptable, because water pressure in reverse will close the gate fully under normal operation.

Trim Adjustment

5- During operation, if the boat tends to "pull" to port or starboard, the flow fins may be adjusted to correct the condition.

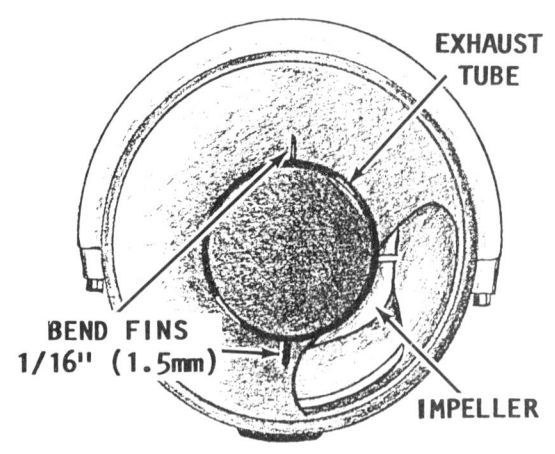

These fins are located at the top and bottom of the exhaust tube.

If the boat tends to "pull" to starboard, bend the trailing edge of each fin approximately 1/16" (1.5mm) toward the starboard side of the jet drive. Naturally, if the boat tends to "pull" to port, bend the fins toward the port side.

12
REMOTE CONTROLS

12-1 DESCRIPTION

The remote control unit allows the helmsperson to control throttle operation and shift movements from a location other than where the outboard unit is mounted.

In most cases, the remote control box is mounted approximately halfway forward (midship) on the starboard side of the boat.

The control unit houses a key switch, a choke switch, a "kill" switch, a neutral safety switch, a warning horn, and the necessary wiring and cable hardware to connect the control box to the outboard unit.

A safety feature is incorporated in the unit. The control arm can be shifted out of the neutral position if and **ONLY** if the neutral lever is squeezed into the control arm. This feature prevents the arm from being accidently moved from neutral into either forward or reverse gear. Unintentional movement of the shift lever could be dangerous, resulting in personal injury to the operator, passengers, or the boat.

Starting from the upright **NEUTRAL** position, when the control arm is moved forward to about 30° from the vertical, the unit shifts into **FORWARD** gear. At this point the throttle plate is fully closed. As the control arm is moved past the 30° position, forward and downward, the throttle will open, until the wide open position, approximately 90° from the vertical, is reached.

To shift into **REVERSE** gear, the control arm is first returned to the full upright position, **NEUTRAL**, momentarily. From the upright position, the control arm is moved aft about 30°, and the unit shifts in **REVERSE** gear. At this point, the throttle plate is fully closed. If the arm is moved further aft and downward, the throttle will be opened until the wide open position is reached at about 60°.

The remote control unit is equipped with a free acceleration lever. This lever is moved up and down -- up to open the throttle and down to close the throttle. This lever is utilized **ONLY** during powerhead startup and when the control arm is in the full upright **(NEUTRAL)** position. When the free acceleration lever is not in the full down, idling position, the control lever cannot be moved from the **NEUTRAL** position.

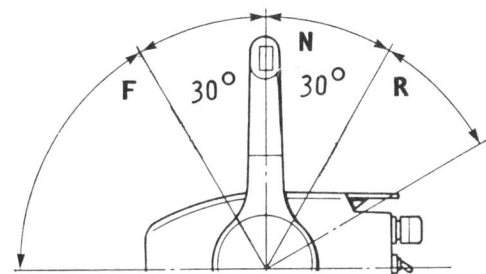

Shift lever positions for the remote control unit, as explained in the text.

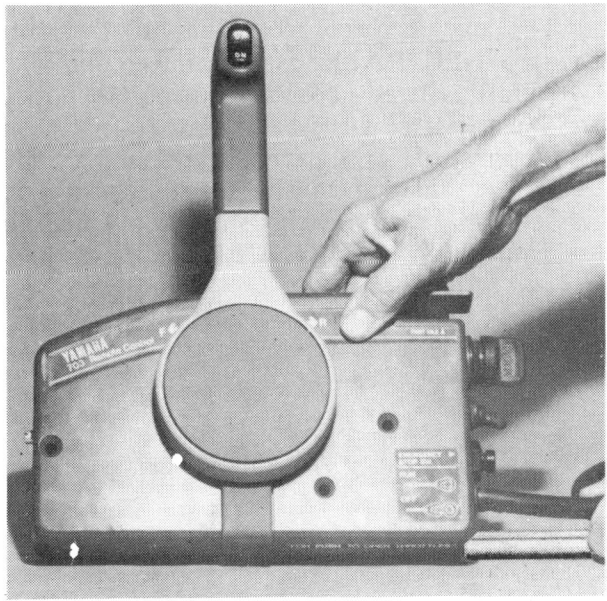

Remote control box removed from the boat and ready for disassembly and service.

12-2 REMOTE CONTROLS

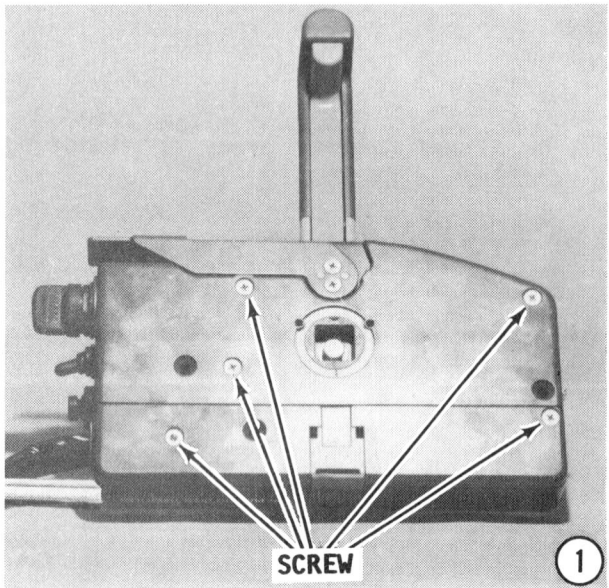

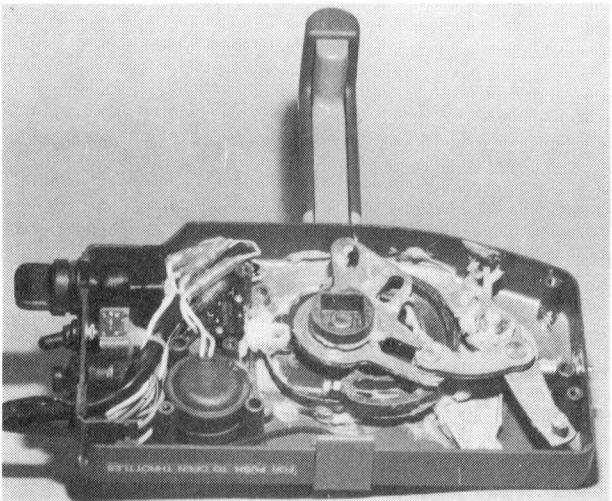

Interior view of an opened control box showing the wires neatly routed and tucked away from moving parts.

Complete, detailed, and illustrated procedures to disassemble, service, and assemble the control box are included in this chapter.

12-2 DISASSEMBLING

The following procedures pickup the work after the control has been disconnected, removed from the boat and placed on a suitable work surface. Removal is accomplished by first disconnecting the throttle and shift cables at the powerhead, and then disconnecting the main wiring harness at the quick disconnect fitting. Next, the three mounting screws are removed securing the control box to the boat.

1- Move the control arm to the full upright **(NEUTRAL)** position. Remove the five Phillips head screws securing the upper and lower parts of the back plate.

2- Pry off the circlip retaining the throttle cable end to the throttle arm and the circlip retaining the shift cable end to the shift arm. **TAKE CARE** not to lose

these two small circlips. Lift off the cable ends from the arms and be **CAREFUL** not to alter their length at the turnbuckles. Remove both cable ends from the box.

WORDS FROM EXPERIENCE

One of the most difficult tasks during assembling of the control unit is attempting to return all the wires back into their original positions.

Some wires are tucked neatly under switches, others are routed into neat bundles secured with plastic retainers, some are looped and double back, but "believe-it-or-not", with patience and some good words,

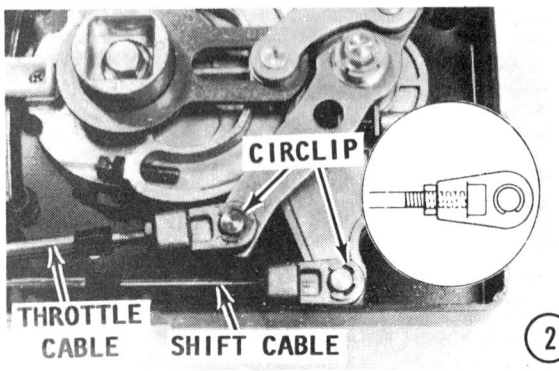

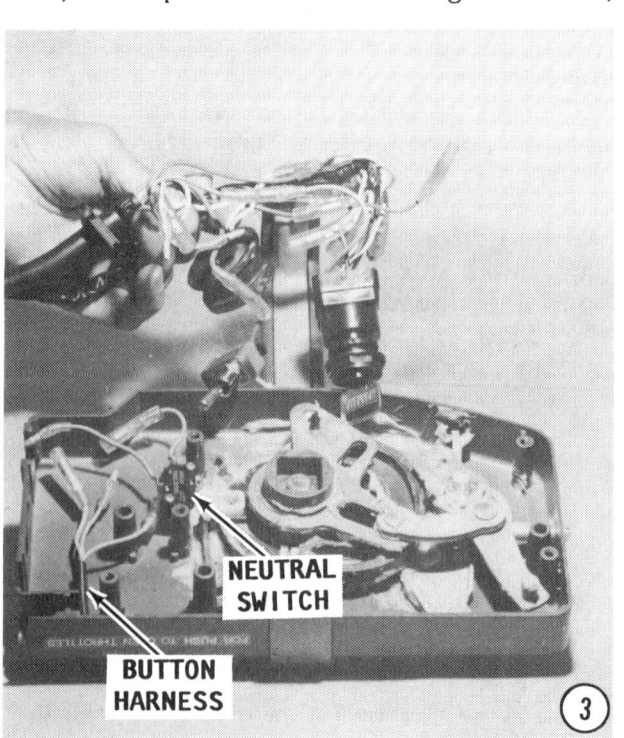

DISASSEMBLING 12-3

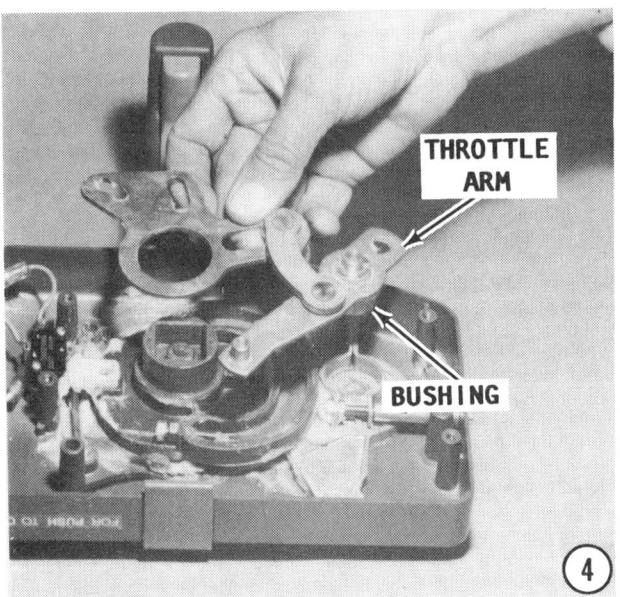

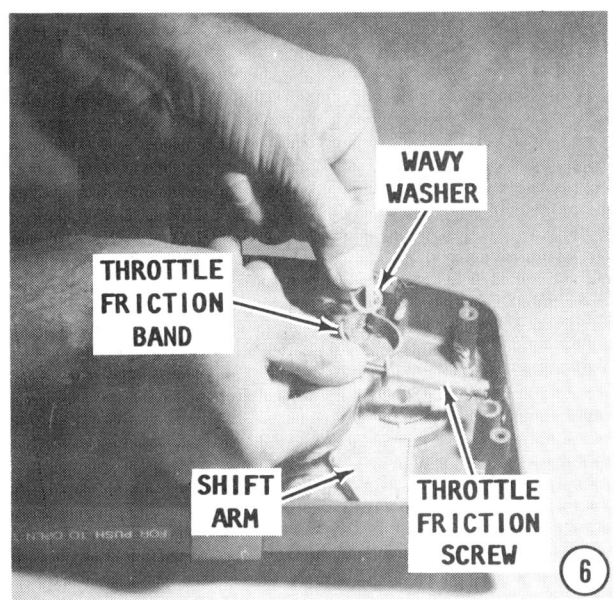

they all fit into one side of the box away from moving parts.

THEREFORE, it would be most advantageous to take a Polaroid picture of the unit, **NOW**, as an aid during assembling.

3- Remove the retaining nuts on the main key switch, the electric choke switch, and the "kill" switch on the side of the control box. Ease these three switches out of their grommets and holes. Disconnect the following leads at their quick disconnect fittings:

The Pink and Yellow leads to the neutral safety switch, and for models with trim/tilt.

The Yellow, Green, and Black leads encased in a small harness leading to the **UP/DOWN** button on the upper control arm.

The entire harness, switches, and horn may now be lifted free of the control box, as shown. The horn is not secured with hardware, but is simply retained between four bosses.

4- Lift out the throttle arm, consisting of three pieces hinged together. A plastic bushing will probably remain on the underneath side of the throttle arm. This bushing was indexed with the throttle friction bands.

5- Remove the two Phillips head screws securing the neutral switch, and then remove the switch. Remove the Phillips head screw and switch arm from the gear.

6- Lift the two halves of the throttle friction band up and out of the control box. The throttle friction screw, with the circlip attached, will come away with the band.

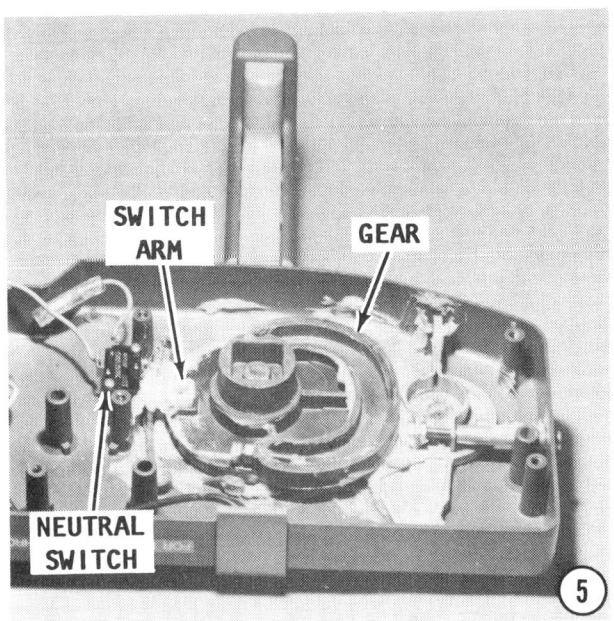

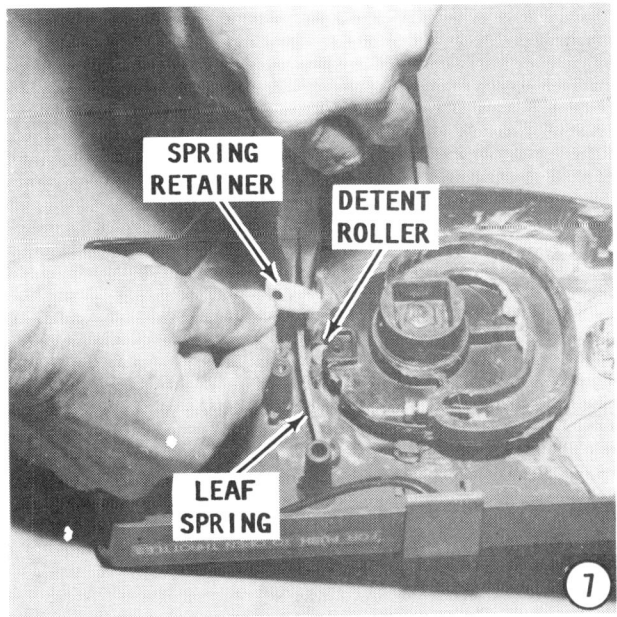

12-4 REMOTE CONTROLS

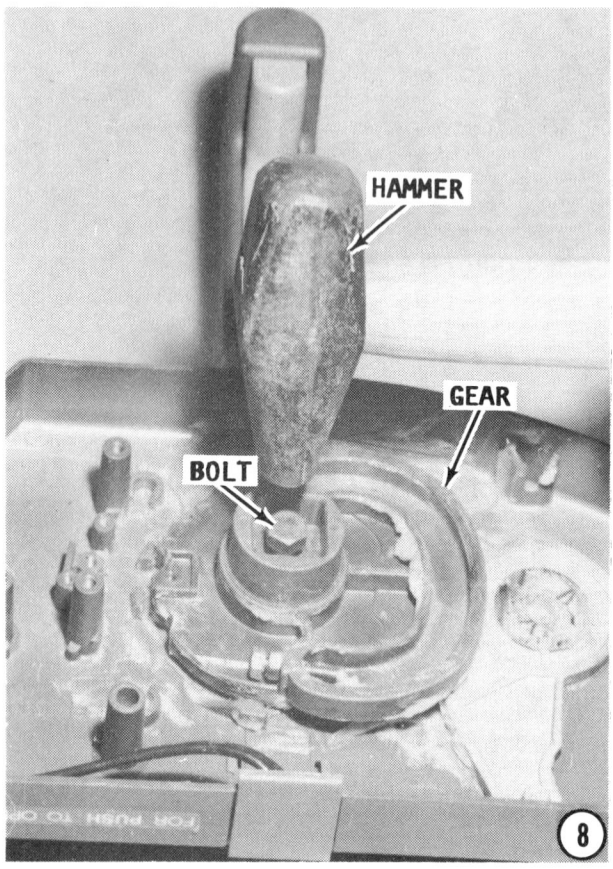

Lift out the small wavy washer from the center of the shift arm.

7- Remove the Phillips screw securing the spring retainer, and then remove the spring retainer. Grasp one end of the leaf spring pack with a pair of needle nose pliers.

Pull up on the pack and at the same time allow the pack to straighten to release tension on the springs. Lift out the detent roller from under the gear.

8- Loosen the bolt in the center of the gear 1/2 turn only. After the center bolt has been loosened, lightly tap on the bolt to free it from the gear. **NOW**, remove the bolt.

9- Turn the control box over. Three things are now to be performed, almost simultaneously. Support the gear -- now underneath -- with one hand and at the same time unsnap the bottom harness cover from the front plate, as the control arm is lifted from the box. Remove the attaching screws, and then the neutral position plate.

10- Support the gear and at the same time turn the control box over, and then lift the gear free of the box.

11- Lift off the shift arm, a large flat washer, and two bushings from the front

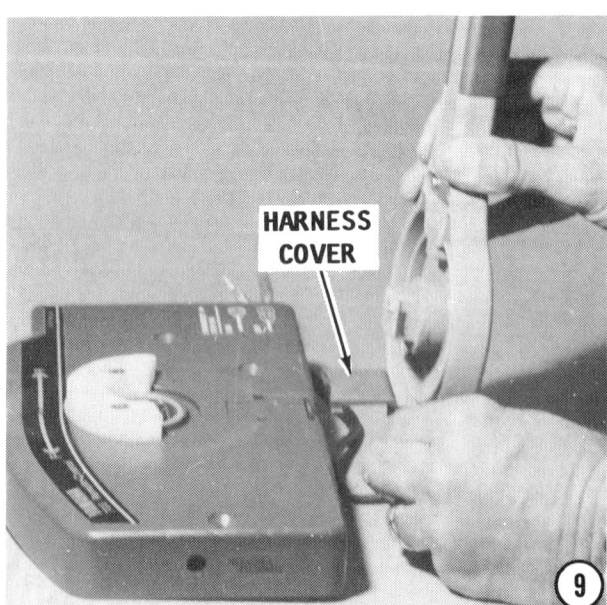

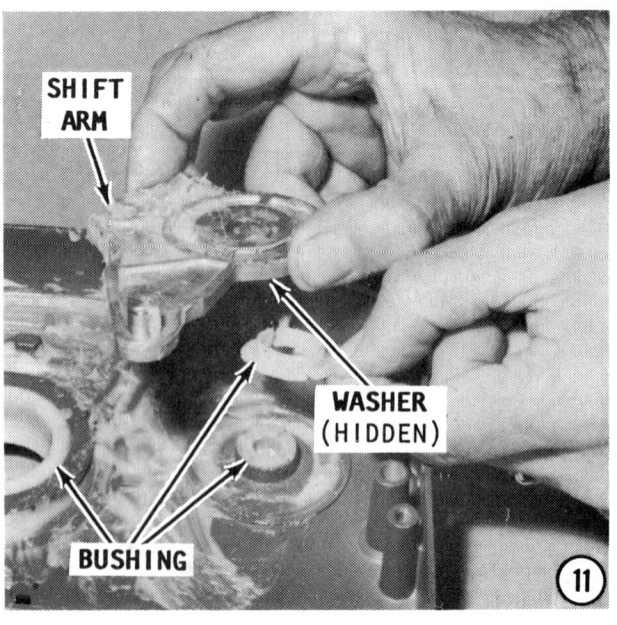

DISASSEMBLING 12-5

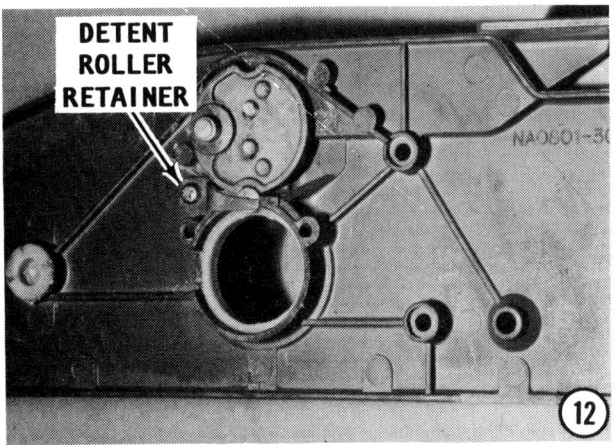

plate. This step concludes the disassembly of the inside of the front plate.

Acceleration Lever Removal

12- Place the upper back plate on the work bench with the lever facing down. Remove the Phillips screw and lift off the detent roller retainer.

13- Pry out the detent roller from the free acceleration disc.

14- Turn the back plate over and remove the two Phillips screws securing the free acceleration lever to the disc. Lift off the lever and the wavy washer under the lever.

15- Turn the back plate over again, and then lift off the free acceleration disc and another wavy washer.

16- Remove the neutral position lever retainer and slide the lever from the arm. Remove the small spring between the top of the lever and the top of the arm. (This spring is not visible in the accompanying illustration.)

17- Remove the two small Phillips screws securing the handle to the control arm.

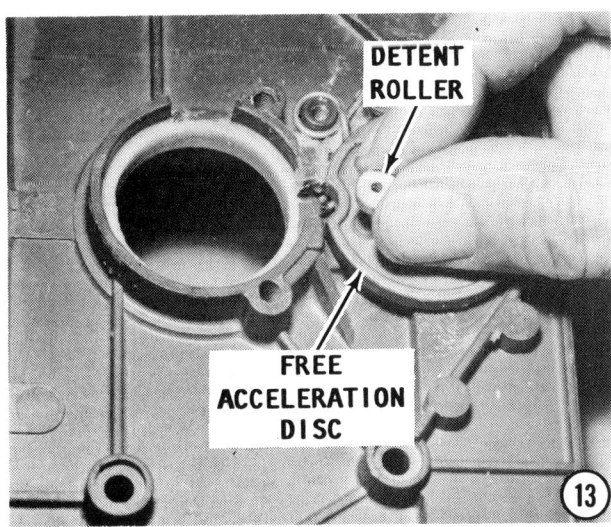

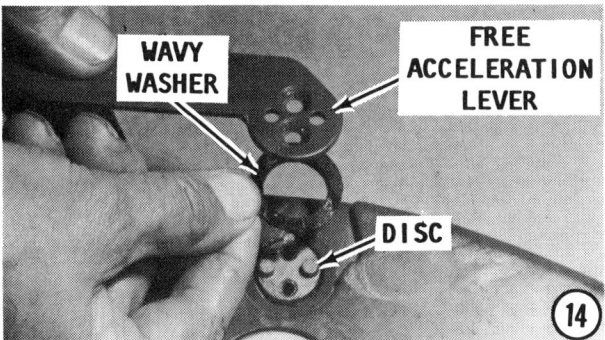

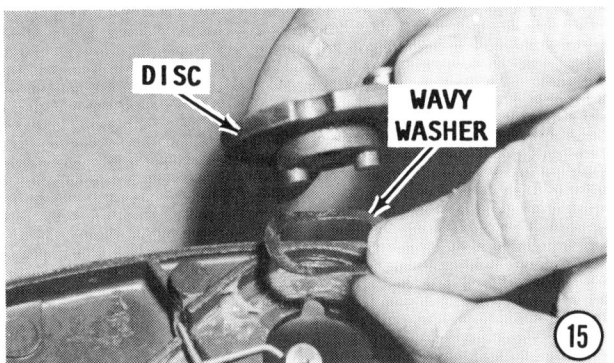

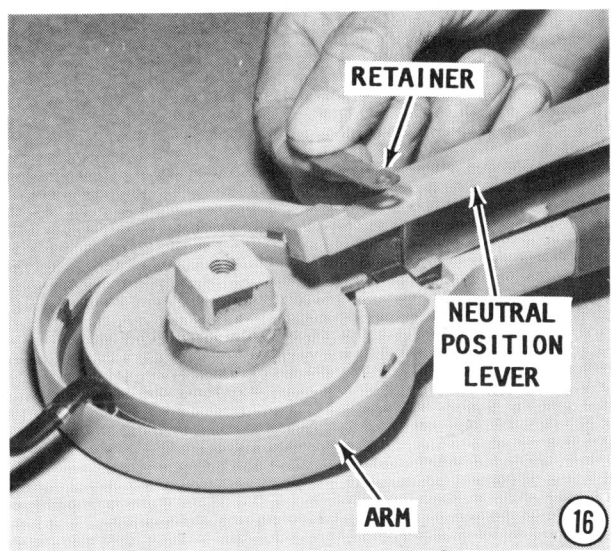

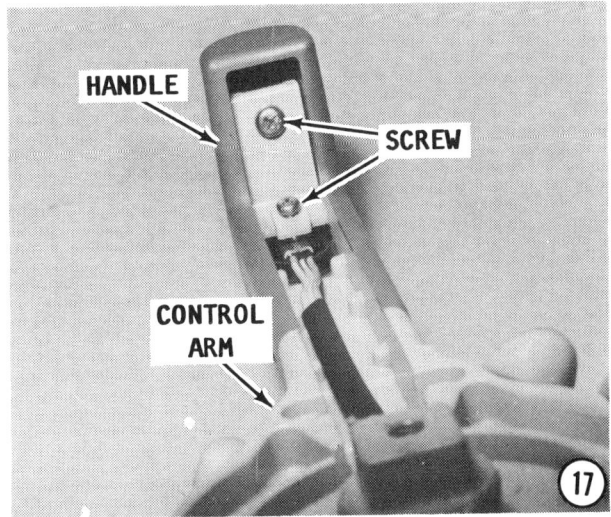

12-6 REMOTE CONTROLS

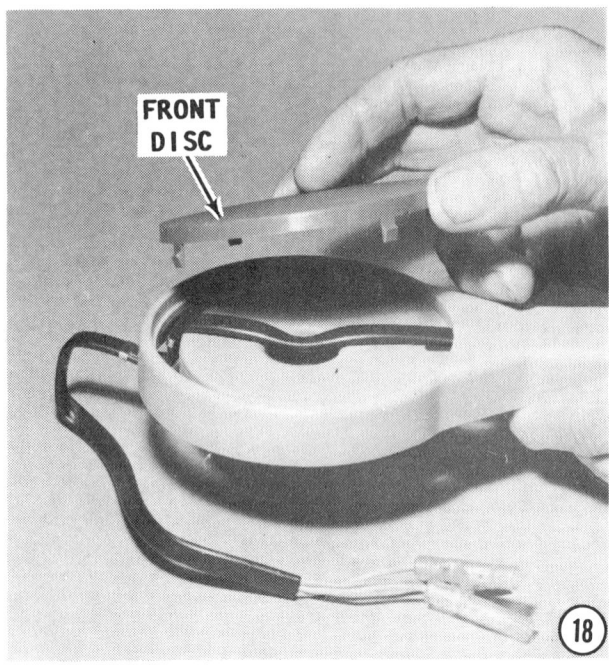

18- Pry up on the front disc and separate the disc from the control lever arm. Unthread the harness, to the **UP/DOWN** button, from the cavity of the arm.

19- Grasp the handle in one hand and the shaft of the control arm in the other. Gently slide them apart about 2 inches (5cm) or until the Trim/Tilt button with the harness attached can be removed from the handle.

12-3 CLEANING AND INSPECTING

Clean all metal parts with solvent, and then blow them dry with compressed air.

NEVER allow nylon bushings, plastic washers, nylon retainers, wiring harness retainers, and the like, to remain submerged in solvent more than just a few moments. The solvent will cause these type parts to expand slightly. They are already considered a "tight fit" and even the slightest amount of expansion would make them very difficult to install. If force is used, the part is most likely to be distorted.

Inspect the control housing plastic case for cracks or other damage allowing moisture to enter and cause problems with the mechanism.

Carefully check the teeth on the gear and shift arm for signs of wear. Inspect all ball bearings for nicks or grooves which would CAUSE them to bind and fail to move freely.

Closely inspect the condition of all wires and their protective insulation. Look for exposed wires caused by the insulation rubbing on a moving part, cuts and nicks in the insulation and severe kinking which could cause internal breakage of the wires.

Inspect the bosses on both ends of the leaf spring. If either end shows signs of failure, the front plate **MUST** be replaced.

Inspect the edges of the cut-out in the neutral position plate. Replace this plate if the corners of the cut-out show any sign of "rounding". If "rounded", a slight pressure

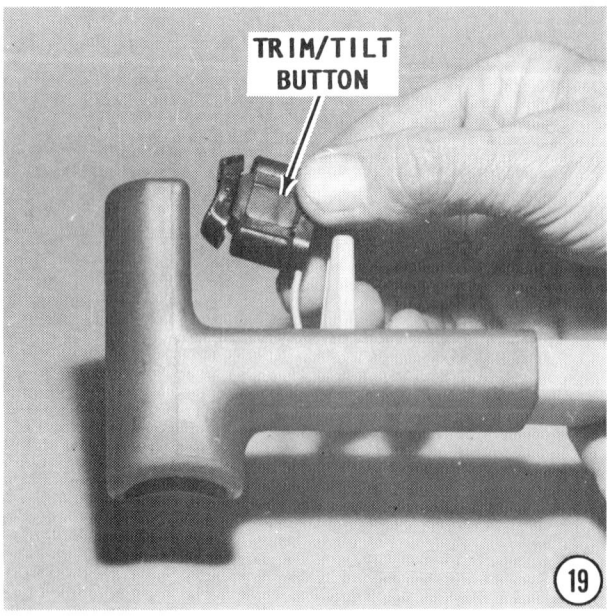

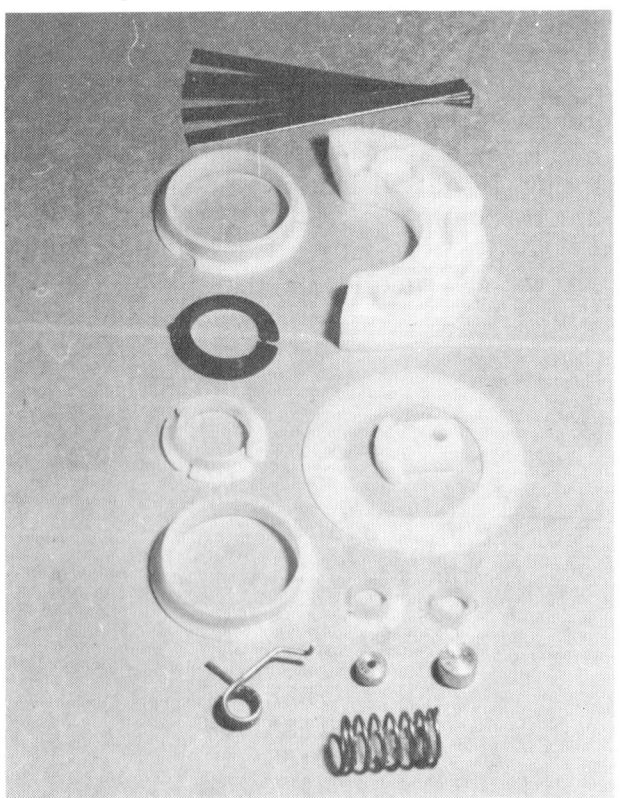

Some, not all, of the small metal and nylon parts from the interior of a remote control unit.

CLEANING & INSPECTING 12-7

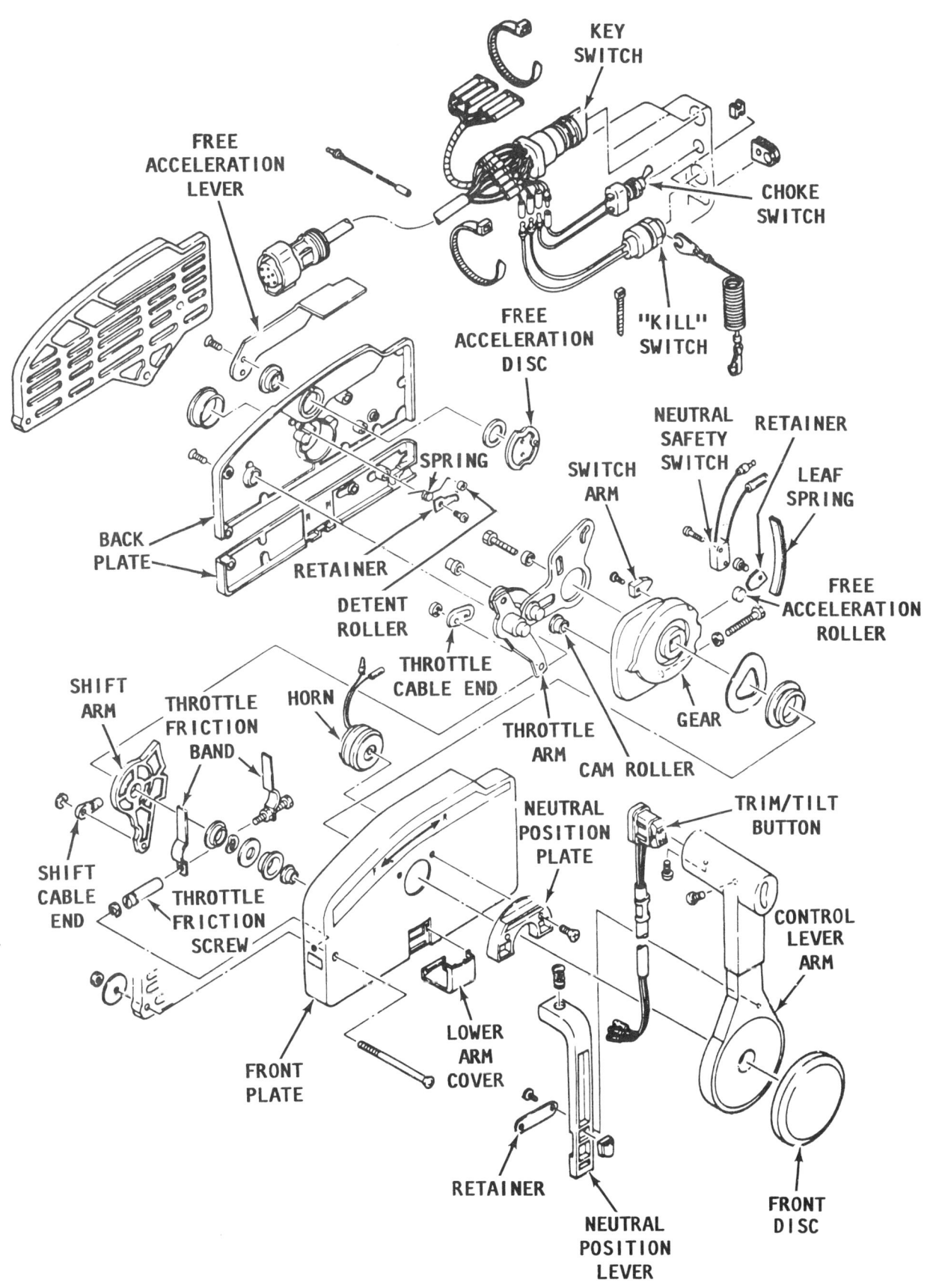

Exploded drawing of the remote control box used as factory furnished equipment for the outboard units covered in this manual. Major parts are identified.

12-8 REMOTE CONTROLS

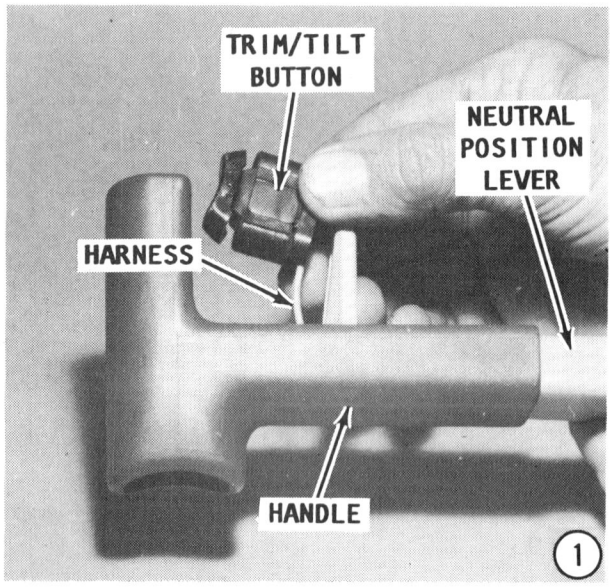

on the neutral position lever could throw the lower unit into gear. Check and double check all components of the neutral position system from the spring at the top of the lever down to the extension at the bottom of the lever which indexes into the cut-out of the neutral position plate.

TESTING ELECTRICAL PARTS

The following electrical parts in the remote control box are all tested in Chapter 9, Section 9-16: The neutral safety switch, low oil warning horn, "kill" switch, choke switch, power tilt/trim button, and the harness connector.

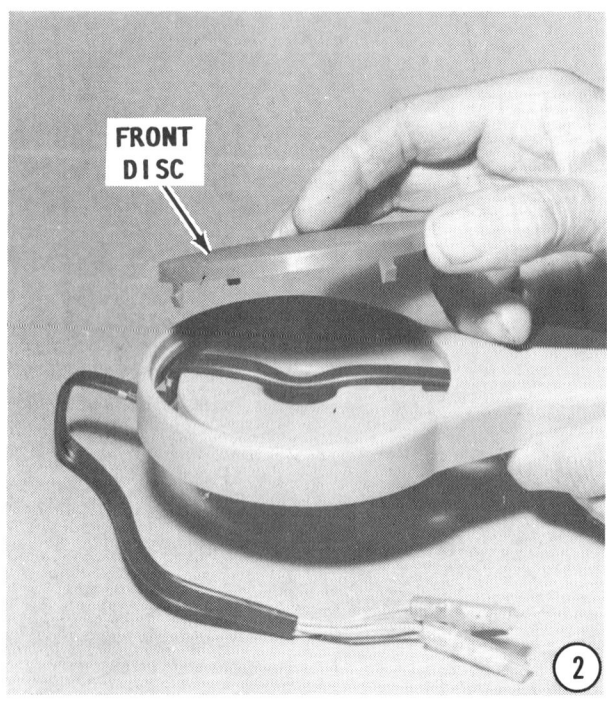

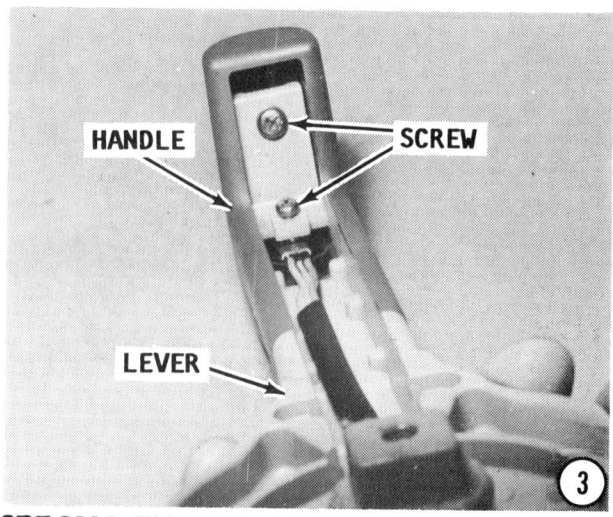

SPECIAL WORDS

Good shop practice dictates a thin coat of Yamalube, or equivalent lubricant be applied to all moving parts as a precaution against the "enemy" moisture. Of course the lubricant will help to ensure continued satisfactory operation of the mechanism.

12-4 ASSEMBLING

The following procedures provide complete detailed instructions to assemble all parts of the remote control unit. If certain areas were not disturbed during disassembling, simply bypass the steps involved and proceed with the work.

Trim/Tilt Button and Harness

1- Feed the harness wires for the power tilt/trim button down from the top in between the handle and the arm. Push the face of the button into the handle and observe the button from the operator's position. The word "UP" should be up and the "DN" should be down. Slide the control lever arm up into the handle.

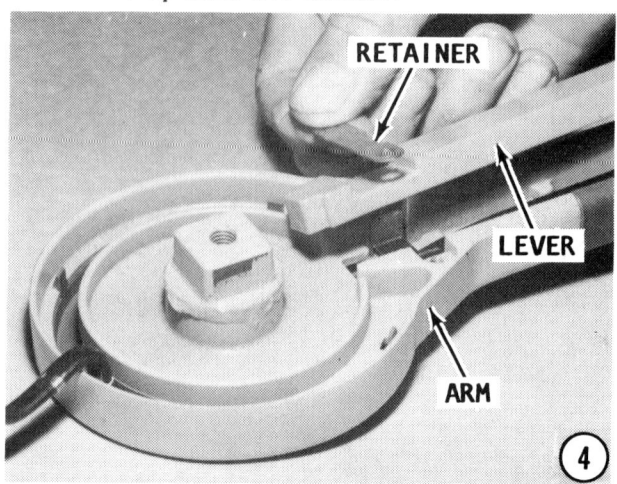

ASSEMBLING 12-9

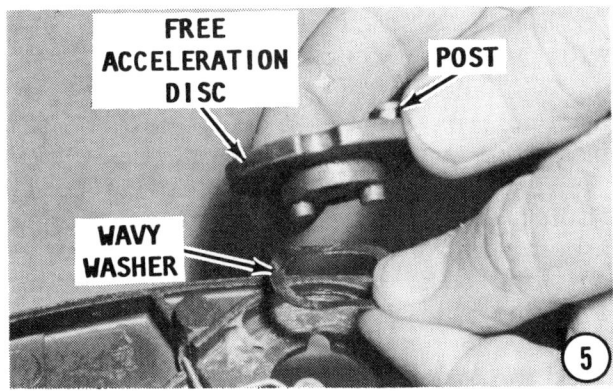

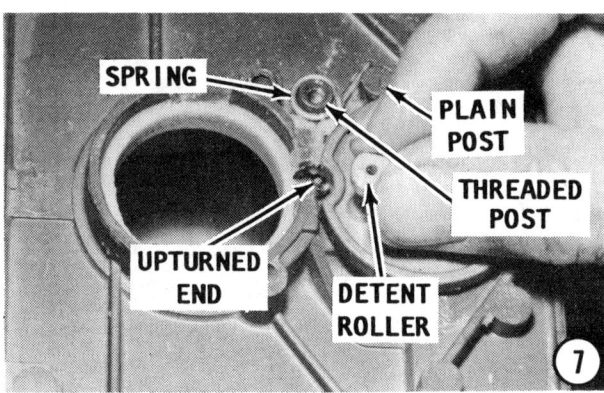

2- Continue to feed the harness wires along the length of the arm -- in and out of the cavity of the arm. Check to be sure the harness lies flat with no kinks. A kink in the wire will damage the harness over a period of time. Snap the front disc over the cavity.

3- Secure the arm to the handle with the two Phillips head screws.

4- Hold the return spring in place in the recess at the top of the neutral position lever and at the same time slide the lever along the length of the arm up into the handle. Install the lever retainer across the lever and secure it to the arm with the two small Phillips head screws.

Free Acceleration Lever Installation

5- Apply just a "dab" of Yamalube to the wavy washer and the free acceleration disc. Place the wavy washer and the free acceleration disc over the smaller hole in the inside of the top back plate. The single large post on the disc must face **AWAY** from the notch cut in the plate for the free acceleration arm.

Hold the washer and disc (the Yamalube will help), in place -- from falling away -- and at the same time, turn the plate over on the work bench.

6- Place another wavy washer and the free acceleration arm over the installed disc.

GOOD WORDS

The arm can only be installed one way because of the groove cut into the face of the upper back plate to accommodate the arm. If the free acceleration disc has been installed correctly in the previous step, the two bevelled holes in the arm will align with the two threaded holes in the disc. Also, the two round holes in the arm will index over the two posts on the disc.

Install and tighten the two Phillips head screws. Turn the plate over again with the inside facing upward.

7- Install the upturned portion of the spring between the installed disc and the center boss, as shown. Hook the coiled part of the spring over the threaded post and at the same time direct the free end of the spring to anchor between the threaded and plain posts.

The detent roller **MUST** be positioned over the upturned end of the spring and be seated into the curved notch of the free acceleration disc. A small screwdriver inserted along the spring under the disc may aid in pushing the end into place to allow the roller to drop down over the spring end and against the disc.

8- Install the retainer over the detent roller. Secure the roller in place with the

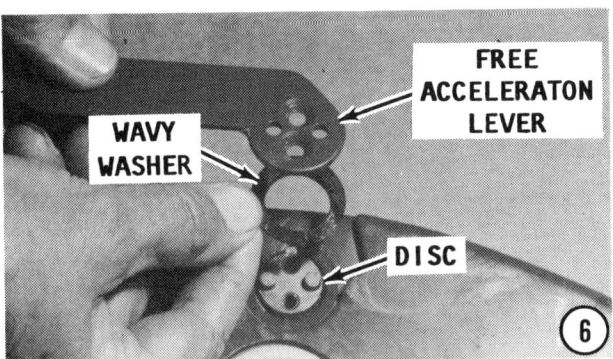

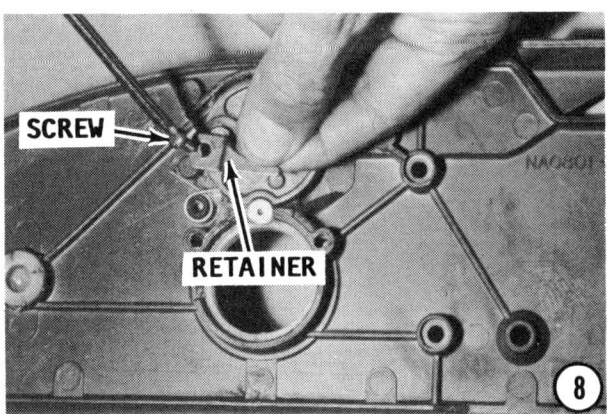

12-10 REMOTE CONTROLS

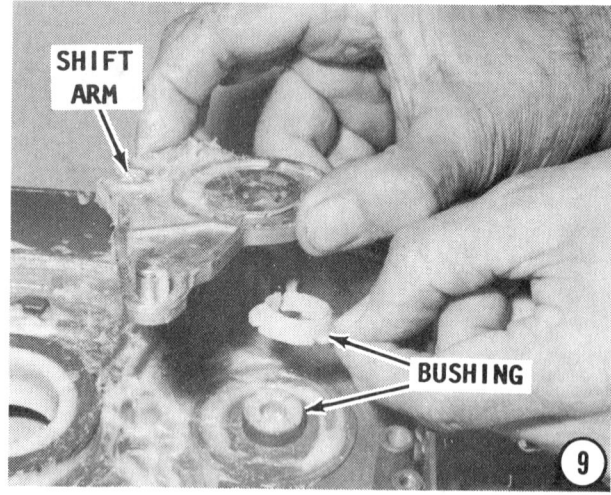

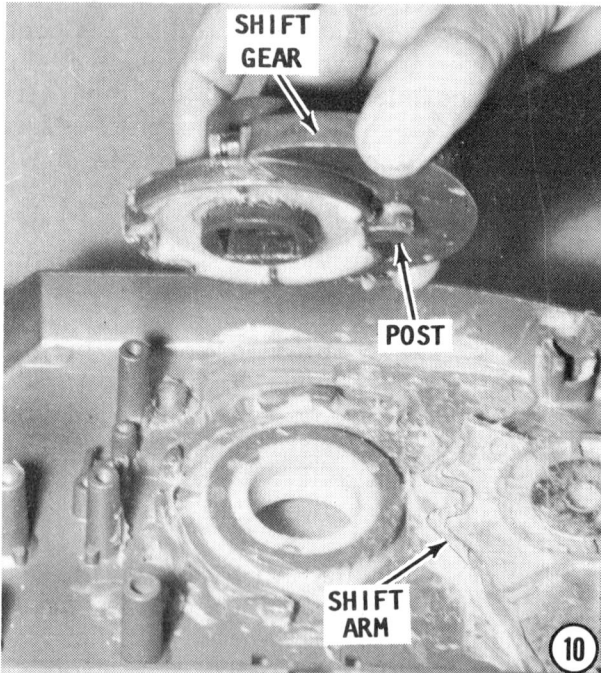

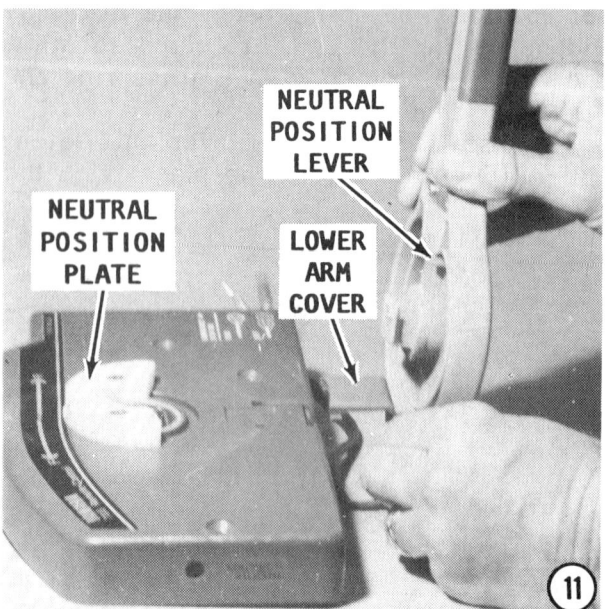

Phillips head screw threaded into the post with the spring around it.

9- Place the small bushing onto the shift arm boss on the inside back plate. Next, place the large bushing onto the same boss. The small and large bushing **MUST** be installed face-to-face. After both bushings are in position, slide the large flat washer onto the boss. Now, install the shift arm over the flat washer with the post for the shift cable facing **UPWARD** -- pointed toward the lower part of the case.

10- Lower the shift gear into the case with the post indexing into the notch in the shift gear. Support the shift gear inside the cover and turn the front plate over.

11- Place the neutral position plate onto the front plate and secure it in place with the two Phillips head screws. Position the control arm over the front plate with the neutral position lever indexed into the cutout of the neutral position plate.

SAFETY WARNING
IF THE NEUTRAL POSITION LEVER IS NOT SEATED PROPERLY, A SLIGHT PRESSURE ON THE HANDLE COULD SHIFT THE LOWER UNIT INTO GEAR —FORWARD OR REVERSE. SUCH ACTION COULD CAUSE SERIOUS INJURY TO CREW, PASSENGERS, OR THE BOAT.

Guide the wire harness into the groove below the arm and snap on the lower arm cover to secure the harness in place.

12- Support the control arm and turn the front plate over. Install the washer and bolt into the square recess of the gear. Tighten the bolt to a torque value of 5.8 ft lbs (8Nm).

13- Slide the free acceleration roller into the center left notch in the shift gear. Grasp the pack of leaf springs with a pair of needle nose pliers. Insert one end of the

ASSEMBLING 12-11

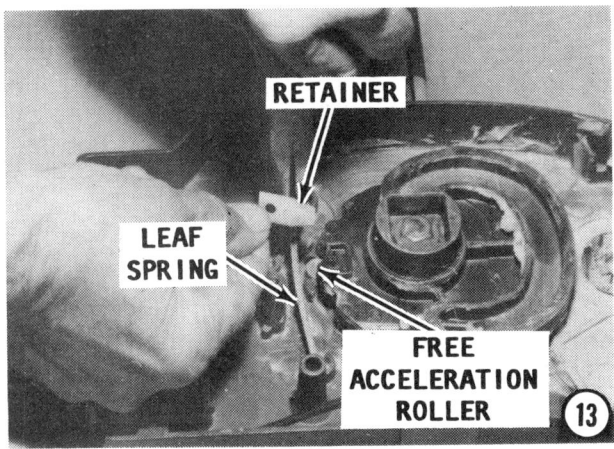

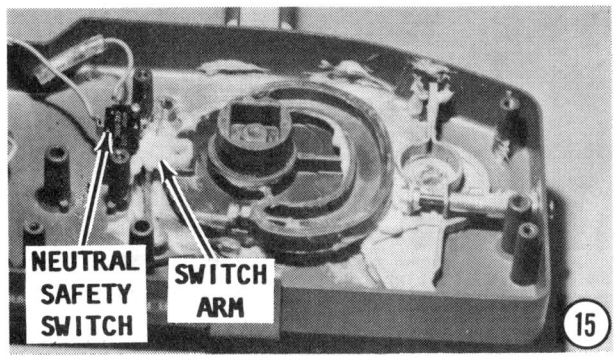

pack into the lower boss. Bend the pack around the roller. Guide the other end into the upper boss. Install the small metal leaf spring retainer onto the threaded post to the left of the leaf spring. Secure the retainer in place with a Phillips head screw.

14- Install the throttle friction band over the shift arm. Move the two ends of the band into the boss on the front plate. Center the opening of the band squarely over the shift arm. Insert a wavy washer into the center of the band on the shift arm.

15- Install the neutral safety switch with the tab on the switch facing **TOWARD** the shift gear. Secure the switch in place with the two Phillips head screws. Position the white plastic switch arm into the slot provided on the shift gear. Install and tighten the Phillips head screw to secure the arm to the gear.

16- Install the throttle arm. Try not to disturb the wavy washer in the center of the throttle friction band when the bushing on the underneath side of the throttle arm indexes with the band.

WORDS OF EXPERIENCE

If you think some of the other assembling procedures were tricky on this unit, stand by! This next one will put you to the test.

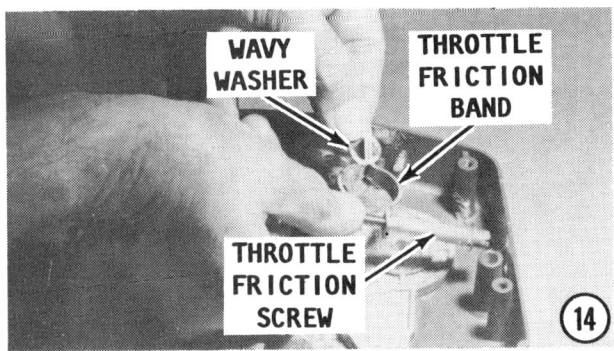

17- Connect the neutral safety switch leads, Pink and Yellow (male) at the quick disconnect fittings -- color to color. Models with power trim/tilt: connect the button harness leads Yellow (female), Green and Black at the quick disconnect fittings -- color to color. Position the horn between the four tall posts next to the neutral safety switch. Install the main key switch, the electric choke switch, and the "kill" switch into their respective openings on the side of the remote control box. Arrange the mess of wires neatly.

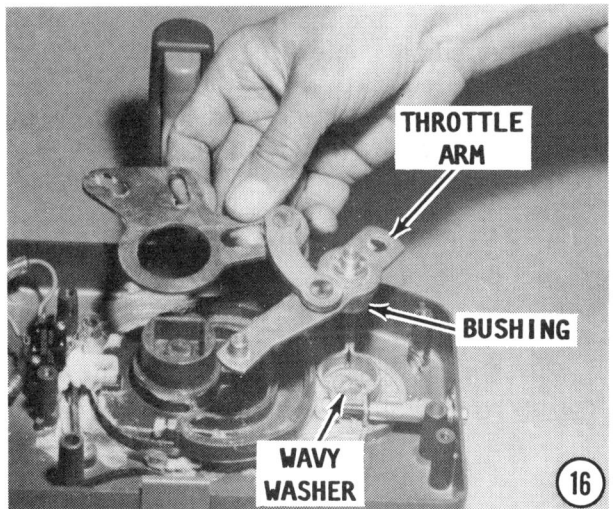

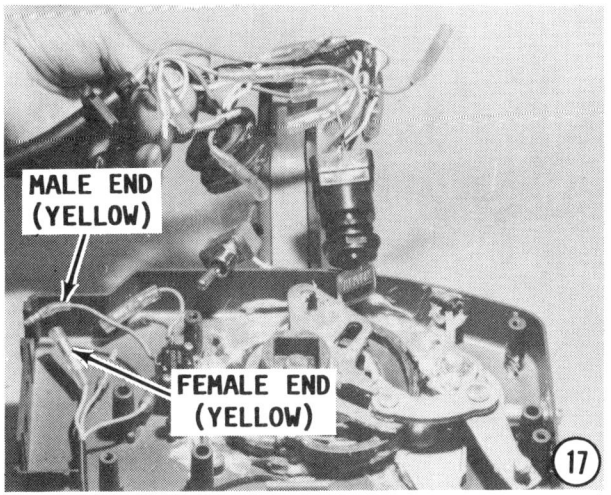

12-12 REMOTE CONTROLS

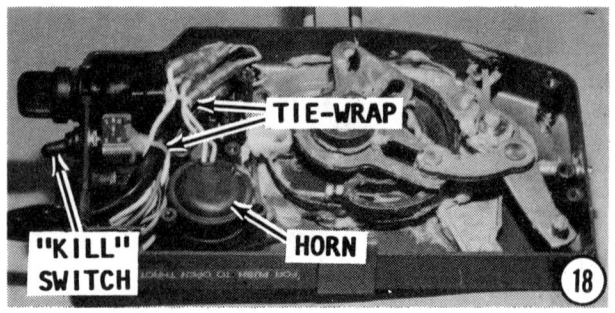

REMEMBER, some wires are tucked neatly under switches, others are routed into neat bundles secured with plastic retainers, and some are looped then doubled back. "Believe-it-or-not", with patience and some good words, they will all fit into one side of the box. The wires MUST be clear of moving parts.

18- Secure the electrical switches to the side of the box with the retaining nuts. Slide the rubber boot over the choke switch. If possible, secure a "tie-wrap" around the bundle of wires to prevent loose wires from rubbing against and interfering with moving parts.

19- Thread the cable joint 0.3" (8mm) onto the shift cable. Hold the joint to prevent it from rotating and at the same time hook the cable end onto the shift arm post. Secure the cable joint with the restraining circlip.

Install the throttle cable in the same manner as the shift cable. Bring both halves of the back plate together with the front plate. Secure it all together with the five Phillips head screws.

20- Check operation of the remote control lever to and between each of the three positions -- neutral, forward, and reverse. The lever should move smoothly and with a definite action. The shift cable and the throttle cable should move in and out of the sheathing without any indication of binding. The movement can be checked at the powerhead end of the cable.

Control Box Installation in Boat

Guide the cables along a selected path and secure them in place with the retainers. DO NOT bend any cable into a diameter smaller than 16" (40cm).

Connect the shift cable to the shift lever on the powerhead. Insert the outer throttle cable wire into the throttle cable bracket. Connect the cable joint onto the throttle control attachment.

DO NOT confuse the cables. If in doubt, operate the control lever on the control box. The cable moving first is the shift cable. Tilt the outboard unit from full up to the full down position, and from hard over starboard to hard over port, to verify the cables move smoothly without binding, bending, or buckling.

Align the mark on the wire harness plug with the mark on the coupler plug, and then connect the plugs.

Install the safety tether to the back of the "kill" switch button.

REMEMBER, the powerhead will not start without the tether in place.

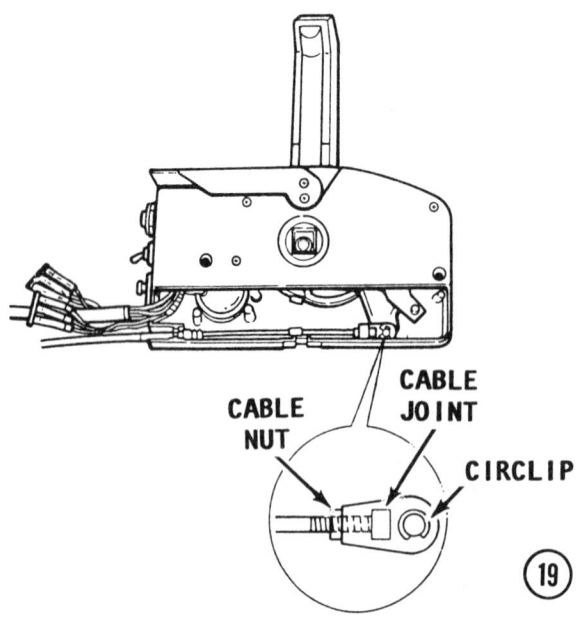

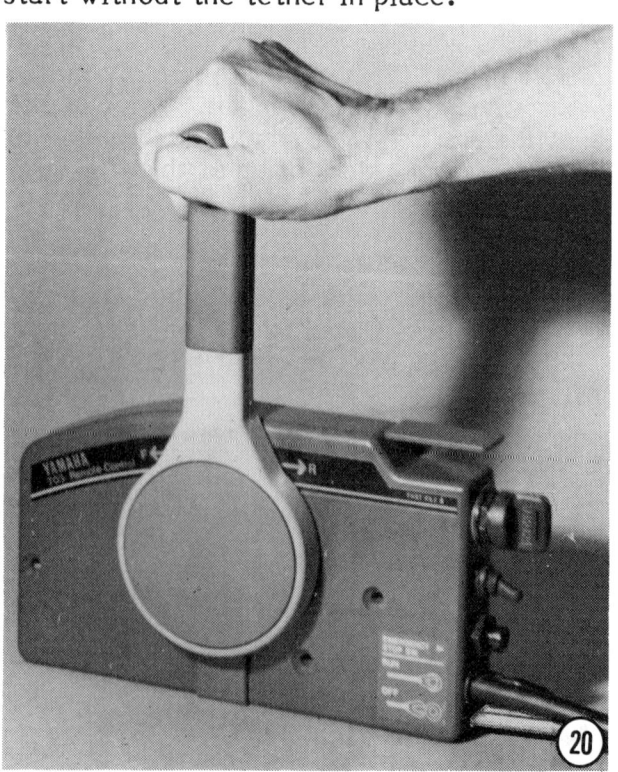

APPENDIX

METRIC CONVERSION CHART

LINEAR
inches	X 25.4	= millimetres (mm)
feet	X 0.3048	= metres (m)
yards	X 0.9144	= metres (m)
miles	X 1.6093	= kilometres (km)
inches	X 2.54	= centimetres (cm)

AREA
$inches^2$	X 645.16	= $millimetres^2$ (mm^2)
$inches^2$	X 6.452	= $centimetres^2$ (cm^2)
$feet^2$	X 0.0929	= $metres^2$ (m^2)
$yards^2$	X 0.8361	= $metres^2$ (m^2)
acres	X 0.4047	= hectares (10^4 m^2) (ha)
$miles^2$	X 2.590	= $kilometres^2$ (km^2)

VOLUME
$inches^3$	X 16387	= $millimetres^3$ (mm^3)
$inches^3$	X 16.387	= $centimetres^3$ (cm^3)
$inches^3$	X 0.01639	= litres (l)
quarts	X 0.94635	= litres (l)
gallons	X 3.7854	= litres (l)
$feet^3$	X 28.317	= litres (l)
$feet^3$	X 0.02832	= $metres^3$ (m^3)
fluid oz	X 29.60	= millilitres (ml)
$yards^3$	X 0.7646	= $metres^3$ (m^3)

MASS
ounces (av)	X 28.35	= grams (g)
pounds (av)	X 0.4536	= kilograms (kg)
tons (2000 lb)	X 907.18	= kilograms (kg)
tons (2000 lb)	X 0.90718	= metric tons (t)

FORCE
ounces - f (av)	X 0.278	= newtons (N)
pounds - f (av)	X 4.448	= newtons (N)
kilograms - f	X 9.807	= newtons (N)

ACCELERATION
$feet/sec^2$	X 0.3048	= $metres/sec^2$ (m/S^2)
$inches/sec^2$	X 0.0254	= $metres/sec^2$ (m/s^2)

ENERGY OR WORK (watt-second - joule - newton-metre)
foot-pounds	X 1.3558	= joules (j)
calories	X 4.187	= joules (j)
Btu	X 1055	= joules (j)
watt-hours	X 3500	= joules (j)
kilowatt - hrs	X 3.600	= megajoules (MJ)

FUEL ECONOMY AND FUEL CONSUMPTION
miles/gal	X 0.42514	= kilometres/litre (km/l)

Note:
235.2/(mi/gal) = litres/100km
235.2/(litres/100 km) = mi/gal

LIGHT
footcandles	X 10.76	= $lumens/metre^2$ (lm/m^2)

PRESSURE OR STRESS (newton/sq metre - pascal)
inches HG (60 F)	X 3.377	= kilopascals (kPa)
pounds/sq in	X 6.895	= kilopascals (kPa)
inches H2O (60° F)	X 0.2488	= kilopascals (kPa)
bars	X 100	= kilopascals (kPa)
pounds/sq ft	X 47.88	= pascals (Pa)

POWER
horsepower	X 0.746	= kilowatts (kW)
ft-lbf/min	X 0.0226	= watts (W)

TORQUE
pound-inches	X 0.11299	= newton-metres (N·m)
pound-feet	X 1.3558	= newton-metres (N·m)

VELOCITY
miles/hour	X 1.6093	= kilometres/hour (km/h)
feet/sec	X 0.3048	= metres/sec (m/s)
kilometres/hr	X 0.27778	= metres/sec (m/s)
miles/hour	X 0.4470	= metres/sec (m/s)

TEMPERATURE

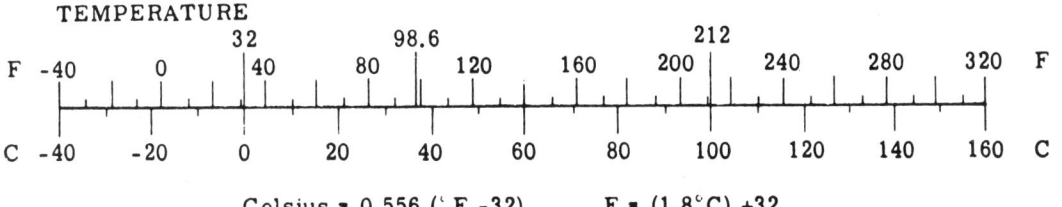

Celsius = 0.556 (°F -32)   F = (1.8°C) +32

ENGINE SPECIFICATIONS AND

MODEL	CYL.	H.P.	CU. CM. DISPL.	CU IN DISPL.	OPERATING RANGE RPM	BORE INCHES	STROKE INCHES
1984							
115	V4	115	1730	105.5	4500-5500	3.54	2.68
150	V6	150	2596	158.4	4500-5500	3.54	2.68
175	V6	175	2596	158.4	4500-5500	3.54	2.68
200	V6	200	2596	158.4	4500-5500	3.54	2.68
V6 Sp	V6	220	2596	158.4	4800-5800	3.54	2.68
1985							
115	V4	115	1730	105.5	4500-5500	3.54	2.68
150	V6	150	2596	158.4	4500-5500	3.54	2.68
175	V6	175	2596	158.4	4500-5500	3.54	2.68
200	V6	200	2596	158.4	4500-5500	3.54	2.68
V6 Sp	V6	220	2596	158.4	4800-5800	3.54	2.68
1986							
115	V4	115	1730	105.5	4500-5500	3.54	2.68
150	V6	150	2596	158.4	4500-5500	3.54	2.68
Pro V 150	V6	150	2596	158.4	4500-5500	3.54	2.68
175	V6	175	2596	158.4	4500-5500	3.54	2.68
200	V6	200	2596	158.4	4500-5500	3.54	2.68
V6 Sp	V6	220	2596	158.4	4800-5800	3.54	2.68

TUNE-UP ADJUSTMENTS

SPARK PLUG NGK	PLUG GAP INCH	FULL ADV. TIME	TIMING AT IDLE	PICK UP TIMING	CARB. PILOT SCREW (TURNS OUT)	IDLE RPM
1984						
B-8HS10	0.039	23°BTDC	5°ATDC	4°ATDC	1 5/8	750
B-7HS10	0.039	24°BTDC	5°ATDC	4°ATDC	Note 1	800
B-8HS10	0.039	22°BTDC	5°ATDC	4°ATDC	Note 2	800
B-8HS10	0.039	22°BTDC	5°ATDC	4°ATDC	Note 3	800
BR-8HS10	0.039	Note 4	Note 4	Note 4	1¾	800
1985						
B-8HS10	0.039	23°BTDC	5°ATDC	4°ATDC	1½	750
B-7HS10	0.039	24°BTDC	7°ATDC	6°ATDC	1 1/8	800
B-8HS10	0.039	22°BTDC	7°ATDC	6°ATDC	Note 5	800
B-8HS10	0.039	22°BTDC	7°ATDC	6°ATDC	Note 6	800
BR-8HS10	0.039	Note 4	Note 4	Note 4	1½	750
1986						
B-8HS10	0.039	25°BTDC	5°ATDC	4°ATDC	Note 7	750
B-7HS10	0.039	24°BTDC	7°ATDC	6°ATDC	Note 8	800
B-7HS10	0.039	28°BTDC	7°ATDC	6°ATDC	Note 9	800
B-8HS10	0.039	22°BTDC	7°ATDC	6°ATDC	Note 5	800
B-8HS10	0.039	22°BTDC	7°ATDC	6°ATDC	Note 6	800
BR-8HS10	0.039	Note 4	Note 4	Note 4	Note 10	750

NOTE: See Appendix Page A-6 for special timing notes called out in this table.

ENGINE SPECIFICATIONS AND

MODEL	CYL.	H.P.	CU. CM. DISPL.	CU IN DISPL.	OPERATING RANGE RPM	BORE INCHES	STROKE INCHES
1987							
115	V4	115	1730	105.5	4500-5500	3.54	2.68
150	V6	150	2596	158.4	4500-5500	3.54	2.68
Pro V 150	V6	150	2596	158.4	4500-5500	3.54	2.68
175	V6	175	2596	158.4	4500-5500	3.54	2.68
200	V6	200	2596	158.4	4500-5500	3.54	2.68
V6 Sp	V6	220	2596	158.4	4800-5800	3.54	2.68
Jet Drive 115	V4	115	1730	105.5	4500-5500	3.54	2.68
Jet Drive 200	V6	200	2596	158.4	4500-5500	3.54	2.68
1988							
115	V4	115	1730	105.5	4500-5500	3.54	2.68
130	V4	130	1730	105.5	4500-5500	3.54	2.68
150	V6	150	2596	158.4	4500-5500	3.54	2.68
Pro V 150	V6	150	2596	158.4	4500-5500	3.54	2.68
175	V6	175	2596	158.4	4500-5500	3.54	2.68
200	V6	200	2596	158.4	4500-5500	3.54	2.68
V6 Excel	V6	225	2596	158.4	4800-5800	3.54	2.68
Jet Power 80	V4	115	1730	105.5	4500-5500	3.54	2.68
Jet Power 140	V6	200	2596	158.4	4500-5500	3.54	2.68

TUNE-UP ADJUSTMENTS

SPARK PLUG NGK	PLUG GAP INCH	FULL ADV. TIME	TIMING AT IDLE	PICK UP TIMING	CARB. PILOT SCREW (TURNS OUT)	IDLE RPM
1987						
B-8HS10	0.039	25°BTDC	5°ATDC	4°ATDC	Note 7	750
B-7HS10	0.039	24°BTDC	7°ATDC	6°ATDC	Note 8	800
B-7HS10	0.039	28°BTDC	7°ATDC	6°ATDC	Note 9	800
B-8HS10	0.039	22°BTDC	7°ATDC	6°ATDC	Note 5	800
B-8HS10	0.039	22°BTDC	7°ATDC	6°ATDC	Note 6	800
BR-8HS10	0.039	Note 4	Note 4	Note 4	Note 10	750
B-8HS10	0.039	25°BTDC	5°ATDC	4°ATDC	Note 7	750
B-8HS10	0.039	22°BTDC	7°ATDC	6°ATDC	Note 6	800
1988						
B-8HS10	0.039	25°BTDC	5°ATDC	4°ATDC	Note 7	750
B-9HS10	0.039	22°BTDC	5°ATDC	4°ATDC	7/8	750
B-7HS10	0.039	24°BTDC	7°ATDC	6°ATDC	Note 8	800
B-7HS10	0.039	28°BTDC	7°ATDC	6°ATDC	Note 9	800
B-8HS10	0.039	22°BTDC	7°ATDC	6°ATDC	Note 5	800
B-8HS10	0.039	22°BTDC	7°ATDC	6°ATDC	Note 6	800
BR-8HS10	0.039	Note 4	Note 4	Note 4	Note 10	750
B-8HS10	0.039	25°BTDC	5°ATDC	4°ATDC	Note 7	750
B-8HS10	0.039	22°BTDC	7°ATDC	6°ATDC	Note 6	800

NOTE: See Appendix Page A-6 for special timing notes called out in this table.

ENGINE SPECIFICATIONS

TIMING NOTES

Note 1 150hp Serial Numbers 400101 to 400185 and Serial Numbers 750101 to 750150, 2-3/8 pilot screw turns out.

Note 2 175hp Serial Numbers 400101 to 400163 and Serial Numbers 750101 to 750152, 2-1/8 pilot screw turns out.

Note 3 200hp Serial Numbers 400101 to 400161 and Serial Numbers 750101 to 750153, 2-1/8 pilot screw turns out.

Note 4 Computer controlled, therefore no adjustment necessary.

Note 5 Pilot screw turns out for single bowl Teikei carburetors: Port carburetor 1-3/4, Starboard carburetor 1-1/8.

  Pilot screw turns out for dual bowl Nikki carburetors: Port and Starboard carburetors 1-3/8.

Note 6 Pilot screw turns out for single Teikei carburetors: Port carburetor 1-1/2, Starboard carburetor 1-1/8.

  Pilot screw turns out for dual Nikki carburetors: Port carburetor 1-5/8, Starboard carburetor 7/8.

Note 7 Pilot screw turns out for single bowl Teikei carburetors: Port carburetor 2-1/8, Starboard carburetor 1-7/8.

  Pilot screw turns out for dual bowl Nikki carburetors: Port and Starboard carburetors 5/8.

Note 8 Pilot screw turns out for single bowl Teikei carburetors: Port and Starboard carburetors 1-1/8.

  Pilot screw turns out for dual bowl Nikki carburetors: Port and Starboard carburetors 1-1/4.

Note 9 Pilot screw turns out on Port carburetor 1-1/2, on Starboard carburetor 1.

Note 10 Pilot screw turns out for single Teikei carburetors: Port and Starboard carburetors 1-1/2.

  Pilot screw turns out for dual Nikki carburetors: Port carburetor 1-5/8, Starboard carburetor 3/4.

WIRE IDENTIFICATION A-7

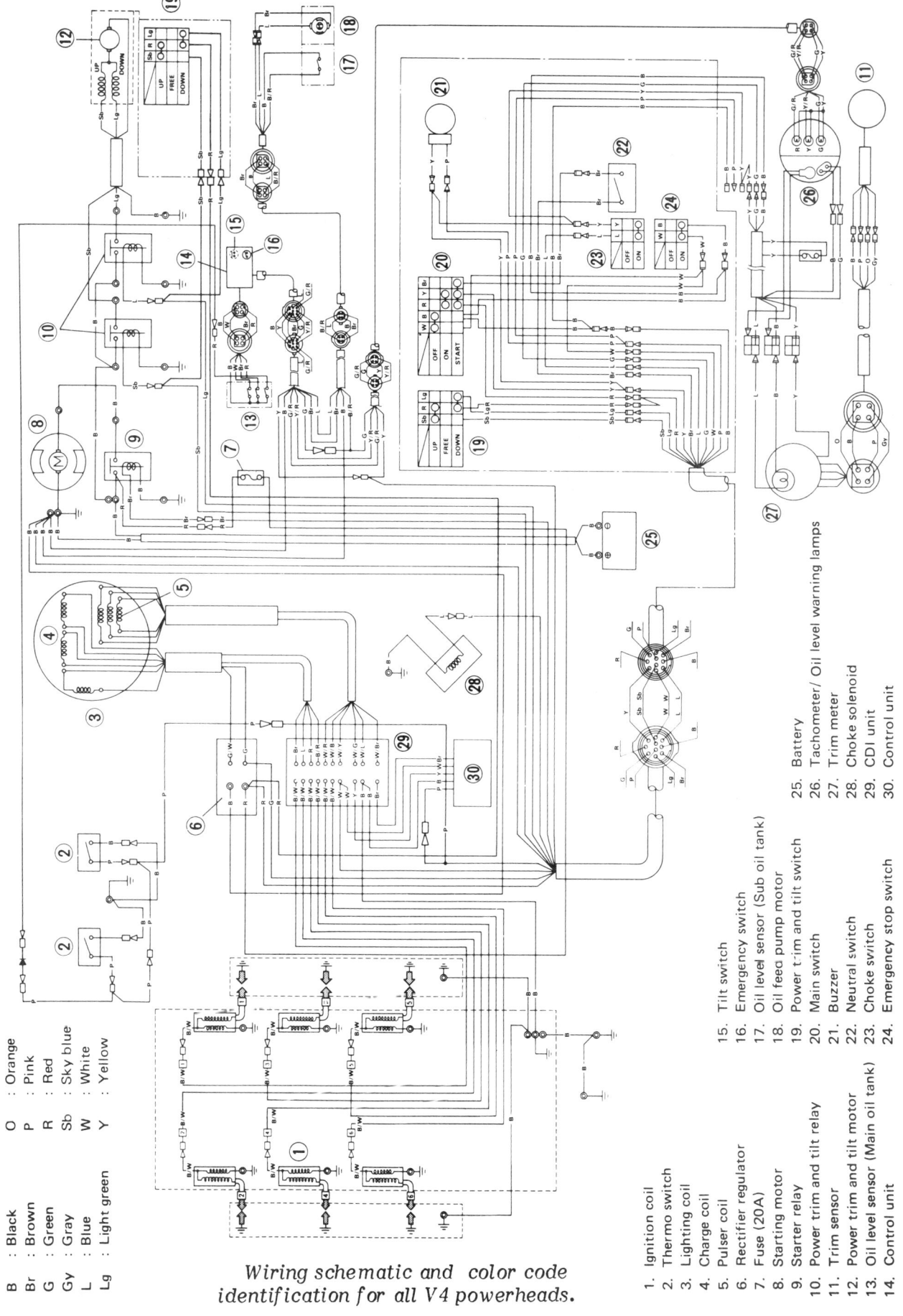

Wiring schematic and color code identification for all V4 powerheads.

B : Black
Br : Brown
G : Green
Gy : Gray
L : Blue
Lg : Light green
O : Orange
P : Pink
R : Red
Sb : Sky blue
W : White
Y : Yellow

1. Ignition coil
2. Thermo switch
3. Lighting coil
4. Charge coil
5. Pulser coil
6. Rectifier regulator
7. Fuse (20A)
8. Starting motor
9. Starter relay
10. Power trim and tilt relay
11. Trim sensor
12. Power trim and tilt motor
13. Oil level sensor (Main oil tank)
14. Control unit
15. Tilt switch
16. Emergency switch
17. Oil level sensor (Sub oil tank)
18. Oil feed pump motor
19. Power trim and tilt switch
20. Main switch
21. Buzzer
22. Neutral switch
23. Choke switch
24. Emergency stop switch
25. Battery
26. Tachometer/ Oil level warning lamps
27. Trim meter
28. Choke solenoid
29. CDI unit
30. Control unit

A-8 APPENDIX

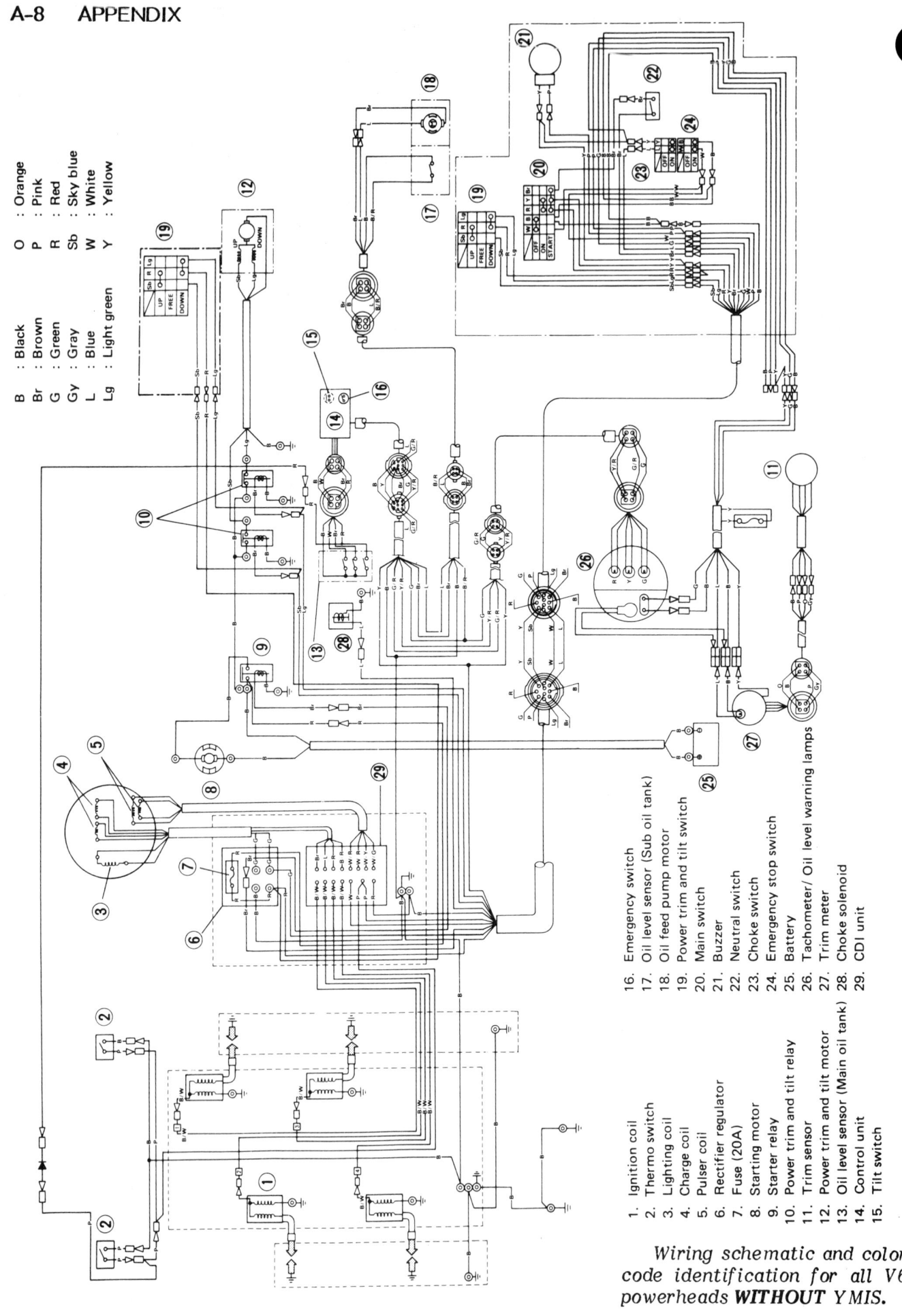

Wiring schematic and color code identification for all V6 powerheads WITHOUT YMIS.

WIRE IDENTIFICATION A-9

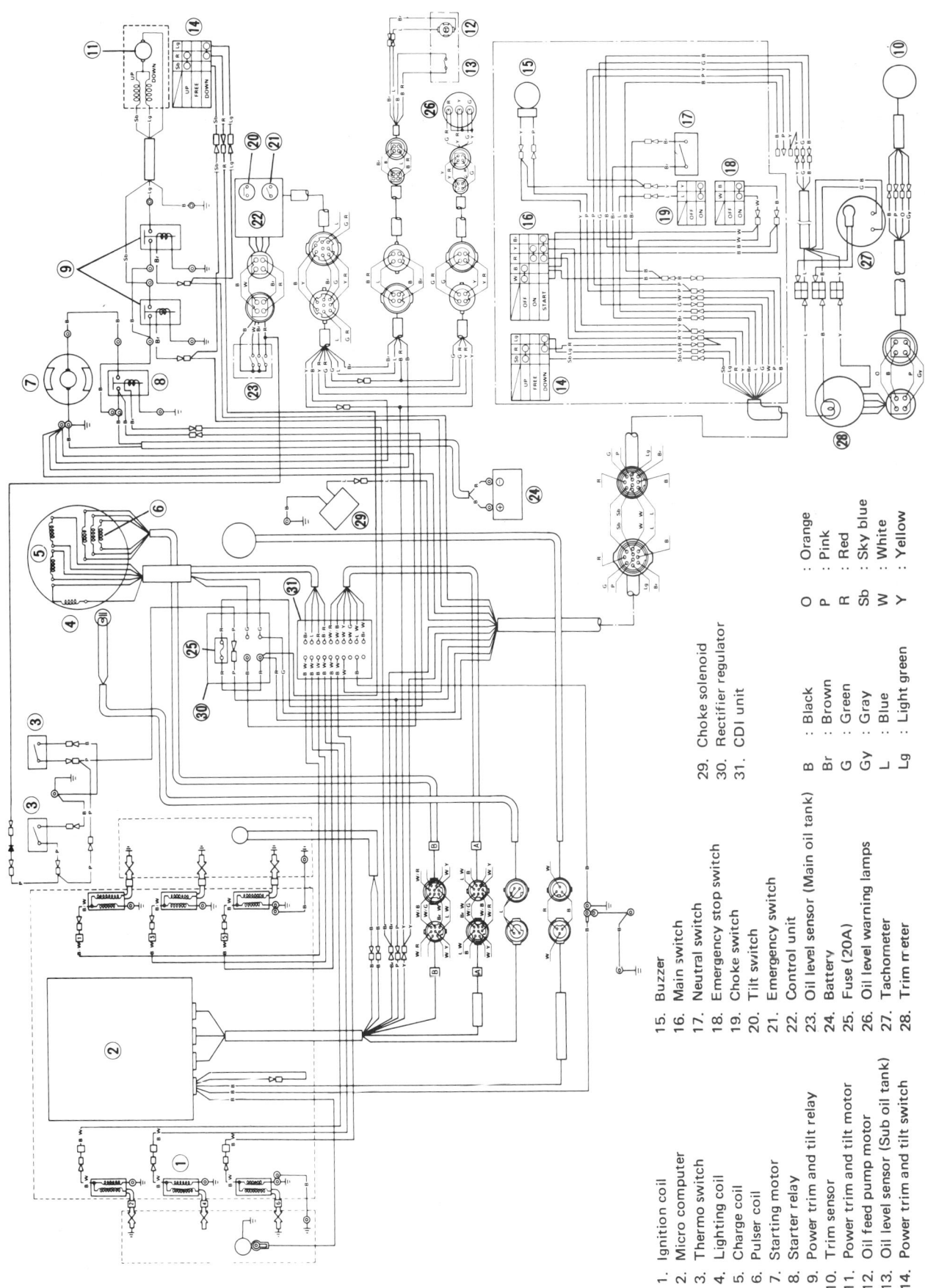

*Wiring schematic and color code identification for V6 powerheads **WITH YMIS**.*

Other Seloc Marine Manuals

New titles are constantly being produced and the updating work on existing manuals never ceases.
All manuals contain complete detailed instructions, specifications, and wiring diagrams.